Applied Multiple Regression/Correlation Analysis for Aviation Research

Applied Multiple Regression/Correlation Analysis for Aviation Research describes and illustrates multiple regression/correlation (MRC) analysis in an aviation context, including flight instruction, airport design, airline routes, and aviation human factors research. Structured in four parts, the book first reviews the major concepts of bivariate correlation and regression and then extends the bivariate case to two, four, and *k* predictors coupled with discussions on statistical inference, underlying assumptions, and regression diagnostics relative to MRC analysis. The book then builds on this foundation by presenting MRC variable selection strategies (simultaneous, hierarchical, and statistical regression), analyzing sets of predictors, and introducing coding strategies for nominal predictors. The book concludes by presenting how MRC can be used to conduct an analysis of covariance (ANCOVA), interactions, mediation analysis, and binary logistic regression.

Throughout the presentation, the book provides a balance between procedural knowledge as well as conceptual understanding. Detailed guided examples are presented at the end of each chapter that apply the topics and concepts of the chapter from the perspective of conducting a research study. The analytic strategies demonstrated via these guided examples are clearly explained, enabling readers to understand, conduct, and report results correctly. Pedagogical features associated with each chapter include a set of student learning outcomes and an end-of-chapter package that consists of a summary of the chapter's main topics/concepts, a list of key terms, and review exercises, including multiple-choice items and a research scenario with data for students to analyze.

Tailored to the needs of aviation students, *Applied Multiple Regression/Correlation Analysis for Aviation Research* is the ideal textbook for research-oriented graduate aviation programs such as a thesis-based master's degree or doctoral program that require knowledge of advanced statistical strategies for analyzing research data.

Michael A. Gallo, Ph.D., is a Professor Emeritus in the College of Aeronautics at Florida Institute of Technology, USA.

Ulreen O. McKinney, Ph.D., is an Associate Professor in the College of Aeronautics at Florida Institute of Technology, USA.

Applied Multiple Regression/Correlation Analysis for Aviation Research

Michael A. Gallo
Florida Institute of Technology

Ulreen O. McKinney
Florida Institute of Technology

Routledge
Taylor & Francis Group
LONDON AND NEW YORK

Designed cover image: © Getty Images

First published 2026
by Routledge
4 Park Square, Milton Park, Abingdon, Oxon OX14 4RN

and by Routledge
605 Third Avenue, New York, NY 10158

Routledge is an imprint of the Taylor & Francis Group, an informa business

British Library Cataloguing-in-Publication Data
A catalogue record for this book is available from the British Library

ISBN: 978-1-032-82912-8 (hbk)
ISBN: 978-1-032-82911-1 (pbk)
ISBN: 978-1-003-50693-5 (ebk)

DOI: 10.4324/9781003506935

Typeset in Arial
by Michael A. Gallo

Access the Support Material: www.routledge.com/9781032829111

Publisher's Note
This book has been prepared from camera-ready copy provided by the authors.

To my academic triad (in reverse alphabetical order)

William M. Setek, Jr., Robert B. Nenno (Late), & Lawrence G. Gilligan

Thank you for your guidance, wisdom, encouragement, inspiration,
and resolute support for the past 52 years.

—M.A.G.

To my parents

Ulric O. E. Jones & E. Doreen Jones

All is well with my soul.
You are my peace and my strength in the vicissitudes of life.
The memory of your sacrifices,
your loving kindness and wisdom live forever in my heart and mind.

—U.O.M.

Contents

Part B
Establishing the Foundation of MRC 55

Figures

Tables

Preface

Overview

Applied Multiple Regression/Correlation Analysis for Aviation Research presents a general introduction to MRC within an aviation context. The need for this book became apparent when one of the authors transitioned from teaching MRC in the behavioral sciences to teaching MRC in the aviation sciences: Although the statistical concepts were the same, their respective applications were quite different. As a result, we wrote this book to help make the study of MRC more meaningful and accessible for graduate students pursuing their master's or doctoral degree in an aviation discipline by relating the concepts of MRC to applications aligned to aviation research. Our presentation is straightforward, the approach is intuitive, and the topics are developed completely and explained from both conceptual and procedural perspectives. All data analyses were performed using JMP (http://www.jmp.com) and JASP (https://jasp-stats.org/), the latter of which is a free and open-source project supported by the University of Amsterdam and informally referred to as "SPSS-lite." The book is designed for use in a one-semester course, and it is assumed that students using it have the necessary prerequisite knowledge commensurate with what is conventionally taught in a traditional first-level statistics course.

Organizational Structure

We organized the material in this book around four parts.

- Part A, *Fundamental Concepts of Correlation and Regression*, consists of two review chapters. Chapter 1 reviews the basic concepts of bivariate correlation, and Chapter 2 reviews the basic concepts of bivariate regression. The material presented in these chapters is designed to serve as a refresher and is not to be considered exhaustive. Various parts of these chapters first appeared in *Fundamentals of Statistics for Aviation Research*, which was coauthored by the first author, and are reprinted in the current book with permission.
- Part B, *Establishing the Foundation of MRC*, consists of three chapters that present the basic concepts and methods of MRC that serve as the cornerstone for all subsequent chapters. Chapter 3 begins the transition from bivariate regression to multiple regression involving two IVs. Chapter 4 extends this discussion to four IVs and then generalizes to *k* IVs. Chapter 5 introduces preliminary data screening activities that include the underlying assumptions of multiple regression as well as various regression diagnostics.

- Part C, *Data Analytic Strategies for MRC*, consists of three chapters that demonstrate the robustness of multiple regression. Chapter 6 introduces the variable selection approaches of simultaneous, hierarchical, and statistical (stepwise) regression, including their respective advantages, disadvantages, and applicability. Chapter 7 extends this discussion from selecting and analyzing individual IVs to selecting and analyzing sets of IVs. Chapter 8 presents the coding strategies of dummy, effects (unweighted and weighted), and contrast coding for analyzing variables measured on a nominal scale with three or more groups.
- Part D, *Additional Applications of MRC*, consists of two chapters that broaden the scope of multiple regression. Chapter 9 introduces the concepts of Analysis of Covariance (ANCOVA) and interactions (moderation) from a multiple regression perspective, and Chapter 10 focuses on mediation and binary logistic regression.

The last section of each chapter consists of a guided example that demonstrates how to apply that chapter's statistical strategy to analyze data from a research study. These research studies—as well as those presented in the examples throughout the book—reflect actual past studies that we or our students conducted. However, the corresponding data sets were modified either for instructional purposes or because ethical guidelines of human-subject research prohibited using the actual data outside the initial research setting.

Each chapter concludes with a packaged set of review/study material. A *Chapter Summary* consists of a section-by-section bullet-item list that summarizes the salient aspects of the chapter. *Key Terms and Concepts* contain a vocabulary list of words or phrases central to the material presented. *Review Exercises* consist of 10 multiple-choice items followed by a research study and corresponding data set similar to the chapter's Guided Example. *References* contain the bibliographic information for the in-text chapter citations. Three appendixes and an index complete the book. Appendix A contains a review of the underlying concepts of research design and statistics, Appendix B contains statistical tables, and Appendix C contains the correct responses to the end-of-chapter multiple-choice review exercises.

A Word on Style and Nomenclature

We mostly follow the guidelines of the seventh edition of the *Publication Manual of the American Psychological Association* (see https://doi.org/10.1037/0000165-000), which our students use to prepare their thesis or dissertation. These include but are not limited to: (a) writing in the first-person; (b) following the general guidelines for and exceptions to writing numbers in text, including no leading zeros for decimal fractions when the statistic cannot be greater than 1 or smaller than –1 (e.g., $r = .25$ *not* $r = 0.25$); and (c) using the Oxford comma. One topic we differ from APA is rounding. Although APA guidelines suggest rounding results to two or three decimal places (which we mostly do), this is not always feasible for certain contexts. For example, if the regression coefficient for flight hours is $B = 0.000573$, rounding to two decimal places does not provide an accurate interpretation of the slope. Because flight time generally is reported in thousands of hours (or tens of thousands of hours), we round this coefficient to either three or four decimal places to make the interpretation more meaningful. So, if $B = 0.000573$: "For every 1,000-hour (or 10,000-hour) increase in flight time, Y increases, on average, by approximately 0.6 (or 5.7) units." We also apply the conventional rounding rule, although there might be some instances where we absentmindedly "rounded to even," for which we offer a sincere mea culpa. Lastly, when conducting an a priori power analysis, we primarily consult Daniel Soper's online calculators (https://www.danielsoper.com/statcalc/default.aspx). As a result, readers who use a different source (e.g., G*Power) might get different results.

As for nomenclature, one instance is we use informal labels to refer to different categories of pilots. For example, "Part 121 pilots" refer to pilots working for an airline flying under Part 121, "GA pilots" refer to pilots who have at most a private pilot license (PPL), and "airline transport pilot" is a pilot who holds an ATP license. We also use an uppercase N for sample sizes and a lowercase n for group sizes. For example, a sample of $N = 100$ consists of four groups each of size $n = 25$. We also do not provide operational definitions of variables, terms, or phrases, so readers will need to provide this information.

Acknowledgements

Many people contributed considerably, either directly or indirectly, in the preparation of this book. First, and foremost, we are indebted to the late Jacob Cohen whose published works, particularly *Applied Multiple Regression/Correlation Analysis for the Behavioral Sciences*, have influenced us since 1988. In some ways, one might consider our book as "Cohen-lite" applied to the aviation sciences. A special affirmation also is conveyed to Phil Horton who introduced the first author to and fostered his interest in MRC. Third, we are grateful to our editor, Andrew Harrison, who gave us the opportunity to write this book, and his colleague, Zoe Everitt-Spence, who assisted Andrew on the project. Fourth, we acknowledge the generosity of Brooke Wheeler and Isaac Silver for permitting us to include various passages and line art from *Fundamentals of Statistics for Aviation Research*. Fifth, a note of appreciation is extended to the collective group of our former and current master's and doctoral students at Florida Institute of Technology whose comments and research projects helped shape our exposition and motivated the research studies presented in this book. A sincere "thank you" is extended to our colleague in the College of Aeronautics, Vivek Sharma, who read and commented on various drafts of the manuscript and checked the analyses. A personal note of gratitude is also expressed to Alan Tirpak and his daughters, Danica and Mila, for taking the 'vette out for a spin during extensive writing periods to keep the tires from getting flat spots. Last but not least, this book would not have been completed without the support from our respective families. Thank you Janie, Carl, and Nasserie for giving us the space and time to bring this project to fruition. The substance of this book is manifested in your patience, encouragement, and love. *Amor vincit omnia.*

Part A

Fundamental Concepts of Correlation and Regression

Chapter 1
Bivariate Correlation

Student Learning Outcomes

After studying this chapter, you will be able to do the following:

1. Examine a scatter plot and line of best fit and discern the characteristics of the corresponding correlation.
2. Determine and interpret the Pearson *r*, the 95% confidence interval, and the coefficient of determination (r^2).
3. Conduct and interpret the results of a hypothesis test of the Pearson *r*.
4. Conduct a bivariate correlation analysis for a research study by engaging in the activities presented in Section 1.7.

1.1 Chapter Overview

This chapter presents statistical procedures for analyzing research data involving a single sample or group and focuses exclusively on *bivariate correlation*, which examines the relationship between two variables. We begin by visualizing relationships graphically via *scatter plots* and then present the *Pearson r*, which is a single numerical metric called a *correlation coefficient* that denotes the strength and direction of a relationship. We follow this with a discussion of various statistical aspects of correlation, including *confidence intervals* and *hypothesis testing* involving the correlation coefficient as well as the applications of correlational studies. We conclude the chapter with a guided example that demonstrates the application of correlation to a correlational research study.

1.2 The Concept of Correlation

Correlational Study vs. Correlation Analysis

From a research design perspective, a correlational study involves examining the relationship between two or more research factors from a single entity to determine if the variables share something in common, and if they do, we say that the variables are correlated (or *co-related*). For example, in the context of aviation research, the director of a flight school might be interested in determining if there is a relationship between flight students' age and the time it takes for them to achieve their private pilot license (PPL). Other examples include examining the relationship between the number of flight hours as pilot-in-command (PIC) and aviation safety incident rate per 10,000 flight hours, employment and commercial passenger enplanements at an airport, the number of aircraft missed approaches and lower-level wind velocities, fatigue or stress and the number of maintenance errors committed by aviation maintenance technicians (AMTs), and airline profitability and the cost of jet fuel. Because a correlational study is considered observational (see Appendix A), data sources are often referred to as *observational units* and may include but are not limited to groups

DOI: 10.4324/9781003506935-2

of individuals; events; physical and psychological phenomena; physiological, demographic, and socioeconomic information; and periods.

As an example of a correlational study, let's assume a researcher is reviewing plans for airports located within the Federal Aviation Administration (FAA) Southern region and seeks to understand the relationship between the population of airport service areas and the number of annual operations at various airports in the region. The corresponding research question for this hypothetical example follows (operational definitions of key terms/variables are left to the reader):

What is the relationship between airport service area population and annual operations in the FAA Southern region?

Note that this study has two variables—annual airport operations and airport service area population—and the data source consists of one observational unit (airports). To answer this RQ, we conduct a correlation analysis, or simply, correlation. A correlation analysis (a) involves the observation or identification of relationships between variables that covary and (b) provides a means by which these relationships may be described and summarized. From a practical standpoint, a correlation analysis is useful because it can be an inexpensive and expedient way of studying variables of interest. A correlation analysis that examines the relationship between two variables is called bivariate correlation, and the distribution of values or scores on the two variables is called a bivariate distribution. A bivariate distribution implies that observational units have scores on each of the two variables. These data are termed bivariate data and are treated as *pairs of scores* instead of individual scores.

An example of a bivariate distribution is provided in Table 1.1, which contains archival data associated with the RQ for the current example. These data were extracted from the Federal Aviation Administration (FAA, 2024) OPSNET and the U.S. Census Bureau (2024) for a random sample of N = 38 airports located in the FAA Southern region and are also provided in the file "Ch_1 Annual Ops vs. Population" as a convenience for the reader. From Table 1.1, note that each airport has a paired score (x, y), where X = Airport Service Area Population is the independent variable (IV) and Y = Annual Operations is the dependent variable (DV). For example, Airport 10's paired observation is (335509, 34695), and Airport 30's paired observation is (332285, 101366).

When working with bivariate data, if there is a clear understanding that one variable is an IV and the other is a DV, then we identify X as the IV and Y as the DV. For our running example, we label airport service area population as the IV and annual airport operations as the DV because it is more meaningful to examine this relationship from the perspective of service area population influencing annual operations. On the other hand, if there is no such understanding—for example, the length of the longest runway and elevation of the airport—then it does not matter which variable is identified as X or Y.

When conducting a correlation analysis, our primary interest is examining systematic patterns of association that are encountered between variables or exist in the natural environment. For example, if high values of one variable are related to high values of another variable, then the variables are *positively related*. Two examples of a positive relationship are (a) the number of aircraft missed approaches versus lower-level wind velocities, and (b) employment versus commercial passenger

TABLE 1.1 *Airport Service Area Population and Annual Operations at Airport Sites in the FAA Southern Region*

Site	*X*	*Y*	Site	*X*	*Y*
1	202518	36132	20	47632	35412
2	413210	60084	21	192122	31494
3	217702	104847	22	382295	44363
4	694144	234964	23	231840	50716
5	367804	66999	24	766757	88907
6	406222	122886	25	936250	127167
7	354081	27184	26	153159	15317
8	166998	162816	27	74125	86686
9	105882	72453	28	937166	229451
10	335509	34695	29	413210	77257
11	770577	114863	30	332285	101366
12	328297	185697	31	197938	22094
13	523542	63036	32	101511	16753
14	208080	56407	33	65418	25224
15	537174	91134	34	759297	161371
16	523542	53881	35	766757	175666
17	523542	41191	36	974996	98298
18	231840	23516	37	601942	85130
19	159551	26995	38	158348	35816

Note. $N = 38$. X = Airport Service Area Population, Y = Annual Operations.

enplanements. Alternatively, if high values of one variable are related to low values of another variable, then the variables are *negatively related.* Two examples of a negative relationship are (a) the number of flight hours as PIC versus aviation safety incident rate per 10,000 flight hours, and (b) airline profitability versus the cost of jet fuel. The reader will note that these examples generally reflect *linear relationships.*

It also is possible for two variables to have a *nonlinear relationship*, including parabolic, logarithmic, and exponential. For example, we would expect the relationship between fuel burn rate (or fuel efficiency) and the true airspeed of an aircraft to be nonlinear. There also is the possibility that there is no systematic relationship between variables. For example, we would not expect commercial passenger enplanements to be systematically associated with airport elevation. Figure 1.1 contains a graphical illustration of general forms of bivariate relationships.

Visualizing Relationships via Scatter Plots

Although distribution tables such as Table 1.1 help us organize bivariate data, they rarely reveal potential relationships, patterns, or trends between variables. For example, examining the data presented in Table 1.1, it is almost impossible to detect if the relationship between X and Y reflects any of the forms given in Figure 1.1, or if there appears to be a pattern or trend between the variables. A better way to visualize the relationship between paired data from a correlation is via a scatter plot. A scatter plot consists of each *ordered pair* (x, y) plotted on the X–Y coordinate system where x scores are placed along the horizontal axis and y scores are placed along the vertical axis. Most statistics programs provide scatter plots as part of the results of a correlation analysis. The scatter plot for the bivariate distribution given in Table 1.1 involving annual operations and population among airports is illustrated in Figure 1.2.

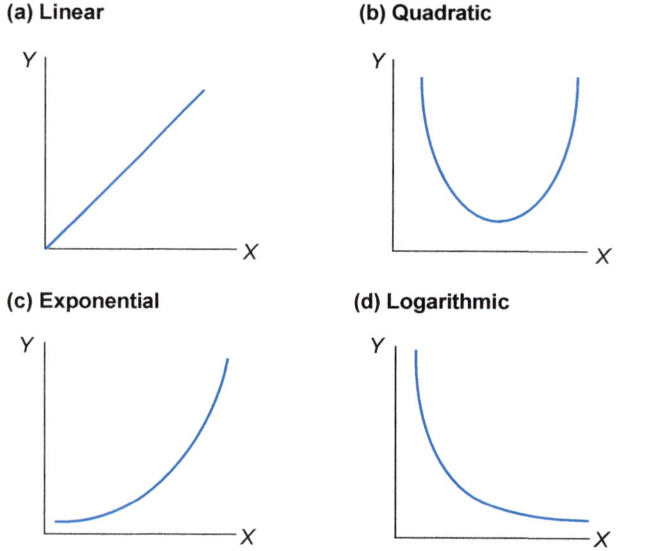

FIGURE 1.1 General forms of various bivariate relationships. (*Source:* Gallo et al., 2023, p. 134. Reprinted with permission.)

As illustrated in Figure 1.2(b), after plotting the paired observations, we draw an imaginary line, called the line of best fit, through the middle of the cluster of points. This line, which we will discuss in more detail in Chapter 2, represents the center of the relationship similar to the way a mean and median represent the center of a normal distribution and helps identify the general trend of the data. For example, if the line of best fit "rises" from left to right, then the values of X and Y tend to increase and decrease together. This appears to be the general trend for the annual airport operations and airport service area population data: high annual operations tend to be associated with high population numbers, and low annual operations tend to be associated with low population numbers. On the other hand, if the line of best fit "falls" from left to right, then the general trend is for high values of X to be associated with low values of Y, and vice versa.

It is important to understand that the line of best fit is not necessarily a "line," per se. If the paired data form a nonlinear relationship, then the line of best fit will follow the trend of the data. As an example, consider an extension to our current example that includes median household income ($), which we acquired from the U.S. Census Bureau (2024) database and manipulated for instructional purposes. A copy of this data set is provided in the file, "Ch_1 Annual Ops vs. Median Income." As illustrated in the scatter plot given in Figure 1.3, the relationship between Y = Annual Operations and X = Median Income is nonlinear (i.e., quadratic). Notice how the "line of best fit" follows the trend of the data even though it is not linear.

Characteristics of Relationships

The relationship between variables may be described by three primary characteristics: the *direction* of the relationship, the *form* of the relationship, and the *strength* of the relationship. A brief discussion of each characteristic follows.

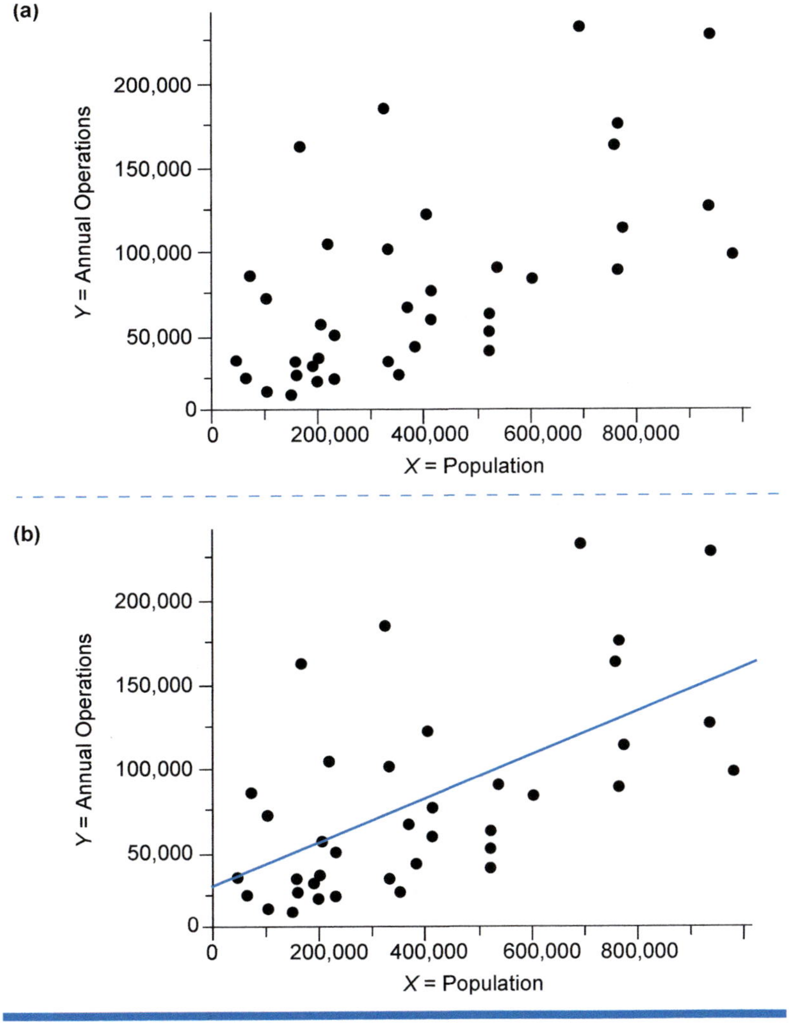

FIGURE 1.2 A bivariate scatter plot of the raw data in Table 1.1.

The Direction of a Relationship. As noted earlier, the direction of a relationship can be positive or negative. In a positive correlation, the variables tend to move in the same direction: they increase or decrease together. A positive relationship is characterized graphically in a scatter plot by a line of best fit that "rises" from left to right. A positive correlation also is referred to as a direct relationship. In Figure 1.2, the correlation between annual operations and service area population is positive because scores on both variables increase together.

In a negative correlation, the two variables tend to move in opposite directions: as one increases, the other decreases. A negative relationship is characterized graphically in a scatter plot by a line of best fit that "falls" from left to right. A negative correlation also is referred to as an inverse relationship. An example of an inverse relationship is illustrated in Figure 1.4, which is a scatter plot of the relationship

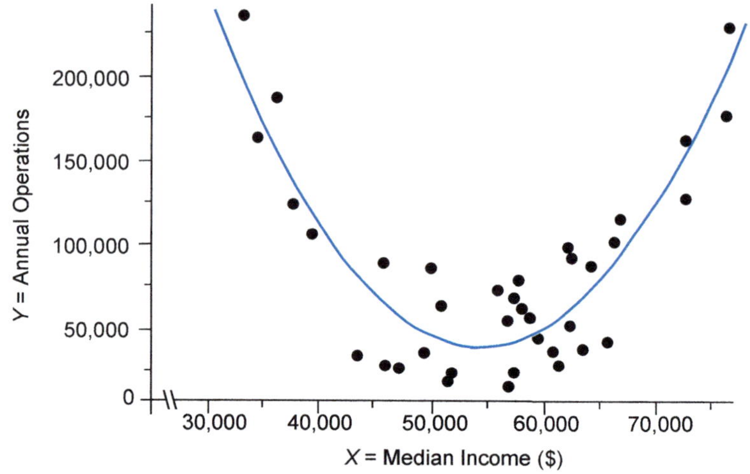

FIGURE 1.3 Example of a nonlinear relationship.

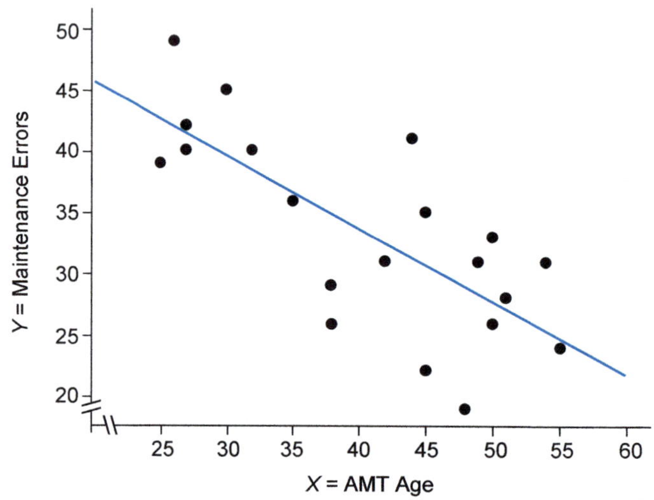

FIGURE 1.4 Scatter plot of the number of maintenance errors versus AMTs' age.

between *Y* = Maintenance Errors and *X* = Aviation Maintenance Technicians' (AMTs) Age: As AMTs' age increases, the number of maintenance errors they commit decreases (Uhuegho, 2017). A copy of this data set is provided in the file, "Ch_1 Maintenance Errors vs. Age" as a courtesy to the reader.

The Form of a Relationship. The form of a relationship refers to whether the relationship is linear or nonlinear. Although our discussion focuses on linear relationships, it also is possible to have nonlinear relationships as illustrated previously in Figure 1.3. The relationship initially is negative with increasing median household incomes associated with decreasing annual airport operations. As median household incomes continue to increase, they reach a point, with income between $55,000 and $60,000, where then the relationship becomes positive with higher median incomes associated with higher number of annual operations. Thus, the relationship *Y* has with *X* changes from negative to positive, resulting in high annual airport operations being associated with low *and* high median household incomes. In general, when the rate at which *Y* increases or decreases changes as *X* changes, the scatter plot has a "curved" pattern as opposed to a "straight" line pattern, and therefore nonlinear relationships are commonly called *curvilinear relationships*.

The Strength of a Relationship. Scatter plots can also provide a visual representation of the strength of a relationship, which is a function of how consistent the association is between variables. There are three general levels of consistency: *perfect*, *zero*, and *intermediate*, which is somewhere between zero and perfect. Examples of these are illustrated in Figure 1.5 and summarized below.

Perfect Correlation. A perfect correlation means that for every unit increase in *X*, there is a consistent and predictable increase in *Y*. When viewed from a scatter plot, all the data points of a perfect correlation fall on the line of best fit as shown in Figures 1.5(a) and 1.5(b). A perfect correlation may be positive or negative. It is rare to find this type of correlation in bivariate data.

Zero Correlation. A zero correlation means there is no consistency between variables. When viewed from a scatter plot, the data points are scattered randomly throughout the graph and appear as a cloud of points without any discernible pattern or trend such as a line or curve. An example of a zero (or near zero) correlation is shown in Figure 1.5(f), which is between pilot height and the number of FAA ratings. The line of best fit for a zero relationship is horizontal.

Intermediate Correlation. An intermediate correlation is neither perfect nor zero and indicates that as values of *X* increase, values of *Y* increase but they also might decrease. Thus, changes in *Y* with respect to changes in *X* are neither consistent nor predictable. When viewed from a scatter plot all the data points will not fall on the line of best fit. They instead will vary in how "close" they are to the line of best fit. In some cases, the points might be very close to the line as shown in Figure 1.5(c). In other cases, the points might be more spread out as shown in Figures 1.5(d) and 1.5(e).

1.3 The Pearson *r*

A correlation analysis generates a numerical index called the correlation coefficient, which is derived from the raw data. Similar to the mean and standard deviation, which respectively describe the center and spread of a distribution, a correlation coefficient is a statistical measure that numerically describes the strength of the relationship between sets of observations. One such index is the Pearson product moment correlation coefficient, denoted *r*, or more simply, the Pearson *r*. The Pearson *r* measures linear relationships between variables measured on an interval or ratio scale where the variables have a normal distribution for any one value of another variable. Thus, the Pearson *r* is not an appropriate measure for variables that have a nonlinear relationship, and it is not appropriate for variables measured on a nominal or ordinal scale.

The Pearson *r* describes the direction and strength of the relationship between two variables and can vary between –1.0 and +1.0. If $r > 0$, then the direction is positive; if $r < 0$, then the direction is negative; if $r = 0$, then there is no relationship,

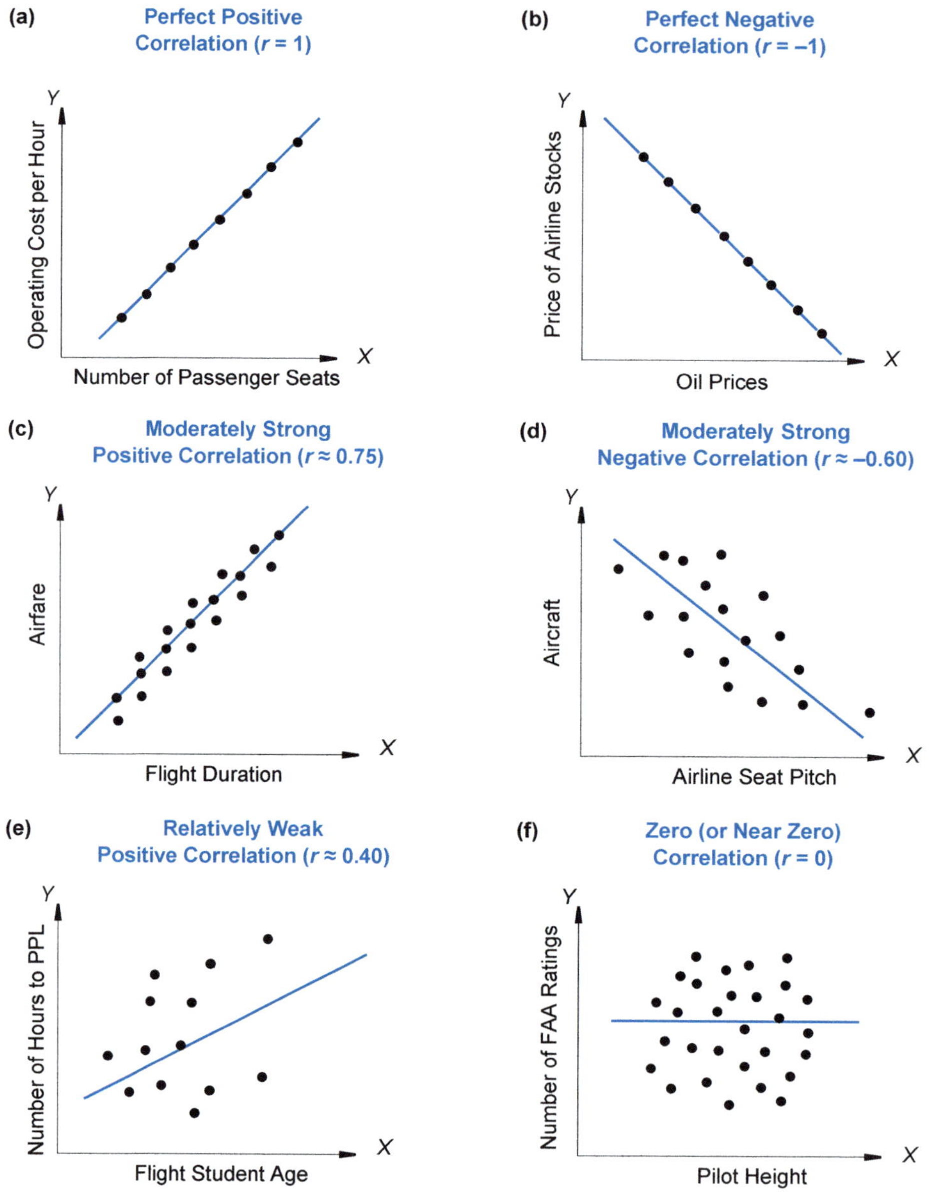

FIGURE 1.5 Scatter plots of six different data sets showing different degrees of linear relationship. (*Source*: Gallo et al., 2023, p. 138. Reprinted with permission.)

which means there is no consistency between the two variables; if $r = 1$ or if $r = -1$, then the relationship is perfect, which means for every unit increase in one variable, there is a consistent and predictable increase in the second variable. Furthermore, as r approaches 0 ($r \rightarrow 0$), the strength of the relationship gets weaker, and as r approaches +1 ($r \rightarrow 1$) or –1 ($r \rightarrow -1$), the strength of the relationship gets stronger. This is graphically illustrated in Figure 1.6. In practice, we do not calculate r manu-

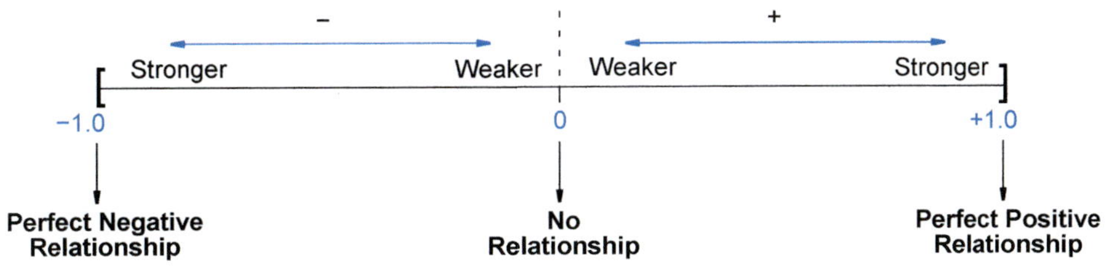

FIGURE 1.6 The range and structure of the Pearson *r*. (Adapted from Kachigan, 1991, p. 126.)

ally but instead rely on our statistics program to provide *r* as part of the output of a correlation analysis.

In addition to the Pearson *r*, several other correlation coefficients are used to measure relationships based on the measurement scale used: (a) *Spearman's rho coefficient* is used when the data are measured on an ordinal scale (i.e., ranked data); (b) the *point-biserial coefficient* is used when one variable consists of numerical data and the second variable is dichotomous; and (c) the *phi-coefficient* is used when both variables are dichotomous. Although the concept of correlation also applies to these coefficients, we restrict our discussion to the Pearson *r*.

1.4 Statistical Aspects of Correlation

Statistical Significance of r

The concept of statistical significance involves determining whether the results obtained from analyzing sample data are truly reflective of the parent population. In the context of correlation, a statistically significant correlation coefficient implies that the *r* value calculated from the sample data, denoted r^*, is most likely an accurate representation of the true measure of association between the variables in the parent population, instead of being due to chance. In other words, when we report a statistically significant correlation coefficient, we are implying that it reflects a true positive or negative relationship in the population and not a zero relationship. It also implies that if we were to randomly draw different samples of the same size from the same population and calculate the respective r^* values for each sample, then these r^* values would be similar.

To determine if a sample *r* is statistically significant, we compare it to a critical *r* value (r_{Critical} or r_C), which corresponds to the sample size and a preset alpha level. (*Note*: We assume the reader is familiar with the concept of *level of significance*, α, from a prior statistics course.) A table of critical *r* values is provided in Appendix B/Table B.2. Using this table for our current example where $N = 38$ ($df = 36$) and $\alpha = .05$, the critical *r* value for a two-tailed test is $r_C = \pm.320$. Because we have a two-tailed test, this means that the critical region will be to the right of $r = .320$ and to the left of $r = -.320$. If the calculated *r* value (r^*), falls in either region, then the correlation is significant and is either positive or negative depending on the direction of r^*. On the other hand, if r^* falls between these boundary values, then r^* is not significant, and there is no significant correlation in the population. As noted previously, the Pearson *r* for the data given in Table 1.1 is $r^* = .6044$. Because $r^* > r_{\text{Critical}}$, it falls in

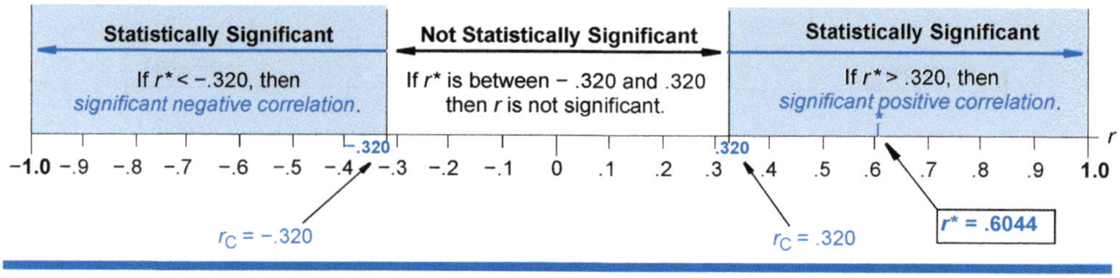

FIGURE 1.7 Illustration of critical r values for a two-tailed test of the correlation coefficient, r, based on a sample size of $N = 38$, $\alpha = .05$, and applied to the bivariate data given in Table 1.1. (Adapted from Gallo et al., 2023, p. 143.)

the critical region and thus the relationship between annual operations and airport service area populations is statistically significant. This is illustrated in Figure 1.7.

Effect of Outliers

An outlier is an extreme data point that is not consistent with other data points in a sample. Outliers may be errors or rare cases and can have a profound effect on the results of any data analysis, particularly in small sample sizes. This is illustrated in Figure 1.8. Because of the impact outliers can have on correlation, one of the assumptions of a correlation analysis is "no outliers," which means we must conduct an outlier analysis. There are several ways to conduct an outlier analysis. For our running example of airport service area population versus annual operations, we follow Stevens' (2002) approach to detecting outliers, which is based on the standard normal distribution (see Appendix A).

The reader might recall that according to the *empirical rule*, 99.7% of the data in a normally distributed variable lies within three standard deviations of the mean. Therefore, Stevens (2002) posited that an outlier in a normally distributed variable will have a standardized (z) score greater than $z = 3$ or less than $z = -3$ because such a score is unlikely to occur. Stevens further noted that even for *any* type of distribution this threshold is reasonable, although he suggested extending the rule to $|z| > 4$ (Stevens, 2002, p. 16).

For our running example, when we transform both the annual operations and population data in Table 1.1 to z scores, all the z scores for both variables are less than 2.7 and hence based on Stevens' outlier approach, the data set is free of outliers. The reader is informed that there are several other approaches to conducting an outlier analysis, including using *box plots*, the *interquartile range*, and *Jackknife distances*, and therefore it is important to report what approach is used because the results might vary. In practice, if outliers are present in a sample, rather than simply delete them to be compliant with the assumption, we perform two independent analyses: one in the presence and one in the absence of outliers. We also determine if the outliers are rare cases or contaminants. We then decide whether to include or exclude the outliers in the final analysis and inform the reader of our decision and corresponding reason(s).

The Coefficient of Determination

When we square the Pearson r, the result is an index called the coefficient of determination, denoted r^2, which represents the percentage of variance in one variable

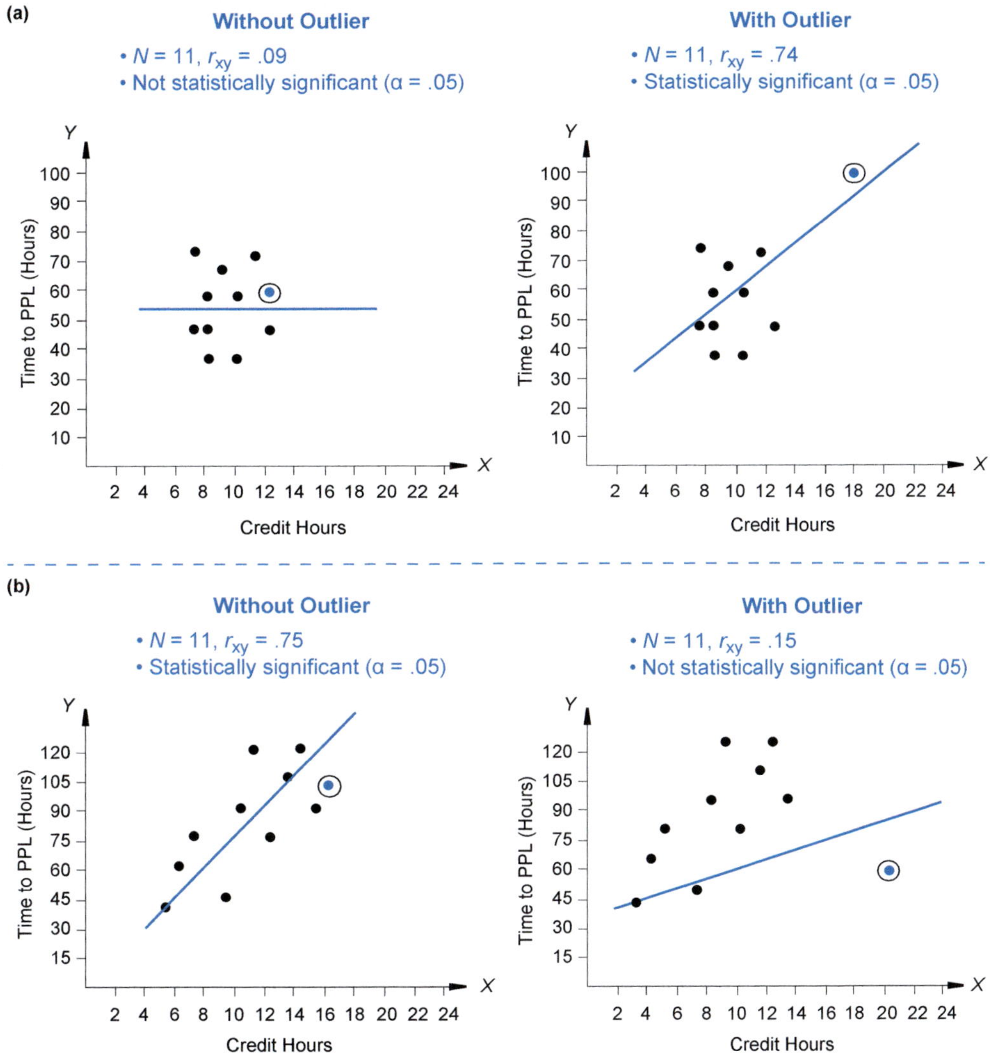

FIGURE 1.8 The effect of outliers on correlation. (a) Outliers can either inflate significance or (b) mask significance. (*Source*: Gallo et al., 2023, p. 144. Reprinted with permission.)

that is associated (or is common) with the percentage of variance in a second variable. For example, given $r = .6044$, which is the correlation between airport service area population and the number of annual aircraft operations from our running example, $r^2 = .3653$ and is interpreted as:

Approximately 36.5% of the variance in annual aircraft operations is explained by the variance in airport service area population.

We also can interpret the coefficient of determination from a prediction perspective:

If we know the airport service area population (X), then we have approximately 36.5% of the information needed to perfectly predict the number of annual aircraft operations (Y).

In a generic sense, we may think of r^2 as a measure of the amount of information one variable carries about another variable. Furthermore, because we are squaring r, a negative and positive correlation of the same magnitude are equivalent in their respective prediction capacity. So, if $r = .8$ and $r = -.8$, then $r^2 = .64$, and both correlation coefficients yield the same predictive information, namely, "X provides 64% of the information needed to perfectly predict Y." Note that r^2 can range from 0 to 1.00, and if $r^2 = 1.00$, then $r = 1.00$, which reflects perfect correlation. This means that if the correlation between given two variables is $r = 1$, then the scores from one variable can be used to *perfectly predict* the scores on the other variable.

1.5 Statistical Inferences Involving the Pearson *r*

Statistical inference can be approached directly via confidence intervals or indirectly via hypothesis testing. When applied to the context of bivariate correlation, the focus is the Pearson r, which is a sample statistic, and the corresponding population parameter, which is the Greek letter for "r," namely, rho, symbolized as ρ.

Confidence Intervals

The reader might recall from a previous statistics course that a confidence interval (CI) is based on a sample statistic and consists of a range of probable values that the corresponding population parameter can be assigned to estimate the "true" value of the population parameter. When a probability factor such as 95% is associated with a CI, then the range of probable values has a specific degree of certainty. For example, if the 95% CI for a mean score ranges from 13 to 19, then we would say there is a 95% chance that the population mean is somewhere between 13 and 19. This is denoted as 95% CI $\mu = [13, 19]$, where 13 is the *lower limit* and 19 is the *upper limit* of the interval. The general form of a confidence interval is

$$Population\ Parameter = Point\ Estimate \pm (SE)\ (t_{\text{Critical}})$$

When applied to correlation, the population parameter is rho (ρ), the point estimate is the Pearson r, SE is the standard error of the correlation, and t_{Critical} is the t value that denotes the boundary between regions of significance and nonsignificance. Furthermore, the general from for reporting the 95% CI for correlation is

$$95\%\ \text{CI}\ \rho = [a, b]$$

where a is the lower limit and is equal to $r - (SE)\ (t_{\text{Critical}})$, and b = the upper limit and is equal to $r + (SE)\ (t_{\text{Critical}})$. The reader is advised, though, that these calculations are not straightforward because correlation generally involve scores from different distributions. For example, in our running example, the measure of annual operations at an airport is quite distinct from a measure of population. To address this issue of disparate distributions, a statistical procedure called the Fisher z transformation is performed. In practice, statistics programs automatically provide the CI for the correlation coefficient, so it is not necessary to perform a z transformation.

For our purposes, we will focus on interpreting the CI for the correlation coefficient. As an example, as noted previously, the correlation between airport service area population and annual operations data given in Table 1.1 is $r = .6044$. As part of the output from our statistics program, the corresponding 95% CI $\rho = [.3530, .7745]$, which is interpreted as:

95% of the time, we can expect the relationship between airport service area population and annual airport operations in the FAA Southern region to be between .3530 and .7745.

An alternative interpretation is:

If we were to randomly select 100 samples of size N = 38 from the same population of airports in the FAA Southern region, then we would expect 95 of these 100 samples to have a correlation coefficient between r = .3530 and r = .7745.

When reporting a CI, it is informative to comment on the CI's precision, which is a function of the CI's width. If the width is relatively narrow, then we say that the CI provides an accurate estimate of the population parameter, but if the width is relatively broad, then the CI does not provide an accurate estimate of the parameter. This concept of how accurate a CI is in estimating the population parameter is called accuracy in parameter estimation (AIPE). When commenting on a CI's AIPE, some judgment needs to be applied. For example, with respect to our current example, the CI's width is .7745 − .3530 = .4215, which is a relatively wide interval. Because the precision is relatively low, this means that the AIPE also is low. Therefore, the sample of size N = 38, provides a relatively poor estimate of the true relationship between airport service area population and annual operations in the target population of airports in the FAA Southern region.

Hypothesis Testing

Independent of confidence intervals, we also can conduct a hypothesis test to acquire information about the population parameter. Unlike CIs, though, which provide a direct approach to statistical inference, a hypothesis test is an *indirect approach* to statistical significance. The reader might recall from a previous statistics course that to conduct a hypothesis test we: (a) first state a hypothesis about a targeted population parameter; (b) establish the test criteria that will be used to determine statistical significance; (c) collect data from a representative sample of the population relative to the hypothesis, check the corresponding assumptions associated with the statistical strategy being used, statistically analyze these data, and report the results; and (d) based on the results of data analysis, use the resulting sample statistic to either support or refute the hypothesis, and then make a concluding statement relative to the RQ. Applying these steps to bivariate correlation:

1. The population parameter is rho (ρ), and we hypothesize that the correlation between two variables in the population is zero ($\rho = 0$), which is the *null hypothesis*. The *alternative hypothesis* is that the relationship will be something other than zero, which means it could be "not zero" ($\rho \neq 0$), positive ($\rho > 0$), or negative ($\rho < 0$), depending on the researcher's prediction based on theory, prior research, preliminary studies, or personal expertise. The reader might recall that $\rho \neq 0$ leads to a two-tailed test of the hypothesis, whereas $\rho > 0$ or $\rho < 0$ leads to a one-tailed test.
2. Establish the test criteria, which consist of the level of significance and corresponding critical r value. The level of significance is traditionally based on an alpha level of $\alpha = .05$, and we may use either Appendix B / Table B.1 or Appendix B / Table B.2 to determine the critical boundary value(s).

3. Select a representative sample from the parent population, collect data from this sample, and analyze the data via a correlation analysis.
4. Determine if the sample correlation coefficient, r, reflects more than a chance deviation from the hypothesized population correlation coefficient, ρ. If it does, then we can infer that the relationship observed in the sample also exists in the population.

This procedure is illustrated next using the data set given in Table 1.1. The reader will recall that the overall RQ is, "What is the relationship between airport service area population and the number of annual operations at airports in the FAA Southern region?"

1. **Establish the null and alternative hypotheses.** The null hypothesis, denoted H_0, always claims "no significant relationship," whereas we have a choice for the alternative hypothesis, denoted H_1. For our running example, we believe, based on informal observations and personal experience, that the relationship will be positive.

   H_0: $\rho = 0$. There is no significant relationship between airport service area population and the number of annual operations at airports within the FAA Southern region.

   H_1: $\rho > 0$. There is a significant, positive relationship between airport service area population and the number of annual operations at airports within the FAA Southern region.

2. **Establish the test criteria.** This involves the following three components:

   Level of Significance (α). We set alpha to $\alpha = .05$.

   Test Statistic. The test statistic may be either r or t. We will use r.

   The Critical Boundary. The critical r value from Appendix B / Table B.2 for a one-tailed test to the right with $N = 38$, $df = 36$, and $\alpha = .05$ is $r_{\text{Critical}} = .271$.

3. **Collect data, check assumptions, run the analysis, and report the results.**

   Collect Data. The data are given in Table 1.1 as well as in the file "Ch_1 Annual Ops vs. Population."

   Check Assumptions. There are five assumptions for bivariate linear correlation, and we will assume all assumptions have been met.

   - The relationship is linear.
   - The variables are continuous (i.e., they are measured on an interval or ratio scale).
   - The paired data are from a random sample and independent of each other.
   - There are no outliers.
   - Both variables should be approximately normally distributed.

   Run the Analysis and Report the Results. When we run the analysis using our statistics program, as previously reported, $r^* = .6044$, and the corresponding r critical is $r_c = .271$.

4. **Reject or fail to reject H_0 and make a concluding statement relative to the RQ.**

   Decision. The decision is to *reject the null hypothesis* because $r^* = .6044$ is greater than r critical, $r_c = .271$ and hence lies in the critical region.

Concluding Statement. Relative to the preset alpha level of $\alpha = .05$, the sample data provide sufficient evidence of a significant positive, linear relationship between airport service area population and the number of annual operations at airports in the FAA Southern region: As airport service area population increases, annual operations also increase.

1.6 Additional Considerations

Correlation and Causation

Correlational studies are *observational* and do not involve manipulation and control (see Appendix A). Although there are many instances of variables that might seem to be plausibly related to other variables, they fail as causal explanations because there is no manipulation of the independent variable, and there is no control of extraneous variables that could provide alternative explanations for the results. This means that for our running example, we cannot conclude that the increase in the airport service area population *caused* annual operations to increase. It is possible that the observed relationship could be explained by one (or more) alternative explanations even though the two variables are significantly correlated. For example, it is plausible that the significant relationship is due to a third variable such as employment. As employment opportunities in the airport's service area increase, so does the population due to migration. Thus, it is conceivable that the two variables might not be related at all, but instead are separate results of a third variable: employment. This example demonstrates the concept of common cause.

Another alternative explanation is the concept of reverse causality, which is when the presumed cause (X) and the corresponding effect (Y) are reversed. For example, if there is a significant relationship between pilots' consumption of high-sugar diets and depression, it might not be that high-sugar diets cause depression in pilots, but instead, it is depression that causes pilots to consume high-sugar diets.

The bottom line is that an observed relationship between variables does *not* imply and must *not* be interpreted as a cause-and-effect relationship regardless of the strength of the relationship. In short: ***Correlation does not imply causation***. For a good demonstration of this point, the reader is directed to Vigen (n.d.), which presents several examples of relationships involving correlations of $|r| > .80$.

The Utility of Correlation Analysis

As noted by Kachigan (1991), various fields of study have key variables that need to be explored and better understood. The aviation discipline has an extensive set of important variables that overlap with several other fields of study, including but not limited to, econometrics, physiology, psychology and human factors, finance, airport and airline management, meteorology, and environmental science. Correlation analysis is useful in the aviation industry because it helps discover and utilize associations between variables. For example, in bivariate correlation, we may want to determine if an airline's organizational culture has a relationship with flight attendants' career satisfaction. We can then apply the results of this study to other pursuits if a relationship is found. The results of a correlation analysis also are often the forerunner to an experimental study, used to confirm or generate a hypothesis, and provide the ability to predict scores and determine the reliability of scores on a variable (Kachigan,

1991). More importantly, it is the foundation of other more advanced analyses addressed in subsequent chapters.

1.7 Using Bivariate Correlation in Research: A Guided Example

We now present a guided example of how to conduct a correlation analysis for a research study that involves two variables, one independent and one dependent. The research description for the study is outlined below.

> Inspired by Fernandez-Juricic et al. (2011), an aviation researcher conducted a follow-up study of the relationship between aircraft tail color schemes and the number of avian strikes, more commonly referred to as bird strikes, over the past 3 decades. The study focused on wildlife incidents involving Boeing B737 series aircraft at small, medium, and large hub airports in the U.S. The corresponding data set, which contains hypothetical data, is provided in the file "Ch_1 Guided Example Data," and the variables are defined as follows:
>
> - X = Aircraft Tail Color represents the RGB score. This score is the mean value of the red, blue, and green colors painted on the aircraft tail.
> - Y = Bird Strike Rate represents the number of bird strikes per 10,000 departures. For example, a bird strike rate of 0.124 means that 124 bird strikes occurred for every 10 million departures.

The guided example is presented in five parts: Research Questions and Statistical Hypotheses, Sample Size Planning, Data Analysis, Interpretation of Results, and Post Hoc Considerations.

Research Question and Statistical Hypotheses

The research question and corresponding null/alternative hypotheses are as follows:

Research Question. What is the relationship between aircraft tail color scheme and the occurrence of bird strikes at U.S. airports?

Hypotheses. The null and alternate statistical hypotheses are:

H_0: $\rho = 0$. There is no significant relationship between aircraft tail color schemes and bird strike rates.

H_1: $\rho \neq 0$. There is a significant relationship between aircraft tail color schemes and bird strike rates. We are uncertain of the direction, though.

Sample Size Planning

We use Appendix B/Table B.3 to determine the minimum sample size needed. To use this table, we need to know power, which is set at .80, and r, which is what we expect the correlation coefficient to be in the population relative to the alternative hypothesis. We will assume a large effect based on the findings of Fernandez-Juricic et al. (2011). According to Cohen (1988, pp. 79–80): $|r| = .10$ is a small effect, $|r| = .30$ is a medium effect, and $|r| = .50$ is a large effect. However, we do not assume the directionality of the association Fernandez-Juricic et al. observed because we are uncertain how the association between aircraft tail colors and bird strike rates may differ from that with aircraft fuselage color schemes. Therefore, for a large effect, using a two-tailed test for α = .05, the corresponding sample size is N^* = 28, which is the size of the given sample.

Data Analysis

The Test Criteria. We set alpha to α = .05 and use the Pearson *r* as our test statistic. To determine the critical value for the Pearson *r*, we use Appendix B / Table B.2 with respect to a two-tailed test, *N* = 28, and α = .05. This yields a critical *r* value of r_C = .374.

Check Assumptions. There are five assumptions: linearity, continuous variables, paired observations, no outliers, and normality.

- Linearity. As shown in Figure 1.9, the scatter plot and line of best fit confirm the linearity assumption.
- Continuous variables. Both variables are continuous. The aircraft tail color schemes measured by RGB values and bird strike rates are ratio scale measures.
- Paired observations. Every data point represents a paired response independent of each other. This condition is met in the data set.
- No outliers. We use a bivariate quantile range screening technique to detect outliers. For this analysis, outliers were set as values three times the interquartile range past the lower and upper quantiles. The results from our statistics program confirm the data set is free of outliers.
- Normality. To check the normality of each distribution, we use our statistics program to report the results of the Shapiro-Wilk Goodness of Fit test, *W*, for each variable. The null hypothesis of Shapiro-Wilk assumes a normal distri-

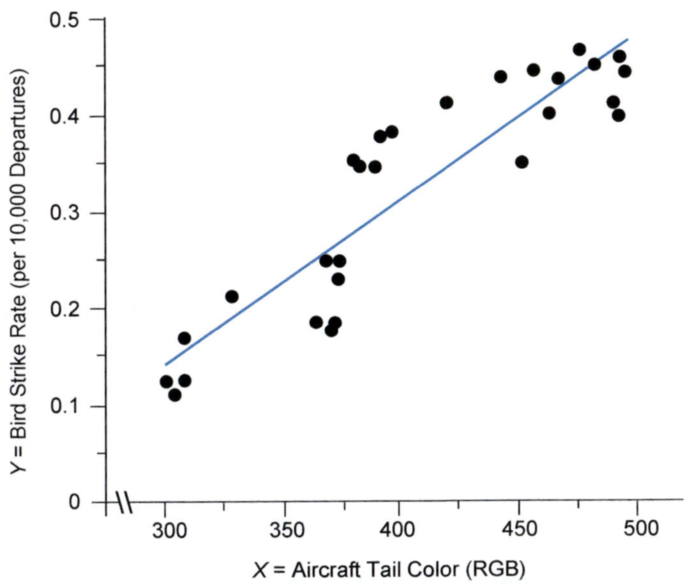

FIGURE 1.9 Visual confirmation of the linearity assumption for Chapter 1's Guided Example.

bution, which means if H_0 is rejected, then the data do *not* form a normal distribution. When applied to the Guided Example, the results are: (a) Y = Bird Strike Rate, $W = .886$, $p = .0056$; and (b) X = Aircraft Tail Color, $W = .920$, $p = .0338$. Thus, neither variable is normally distributed. As an alternative, some statistics programs also report the Shapiro-Wilk Test for Bivariate Normality. For the current example, this test yields $W = .937$, $p = .046$, which also indicates that the distributions are not normal. Nevertheless, we will continue with the analysis but advise the reader of this outcome. (*Note*: Another test for normality uses the concept of *residuals*, which is discussed in subsequent chapters. When the current example's normality assumption is examined from the perspective of residuals, the normality assumption is indeed met.)

We now conduct a bivariate correlation analysis on the data set, which yields the following results: $r^* = .9083$, $p < .0001$, 95% CI $\rho = [.8095, .9571]$, $r^2 = .825$.

Interpretation of Results

Because the calculated r value, r^*, is greater than r_{Critical}, $.9083 > .374$, we reject the null hypothesis and conclude there is a significant, positive relationship between aircraft tail color schemes and bird strike rates involving Boeing B737 series aircraft: As aircraft tail color schemes become brighter (increasing RGB values), the number of bird strikes that occur every 10,000 departures increases significantly.

Post Hoc Considerations

We now interpret the effect size, the 95% confidence interval, and power of the study.

Effect Size. Recall that the coefficient of determination, r^2, is used to measure the effect size. As noted above, this is $r^2 = .825$ and may be interpreted from either an explained variance or prediction perspective:

- Explained variance. $r^2 = .825$ indicates that 82.5% of the variance in bird strike rates is explained by the variance in aircraft tail color schemes (and vice versa).
- Prediction. $r^2 = .825$ indicates that if we know the aircraft tail color scheme, then we have 82.5% of the information needed to perfectly predict the bird strike rate.

95% CIs. As reported above, the 95% confidence interval is [.8095, .9571]. Therefore, 95% of the time we can expect the true relationship between bird strike rates and aircraft color scheme in the population to range between $r = .8095$ and $r = .9571$. Alternatively, if we were to randomly select 100 samples of size $N = 28$ from the same population, then 95 of these samples would have a calculated r value between $r = .8095$ and $r = .9571$. Furthermore, given H_0: $\rho = 0$, because 0 is not within this interval, we reject H_0. We also point out that the width of the interval is $.9571 - .8095 = .1476$, which is fairly narrow and infers that the corresponding AIPE is relatively good. Therefore, the sample data provide a good estimate of the true relationship between the targeted variables in the population.

Power (Post Hoc). We use Appendix B / Table B.3 for a two-tailed test to determine the approximate post hoc power of the study. Because $r^* = .9083$, we focus on the $r = .90$ column and then scroll down to find an entry as close to the sample size of $N = 28$, which is the very last one where $N = 11$. This corresponds to power = .99. Because the sample size is greater than 11, the power of the study is greater than .99. Therefore, there is a greater than 99% probability that we made the correct decision to reject H_0, and there is a greater than 99% probability that the large effect found in the sample truly exists in the population.

Chapter 1 Summary

1.2 The Concept of Correlation

- A correlational study examines the relationship between variables and involves collecting data on two or more variables from a single group or observational unit.
- A correlation analysis is a statistical procedure for identifying, describing, and measuring the relationship between variables. If the analysis involves exactly two variables, it is called bivariate correlation.
- A scatter plot is a graphical illustration of a bivariate relationship and consists of a set of ordered pairs (x, y), which represent the paired data for each observation. A line of best fit is used in tandem with a scatter plot to help identify the general trend of the relationship.

1.3 The Pearson *r*

- The characteristics of a relationship include its direction (positive or negative), form (linear or nonlinear), and strength.
- To quantify the strength of a relationship, we examine the corresponding correlation coefficient, which is a numerical index derived from the raw data. The Pearson *r* is a numerical index used to measure linear relationships.
- The Pearson *r* can vary between −1 and +1: $r = 1$ or $r = -1$ is a perfect relationship, $r = 0$ is a zero relationship, and $-1 < r < 1$ is an intermediate relationship. The corresponding population parameter is rho (ρ).

1.4 Statistical Aspects of Correlation

- The concept of statistical significance involves determining whether the results obtained from analyzing sample data are truly representative of the parent population. A statistically significant correlation coefficient implies that the *r* value calculated from the sample data is most likely an accurate representation of the true measure of association between the variables in the parent population, instead of being due to chance.
- Outliers can mask a significant relationship or inflate a nonsignificant relationship and therefore need to be examined prior to conducting a correlation analysis.
- The coefficient of determination, r^2, is an effect size metric that can be interpreted from the perspective of explained variance or prediction.

1.5 Statistical Inferences Involving the Pearson *r*

- Statistical inference can be approached directly via confidence intervals or indirectly via hypothesis testing.

- A CI's AIPE is a function of the width of the CI. This width relates to its precision and informs us of how accurate the CI is in estimating the population parameter.
- A hypothesis test of a bivariate correlation involves presenting null and alternative hypotheses, establishing the test criteria, collecting data from a representative sample, checking the data set for compliance with the assumptions of correlation, running the analysis, reporting the results, rejecting or failing to reject the null hypothesis relative to the test criteria, and making a concluding statement with respect to the RQ.

1.6 Additional Considerations

- The aviation discipline has an extensive set of important variables that overlap with several other fields of study. A correlation analysis is useful in the industry because it allows us to discover and utilize associations between variables.
- A correlation analysis can be the forerunner to an experimental study, used for the confirmation or generation of a hypothesis, and provide the ability to predict scores on a variable.
- A significant correlation coefficient does *not* imply causation because it is possible that the relationship could be explained by common cause, reverse causality, or other factors that were not considered.

Key Terms and Concepts

Accuracy in parameter estimation
Bivariate correlation
Bivariate data
Bivariate distribution
Cause-and-effect relationship
Coefficient of determination (r^2)
Common cause
Confidence interval (CI)
Correlation
Correlation analysis
Correlation coefficient
Critical *r* value
Curvilinear relationship
Direct relationship
Direction of a relationship
Fisher *z* transformation
Form of a relationship
Hypothesis test
Intermediate correlation
Inverse relationship
Line of best fit
Negative correlation
Pearson product moment correlation coefficient
Pearson *r*
Perfect correlation
Positive correlation
Reverse causality
Rho (ρ)
Scatter plot
Shapiro-Wilk Goodness of Fit Test
Shapiro-Wilk Test for Bivariate Normality
Standardized scores
Strength of a relationship
t statistic
Zero correlation

Review Exercises

A. Checking Conceptual Knowledge

In 1–10, choose the best answer among the choices provided.

1. Which of the following statements is true about bivariate correlation?
   a. Bivariate correlation measures the strength and direction of the relationship between two variables.
   b. Bivariate correlation determines the causality between two variables.
   c. Bivariate correlation is used to assess the variance within a single variable.
   d. Bivariate correlation can only be applied to categorical data.
2. An outlier in a correlation analysis is a score that
   a. makes $r = 1$. b. makes $r = 0$.
   c. increases or decreases r. d. is usually of no concern.
3. When compared to r, the coefficient of determination, r^2
   a. is always larger than r.
   b. can be larger or smaller than r.
   c. is never greater than r.
   d. is negative if r is negative.
4. Which of the following correlation coefficients has the least predictive value?
   a. $r = .68$ b. $r = .21$ c. $r = -.63$ d. $r = -.78$
5. One advantage a scatter plot has over the Pearson r is that a scatter plot can confirm
   a. that the form of the relationship is linear.
   b. whether the relationship between the variables is positive, negative, or zero.
   c. which variable is the IV, and which variable is the DV.
   d. that all the variables are being considered.
6. An aviation safety study assessed the relationship between aircraft performance metrics and operational factors. The study determined a relationship between fuel efficiency and the weight of the payload carried, with a coefficient of determination $r^2 = .73$. What does $r^2 = .73$ signify in this context?
   a. 73% of the aircraft's fuel efficiency is independent of its payload.
   b. 73% of the bivariate data set fell within the expected range of fuel efficiency.
   c. 73% of the variability in aircraft fuel efficiency can be explained by the variability in the weight of the payload carried.
   d. 73% of the payload carried influences operational factors.
7. An aviation researcher examines the relationship between the altitude at which a commercial jet starts its descent and the time it takes to reach the airport. The researcher reports a strong positive correlation between altitude and time. What can be inferred from this result?
   a. As the altitude at which the jet starts its descent increases, the time it takes to reach the airport decreases.
   b. As the altitude at which the jet starts its descent increases, the time it takes to reach the airport also increases.
   c. There is no relationship between the altitude at which the jet starts its descent and the time it takes to reach the airport.
   d. As the time it takes to reach the airport increases, the altitude at which the jet starts its descent decreases.

8. Researchers calculated the 95% confidence interval for the relationship between fuel consumption and trip length for a Piper Pilot 100 training aircraft. The 95% CI for the correlation coefficient r = [.6734, .6921]. What can be concluded from this CI?
   a. There is a 95% probability that the relationship between fuel consumption and trip length is $r < .6734$.
   b. 95% of the time we can expect the true relationship between aircraft fuel consumption and trip length to range between .6734 and .6921.
   c. There is a 95% probability that the relationship between fuel consumption and trip length is less than $r > .6921$.
   d. The relationship between fuel consumption and trip length is exactly r = .6828 with 95% confidence.
9. In an aviation study examining the relationship between aircraft distance traveled and fuel consumption, a hypothesis test is conducted to determine if there is a significant correlation between these variables. The null hypothesis states there is no significant correlation between aircraft distance traveled and fuel consumption. The level of significance was set to $\alpha = .05$ and the study determined a correlation coefficient $r = .42$. If the p value obtained from the test is $p = .03$, what does this imply about the correlation?
   a. There is no significant correlation between aircraft distance traveled and fuel consumption.
   b. There is a significant positive correlation between aircraft distance traveled and fuel consumption.
   c. There is a significant inverse correlation between aircraft distance traveled and fuel consumption.
   d. There is a significant correlation between aircraft distance traveled and fuel consumption only if the p value is less than .01.
10. An analysis is conducted to identify potential outliers in an aviation study investigating the correlation between commercial passenger enplanements and effective buying income. Suppose one data point is found significantly distant from the general trend of the data. How should this outlier be considered in the context of the bivariate correlation analysis?
    a. The outlier is only relevant if it falls outside the 95% CI of the data.
    b. The outlier should be included as it will provide additional data points to better understand the relationship.
    c. The presence of an outlier means that the bivariate correlation is not valid and should be disregarded entirely.
    d. The outlier should be removed from the analysis to ensure the correlation results are not masked or inflated.

B. Applying Procedural and Conceptual Knowledge

Use the following research description and the corresponding data set to conduct the activities given in Parts B1–B5. See Section 1.7 for guidance.

Motivated by Hunter (2006), coupled with the trend in increasing cases of air rage following the worldwide relaxation of COVID-19 travel restrictions, a study examined the relationship between X = Pent-up Demand related to the desire to travel and Y = Propensity for Air Rage in airline passengers.

The study used a researcher-constructed questionnaire designed to measure the domains of pent-up demand and air rage based on the literature. Pent-up demand was defined as scores on the Pent-up Demand Index (PDI), which consisted of 10 items, and air rage was defined as scores on the Air Rage Index (ARI), which consisted of 20 items. Both variables were scored on a traditional 5-point Likert response scale with 1 = Strongly Disagree and 5 = Strongly Agree. Thus, aggregate PDI scores could range from 10 to 50, with higher scores indicating a greater desire to travel following the relaxation of COVID-19 travel restrictions, and aggregate ARI scores could range from 20 to 100, with higher scores indicating a higher propensity toward air rage. A copy of the data set, which contains hypothetical data, is provided in the file "Ch_1 Review Exercises Data-Part B."

B1. Research Questions and Hypotheses. Write the overall research question and corresponding null and alternative hypotheses.

B2. Sample Size Planning. Conduct an a priori power analysis to determine the minimum sample size needed. Compare this result to the size of the given data set and explain what impact the size of the given sample will have on the results relative to the minimum size needed.

B3. Data Analysis. Conduct a hypothesis test of the correlation coefficient for the given data set and report the results.

B4. Interpretation of Results. Interpret the results relative to the research question and corresponding null hypotheses stated in Part B1.

B5. Post Hoc Considerations. Calculate and comment on the effect size, 95% CI, and power. Present a plausible explanation for the results.

References

Cohen, J. (1988). *Statistical power analysis for the behavioral sciences* (2nd ed.). Lawrence Erlbaum Associates.

Federal Aviation Administration (2024). *The operations network (OPSNET).* https://aspm.faa.gov/opsnet/sys/Main.asp

Fernandez-Juricic, E., Gaffney, J., Blackwell, B. F., & Baumhardt, P. (2011). Bird strikes and aircraft fuselage color: A correlational study. *The Human-Wildlife Interactions*, *5*(2), 224–234.

Gallo, M. A., Wheeler, B. E., & Silver, I. M. (2023). *Fundamentals of statistics for aviation research.* Routledge.

Hunter, J. (2006). A correlational study of how airline customer service and consumer perception of airline customer service affect the air rage phenomenon. *Journal of Air Transportation*, *11*(3), 78–109.

Kachigan, S. (1991). *Multivariate statistical analysis: A conceptual introduction* (2nd ed.). Radius.

Stevens, J. (2002). *Applied multivariate statistics for the social sciences* (4th ed.). Lawrence Erlbaum Associates.

Uhuegho, K. O. (2017). *Examining the safety climate of U.S. based aviation maintenance, repair, and overhaul (MRO) organizations*. [Doctoral dissertation, Florida Institute of Technology]. Scholarship Repository. https://repository.fit.edu/etd/37/

United States Census Bureau (2024). *Population and people*. https://data.census.gov/table

Vigen, T. (n.d.). *Spurious correlations*. https://www.tylervigen.com/spurious-correlations

Chapter 2
Bivariate Linear Regression

Student Learning Outcomes

After studying this chapter, you will be able to do the following:

1. Determine the equation of a regression line.
2. Determine and interpret the regression coefficient, *B*, the regression constant, B_0, and the 95% confidence interval of the regression coefficient.
3. Conduct and interpret the results of a hypothesis test of the regression coefficient.
4. Conduct a bivariate linear regression analysis for a research study by engaging in the activities presented in Section 2.6.

2.1 Chapter Overview

In this chapter, we introduce *bivariate linear regression* as an extension of bivariate correlation involving two variables. This statistical strategy describes the linear relationship between variables via a mathematical model called a *regression equation*, and then uses this equation to predict scores on a dependent variable (*Y*) from scores on an independent variable (*X*). We introduce the regression equation numerically and graphically and examine the two main terms of the equation: the *regression coefficient* and *regression constant*. We follow this with a discussion on various statistical aspects of regression, including confidence intervals and hypothesis testing involving the regression coefficient, and present information related to the accuracy of the prediction and the practical applications of regression analyses. We conclude the chapter with a guided example that demonstrates the application of bivariate linear regression in a research study that involves one independent variable and one dependent variable.

2.2 The Basics of Linear Regression

Overview

In Chapter 1, we learned that bivariate correlation involves examining relationships between paired scores from two variables. When combined with scatter plots and the Pearson *r*, the results of a correlation analysis provide information about the form of the relationship (e.g., linear vs. nonlinear), the direction of the relationship (e.g., positive, negative, or zero), and the strength of the relationship. Although informative, these results are limited in their application.

The utility of correlation, however, is enhanced when it is associated with regression. This enhancement comes in two forms. First, a regression analysis can be used to confirm or disprove the relationship between variables, and second, it can further describe the relationship between variables via an equation that could be used to predict *Y* from *X*. With respect to the former, consider the study by Ali and

DOI: 10.4324/9781003506935-3

Malik (2022) who examined the relationship between the "Big Five" personality traits and safety attitudes among aviators in Pakistan. They reported that extraversion (r = .571) and conscientiousness (r = .600) were significantly correlated with safety attitudes. They also reported that agreeableness (r = .225) and neuroticism (r = –.282) also had significant relationships with safety attitudes but these relationships were "marginal" (pp. 10–11). The results of an ensuing regression analysis confirmed that the two former traits were indeed significant predictors of safety attitudes among pilots, but the latter two were not. Thus, the two analyses complemented each other: They identified the nature of the relationships between the variables as well as where there was no systematic pattern of association.

With respect to prediction, recall from Chapter 1's Guided Example that the relationship between avian strike rate and aircraft tail color was r = .9083, which is a very strong, positive relationship. Recall further that r^2 = .825, which indicates that 82.5% of the variance in avian strike rate is being explained by aircraft tail color. If we were to now associate these results with regression, we would get a regression equation, called a *mathematical model*, which could be used to predict the avian strike rate based on aircraft tail color. Notice how these two sets of results complement each other and provide a more complete picture of the targeted relationship. Not only do we know how much of the variance in the avian strike rate is being explained by the aircraft tail color, but we now will have an equation that can predict this rate by the aircraft tail's mean RGB color. The greater the association between variables, the greater the predictive power of the regression.

In a bivariate regression analysis, the primary objectives are to (a) determine if a relationship exists between two variables, (b) mathematically model the nature of that relationship in the form of a regression equation, and (c) assess how accurate the mathematical model is in predicting the DV. For the remaining sections of the current chapter, we pursue these three objectives relative to bivariate linear regression to establish the foundation for linear regression. In subsequent chapters, we extend this discussion to *multiple regression/correlation*, which involves a single, continuous DV and multiple IVs. Although we acknowledge that nonlinear relationships also can be examined via linear regression (e.g., see Cohen et al., 2003 and Kachigan, 1991), we restrict our discussion to linear relationships throughout the remainder of this book.

The Equation of a Line

Recall from Chapter 1 that we superimposed the line of best fit in the "center" of the scatter plots to help identify the form of the relationship between the targeted variables. In regression, the line of best fit is critical because the equation of this line is used as the mathematical model. However, without an objective criterion for determining the center, the placement of the line is an educated guess and subjective. Therefore, to eliminate any uncertainty about where the line of best fit should be placed, we need a procedure that always defines a line that is indeed the "best fit" for a given bivariate distribution. As such, finding the equation for the best fitting straight line, called the regression line, is critical for linear regression because we use the equation, called the regression equation, to make predictions.

In linear regression, we use the slope-intercept form of a linear equation as the mathematical model for making predictions. The reader might recall from algebra that

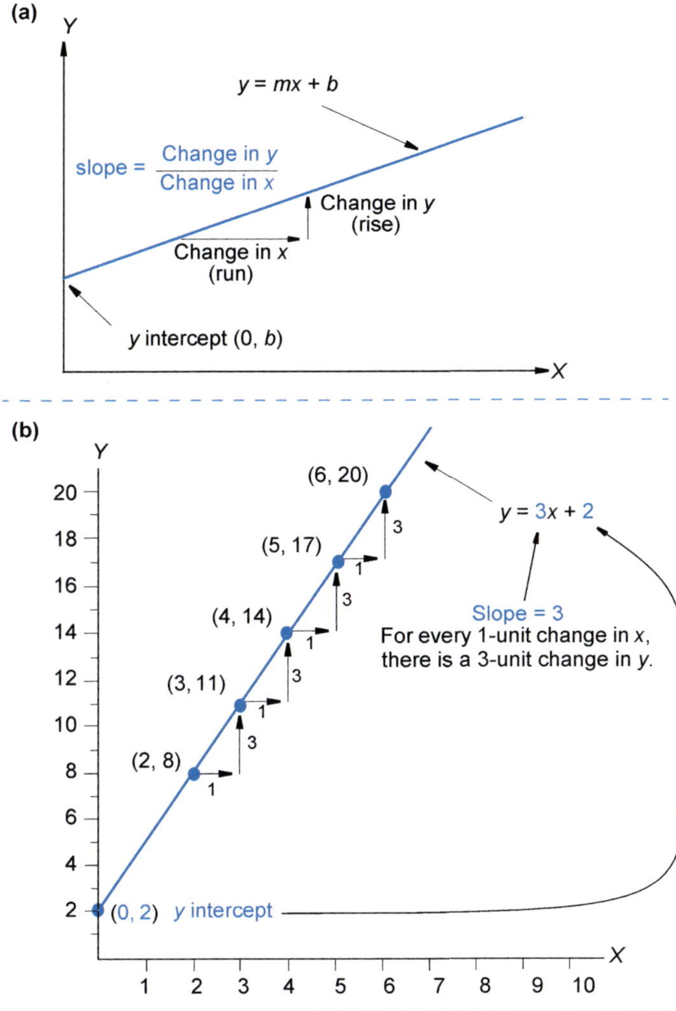

FIGURE 2.1 (a) Illustration of the general characteristics of a linear equation given in slope-intercept form, and (b) a specific application of the slope-intercept form of a linear equation. (*Source*: Gallo et al., 2023, p.161. Reprinted with permission.)

the slope-intercept form of the equation of a line is $y = mx + b$, where m is the slope of the line and b is the y-coordinate of the y intercept, (0, y), which is where the line crosses the y axis. This is illustrated in Figure 2.1(a). Note that slope is defined as the ratio of the change in vertical distance (Δy), called the "rise," to the change in horizontal distance (Δx), called the "run."

$$m = \text{slope} = \frac{\text{rise}}{\text{run}} = \frac{\text{Change in } y}{\text{Change in } x}$$

An application of the slope-intercept form is illustrated in Figure 2.1(b) for the equation $y = 3x + 2$. Note that the slope is equal to 3, which is $\frac{3}{1}$, and the y intercept is the point (0, 2). Therefore, for every 1-unit change in x there is a corresponding 3-unit change in y. We can confirm that the slope is $m = 3$ via the coordinates of each point. As illustrated in Figure 2.1(b), for each 1-point increase in x, there corresponds a 3-point increase in y. When applied to the context of bivariate linear regression, we alter the notation of the slope-intercept form slightly to represent the general form of the regression equation:

$$\hat{Y} = BX + B_0$$

From this representation, note the following:

- Instead of using Y, we use $\hat{Y}$ (pronounced Y hat) to signify that the equation is a prediction equation because we will use it to predict y values from x values.
- The slope, m, is denoted B and is called the regression coefficient.
- The y coordinate of the y intercept, b, is denoted B_0 and is called the regression constant.

A common procedure used to determine the regression line is called the least squares criterion, which minimizes the distance between the regression line and the points that represent the paired observations. For example, Figure 2.2 contains the scatter plot and line of best fit for Chapter 1's Guided Example where X = Aircraft Tail Color and Y = Bird Strike Rate. Note that Point 1, which corresponds to the observation (483, .451), falls directly on the line and thus has no deviation from the line; Point 2, which corresponds to the observation (330, .211), has a small amount of deviation; and Point 3, which corresponds to the observation (420, .412), has a considerable amount of deviation. The least squares criterion method produces a regression line that results in the smallest sum of squared deviations. This leads to a line where the distance each point is from the line is minimal. A regression analysis that uses the least squares criterion method is called ordinary least squares (OLS) regression.

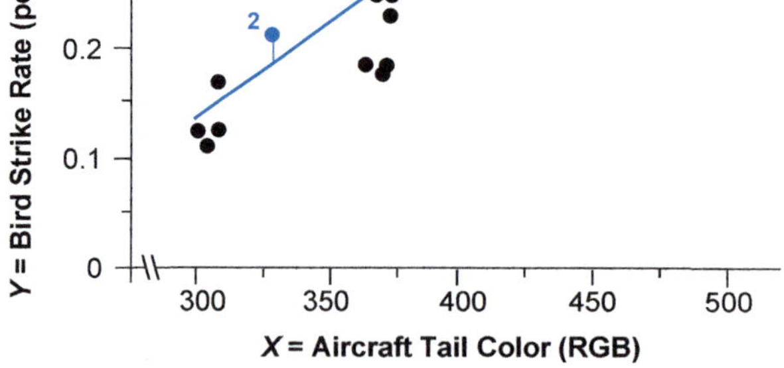

FIGURE 2.2 Illustration of the deviation between observed points and their corresponding predicted points on the regression line for Chapter 1's Guided Example. (Adapted from Kachigan, 1991, p. 165; Gallo et al., 2023, p. 163.)

Following is an example of a real-world setting that demonstrates the application of bivariate regression.

> Selçuk (2018) examined the relationship between airlines' average air fare in U.S. dollars ($) and route length, which was defined as the distance in miles between the origin and destination airports of city-pair markets. The airlines were a combination of legacy, low-cost, and hybrid carriers, and the city-pair markets reflected the top 1,000 in the U.S. for the total number of passengers each year from 2005 through 2015. The objective was to derive an equation that could be used to predict average air fares based on route length. A copy of the data set, which was modified for instructional purposes, is provided in the file, "Ch_2 Route Length-Average Air Fare Data."

The corresponding research question follows (operational definitions of key terms and variables are left to the reader), and a summary of the results of a bivariate regression analysis is provided in Table 2.1.

What is the effect of route length on the average air fares of major U.S. airlines between city-pair markets?

Observe from Table 2.1 that the regression coefficient is $B = 0.056$ and the regression constant is $B_0 = 125.93$. When expressed in the form of $y = mx + b$, the regression equation is $\hat{Y} = 0.056X + 125.93$. Note that we round the regression coefficient to three decimal places because the route lengths range from approximately 200 miles to 2,500 miles, and we round the regression constant to two decimal places because air fares are given in U.S. dollars. Further note from Table 2.1, that $r = .637$ and $r^2 = .405$. Thus, 40.5% of the variance in average air fares is being explained by

TABLE 2.1 *Summary of Bivariate Regression Analysis for the Route Length–Air Fare Data*

Model Summary

r	.637
r^2	.405
Adjusted r^2	.402
Standard Error of the Estimate	30.86
Mean of Response	178.94
Sample Size	183

Coefficients (Parameter Estimates)

Term	*B*	*SE*	95% CI	*t*(181)	*p*
Constant	125.934	5.290	[115.50, 136.37]	23.81	< .0001
X = Route Length	0.056	0.005	[0.05, 0.07]	11.11	< .0001

ANOVA

Source	*SS*	*df*	*MS*	*F*	*p*
Regression	117475.41	1	117475.41	123.43	< .0001
Residual (Error)	172389.43	181	952.43		
Total	289864.84	182			

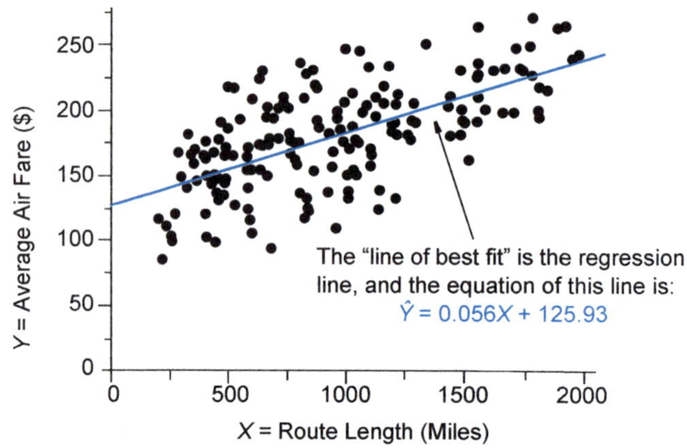

FIGURE 2.3 Graph of the regression line $\hat{Y} = 0.056X + 125.93$.

the route length, which is significant, $F(1, 181) = 123.34$, $p < .0001$. A scatter plot of the relationship with the line of best fit—that is, the regression line—is provided in Figure 2.3. Notice how the combination of correlation and regression provide a more complete picture of the relationship between these variables. Also note that the *y-intercept* is the point (0, 125.93), and the slope is $B = 0.056$, which is equivalent to the fraction 56/1000. Therefore, when the distance between city-pairs is 0 miles, average air fares start at $125.93. From that point forward, for every 1,000 miles in route length, the predicted average air fare is found graphically by moving to the *right* 1,000 miles on the *x* axis, and *up* $56 on the *y* axis. This is consistent with the interpretation of the slope: "For every 1,000-unit change in *X* there is a 56-unit change in *Y*." This also could be interpreted proportionally by progressively reducing the fractional slope:

- "For every 500-mile increase in route length, average air fares increase by $28."
- "For every 250-mile increase in route length, average air fares increase by $14."
- "For every 125-mile increase in route length, average air fares increase by $7."

Understanding the Regression Equation Relative to the Scatter Plot

Two distinct sets of points can be observed in a graph of a regression line accompanying a scatter plot. One set consists of the actual cluster of data points from the sample that makes up the scatter plot, which are the observed scores. The second set consists of the points that lie on the regression line, which are known as the predicted scores because the equation that corresponds to the regression line is used to predict *Y* scores from *X* scores. For example, in the scatter plot shown in Figure 2.4, each *x* value corresponds to two points. The first point, (x, y), is an observed score, and the second point, $(x, \hat{y})$, is the corresponding predicted score. It is the predicted *y* value given *x*, which is obtained from the regression equation by substituting the numerical value of *x* into the regression equation and solving for $\hat{y}$.

If the difference between y and $\hat{y}$ is 0 (i.e., $y - \hat{y} = 0$), then the predicted score matches the observed score, which means that the observed score falls on the regression line and there is no error in the prediction. One such point is depicted in Figure 2.4. If the difference between y and $\hat{y}$ is positive (i.e., $y - \hat{y} > 0$), then the

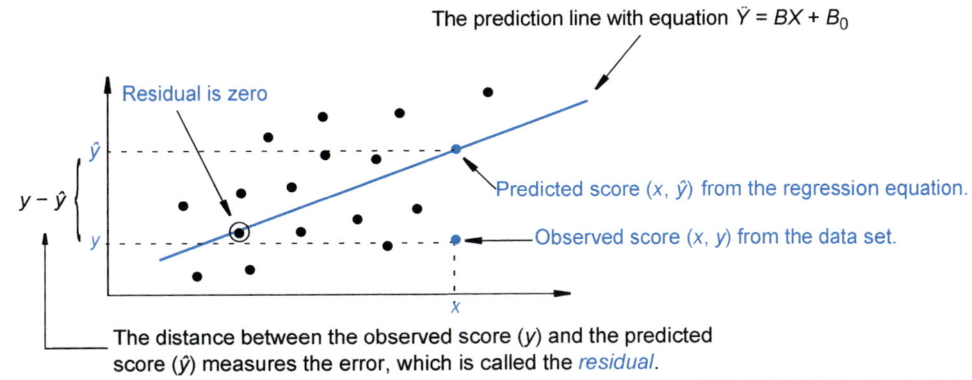

FIGURE 2.4 Illustration of a regression line highlighting the presence of two points that correspond to any given x value: the observed score (x, y), and the predicted score $(x, \hat{y})$. (Source: Gallo et al., 2023, p. 166. Reprinted with permission.)

observed score lies *above* the regression line, and if $y - \hat{y} < 0$, then the observed score lies *below* the regression line. The nonzero difference between y and $\hat{y}$ is the amount of *error* that exists between an observed score and its corresponding predicted score. As noted in Figure 2.4, this difference corresponds to the vertical distance between the observed and predicted scores. This error is called the residual and is denoted $y - \hat{y}$.

To illustrate this concept more concretely, let's examine the scatter plot, regression line, and regression equation for the data set involving annual operations vs. service area population, which we presented initially in Chapter 1 / Table 1.1 as well as in the file, "Ch_1 Annual Ops vs. Population." The corresponding scatter plot from Figure 1.2 is replicated in Figure 2.5 as a courtesy to the reader, and an extension of Table 1.1 that includes the predicted Y values and the residuals is provided in Table 2.2 (p. 35). Note that the regression equation is $\hat{Y} = 0.1320122X + 28592.968$. Let's now focus on Case 6, which is point A in Figure 2.5. The observed score for this airport is (406222, 122886), which means that this airport's service area population is $X = 406222$ and the number of annual operations is $Y = 122886$. To get the corresponding predicted score, we substitute 406222 into the regression equation and solve for $\hat{Y}$:

$$\begin{aligned}\hat{Y} &= 0.1320122X + 28592.968\\ &= 0.1320122(406222) + 28592.968\\ &= 53626.25991 + 28592.968\\ &= 82219.22791\\ &= 82219\end{aligned}$$

Therefore, based on the sample data, we predict that an airport with a service area population of 406,222 will have 82,219 annual operations. Note that the residual for this prediction is $Y - \hat{Y} = 122{,}886 - 82{,}219 = 40{,}667$. Because this difference is positive, the observed score lies *above* the regression line. Further, note that this difference of 40,667 represents how far "off" the predicted Y value is compared to the observed Y value. This is illustrated in Figure 2.5. It should be noted that when we

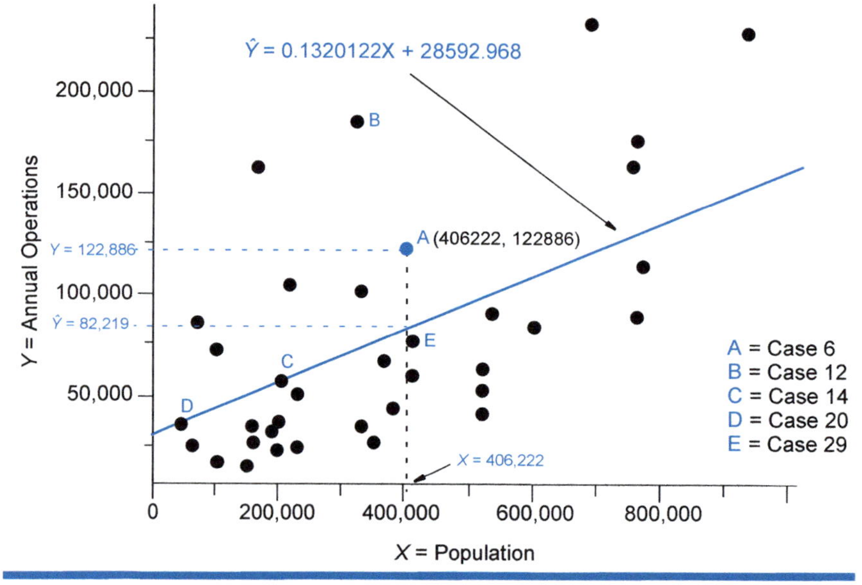

FIGURE 2.5 Scatter plot of the data set from the file, "Ch_1 Annual Ops vs. Population." (See also Table 2.2.)

calculated the predicted $\hat{Y}$ value using the regression equation, we did not round the regression coefficient or constant because the rounding error in the final score would have been too large.

Also observe from Figure 2.5 that the closer an observed score is to the regression line, the smaller the error will be. For example, the residual for Case 29 (point E) is –5885. Similarly, the farther away an observed score is from the regression line, the larger the error will be. For example, the residual for Case 12 (point B) is 113,765. However, if an observed score falls directly on the regression line, then the prediction for that case will be "perfect" because the observed and predicted scores would be equal, and the residual would be 0. Be wary of graphical illustrations, though. For example, although Case 14 (point C) and Case 20 (point D) in Figure 2.5 appear to fall directly on the regression line, their respective residuals are 344.94 and 531.03, neither of which is 0. This is because of the scales used for the axes coupled with the thickness of the line and points. Nevertheless, their respective residuals are very close to 0 relatively speaking.

Let's now extend this concept of perfect prediction for a single case to perfect prediction for *all* cases in a sample. From a graphical perspective, perfect prediction means that all the observed scores fall exactly on the regression line as was illustrated in Figures 1.5(a) and 1.5(b) in Chapter 1. It also implies that perfect correlation ($r = 1$ or $r = -1$) leads to perfect prediction. As the amount of error increases, though, the correlation approaches zero ($r \to 0$) and the gap between the predicted and observed scores increases, which leads to less accurate predictions. So, returning to Case 6 of our running example, we predicted that an airport with a service area population of 406,222 would have 82,219 annual operations, which is 40,667 more operations than what was observed. This prediction is not accurate because the correlation between the two variables reported in Chapter 1 ($r = .6044$) is not perfect.

TABLE 2.2 *Airport Service Area Population and Annual Operations at Airport Sites in the FAA Southern Region*

Site	X	Y	$\hat{Y}$	$Y-\hat{Y}$	Site	X	Y	$\hat{Y}$	$Y-\hat{Y}$
1	202518	36132	55328	−19196	20	47632	35412	34881	531.03
2	413210	60084	83142	−23058	21	192122	31494	53955	−22461
3	217702	104847	57332	47515	22	382295	44363	79061	−34698
4	694144	234964	120228	114736	23	231840	50716	59199	−8483
5	367804	66999	77148	−10149	24	766757	88907	129814	−40907
6	406222	122886	82219	40667	25	936250	127167	152189	−25022
7	354081	27184	75336	−48152	26	153159	15317	48812	−33495
8	166998	162816	50639	112177	27	74125	86686	38378	48308
9	105882	72453	42571	29882	28	937166	229451	152310	77141
10	335509	34695	72884	−38189	29	413210	77257	83142	−5885
11	770577	114863	130319	−15456	30	332285	101366	72459	28907
12	328297	185697	71932	113765	31	197938	22094	54723	−32629
13	523542	63036	97707	−34671	32	101511	16753	41994	−25241
14	208080	56407	56062	344.94	33	65418	25224	37229	−12005
15	537174	91134	99506	−8372	34	759297	161371	128829	32542
16	523542	53881	97707	−43826	35	766757	175666	129814	45852
17	523542	41191	97707	−56516	36	974996	98298	157304	−59006
18	231840	23516	59199	−35683	37	601942	85130	108057	−22927
19	159551	26995	49656	−22661	38	158348	35816	49497	−13681

Note. $N = 38$. X = Airport Service Area Population, Y = Annual Operations, $\hat{Y}$ = Predicted y values from the regression equation, $\hat{Y} = 0.132X + 28592.97$, and $(Y - \hat{Y})$ = the residual, which is the difference between the observed y value and the corresponding predicted y value. Cases 14 and 20 are given to two decimal places to match the in-text discussion.

When using a regression equation to make predictions, it also is important to be sensitive to the concept of the interval of predictability, which restricts predictions to the range of X values of the observed scores. For example, because the population range in our running example is between 48,000 and 1 million, we should restrict any predictions of annual operations to this range. As noted in our discussion of correlation, the prediction only applies to airports within the FAA Southern region with airport service area populations between 48,000 and 1 million. Making predictions outside this interval is risky because we do not have any data to represent that part of the relationship. It should be noted that the common industry practice of forecasting aviation demand using econometric regression analyses is also risky when predictions are made based on changes in an IV that represent data outside the range used to model the relationship with aviation demand. For example, the assumption that a regression model for predicting passenger enplanements at an airport using historical per capita income data is representative of the future would be risky if the future years' per capita income data are outside the range used to generate the regression model.

Interpreting a Regression Equation

The interpretation of a regression equation involves interpreting the regression coefficient, B, and the regression constant, B_0. To illustrate how this is done, we will use the regression equation for the population–annual operations study, $\hat{Y} = 0.13201X + 28593$.

Interpreting the Regression Coefficient *(B)*. To understand how to interpret a regression coefficient, it is helpful to begin with a generic interpretation and then progress toward an interpretation that is relative to the research setting. It also is helpful to recall that the regression coefficient is the slope of the regression line.

(a) A generic interpretation of the regression coefficient B:

For every 1-unit change in X, there will be,
on average, an estimated change of B units in Y.

(b) We refine this generic interpretation to include the regression coefficient, $B = 0.13201$:

For every 1-unit change in X, there will be,
on average, an estimated change of 0.13201 units in Y.

(c) We refine the previous interpretation even more to reflect the context of the given setting where X = population and Y = annual operations:

For every 1-person increase in an airport's service area population, there will be,
on average, an estimated increase of 0.13201 operations annually.

(d) Finally, to interpret it more realistically in the context of the given research setting where the population is expressed in units of *one hundred thousand*, we multiply "1-person" by a factor of 100,000, which means we also multiply B by a factor of 100,000:

For every 100,000-person increase in an airport's service area population,
there will be, on average, an estimated increase of 13,201 operations annually.

More succinctly: For every 100,000-person increase in the service area population of an airport in the FAA Southern region, we predict that the number of annual operations will increase on average by approximately 13,201 operations. Note that this is nothing more than interpreting the slope of a line, except now we are doing so with respect to a specific application.

Interpreting the Regression Constant (B_0). The regression constant is equal to the y value when $x = 0$. In the regression equation for the population–annual operations study, $\hat{Y} = 0.13201X + 28593$, the regression constant is $B_0 = 28593$ and is interpreted as follows:

When an airport's service area population is 0, we predict that,
on average, the number of annual operations will be 28,593.

In the context of the current study, B_0 is meaningless because an airport's service area population will never be zero, and if it were, the airport would be inactive or not operational. So, we would conclude that the regression constant is meaningless in the context of the given research setting. This is an important observation: Although the regression coefficient is always interpretable, sometimes a regression constant is not interpretable in a particular context. Thus, it is incumbent for us to determine if the constant is meaningful for the given research setting.

2.3 Statistical Aspects of Regression

Statistical Significance of Regression

The concept of statistical significance with respect to regression involves deciding if the sample regression coefficient, B, holds in the parent population. Because B is the slope of a regression equation, if there is no relationship between X and Y, then we would expect the slope to be zero, which means that the regression line is horizontal. On the other hand, if the relationship between X and Y is positive or negative, then the slope will be positive or negative, respectively, and the regression line will rise from left to right ($B > 0$) or fall from left to right ($B < 0$). To test the significance of the regression coefficient, we use the t statistic with $df = n - 2$ and we consult Appendix B/Table B.1 to determine the critical t value. This is demonstrated later in this section when we discuss hypothesis testing involving the regression coefficient.

Accuracy of a Prediction

A regression equation enables us to make predictions but it does not tell us anything about the accuracy of the predictions. Because a regression equation is derived from sample data, the slope, B, is considered a sample statistic and therefore is subject to *sampling error* just like any other sample statistic such as the mean. As a result, B is an estimate of the corresponding population parameter, β (beta), which is the Greek letter for "b." So, if we use a regression equation derived from sample data to predict y from x in the population, how accurate is the prediction?

To measure the accuracy of a prediction, we use the standard error of estimate ($SE_{\text{Residuals}}$), which is part of the output of a regression analysis. The standard error of estimate provides a measure of the distance between a regression line and the actual data points in the population, and it uses the standard deviation of the residuals from the sample. This is why we denote the subscript of the standard error of estimate as "residuals" ($y - \hat{y}$). It is the estimated population standard deviation of the residuals from estimating y from x and therefore is an indicator of the variability of errors, which gives us an idea of the accuracy of a prediction.

From the results of the regression analysis for the route length–average air fare example given in Table 2.1, the standard error of estimate is 30.86. This means that if we use the corresponding regression equation, $\hat{Y} = 0.056X + 125.93$, to predict an average air fare based on the route length, the predicted average air fare will be "off," on average, by about \$30.86. For example, if a route length is 1,000 miles, then

$$\begin{aligned}\hat{Y} &= 0.056X + 125.93\\ &= 0.056(1000) + 125.93\\ &= 56 + 125.93\\ &= 181.13\end{aligned}$$

Thus, the predicted average air fare is \$181.13, but the prediction is "off," on average, by about \$30.86. This result is reasonable given that the corresponding correlation coefficient is $r = .637$ and the coefficient of determination is $r^2 = .405$, which means that if we know the route length, then we have 40.5% of the information needed to perfectly predict the average air fare.

It is important to observe that although this metric is called the "standard error of estimate," it is *not* the same concept as the "standard error of the mean." A more appropriate term for standard error of estimate would be standard deviation of regression because it represents the standard deviation of Y for each corresponding X. Many statistical programs report the standard error of estimate as root mean square error (*RMSE*) where "root" is for square root, "mean" indicates we are taking the average squared deviations when we divide by df, and "error" reflects that this represents the "average error" we get when we use the regression equation to predict y from x.

2.4 Statistical Inferences Involving the Regression Coefficient

Statistical inference can be approached directly via confidence intervals (CIs) or indirectly via hypothesis testing. When applied to bivariate linear regression, the focus shifts to the regression coefficient, *B*, and the corresponding population parameter, β. We now extend our discussion of CIs and hypothesis testing from correlation in Chapter 1 to include regression.

Confidence Intervals

As discussed in Chapter 1, the general form of a CI is:

$$\textit{Population Parameter} = \textit{Point Estimate} \pm (SE)\,(t_{\text{Critical}})$$

When applied to regression, the point estimate is the regression coefficient, *B,* and *SE* is the standard error of the regression coefficient (SE_B), which is provided by the statistics program. Therefore:

$$95\%\ \text{CI for } \beta = [a, b]$$

where *a* is the lower limit, which is equal to $B - (SE_B)(t_{\text{Critical}})$, and *b* is the upper limit, which is equal to $B + (SE_B)(t_{\text{Critical}})$. In practice, we do not calculate these limits but instead rely on our statistics program to determine them.

As an example, let's revisit the regression analysis for the data set given in the file, "Ch2_Route Length-Average Air Fare Data." As reported in Table 2.1, the output for the 95% CI from our statistics program is [0.04626, 0.06626]. Interpreting this CI, we would report that with 95% certainty, in the population of the city-pair markets of U.S. airlines, the slope of the regression line that reflects the effect route length (*X*) has on average air fares (*Y*) is between 0.04626 and 0.06626. More concretely:

In the population of city-pair markets of U.S. airlines, 95% of the time we can expect that for every 1,000-mile increase in route length, the average air fare will increase, on average, between $46.26 and $66.26.

An alternative interpretation is:

If we were to randomly select 100 samples of size N = 183 from the same population of city-pair markets of U.S. airlines, then 95 of these samples will have a regression coefficient between B = 0.04626 and B = 0.06626.

Focusing on the precision of this CI, note that the width is 0.06626 – 0.04626 = 0.02, which is a moderate interval in this context. Because the precision is moderate, the accuracy in parameter estimation (AIPE) is also moderate (remember, this is a judgment call). To consider the impact of this CI's width, we calculate the range of the predicted air fares for the mean route length (M_X = 942 miles, rounded to nearest whole number) by replacing *B* with the CI's lower limit and upper limits and letting *X* = 942:

- Lower limit. $\hat{Y} = aX + 125.93 = (.04626)(942) + 125.93 = 43.57692 + 125.93 =$ 169.50692
- Upper limit. $\hat{Y} = bX + 125.93 = (0.06626)(942) + 125.93 = 62.41692 + 125.93 =$ 188.34692

Thus, the average air fare is predicted to range between $169.51 and $188.35, a difference of $18.84, which is not substantial when purchasing an airline ticket. So, a sample of size *N* = 183 provides a good estimate of the effect a city-pair's route length has on average air fares. Furthermore, because this CI does not include 0, which would reflect a zero slope (i.e., horizontal regression line), if we were to set up a hypothesis test that investigated the null hypothesis, H_0: β = 0, then based on the 95% CI, we would reject H_0 because 0 is not in the interval.

Hypothesis Testing

To conduct a hypothesis test involving bivariate linear regression, we follow the same procedure outlined in Chapter 1 except now the focus is on the regression coefficient, *B*, and its corresponding population parameter, β, instead of the correlation coefficient, *r* and ρ.

1. The population parameter is beta (β), and the *null hypothesis* is that the regression coefficient in the population is zero (β = 0), which is a horizontal line with zero slope. The *alternative hypothesis* is that the slope of the regression line in the population is "not zero" (β ≠ 0), positive (β > 0), or negative (β < 0), depending on the researcher's prediction. The reader might recall that β ≠ 0 leads to a two-tailed test, and β > 0 or β < 0 leads to a one-tailed test.
2. Establish the test criteria, which consist of the level of significance and corresponding critical *t* value. The level of significance is traditionally based on an alpha level of α = .05, and we use Appendix B / Table B.1 to determine the critical boundary value(s).
3. Select a representative sample from the parent population, collect data from this sample, and analyze the data via a regression analysis.
4. Determine if the sample regression coefficient, *B*, reflects more than a chance deviation from the hypothesized population regression coefficient, β. If it does, then we can infer that the slope of the regression line observed in the sample also exists in the population.

This procedure is illustrated next using the data set for the population–annual operations data, which was presented initially in Chapter 1 (see Table 1.1 and the file, "Ch_1 Annual Ops vs. Population") and then extended in Table 2.2 as part of our development for regression. Recall from this earlier discussion that the regression equation is $\hat{Y} = 0.1320122X + 28592.968$ and the corresponding scatter plot with the line of best fit is provided in Figure 2.5. The reader might also recall from Chapter 1 that the overall RQ for the correlation analysis was, "What is the relationship between airport service area population and the number of annual operations at airports in the FAA Southern region?" Modifying this RQ to reflect a regression analysis, the RQ for the current example is, "What is the effect of service area population on the annual operations of airports in the FAA Southern region?"

1. **Establish the null and alternative hypotheses.** The null hypothesis always claims, "no significant effect," whereas we have a choice for the alternative hypothesis. For the current example, and applying the position established in Chapter 1, we believe that the relationship will be positive, which means that the slope of the line will be positive.

   H_0: β = 0. The population of an airport's service area in the FAA Southern region does not have a significant effect on its annual operations (the slope of the regression line is zero).

   H_1: β > 0. The population of an airport's service area in the FAA Southern region has a significant, positive effect on its annual operations (the slope of the regression line is positive).

2. **Establish the test criteria.** This involves the following three components:
   Level of Significance (α). We set alpha to $\alpha = .05$.
   Test Statistic. The test statistic is *t*.
   The Critical Boundary. The critical *t* value from Appendix B / Table B.1 for a one-tailed test to the right with $N = 38$, $df = 36$, and $\alpha = .05$ is $t_{\text{Critical}} = 1.688$.
3. **Collect data, check assumptions, run the analysis, and report the results.**
   Collect Data. The data are given in Tables 1.1 and 2.2, as well as in the file "Ch_1 Annual Ops vs. Population."
   Check Assumptions. There are four assumptions for bivariate linear regression, and the single best way to confirm them is by examining residual plots involving the DV. We will demonstrate this in Section 2.6 when we present a guided example on how to conduct a bivariate regression analysis for a research study. For the current discussion, we will assume all assumptions are met.
   - Linearity. The relationship between *X* and *Y* must be linear.
   - Constant variance or *homoscedasticity of the residuals.* For any value of *X*, the variance of the residuals around the regression line in the population is assumed to be constant. This idea of "equal or constant variance of the residuals" is known as homoscedasticity (*homo* = same; *scedastic* = scattered).
   - Independence of the residuals. The residuals of the observations must be independent of one another. In other words, there must be no relationship among the residuals for any subset of cases in the analysis. This assumption will be met for any random sample.
   - Normality of the residuals. For any value of the IV, the residuals around the regression line are assumed to have a normal distribution.

   Run the Analysis and Report the Results. As reported in Table 2.3, the regression model is significant, $r^2 = .365$, $F(1, 36) = 20.7243$, $p < .0001$, and the regression equation is $\hat{Y} = 0.13201X + 28592.968$.
4. **Reject or fail to reject H_0 and make a concluding statement relative to the RQ.**
   Decision. The decision is to *reject* H_0 because the calculated t^* is $t(36) = 4.55$, which is greater than $t_{\text{Critical}} = 1.688$ and therefore lies in the critical region.
   Concluding Statement. Relative to the preset significance level of $\alpha = .05$, the sample data provide sufficient evidence that the slope of the regression line in the population differs significantly from zero. Based on the results of data analysis, airport service area population has a significant and positive effect on the annual operations of airports located in the FAA Southern region. More specifically, given a slope of $B = 0.13201$, for every 100,000-person increase in annual population, airport operations increase, on average, by approximately 13,201. This effect of population on annual operations is significant, $t(36) = 4.55$, $p < .0001$.

2.5 Correlation vs. Regression and Applications of Regression Analysis

A question often asked is, "Which statistical strategy—correlation or regression—is the better approach to use to examine linear relationships between two variables? To answer this question, it is helpful to briefly review the two concepts.

TABLE 2.3 *Summary of Bivariate Regression Analysis for Airport Service Area Population and Annual Operations*

Model Summary

r	.604
r^2	.365
Adjusted r^2	.348
Standard Error of the Estimate	47212.90
Mean of Response	81270.21
Sample Size	38

Coefficients (Parameter Estimates)

Term	*B*	*SE*	95% CI	*t*(36)	*p*
Constant	28592.968	13876.42	[450.28, 56735.66]	2.06	.0466
X = Population	0.1320122	0.028998	[0.07, 0.19]	4.55	< .0001

ANOVA

Source	*SS*	*df*	*MS*	*F*	*p*
Regression	4.6196e+10	1	4.62e+10	20.7243	< .0001
Residual (Error)	8.0246e+10	36	2.2291e+9		
Total	1.2644e+11	37			

Note. *N* = 38. "e+" denotes scientific notation. For example, 4.6196e+10 = 4.6196 × 10^{10}, which is 46,196,000,000.

- Correlation describes the *extent* of the relationship between two variables by quantifying the direction and strength of the relationship via the correlation coefficient, *r*. Furthermore, the square of *r*, namely, r^2, which is the coefficient of determination, provides a measure of explained variance, which indicates the amount of variance in one variable that is being explained by a second variable.
- Regression describes the *nature* of the relationship between two variables by quantifying the amount of change in one variable relative to changes in a second variable via the regression coefficient, *B*, which is the slope of the regression line. Furthermore, regression specifies one variable as the IV and the other variable as the DV, and unlike correlation, this designation is *not* interchangeable. A regression analysis also yields r^2 as well as a mathematical model of the form, $\hat{Y} = BX + B_0$, which is the equation of the regression line that can be used for prediction.

Thus, neither analysis is "better" than the other. The two approaches simply describe the linear relationship between variables from different perspectives, with *r* as the key index for correlation, and *B* as the key index for regression. The former asks about the relationship, whereas the latter is concerned with the effect the IV has on the DV (via its slope) and the regression model is used for prediction. Of the two approaches, though, regression, as Kachigan (1991, p. 186) observed, is more beneficial because of its ability to (a) predict scores on a DV with the knowledge of scores on an IV and (b) assess the amount of variance in the DV that can be explained by changes in the IV. Furthermore, because a regression analysis also provides everything that is produced from a correlation analysis, a regression analysis is often referred to by the catchall term, *regression/correlation*.

Kachigan (1991, pp. 186–189) also highlighted the ubiquity of regression by presenting several diverse examples where regression may be applied. This ubiquity is also present within the aviation discipline. For example, as noted in Chapter 1, applications of interest include econometrics, physiology, psychology and human factors, finance, airport and airline management, meteorology, and environmental science. These broad areas have several criterion and predictor variables of interest that may be studied. In aviation demand forecasting, econometric analyses include IVs such as population, employment, effective buying income, and retail sales. In airline management, profitability is a DV that can be associated with specific IVs such as organizational culture, aircraft maintenance costs, passenger and employee satisfaction, aircraft comfortability, fleet mix, load factors, network configuration, jet fuel costs, and product differentiation. In aviation psychology and human factors, variables of interest include flight crew self-efficacy, hazardous attitudes, fatigue, stress, and anxiety, as well as a lack of communication, distraction, complacency, and a lack of resources during aircraft operations. These variables are just a few that could be possible predictor variables within the aviation discipline.

For example, focusing on the airline industry, if an airline is interested in increasing profitability, it must first identify which variables are related to profitability, and this could be accomplished via a regression analysis. However, the airline also would need to specify the unit of analysis such as the route type (domestic or international), city-pair market, or fleet type before proceeding with the analysis. Although these variables are under an airline's control, other IVs are not such as airport landing fees and the cost per enplaned passenger. Global oil prices and jet fuel costs also are generally not under an airline's control but can be mitigated through fuel hedging. However, even in the instances where an airline would have no direct, or limited, control over an IV, a regression analysis still would be beneficial in providing information about future profitability when selecting routes to serve, markets to enter or exit, aircraft fleet type to operate, and so on. Additionally, those variables under an airline's control could be manipulated to test if they do indeed result in a change in profitability.

The airline could also approach the study of profitability from other perspectives. It could study crew bases or groups of employees and include other IVs such as flight crew experience and employee satisfaction. The airline may also consider time periods as its unit of analysis wherein profitability would be related to IVs such as cycles in the global economy (growth, crises, and recessions), seasonal changes in passenger demand, the time elapsed since newer aircraft were introduced to the fleet, and when aging aircraft retire. Returning passengers or those in a loyalty program could be the unit of analysis wherein passenger satisfaction would be an IV that provides information about factors that increase profitability.

As illustrated in the airline example, the choice of how to approach a study by identifying a suitable DV and IV is critical because several different bivariate regression models could yield information about airline profitability. Regression analyses go beyond creating and applying a regression equation to a set of data that conforms to the assumptions for the analysis. As discussed in the preceding sections, interpretations of the regression equation and regression coefficient and other statistical aspects of the analysis determine how useful these mathematical models are when applied in practice (Kachigan, 1991).

2.6 Using Bivariate Linear Regression in Research: A Guided Example

We now present a guided example of a research study that involves a bivariate linear regression analysis. The study is reminiscent of Jones et al. (1999) involving severe weather at a busy airport and focuses on examining the effect low-level (1,000 meters) wind speeds have on the onset of warm season thunderstorms at an airport in central Florida. The research objective is to generate a prediction equation that could be used to forecast within a 60-minute window the onset time of thunderstorms based on the wind speeds at 1,000 meters in the airport's vicinity. The existing forecast capability lies within a 2- to 3-hour time frame. Therefore, the ability to predict the timing of thunderstorms more precisely could improve planning for both commercial and general aviation operations. The data set, which contains actual observed data modified for instructional purposes, is provided in the file "Ch_2 Guided Example Data," and the variables are defined as:

- X = Wind Speed (knots) represents the 1,000-meter wind speed measured in nautical miles per hour as determined from radiosonde observations in the airport's vicinity at 1200 hours Zulu time (Z). (*Note*: 1200 Z is 7 AM Eastern Standard Time in the U.S.)
- Y = Onset Time of Thunderstorms is the time in Zulu hours during which thunderstorms commenced at the airport, as observed via weather radar.

The guided example is presented in five parts: Research Questions and Statistical Hypotheses, Sample Size Planning, Data Analysis, Interpretation of Results, and Post Hoc Considerations.

Research Question and Statistical Hypotheses

The research question and corresponding null/alternative hypotheses are as follows:

Research Question. What is the effect of wind speed at 1,000 meters on the onset time of thunderstorms at airports located in central Florida?

Hypotheses. The null and alternate statistical hypotheses are:

H_0: $\beta = 0$. Wind speeds at 1,000 meters have no significant effect on the thunderstorm onset time in the population of airports located in central Florida. (The slope of the regression line in the population is zero—horizontal line.)

H_1: $\beta \neq 0$. Wind speeds at 1,000 meters have a significant effect on the thunderstorm onset time in the population of airports located in central Florida. (The slope of the regression line in the population differs significantly from zero—oblique line.)

Sample Size Planning

When conducting an a priori power analysis to determine the minimum sample size for linear regression, it is helpful to understand the following:

- In regression, the effect size index is f^2 and is equal to $\frac{R^2}{1 - R^2}$, where R^2 is the coefficient of determination.
- Cohen (1988, pp. 413–414) provided metric-free effect sizes of "small" ($f^2 = .02$), "medium" ($f^2 = .15$), and "large" ($f^2 = .35$). For the current example, in the absence of any guidance, we posit a medium effect.

To determine the minimum sample size, we consult Soper (2025a) with the following inputs: (a) Anticipated effect size = 0.15, (b) Desired statistical power level = .8, (c) Number of predictors = 1, and (d) Probability level = .05. This returns a minimum sample size of N = 54. Given that the given data set has N = 65 cases, the sample is sufficiently large to have at least an 80% chance of correctly rejecting the null hypothesis for a medium effect.

Data Analysis

Prior to conducting the analysis, we establish the test criteria and check assumptions.

The Test Criteria. We set alpha to α = .05 and use Appendix B/Table B.1 to acquire the critical t value. For the current example where N = 65, df = 63, and α = .05, the table does not have an entry for df = 63, so we use the next *smaller* table value, which is df = 60. Thus, for a two-tailed test, the corresponding critical t value is $t_{\text{Critical}} = \pm 2.00$. A more precise critical t value may be obtained from an online calculator such as Student T-Value Calculator (2025). Consulting this site, $t_{\text{Critical}} = \pm 1.998341$.

Check Assumptions. As noted previously, bivariate linear regression has four assumptions: linearity, homoscedasticity of the residuals, independence of the residuals, and normality of the residuals. These assumptions are commonly examined via residuals plots as demonstrated here. Prior to doing so, though, it is prudent to conduct an outlier analysis because of the potential effect outliers could have on the results. (*Note*: In Chapter 5, we introduce the concept of *preliminary data screening*, which includes performing *regression diagnostics* such as checking for outliers, as well as checking the data set for compliance with the underlying assumptions of regression.)

- Outliers. For the current example, the respective univariate distributions for the two variables are approximately normal, and their respective standard deviations are within 2 *SD* of the mean for all cases indicating there are no outliers. As a complement to this analysis, we also examine the data set for outliers from a multivariate perspective using Jackknife distances, which also did not flag any outliers.
- Linearity. To test the linearity assumption, we examine a scatter plot with the residuals of Y placed on the vertical axis and the observed Y values placed on the horizontal axis as shown in Figure 2.6(a). This scatter plot clearly shows a linear relationship and therefore this assumption is met.
- Homoscedasticity of residuals. To confirm this assumption, we examine a scatter plot with the residuals of Y placed on the vertical axis and the predicted Y values placed on the horizontal axis to determine if there is any discernible pattern. Examples of such patterns include the data "fanning out" to resemble the shape of a "V" on its side or a curvilinear form. As shown in Figure 2.6(b), there is no detected systematic trend. This visual inspection is aided by placing a horizontal line at $y = 0$ to see if the residuals are approximately equally scattered above and below the line without any pronounced pattern. The scatter plot in Figure 2.6(b) clearly shows no patterns and therefore this assumption is met. (*Note*: This scatter plot also can be used to confirm the linearity assumption and is explained in more detail in Chapter 5.)

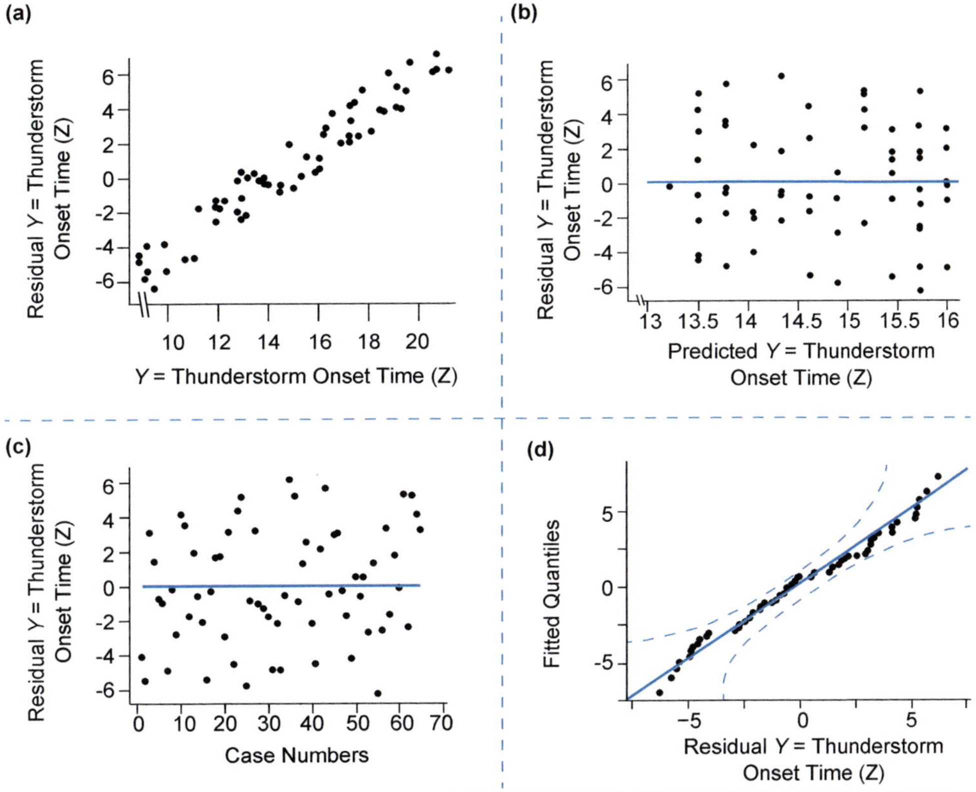

FIGURE 2.6 Scatter plots used to examine the assumptions of regression for Chapter 2's Guided Example.

- Independence of residuals. To confirm this assumption, we examine a scatter plot with the residuals of *Y* placed on the vertical axis and the case numbers, which are the row numbers, placed on the horizontal axis to determine if there is any systematic pattern. This visual inspection is aided by placing a horizontal line at $y = 0$ to see if the residuals are approximately equally scattered above and below the line without any pronounced pattern. As shown in Figure 2.6(c), there is no discernible pattern, and therefore, this assumption is met.
- Normality of residuals. To confirm this assumption, we report the results from the Shapiro-Wilk Goodness of Fit test, *W*, which is a statistical test for normality. The null hypothesis of this test is that the data are normally distributed. For the current example, $W = .971$, $p = .1249$, and therefore this assumption is met. Complementing this test is a *q-q plot*, which is shown in Figure 2.6(d). This plot, which is explained in Chapter 5, consists of a "normal line" placed at a 45-degree angle surrounded by a 95% confidence band. To satisfy the normality assumption, the data points should "hug" the normal line with the majority (if not all) falling within the 95% confidence band, which is the case for Figure 2.6(d).

TABLE 2.4 *Summary of Bivariate Regression Analysis for Chapter 2's Guided Example*

Model Summary

r	–.263
r^2	.069
Adjusted r^2	.054
Standard Error of the Estimate	3.31
Mean of Response	14.75
Sample Size	65

Coefficients (Parameter Estimates)

Term	*B*	*SE*	95% CI	*t*(63)	*p*
Constant	17.96	1.54	[14.89, 21.03]	11.68	< .0001
X = Wind Speed [a]	–0.28	0.13	[–0.54, –0.02]	–2.17	.0340

Note. [a] Wind speed is the velocity of the wind, in knots, at 1,000 meters.

ANOVA

Source	*SS*	*df*	*MS*	*F*	*p*
Regression	51.56	1	51.56	4.70	.0340
Residual (Error)	691.82	63	10.98		
Total	743.38	64			

Note. $N = 65$. Y = Onset time, in Zulu hours, of thunderstorms.

With the assumptions of bivariate linear regression met, we now run the analysis. As reported in Table 2.4, the regression equation is $\hat{Y} = -0.2794X + 17.960$ and the overall regression model is significant, $r^2 = .069$, $F(1, 63) = 4.6952$, $p = .0340$. (Alternatively, $t(63) = -2.17$ vs. $t_{\text{Critical}} = \pm 1.998341$.)

Interpretation of Results

Based on the results of data analysis, we reject the corresponding null hypothesis and conclude that 1,000-meter wind speeds measured at 1200 Z have a significant and inverse effect on the onset time of thunderstorms at airports located in central Florida during the warm season. Given slope $B = -0.2794$, the generic interpretation is: "When observed at an altitude of 1,000 meters, for every 1-unit change in X we can expect (or we predict) on average a B-unit change in Y." Applying this to the given research setting where X = wind speeds at 1,000 meters and Y = the onset time in Zulu hours of thunderstorms:

> *When observed at an altitude of 1,000 meters, for every 1-knot increase in wind speed, the onset time of thunderstorms in Zulu hours is –0.2794.*

More concretely, because (a) there are 60 minutes in an hour and (b) a negative value in the context of time equates to an *earlier* event (e.g., think of the countdown for the liftoff of a rocket):

> *When observed at an altitude of 1,000 meters, for every 1-knot increase in wind speed, thunderstorm activity will begin, on average, (–0.2794)(60) = –16.764, or approximately 17 minutes earlier.*

Post Hoc Considerations

We now present and interpret various aspects of the findings, including the effect size, coefficient of determination, the 95% confidence interval, the standard error of estimate, and post hoc power.

Effect Size. As noted previously, the effect size is $f^2 = R^2 / (1 - R^2)$. As reported in Table 2.4, $r^2 = .069$. Therefore, the effect size is $f^2 = .069/.931 = 0.07$, which according to Cohen (1998) is considered a small effect size.

Coefficient of Determination. Recall that the coefficient of determination, r^2, can be interpreted from either an explained variance or prediction perspective.

- *Explained variance.* $r^2 = .069$ indicates that 6.9% of the variance in thunderstorm onset time is explained by the variance in 1,000-meter wind speeds. Thus, 93.1% of the variance in thunderstorm onset time is explained by other factors.
- *Prediction.* $r^2 = .069$ indicates that if we know the 1,000-meter wind speeds at 1200 Z, then we have 6.9% of the information needed to perfectly predict the thunderstorm onset time at airports in located central Florida.

95% CIs. As reported in Table 2.4, the 95% CI for β is [–0.54, –0.02], which tells us that 95% of the time we can expect the true effect 1,000-meter wind speeds have on thunderstorm onset time to lie within this interval. In other words, 95% of the time we can expect thunderstorm onset times to occur anywhere from (–0.54)(60) = 32 minutes to (–0.02)(60) = 1.2 minutes *earlier* for every 1-knot increase in wind speeds at 1200 Z. Another way of looking at this is to say that if we were to randomly select 100 samples of size $N = 65$ from the same population, then in 95 of the samples, the regression coefficient would be between –0.54 and –0.02. Also recognize that given H_0: $\beta = 0$, because 0 is not within the 95% CI, we reject H_0. Finally, we conclude that based on the width of the 95% CI, the corresponding accuracy in parameter estimation (AIPE) is moderate (it's a judgment call). Thus, the given data set provides a moderate estimate of the true effect 1,000-meter wind speeds have on thunderstorm onset time in the population.

Standard Error of the Estimate (Accuracy of Prediction). Recall that the standard error of the estimate provides a metric of how far "off" on average a predicted value will be using the regression equation to predict *Y* from *X*. Based on the results from our statistics program, $RMSE = 3.31$, which indicates that if we use the regression equation to predict thunderstorm onset time based on 1,000-meter wind speeds, then the predicted time would be "off," on average, by approximately 3.31 hours. This accuracy fails to bring the objective to generate a regression equation that could predict the onset time of thunderstorms based on the 1,000-meter wind speeds in the airport's vicinity within a 60-minute window in reach. Therefore, given the existing forecast capability, which, as noted earlier, is within a 2- to 3-hour time frame, the results do not provide any improvement over the current forecast.

Power (Post Hoc). To determine the approximate power of the study, we consult Soper (2025b) with the following inputs: (a) Number of predictors = 1, (b) Observed $R^2 = .069$, (c) Probability level = .05, and (d) Sample size = 65. This returns a power of .59. Thus, there is a 59% probability that the small effect found in the sample truly exists in the population, and there is a 59% probability that the decision to reject the null hypothesis is correct. The reader will note that although sample size planning indicated that the sample size of the given data set was sufficiently large vs. what was needed ($N = 65$ vs. $N = 54$), this was based on a medium effect of $f^2 = 0.15$. For the current study, a small effect size was found ($f^2 = 0.07$). Consulting Soper (2025a), to find an effect of $f^2 = 0.07$ for $\alpha = .05$, a sample size of $N = 112$ is needed. Thus, the sample size of the given data set was insufficient.

Chapter 2 Summary

2.2 The Concept of Regression

- A regression analysis is used to determine if a relationship exists between two variables, to mathematically model the nature of that relationship in the form of a regression equation, and to assess how accurate the mathematical model is in predicting the DV.
- A regression analysis is often paired with correlation, and the collective results provide a more complete picture of the relationship between variables.

2.3 Bivariate Linear Regression

- Regression is a statistical procedure for describing the nature of the relationship between two variables via a mathematical model called the regression equation.
- The graph of a regression equation is called the regression line, which has the general form, $\hat{Y} = BX + B_0$, where $\hat{Y}$ is the predicted score; B is the regression coefficient, which is the slope and reflects the amount of change in Y relative to changes in X; and B_0 is the regression constant, which is equal to the predicted score when $X = 0$.
- The data points in a scatter plot are observed scores, and the points on the regression line are predicted scores. When a regression line is drawn within a scatter plot, there are two y scores for every x score. One is the observed score (x, y), and the second is the corresponding predicted score $(x, \hat{y})$.
- The difference between an observed score, y, and corresponding predicted score, $\hat{y}$, is the residual (error), which is the vertical distance between the two points $(y - \hat{y})$.
- To use a regression equation for prediction, we substitute an x score into the equation and perform the arithmetic. The result is the predicted score, $\hat{y}$, and its corresponding point $(x, \hat{y})$ falls on the regression line.

2.4 Statistical Aspects of Regression

- The standard error of estimate ($SE_{Residuals}$), which is generally reported as the root mean square error (*RMSE*), provides an index of the accuracy of a prediction.
- When using a regression equation for prediction, we should not use scores of x that are outside the interval of predictability.
- The effect size of a regression analysis is given by $f^2 = R^2 / (1 - R^2)$. Effect sizes of $f^2 = .02$, $.15$, and $.35$ are considered small, medium, and large, respectively (Cohen, 1988).

2.5 Statistical Inferences Involving the Regression Coefficient

- Statistical inference can be approached directly via confidence intervals or indirectly via hypothesis testing.
- A CI for a regression coefficient reflects the lower and upper limits of the slope of a regression line.
- A hypothesis test involving a bivariate linear regression involves presenting null/alternative hypotheses for β, establishing the test criteria, collecting data from a representative sample, confirming that the data set is compliant with the assumptions of regression, running the analysis, reporting the results, rejecting or failing to reject the null hypothesis relative to the test criteria, and making a concluding statement with respect to the RQ.

2.6 Correlation vs. Regression and Applications of Regression Analysis

- Correlation quantifies the direction and strength of a relationship between variables via *r*, whereas regression quantifies the amount of change in one variable relative to changes in a second variable via *B*. Both statistical strategies complement each other and are often referred to collectively as "regression/correlation."
- There are numerous ways in which a regression analysis may be applied in practice. Given the possibility of choosing among alternative IVs to study, a recommended approach is to first identify the unit, group, or object of the study.
- A regression analysis is beneficial in providing information even in instances where the researcher has no direct, or limited, control over an IV.
- The applicability of a regression equation in practice is in its interpretation of the regression coefficient, amount of explained *Y* variance, and accuracy of prediction (*RMSE*).

Key Terms and Concepts

Beta (β)
Effect size (f^2)
Homoscedasticity of the residuals
Independence of the residuals
Interval of predictability
Least squares criterion
Normality of the residuals
Observed scores
Ordinary least squares (OLS) regression
Perfect prediction
Predicted scores
Prediction equation
Regression coefficient (*B*)
Regression constant (B_0)
Regression equation
Regression line
Residual plots
Residuals ($y - \hat{y}$)
Root mean square error (*RMSE*)
Slope
Slope-intercept form
Standard deviation of the regression
Standard deviation of the residuals
Standard error of estimate ($SE_{Residuals}$)
Standard error of the regression coefficient
y intercept
Y-hat ($\hat{Y}$)

Review Exercises

A. Checking Conceptual Knowledge

In 1–10, choose the best answer among the choices provided.

1. The slope of a regression line
   a. is always positive.
   b. means that if the value of *X* increases by 1 unit, then *Y* increases by the value of the slope units.
   c. is the value of the line where the line crosses the *y* axis.
   d. is denoted B_0.
2. The *y* value of the *y* intercept of a regression equation represents the
   a. predicted value of *y*.
   b. average distance between observed and predicted scores.
   c. average amount of change in *y* per unit change in *x*.
   d. predicted value of *y* when $x = 0$.
3. If the slope of a regression equation is $B = 0$, then the line is displayed in a scatter plot as
   a. a vertical line.
   b. a horizontal line.
   c. a line that rises from left to right.
   d. a line that falls from left to right.
4. In a study examining the relationship between flight altitude (in feet) and fuel consumption (in gallons) for a specific aircraft, which statement best describes the purpose of conducting a bivariate regression analysis?
   a. To predict the fuel consumption based on various flight speeds.
   b. To determine if there is a significant correlation between flight altitude and fuel consumption and predict fuel consumption based on flight altitude.
   c. To analyze the effect of weather conditions on flight performance.
   d. To compare fuel efficiency between different aircraft models.
5. Which statement about linear regression is NOT accurate?
   a. A regression line can be drawn for a curvilinear relationship.
   b. A regression line fits the data in the best place possible.
   c. A regression line maximizes the distance between observed and predicted scores.
   d. A regression line visually shows how much *Y* will change as a result of a change in *X*.
6. If the coefficient of determination is $r^2 = .5$, then this means that the regression coefficient
   a. is significant
   b. is negative
   c. is positive
   d. There is not enough information given to answer the question.
7. In a regression analysis examining the impact of passenger load (number of passengers) on flight delay time (in minutes) for commercial flights, what does the regression coefficient for passenger load represent?

Continued on next page ...

a. The average increase in flight delay time for each additional passenger on board.
b. The total delay time for all flights regardless of passenger load.
c. The relationship between flight delays and weather conditions.
d. The airline's overall performance rating based on customer feedback.

8. In a bivariate linear regression study, the researcher reported 95% CI β: [-0.47, 0.69]. What would a test for H_0: β = 0 vs. H_1: β ≠ 0 conclude?
   a. Reject the null hypothesis at α = .05 and all smaller α.
   b. Fail to reject the null hypothesis at α = .05 and all smaller α.
   c. Reject the null hypothesis at α = .05 and all larger α.
   d. Fail to reject the null hypothesis at α = .05 and all larger α.
9. In a bivariate regression analysis examining the relationship between aircraft weight (in pounds) and takeoff distance (in feet), a 95% CI for the regression coefficient is calculated to be (0.5, 1.2). What does this CI indicate?
   a. The takeoff distance will always be between 0.5 and 1.2 feet for any aircraft weight.
   b. The average takeoff distance for all aircraft is guaranteed to be between 0.5 and 1.2 feet.
   c. There is a 95% chance that the slope of the relationship between aircraft weight and takeoff distance in the population falls between 0.5 and 1.2.
   d. The relationship between aircraft weight and takeoff distance is weak and not statistically significant.
10. In a bivariate regression analysis assessing the relationship between flight hours and maintenance costs for a fleet of aircraft, the *RMSE* is calculated to be $2,000. What does this *RMSE* value indicate?
   a. The average maintenance cost for all aircraft is $2,000.
   b. The relationship between flight hours and maintenance costs is significant.
   c. The relationship between flight hours and maintenance costs is not significant.
   d. The predicted maintenance costs deviate from the actual costs on average by $2,000.

B. Applying Procedural and Conceptual Knowledge

Use the following research description and the corresponding data set to conduct the activities given in Parts B1–B5. See Section 2.6 for guidance.

Let's imagine that during the COVID-19 pandemic, when mask mandates were in place in the U.S., aviation researchers investigated if flight attendants' attitudes toward flight safety were influenced by their interactions with passengers. Flight attendants reported their interactions with passengers via a single rating from 1 to 10 where 1 = "The passengers I interact with generally are rude and obnoxious" and 10 = "The passengers I interact with generally are respectful and courteous." Flight attendants also completed Ford et al.'s (2013) 36-item Flight Safety Attitudes Questionnaire—Flight Attendants (FSAQ–FA), which is scored on a traditional 5-point Likert response scale. Thus, scores could range from 36 to 180, with higher scores reflecting

Continued on next page ...

more positive attitudes. Because flight attendants are trained professionals who consider flight safety paramount, we will presume the researchers believed there should not be any effect. However, given the mask-mandated environment, they think there could be some effect but are uncertain of the direction. They also endeavored to develop a prediction equation that could be used in similar situations. The file "Ch_2 Review Exercises Data-Part B" contains a set of hypothetical data relative to this study.

B1. Research Questions and Hypotheses. Write the research question and corresponding hypotheses that pertain to the study.

B2. Sample Size Planning. Conduct an a priori power analysis to determine the minimum sample size needed. Compare this result to the size of the given data set and explain what impact the size of the given sample will have on the results relative to the minimum size needed.

B3. Data Analysis. Conduct a hypothesis test of the regression coefficient of the given data set and summarize the results in a table similar to Table 2.4. Assume Case 19 is an outlier and redo the analysis in the absence of the outlier. Compare the results between the outlier present and the outlier absent. Determine if you will keep or delete the outlier and explain why.

B4. Interpretation of Results. Review the results relative to the research question and corresponding null hypotheses stated in Part B1. Determine and interpret the regression equation, including B and B_0.

B5. Post Hoc Considerations. Determine and interpret the effect size; the coefficient of determination from both explained variance and prediction perspectives; the 95% confidence interval, including its precision and AIPE; the standard error of estimate (*RMSE*), and post hoc power.

References

Ali, M. A. & Malik A. A. B. K. (2022). Personality traits and safety attitude among aviators: a correlational study. *Pakistan Journal of Clinical Psychology*, *21*(2), 3–21.

Cohen, J. (1988). *Statistical power analysis for the behavioral sciences* (2nd ed.). Lawrence Erlbaum Associates.

Cohen, J., Cohen, C., West, S. G., & Aiken., L. S. (2003). *Applied multiple regression/correlation analysis for the behavioral sciences* (3rd ed.). Lawrence Erlbaum Associates.

Ford, J., Henderson, R., & O'Hare, D. (2013). *Flight safety attitudes questionnaire—Flight attendants.* https://doi.org/10.1037/t32118-000

Gallo, M. A., Wheeler, B. E., & Silver, I. M. (2023). *Fundamentals of statistics for aviation research.* Routledge.

Jones, U. O., Witiw, M. R., Cristaldi, A. J., & Sharp, D. W. (1999, January 10–15). *Defining predictors for the timing of warm season thunderstorms at the Orlando International Airport* [Paper presentation]. The American Meteorological Society, 8th Conference on Aviation, Range, and Aerospace Meteorology, Dallas, TX, United States.

Kachigan, S. (1991). *Multivariate statistical analysis: A conceptual introduction* (2nd ed.). Radius Press.

Selçuk, B. (2018). *Survival in the U.S. domestic airline market: Strategies for entry, exit, and air fare competition*. [Doctoral dissertation, Florida Institute of Technology]. Scholarship Repository. https://repository.fit.edu/etd/3/

Soper, D. (2025a). *A-priori sample size calculator for multiple regression*. https://www.danielsoper.com/statcalc/calculator.aspx?id=1

Soper, D. (2025b). *Post-hoc statistical power calculator for multiple regression*. https://www.danielsoper.com/statcalc/calculator.aspx?id=9

Student T-Value Calculator (2025). https://goodcalculators.com/student-t-value-calculator/

Part B

Establishing the Foundation of MRC

Chapter 3
MRC with Two Predictors

Student Learning Outcomes

After studying this chapter, you will be able to do the following:

1. Use Venn diagrams to illustrate various relationships between two IVs and a DV.
2. Interpret the results of an MRC analysis with two IVs that are reported in the parameter estimates table, model summary table, and ANOVA summary table.
3. Prepare and interpret the results of a summary table that consists of the various correlation coefficients and their squares of an MRC analysis with two IVs.
4. Conduct an MRC analysis for a research study with two IVs by engaging in the activities presented in Section 3.5.

3.1 Chapter Overview

The primary objective of this chapter is to introduce the statistical strategy known as *multiple regression/correlation* (*MRC*) by providing a transition from the previous chapters on bivariate correlation and bivariate regression to MRC by focusing exclusively on situations involving two independent variables and a single, continuous dependent variable. We begin by demonstrating the difference between conducting two separate regression analyses—one for each IV—as compared to a single regression analysis with both IVs together. As part of this discussion, we introduce the concepts of the *partial regression coefficient* (B_i), the *multiple correlation coefficient* (R) *and its square* (R^2), the *semipartial correlation coefficient* (sr) *and its square* (sr^2), and the *partial correlation coefficient* (pr) *and its square* (pr^2). To facilitate an understanding of these concepts, we examine an MRC analysis with two IVs and focus on interpreting the results of this analysis in the context of the given research setting. We conclude the chapter with a guided example that demonstrates how to conduct an MRC analysis for a research study that involves two IVs and a single, continuous DV. The reader will note that we delay the discussion of statistical inference in MRC to Chapter 4, which generalizes from two IVs to k IVs.

3.2 Bivariate vs. Multiple Regression: A Conceptual Approach

To introduce the difference between bivariate regression and multiple regression involving two independent variables, we will consider a set of hypothetical data that corresponds to Bell et al. (1995), which investigated the effect of pilot experience on aviation safety. For our discussion, the independent variables are: X_1 = Malfunction History, which is the number of inflight malfunctions pilots self-reported they experienced during their aviation career, and X_2 = Age (in years). The dependent variable is Y = Aviation Safety Index (ASI) scores, which consists of pilots' responses to five safety-related statements that corresponded to behaviors associated with preflight preparation. Responses are scored on a 4-item bipolar adjective scale with "Never"

DOI: 10.4324/9781003506935-5

at one end of the scale (1) and "Always" at the other end of the scale (4). Thus, aggregate scores could range from 5 to 20, with higher the scores reflecting higher levels of aviation safety consciousness. The data set is given in the file "Ch_3 ASI Data." Following the procedures for bivariate linear regression, the analysis of the given data set results in two prediction equations, one for each IV, as follows:

- Regressing Y on X_1, the regression equation is $\hat{Y} = 0.24X_1 + 11.1$.
- Regressing Y on X_2, the regression equation is $\hat{Y} = 0.16X_2 + 9.2$.

In the first equation we regress Y on X_1 without any regard for X_2, and in the second equation, we regress Y on X_2 without any regard for X_1. Treating these IVs separately as we did is acceptable if the correlation between X_1 and X_2 is zero, denoted as $r_{12} = 0$. However, for the given data set, the correlation between X_1 and X_2 is *not* zero: $r_{12} = .3428$.

To demonstrate the distinction between $r_{12} = 0$ vs. $r_{12} \neq 0$, consider Figure 3.1, which uses circles (Venn diagrams) to highlight the relationships among variables. In Figure 3.1(a), the two circles that represent X_1 and X_2 are shown without any overlap to illustrate that the two variables are disjoint—that is, they have nothing in common—and hence their correlation is zero (i.e., $r_{12} = 0$). In Figure 3.1(b), we maintain this disjoint state between X_1 and X_2, but we also introduce a circle to represent the dependent variable, Y. Note that although $r_{12} = 0$, this is not the case for X_1 and Y or for X_2 and Y. As illustrated by the overlapping circles:

- Region *a* represents the intersection between X_1 and Y, which means they share information and hence their correlation is nonzero (i.e., $r_{Y1} \neq 0$).
- Region *b* represents the intersection between X_2 and Y, which means they share information and hence their correlation also is nonzero (i.e., $r_{Y2} \neq 0$).

Because X_1 and X_2 *separately* have something in common with Y but not with each other, the respective relationship each IV has with Y is independent of the other. In Figure 3.1(c), we remove the disjoint state between X_1 and X_2 from Figure 3.1(b) so that all three circles now overlap each other. Note that in this configuration:

- X_1 and X_2 have a *nonzero* relationship (i.e., $r_{12} \neq 0$), which means they have something in common—that is, they share information. This is reflected by regions *c* and *d*.
- X_1 and Y have a nonzero relationship (i.e., $r_{Y1} \neq 0$) and hence share information as reflected in regions *a* and *c*.
- X_2 and Y have a nonzero relationship (i.e., $r_{Y2} \neq 0$) and hence share information as reflected in regions *b* and *c*.

Note further from Figure 3.1(c) that region *c* represents the information X_1 shares with Y as well as the information X_2 shares with Y. Because this information is common to both X_1 and X_2, this shared information is *redundant*.

To gain a better understanding of what we mean by this concept of redundancy, we present Figure 3.2, which depicts a developmental progression of Figure 3.1(c) where $r_{12} \neq 0$, $r_{Y1} \neq 0$, and $r_{Y2} \neq 0$. In Figure 3.2(a), we first show the nonzero relationship between Y and X_1 and shade the region of intersection with vertical lines. We then overlay X_2 onto this configuration and shade the region of intersection between X_2 and Y with horizontal lines. Notice that part of the relationship X_2 has

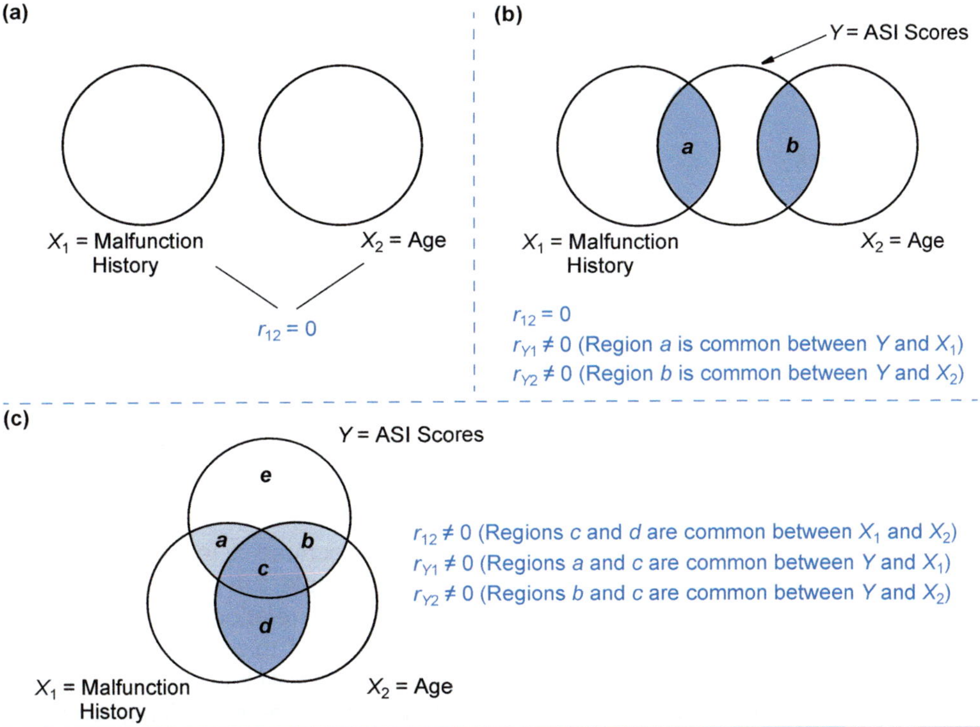

FIGURE 3.1 Graphical representation of the various relationships involving two IVs and a DV.

with *Y* is also part of X_1 as reflected by the "crossed" lines, which is region *c* in Figure 3.1(c). Thus, when X_2 is introduced into a model with X_1 present, some of the information that X_2 shares with *Y* is redundant with that of X_1. In Figure 3.2(b) we reverse the situation. We begin with a nonzero relationship between *Y* and X_2, shade the region of intersection, overlay X_1 onto this configuration, and shade X_1's relationship with *Y*. Similar to Figure 3.2(a), notice that part of the relationship X_1 has with *Y* is also part of X_2. Thus, when X_1 is introduced into a model with X_2 present, some of the information that X_1 shares with *Y* is redundant with that of X_2.

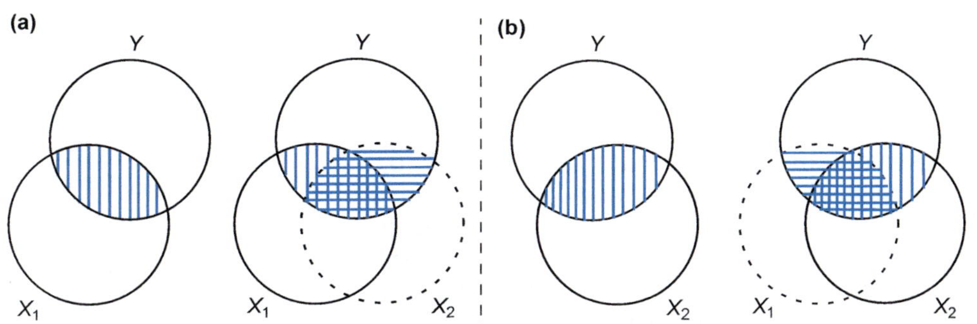

FIGURE 3.2 Graphical representation of the of the concept of redundancy between two IVs and a DV.

TABLE 3.1 *Correlation Matrix for ASI Data*

	Y	X_1	X_2
Y	1		
X_1	.2671	1	
X_2	.5224	.3428	1

Relating this discussion back to our running example of ASI scores, the respective correlations for the given data set are $r_{Y1} = .2671$, $r_{Y2} = .5224$, and $r_{12} = .3428$. These correlations are summarized in a correlation matrix as shown in Table 3.1. Furthermore, because each correlation is strictly between two variables without any consideration for other variables, the respective coefficients are referred to as zero-order correlation coefficients. For the two IVs case, there are three zero-order correlations:

- r_{Y1} is the zero-order correlation between Y and X_1.
- r_{Y2} is the zero-order correlation between Y and X_2.
- r_{12} is the zero-order correlation between X_1 and X_2.

A correlation coefficient is called "zero-order" because it is being calculated without any influence from any other variable. For example, the correlation between Y and X_1 does not take into consideration X_2, and the correlation between Y and X_2 does not take into consideration X_1. One way to think of this is by answering the question, "How many other variables were in the presence of this correlation when it was calculated?" Alternatively: "How many variables were statistically controlled when this correlation was calculated?" The answer to both questions is "zero." We interpret a zero-order coefficient the same as a bivariate correlation coefficient—for example, $r_{Y1} = .2671$ denotes a positive but weak relationship between Y and X_1—and the corresponding r^2 is simply the corresponding coefficient of determination.

Because $r_{Y1} \neq 0$, $r_{Y2} \neq 0$, and $r_{12} \neq 0$, the Venn diagram that corresponds to this data set is the one depicted in Figure 3.1(c), which means that X_1 and X_2 are redundant in terms of carrying information about Y. To account for this redundancy, we need a mathematical model that includes both X_1 *and* X_2, and not two separate regression equations as we did initially. This is presented in Section 3.3. We also need to calculate correlation coefficients that are not zero-order but take into consideration the presence of other variables. This is presented in Section 3.4.

Example 3.1 *Constructing a Venn Diagram for Two IVs and One DV*

Construct a Venn diagram that corresponds to the correlation matrix given in Table 3.2.

Solution

Observe from Table 3.2 that $r_{Y1} = .30$, $r_{Y2} = 0$, and $r_{12} = .08$. Thus: (a) X_1 and X_2 have a nonzero relationship with Y, (b) Y and X_1 have a nonzero relationship, and (c) Y and X_2 are disjoint. The Venn diagram that corresponds to these correlations is given in Figure 3.3.

TABLE 3.2 *Correlation Matrix for Example 3.1*

	Y	X_1	X_2
Y	1		
X_1	.30	1	
X_2	0	.08	1

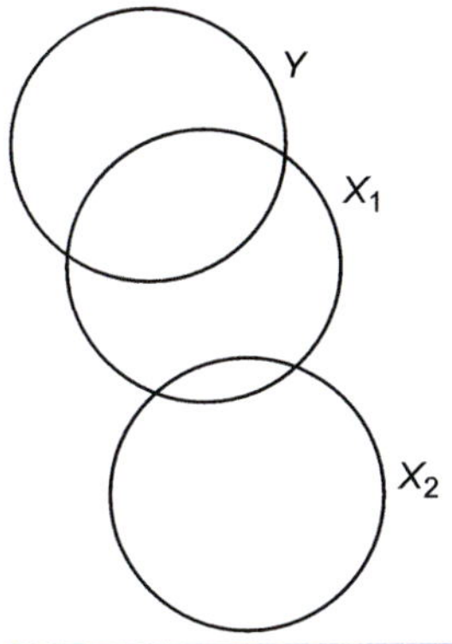

FIGURE 3.3 Venn diagram for solution to Example 3.1.

3.3 Bivariate vs. Multiple Regression: An Empirical Approach

As noted in Section 3.2, when performing a regression analysis involving two IVs, because $r_{12} \neq 0$, we cannot run separate bivariate regression analyses to examine the collective relationship X_1 and X_2 have with Y. Instead, we use a multiple regression approach to capture this relationship and account for the redundancy in Y. To do this we need a mathematical model that includes both X_1 and X_2 together, not separately. This model is given by the following equation:

$$\hat{Y} = B_1X_1 + B_2X_2 + B_0$$

Observe from this equation there are two regression coefficients, B_1 and B_2, which are called partial regression coefficients, and B_0 is the regression constant.

- B_1 represents the effect X_1 has on Y *in the presence of* X_2.
- B_2 represents the effect X_2 has on Y *in the presence of* X_1.

We can think of partial regression coefficients as weights. Thus, B_1 is the weight we give to X_1 in the presence of X_2, and B_2 is the weight we give to X_2 in the presence of X_1. To illustrate this, let's conduct a multiple regression analysis using the Ch_3 ASI Data. Regardless of your statistics software, the results you get for this analysis should resemble the summary information given in the three subtables of Table 3.3. (*Note:* Some software packages provide additional output than what is contained in Table 3.3. We will discuss this additional output later in the chapter.) We now examine each part of Table 3.3.

Examining the Coefficients (Parameter Estimates) Subtable

The Coefficients subtable of Table 3.3 consists of six columns:

- The Term column is a label for each table row.
- The Estimate column contains the partial regression coefficients, B_1 and B_2, and the regression constant, B_0. These B_i coefficients are unstandardized regression coefficients because they result from analyzing "raw scores," which have not been standardized.

TABLE 3.3 *Summary of Multiple Regression Analysis for ASI Data*

Model Summary

R	.530765
R^2	.281712
Adjusted R^2	.213303
Standard Error of the Estimate	2.414916
Mean of Response	13.75
Sample Size	24

Coefficients (Parameter Estimates)

Term	*B*	*SE*	β	*t*(21)	*p*
Constant	8.4939426	2.17515	—	3.90	.0008
X_1 = Malfunction History	0.0885857	0.17493	0.099698	0.51	.6178
X_2 = Age	0.1518039	0.06121	0.488258	2.48	.0217

ANOVA

Source	*SS*	*df*	*MS*	*F*	*p*
Regression	48.03183	2	24.0159	4.1181	.0310
Residual (Error)	122.46817	21	5.8318		
Total	170.50000	23			

- The *SE* column contains the respective standard errors for the B_i values.
- The β column contains the standardized regression coefficients, which result from analyzing raw data that has been standardized. For example, if *z* scores are used to standardize the raw data, then all the IVs will have the same mean of 0 and standard deviation of 1. Standardized regression coefficients are also known as beta coefficients. The regression constant of an MRC analysis based on standardized scores is zero. Some statistics software packages will present the beta coefficients in addition to the B_i values as part of their output.
- The *t* column contains the numerical value of a *t* test that results from testing the significance of the respective B_i values.
- The *p* column contains the respective *p* values corresponding to each *t* test.

Focusing on the B_i estimates, $B_1 = 0.09$, $B_2 = 0.15$, and $B_0 = 8.49$. These B_i values lead to the following unstandardized regression equation:

$$\hat{Y} = B_1X_1 + B_2X_2 + B_0$$

$$\hat{Y} = 0.09X_1 + 0.15X_2 + 8.49$$

The corresponding standardized regression equation is $\hat{Y} = 0.10\beta_1 + 0.49\beta_2 + 0$. We now examine each B_i value in the context of the current example. The reader is encouraged to review from Chapter 2 how to interpret regression coefficients.

B_1 = 0.09. This partial regression coefficient corresponds to X_1 = Malfunction History, which is the total number of inflight malfunctions pilots reported they experienced. Therefore, for every 1-unit increase in the number of inflight malfunctions pilots reported, their ASI scores increase, on average, by 0.09 points. More concretely, for every 10 reported inflight malfunctions, ASI scores increase on average by approximately 1 point (0.9 points). This implies that the more inflight malfunctions pilots reported they experienced, the more safety conscious they became. However, this interpretation is incomplete because the effect malfunction history has on ASI scores is in the presence of age. As a result, the interpretation must include this qualification:

> *Holding age constant*—or *when examined in the presence of age*—for every 1-unit increase in the number of inflight malfunctions pilots reported they experienced, ASI score increase, on average, by 0.09 points.

Recall from the bivariate regression analysis in Section 3.2 where we regressed Y on X_1 without any consideration for age, the regression equation was $\hat{Y} = 0.24X_1 + 11.1$, and $B_1 = 0.24$, which is different than $B_1 = 0.09$. This is because in the multiple regression analysis, X_1 is being analyzed *in the presence of* X_2, and these two variables have a nonzero relationship ($r_{12} = .3428$, see Table 3.1). Therefore, the multiple regression analysis is controlling for the effect of X_2's presence. Note also from the Coefficients subtable in Table 3.3 that in the presence of age, the effect malfunction history has on ASI scores is not statistically significant, $t(21) = 0.51$, $p = .6178$. This also was the case for the bivariate regression, $t(22) = 1.30$, $p = .2071$.

B_2 = 0.15. This partial regression coefficient corresponds to X_2 = Age, Thus, B_2 represents the effect pilots' age has on ASI scores. When interpreting this coefficient, remember that we must qualify the interpretation to include the phrase "in the presence of" or "holding constant" to reflect that the coefficient is being influenced by another variable:

> *Holding malfunction history constant*—or *when examined in the presence of malfunction history*—for every 1-year increase in pilots' age, ASI scores increase, on average, by approximately 0.15 points. More concretely, for every 10-year increase in pilots' age, ASI scores increase, on average, by 1.5 points. This implies that older pilots are more safety conscious than younger pilots.

Comparing $B_2 = 0.15$ to what resulted from bivariate regression ($B_2 = 0.16$) when we did not control for X_1's presence, note that they are nearly the same. In other words, when controlling for the effect of X_1 = Malfunction History in the multiple regression model, the effect X_2 = Age has on ASI scores is nearly the same as when we did not control for X_1's presence. This infers that X_1 has little to no effect on X_2. Note also from the Coefficients subtable in Table 3.3 that in the presence of malfunction history, the effect age has on ASI scores is statistically significant, $t(21) = 2.48$, $p = .0217$. This also was the case for the bivariate regression, $t(22) = 2.87$, $p = .0088$.

B_0 = 8.49. Recall from Chapter 2 that the regression constant in the bivariate case represents the y value of the y intercept, which occurs at the point $(0, y)$. Extending this to multiple regression, B_0 represents the numerical value of the predicted y value (i.e., $\hat{y}$) when all the IVs have been "zeroed out." In other words, we let each $X_i = 0$ in the regression equation. Applying this to the current example, a pilot who reports having experienced 0 inflight malfunctions ($X_1 = 0$) and is 0 years old ($X_2 = 0$) will have, on average, an ASI score of 8.49. Note that in the context of the current example, this interpretation of B_0 is not meaningful: a pilot cannot be 0 years old. If, however, X_2 = Total number of flight hours as pilot-in-command (PIC), then $B_0 = 8.49$ would be meaningful. As a result, before interpreting B_0, we must make sure it is meaningful in the context of the given research setting.

Examining the Model Summary Subtable

The Model Summary subtable in Table 3.3 contains a set of variables that help inform the results of the MRC analysis. A discussion of each entry follows (results rounded to three decimal places).

***R* = .531.** Recall from Chapter 2 that in bivariate correlation, the Pearson *r* is a measure of association between two variables that reflects the strength and direction of the relationship and can vary between –1 and +1. For example, if the correlation between pilots' age and total flight hours is $r = .8$, then the direction of the relationship is positive (as age increases, total flight hours also increase), and the relationship is relatively strong.

In an MRC analysis, the correlation coefficient is called the multiple correlation coefficient or multiple *R*, for short, and is denoted using an uppercase "R" (read: "big R") to distinguish it from the bivariate case. The multiple *R* represents the relationship between *Y* and $\hat{Y}$; that is, $R = r_{Y\hat{Y}}$, where Y = the observed scores and $\hat{Y}$ = the predicted *Y* scores acquired from the regression equation. For example, the MRC analysis for the ASI data resulted in the regression equation $\hat{Y} = 0.09X_1 + 0.15X_2 + 8.49$. Referencing the Ch_3 ASI Data file, note that Case 1 reflects a 38-year-old pilot who reported experiencing 11 inflight malfunctions. Given these data, the corresponding predicted ASI score is:

$$\begin{aligned}\hat{Y} &= 0.09X_1 + 0.15X_2 + 8.49\\ &= 0.09(11) + 0.15\,(38) + 8.49\\ &= 0.99 + 5.7 + 8.49\\ &= 15.18\end{aligned}$$

Thus, the predicted values result from adding the weighted regression coefficients and regression constant as presented in the regression equation. (*Note:* We provide the predicted scores for the ASI data in the last column of the data set given in the Ch_3 ASI Data file as a courtesy.)

Also recognize from its notation that the multiple *R* is a bivariate correlation ($R = r_{Y\hat{Y}}$). As a result, *R* represents the strength of the relationship between an observed score (*y*) and its corresponding predicted score ($\hat{y}$) and is interpreted the same way we interpret the Pearson *r* with one exception: *R* cannot be negative and hence varies between 0 and 1. *R* values that approach 0 indicate very little relationship between the observed and predicted scores, which infers the regression equation is not very good for prediction. *R* values that approach 1, however, indicate a strong relationship, which infers the regression equation is very good for prediction. If $R = 1$, then the relationship is perfect. Thus, for the ASI data, $R = .531$ means that the relationship between the observed scores and their corresponding predicted scores is somewhat "medium": It is neither particularly strong nor particularly weak.

***R* Square = .282.** Recall from Chapter 2 that when we square *r* we get the *coefficient of determination*, which is a measure of explained variance. For example, if the correlation between pilots' age and total flight hours is $r = .8$, then $r^2 = .64$ and is interpreted as: *64% of the variance in age is being explained by the variance in total flight hours*. You also might recall that we can interpret r^2 from a prediction perspective. Using this same example, $r^2 = .64$ would be interpreted as *if you tell me your age, then I have 64% of the information I need to perfectly predict your total number of flight hours.*

In an MRC analysis, the coefficient of determination is called the multiple *R* square, denoted R^2, which is read as big *R* squared, and is analogous to the bivariate r^2. Thus, R^2 represents the total amount of *Y* variance that is attributed to or explained by a group of IVs. Because an MRC analysis involves two or more IVs, we formally denote the overall R^2 with subscripts to reflect the DV and the IVs under discussion. For example, with two IVs, the overall R^2 is denoted as $R^2_{Y \cdot 12}$, where the dependent variable (*Y*) is placed to the left of the dot (•) and the subscripts of the two IVs (X_1 and X_2) are placed to the right of the dot. This notation is read as "Y dot one two" and indicates that R^2 is the result of *Y* being regressed on X_1 and X_2.

As reported in Table 3.3, $R^2_{Y \cdot 12} = .282$, which means that 28.2% of the variance in ASI scores is being explained collectively by pilots' malfunction history and age. We also can interpret this from a prediction perspective: "If we know a pilot's malfunction history and age, then we have 28.2% of the information needed to perfectly predict the pilot's ASI score." Examining this relationship conceptually using Venn diagrams, observe from Figure 3.4 that if we let the circles represent the total variance in a variable—for example, the *Y* circle represents 100% of the variance in ASI scores—then $R^2_{Y \cdot 12}$ corresponds to regions *a* + *b* + *c*. More specifically: $R^2_{Y \cdot 12} = a + b + c = .282$. Thus, the two IVs collectively explain 28.2% of the total variance in ASI scores, which means that 71.8% of the total *Y* variance remains unexplained, which is region *e* in Figure 3.4. We will expand on this concept further in Section 3.4 when we introduce the semipartial and full partial correlation coefficients and their squares.

Adjusted *R* Square = .213. The adjusted R^2 is a modified version of the initial (or "unadjusted") R^2. The adjustment that is made is relative to the number of IVs being analyzed. In MRC, as more variables are included in the analysis, the overall R^2 either will increase or remain the same even if these IVs do not have a significant relationship with *Y*. This makes the overall R^2 biased. The adjusted R^2 "adjusts" for this bias by calculating an R^2 relative to those IVs that have a significant relationship with *Y*. So, when conducting an MRC analysis, the adjusted R^2 should be reported as well. For the current example, the adjusted $R^2_{Y \cdot 12} = .213$, which means that when the overall R^2 has been adjusted for bias, 21.3% of the variance in ASI scores is being explained collectively by pilots' malfunction history and age.

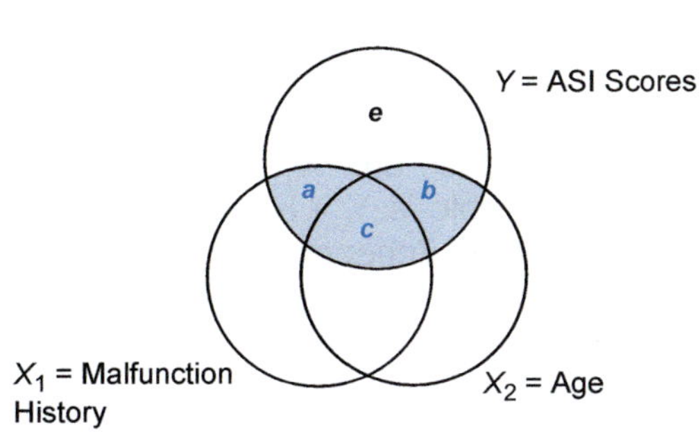

- 100% of *Y* variance = *a* + *b* + *c* + *e*
- $R^2_{Y \cdot 12} = a + b + c = .282$ (amount of *Y* variance explained collectively by X_1 and X_2).
- Therefore, $e = 1 - .282 = .718$ (amount of *Y* variance *not* explained collectively by X_1 and X_2).

FIGURE 3.4 Venn diagram that shows the partitioning of *Y* variance for the MRC analysis involving ASI scores.

Standard Error of the Estimate = 2.41. Recall from Chapter 2 that the standard error of the estimate is a metric that measures the accuracy of a prediction by examining in a scatter plot the distance between the predicted points on a regression line and the actual points that represent the observed scores. Also known as the root mean square error (*RMSE*), this metric indicates how far "off" we will be, on average, when using x to predict y. This metric has the same meaning in MRC except now it reflects the accuracy of predicting the DV from the sum of the weighted regression coefficients and regression constant as dictated by the regression equation. In the current example, $RMSE = 2.41$. This means that if we use pilots' malfunction history and age to predict ASI scores via the corresponding regression equation, then the prediction will be "off," on average, by approximately 2.41 points.

Mean of Response = 13.75. The mean of the response is simply the mean of the DV, independent of any influences by the IVs. For the current example, $M_Y = 13.75$ indicates that the overall the sample's mean ASI score is 13.75.

Sample Size = 24. This metric simply reflects the total number of participants of the study. Thus, for the current example, $N = 24$.

Examining the ANOVA Subtable

The ANOVA subtable in Table 3.3 is a standard summary table used to report the partitioning of variance. In the context of an MRC analysis, the "Regression" row represents the between-groups variance, and the "Residual" row represents the within-groups (error) variance. We use this table to determine if the overall regression model is statistically significant. For the current example, the overall model is indeed significant, $F(2, 21) = 4.1181$, $p = .0310$.

The results provided in the ANOVA subtable serve as an omnibus test for the overall regression model similar to a single-factor ANOVA. Thus, if the overall model is significant, which is the case for the running example, then we interpret the respective regression coefficients. If, however, the overall model is not significant, then any interpretations of the regression coefficients might be spurious because of the possibility of inflated Type I errors. This is consistent with the "protected t test strategy" of ANOVA, which protects against inflated alpha levels.

Example 3.2 *Conducting an MRC Analysis with Two IVs*

The file "Ch_3 Example 3.2 Data" contains a set of hypothetical data that corresponds to Barkhuizen et al. (2002), which examined pilot reaction times. For our discussion, the independent variables are: X_1 = Age (in years) and X_2 = Total Flight Hours. The dependent variable is Y = Reaction Time (in seconds), which reflects the result of pilots' scores on Schepers' (2001) Computerised Information Processing Test Battery. Conduct an MRC analysis using this data set and then respond to Items a–d. (Round to three decimal places and assume $\alpha = .05$.)

a. Determine the overall regression equation.

b. Determine if the overall model is significant.

c. Interpret the regression coefficients and regression constant in the context of the given research setting and determine if they are significant.

d. Determine and interpret the R, R^2, adjusted R^2, and *RMSE* values.

Continued on next page ...

Solution

A summary of the results is provided in Table 3.4.

a. The overall regression equation is $\hat{Y} = 0.272X_1 + 0.003X_2 - 16.625$.

b. The overall model is significant, $F(2, 66) = 25.247$, $p < .0001$.

c. $B_1 = 0.272$. Holding total flight hours constant, for every 1-year increase in age, reaction times increase on average by 0.272 seconds. More concretely, this increase is approximately ¼ of a second, which means that for every 4-year increase in age, reaction times increase by about 1 second. Thus, a 40-year-old pilot will, on average, have a reaction time that is 5 seconds more than a 20-year-old pilot. This result is significant, $t(66) = 4.12$, $p = .0001$.

$B_2 = 0.003$. Holding age constant, for every 1-hour increase in total flight hours, reaction times increase, on average, by 0.003 seconds. More concretely, for every 1,000-hour increase in flight hours, reaction times increase, on average, by 3 seconds. This result is not significant, $t(66) = 1.52$, $p = .1329$.

$B_0 = -16.625$. Zeroing out the IVs (i.e., $X_1 = 0$ and $X_2 = 0$), results in an average reaction time of −16.625 seconds, but this is for a pilot who is 0 years old and has 0 hours of flight time. Also, average pilot reaction times cannot be less than zero in real life. Although this result is significant, $t(66) = -5.54$, $p = .0133$, the interpretation is not meaningful in the current context.

d. $R = .658$. The correlation between the observed reaction times and the corresponding predicted reaction times is .658. This reflects a moderately strong relationship between the observed and predicted scores.

$R^2 = .433$. The coefficient of determination indicates that 43.3% of the variance in reaction times is explained collectively by pilots' age and total flight hours. We

Continued on next page ...

Table 3.4 *Summary of Multiple Regression Analysis for Example 3.2*

Model Summary

R	.658363
R^2	.433442
Adjusted R^2	.416274
Standard Error of the Estimate	4.786152
Mean of Response	9.53
Sample Size	69

Coefficients (Parameter Estimates)

Term	*B*	*SE*	β	*t*(66)	*p*
Constant	−16.62546	6.535114	—	−2.54	.0133
X_1 = Age	0.2723266	0.066115	.515223	4.12	.0001
X_2 = Flight Hours	0.0028609	0.001880	.190328	1.52	.1329

Note. N = 183.

ANOVA

Source	*SS*	*df*	*MS*	*F*	*p*
Regression	1156.6555	2	578.328	25.2465	< .0001
Residual (Error)	1511.8785	66	22.907		
Total	2668.5340	68			

also can interpret this from a prediction perspective: If we know a pilot's age and total flight hours, then we have 43.3% of the information needed to perfectly predict the pilot's reaction time.

Adjusted R^2 = .416. When the overall R^2 has been adjusted for bias, 41.6% of the variance in reaction times is being explained collectively by pilots' age and total flight hours.

RMSE = 4.786. If we use pilots' age and total flight hours to predict reaction times via the corresponding regression equation, then the prediction would be "off," on average, by approximately 4.8 seconds.

3.4 Semipartial and Full Partial Correlation Coefficients and Their Squares

In Section 3.2, we introduced zero-order correlation coefficients for the two IVs case: r_{Y1}, r_{Y2}, and r_{12}. As part of our discussion, we indicated that these coefficients are calculated in the absence of any other variables. We also indicated that these coefficients do not fully inform us about the respective relationships each IV has with Y when $r_{12} \neq 0$ and hence the need for another set of correlation coefficients that control for the presence of other variables in the regression model. In this section, we introduce the *semipartial* and *full partial correlation coefficients* and their squares.

Semipartial Correlation Coefficients and Their Squares

The Squared Semipartial Correlation Coefficient. Recall from our running example that Y = ASI scores, X_1 = Malfunction History, X_2 = Age, and $R^2_{Y \bullet 12}$ = .282. Thus, 28.2% of the variance in ASI scores is being explained collectively by X_1 and X_2. Further recall that the collective contribution X_1 and X_2 make in explaining Y variance includes some redundancy. As illustrated previously in Figure 3.4, this redundancy is captured by region *c*. Thus, part of the information that X_1 makes in explaining Y variance is the same as part of the information that X_2 makes in explaining Y variance, and vice versa. This situation raises the question, "If we know the amount of Y variance being explained collectively by X_1 and X_2, what would be X_1's *unique* contribution to explaining Y variance if X_2's relationship with X_1 is removed from X_1, but not from Y?" Similarly, "If we know the amount of Y variance being explained collectively by X_1 and X_2, what would be X_2's *unique* contribution to explaining Y variance if X_1's relationship with X_2 is removed from X_2, but not from Y?"

Another way of looking at this is to ask, "If we know the amount of information X_2 provides about Y variance without X_1 being present, then how much *additional information* about Y variance would we acquire when X_1 enters the regression model in the presence of X_2?" Similarly, "If we know the amount of information X_1 provides about Y variance without X_2 being present, then how much *additional information* about Y variance would we acquire when X_2 enters the regression model in the presence of X_1?" These questions are answered by the squared semipartial correlation coefficient, sr^2, which identifies the *unique* contribution an IV makes to explaining Y variance in the presence of all other IVs. Figure 3.5 contains a visual demonstration of the concept of the squared semipartial correlation coefficients for X_1 and X_2 with respect to our running example.

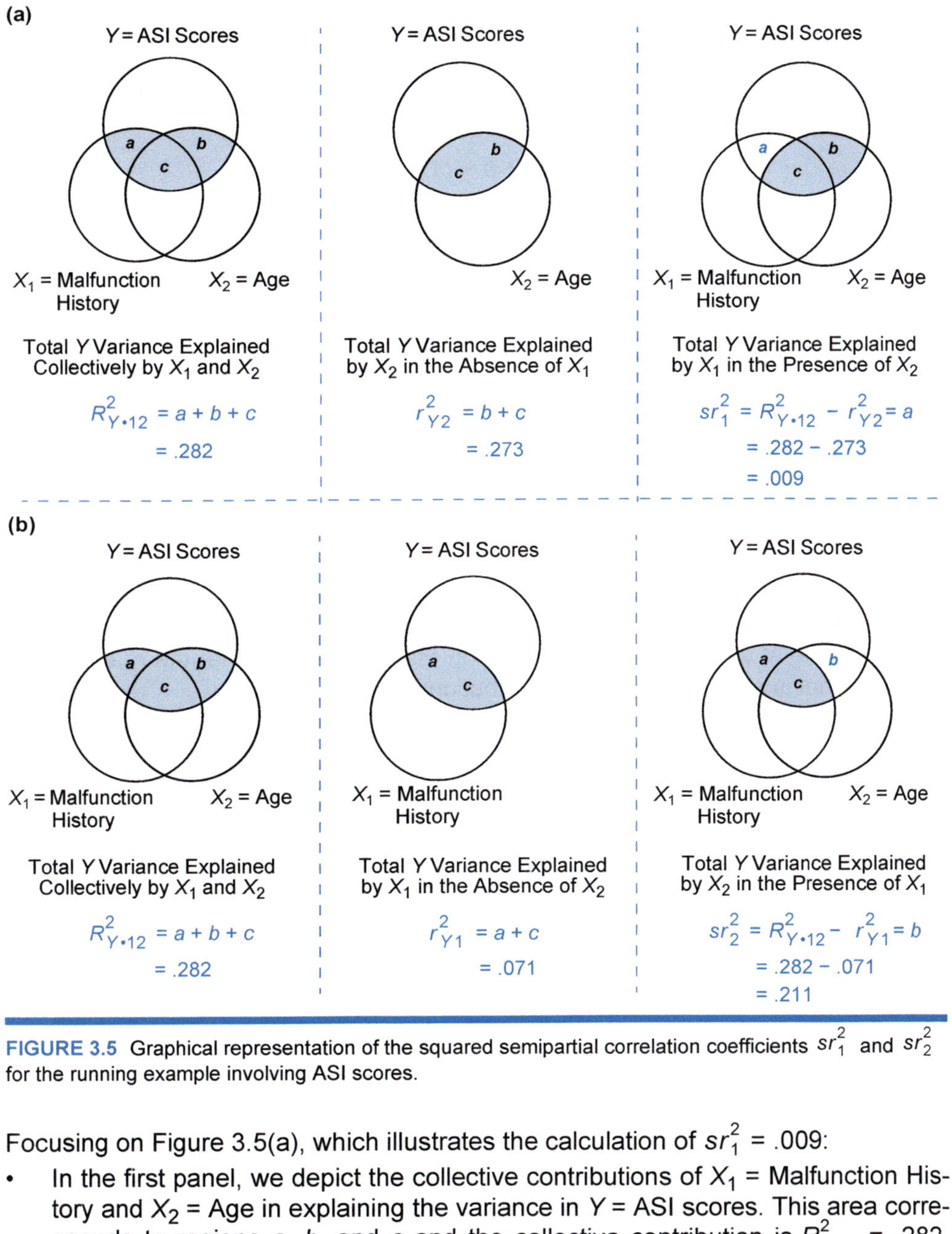

FIGURE 3.5 Graphical representation of the squared semipartial correlation coefficients sr_1^2 and sr_2^2 for the running example involving ASI scores.

Focusing on Figure 3.5(a), which illustrates the calculation of $sr_1^2 = .009$:

- In the first panel, we depict the collective contributions of X_1 = Malfunction History and X_2 = Age in explaining the variance in Y = ASI scores. This area corresponds to regions *a*, *b*, and *c* and the collective contribution is $R^2_{Y\bullet 12} = .282$. Thus, 28.2% of the variance in Y is being explained collectively by X_1 and X_2.
- In the second panel, we present a regression model that involves only Y and X_2. Because no other variables are present, this represents the squared zero-order correlation between Y and X_2, which is represented by regions *b* and *c* and was reported earlier as $r^2_{Y2} = .273$. Thus, 27.3% of the variance in Y is being explained solely by X_2.

- In the third panel, we let X_1 enter the analysis "in the presence of X_2." Note that the only area of Y that X_1 can claim is region *a* because regions *b* and *c* have been claimed by X_2 previously. Thus, region *a* represents the "unique" contribution X_1 makes in explaining Y variance "in the presence of X_2." This region is sr_1^2 and it is equal to the difference between $R^2_{Y \cdot 12}$ and r^2_{Y2}:

$$\begin{aligned} sr_1^2 &= (a + b + c) - (b + c) \\ &= R^2_{Y \cdot 12} - r^2_{Y2} \\ &= .282 - .273 \\ &= .009 \text{ (region } a) \end{aligned}$$

One way to interpret this result is to report that when malfunction history enters the analysis in *the presence of age*, it explains an additional 0.9% of the variance in ASI scores over and above the 27.3% that age provided previously. An alternative approach is to report that less than 1% of the variance in ASI scores is being explained by malfunction history when the effect of age has been removed from malfunction history but not from ASI scores.

This additional contribution to explained Y variance provided by the squared semipartial correlation coefficient is commonly referred to as a variable's unique contribution because it is the additional amount of knowledge gained when an IV enters the analysis in the presence of another IV (or set of IVs). As a result, another interpretation of $sr_1^2 = .009$ is to simply report that the *unique* contribution X_1 makes to explaining Y variance is .009, based on the presumption that the reader understands what is meant by "unique contribution."

Further observe from Figure 3.5(a) that when X_1 enters the model in the presence of X_2 as illustrated in the third panel, the effect X_2 has on Y remains. This is reflected by regions *b* and *c*. However, the effect X_2 has on X_1 is removed because we are not permitting X_1 to claim any part of region *c*. This is why sr_i^2 is called a *semi*partial coefficient. In the current example, we are *partly removing* X_2's effect: It is being removed from X_1 but not from Y.

For completeness, a similar illustration for sr_2^2 is presented in the three panels of Figure 3.5(b). Here, sr_2^2 is equal to the difference between $R^2_{Y \cdot 12}$ and r^2_{Y1}:

$$\begin{aligned} sr_2^2 &= (a + b + c) - (a + c) \\ &= R^2_{Y \cdot 12} - r^2_{Y1} \\ &= .282 - .071 \\ &= .211 \text{ (region } b) \end{aligned}$$

Thus, 21.1% of the variance in ASI scores is being explained by age when the effect of malfunction history has been removed from age but not from ASI scores. In other words, age explains 21.1% of the variance in ASI scores when in the presence of malfunction history, or more simply, the *unique* contribution X_2 makes to explaining Y variance is .211.

Figure 3.6 contains a summary of the complete Venn diagram for the running example that includes the amount of Y variance that is *uniquely* explained by X_1 and X_2 (regions *a* and *b*, respectively), and collectively (region *c*). We also report the amount of unexplained variance (region *e*), which is the amount of knowledge about Y variance that is not explained by X_1 and X_2. For example, 71.8% of the variance in ASI scores is not explained by X_1 = Malfunction History and X_2 = Age, which

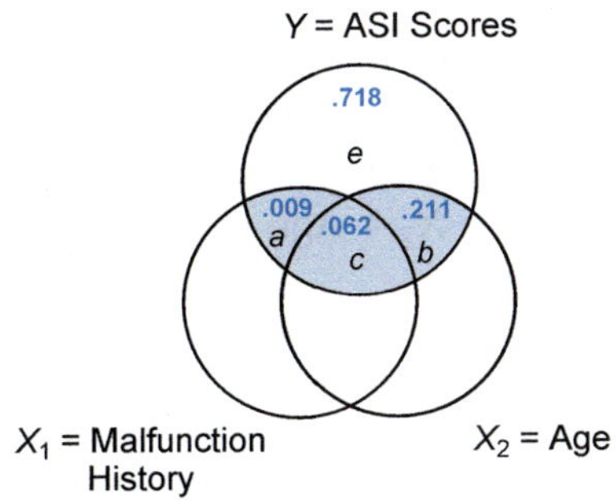

Regions (*a* + *b* + *c*): $R^2_{Y\bullet 12} = .009 + .062 + .211 = .282$. This is the collective contribution to *Y* variance made by X_1 and X_2.

Region *a*: $sr_1^2 = .009$. This is the *unique contribution* X_1 makes to explaining *Y* variance after the effects of X_2 are removed from X_1 but not from *Y*.

Region *b*: $sr_2^2 = .211$. This is the *unique contribution* X_2 makes to explaining *Y* variance after the effects of X_1 are removed from X_2 but not from *Y*.

Region *c*: This is the amount of *shared Y variance* (or *redundancy*) between X_1 and X_2.

$$\begin{aligned}&(a + b + c) - (a + b)\\ &= .282 - (.009 + .211)\\ &= .282 - (.22)\\ &= .062\end{aligned}$$

Region *e*: $1 - R^2_{Y\bullet 12} = 1 - .282 = .718$. This is the amount of *Y* variance *not explained* collectively by X_1 and X_2 (i.e., *unexplained variance*).

Regions (*a* + *b* + *c* + *e*): .009 + .211 + .062 + .718 = 1.00. This is 100% of *Y* variance.

FIGURE 3.6 Summary of complete Venn diagram for the running example involving X_1 = Malfunction History, X_2 = Age, and *Y* = ASI scores.

means that this unexplained variance in ASI scores must be explained by other factors—perhaps total flight hours or hours as PIC. Continuing with Figure 3.6, observe the following:

- Region *c* represents the amount of shared variance between X_1 and X_2. This reflects the amount of redundancy as discussed previously.
- When we add the unique contributions each IV makes in explaining *Y* variance, we do *not* get a full accounting of the overall $R^2_{Y\bullet 12}$:

$$\begin{aligned}R^2_{Y\bullet 12} &\neq sr_1^2 + sr_2^2\\ .282 &\neq (.009 + .211)\\ .282 &\neq .220\end{aligned}$$

The reason the sum of the sr^2s does not equal the overall R^2 is because of region *c*. The shared contribution between X_1 and X_2 does not get applied to either X_1 or X_2 because we partialled out their effects from each other, which leaves this amount of explained variance unaccounted for. This problem can be reconciled by using a multiple regression strategy known as *hierarchical regression*, which involves a prescribed variable entry order with variables entering the analysis one at a time. This strategy, as well as another strategy known as *stepwise regression*, is discussed in Chapter 6.

Although the partial regression coefficients (B_i) are the primary focus of an MRC analysis because they reflect the effect each X_i has on Y, the squared semipartial correlation coefficients provide added-value to the analysis. For example, in the two IVs case, sr_i^2 tell us which IV is the stronger predictor. For our running example, because sr_1^2 = .009 and sr_2^2 = .211, X_2 = Age is a much stronger predictor of ASI scores than X_1 = Malfunction History.

The Semipartial Correlation Coefficient. The semipartial correlation coefficient, denoted *sr*, is the square root of sr^2 and is analogous to the Pearson *r* in that it describes the direction and strength of a relationship between two variables. For the general two IVs case, the semipartial correlation coefficient for X_1, denoted sr_1, represents the relationship between Y and X_1 where the effects of X_2 have been removed from X_1 but *not* from Y. Similarly, the semipartial correlation coefficient for X_2, denoted sr_2, represents the relationship between Y and X_2 where the effects of X_1 have been removed from X_2 but *not* from Y. When examined from the perspective of our running example where Y = ASI scores, X_1 = Malfunction History, and X_2 = Age, the semipartial correlation coefficients rounded to three decimal places obtained from our statistics software are as follows: (*Note:* When taking the square root, the sign of the result follows the sign of the corresponding B_i value. Thus, if B_1 is negative, then sr_1 must be negative.)

- $sr_1 = \sqrt{sr_1^2} = \sqrt{.009} = .095$ means that when the effect of age is removed from malfunction history but not from ASI scores, the relationship between X_1 = Malfunction History and Y = ASI Scores is .095, which is positive but small.
- $sr_2 = \sqrt{sr_2^2} = \sqrt{.211} = .459$ means that when the effect of malfunction history is removed from age but not from ASI scores, the relationship between X_2 = Age and Y = ASI Scores is .459, which is positive and fairly strong.

When the semipartial correlation coefficients are compared to their corresponding zero-order correlation coefficients, note that they are smaller in magnitude. For example: r_{Y1} = .267 but sr_1 = .095, and r_{Y2} = .522 but sr_2 = .459. This is because by controlling for the presence of other variables, we are removing the redundancy between the IVs, which reduces the strength of the relationship.

Full Partial Correlation Coefficients and Their Squares

The Squared (Full) Partial Correlation Coefficient. Another question that emerges from a multiple regression analysis is:

> *When all other variables in a regression model are statistically controlled, how much of the variance in the DV does a targeted variable account for?*

This question is answered by the *squared (full) partial correlation coefficient,* pr^2, and the concept of "statistically controlling" a variable means that the effects of all the other IVs have been removed completely from *both the IV under discussion and the DV*. (*Note:* Although this coefficient is generally known as the squared "full" partial coefficient to contrast it with the squared semipartial correlation coefficient, the word "full" is commonly dropped.) Figure 3.7 contains a visual demonstration of the concept of the squared partial correlation coefficients for X_1 and X_2 with respect to our running example involving ASI scores. Focusing on Figure 3.7(a), which illustrates the calculation of pr_1^2 = .012:

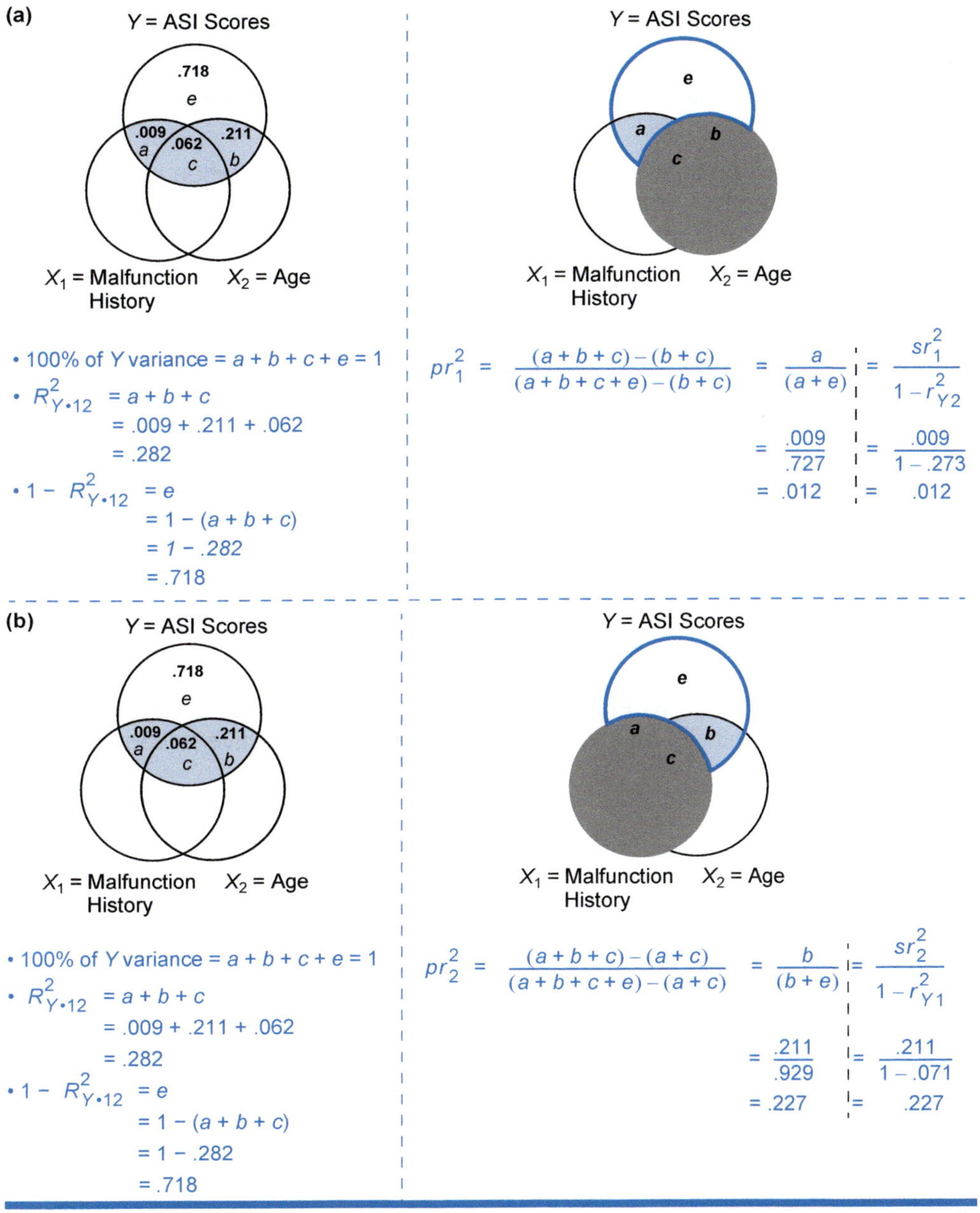

FIGURE 3.7 Graphical representation of the concept of the squared partial correlation coefficients pr_1^2 and pr_2^2 for the running example involving ASI scores.

- In the first panel of Figure 3.7(a), we depict the fully configured Venn diagram with the numerical values of the coefficients embedded in each region.
- In the second panel, we completely blocked out the entire circle representing X_2 to demonstrate graphically what it means to *statistically control for the effect of* X_2: We "fully" or "completely" *remove the effect of* X_2 *from both* X_1 *and* Y. (Think of this by placing your thumb over the entire circle.) This involves deleting

regions *b* and *c*, which is shown by numerically subtracting ($b + c$) in both the numerator and denominator of the equation given below the Venn diagram. Note that by eliminating region *b*, part of X_2's effect on *Y* is removed, and when eliminating region *c*, the rest of X_2's effect on *Y* is removed, and this also removes X_2's effect on X_1.

Further note from the second panel of Figure 3.7(a), that when we remove the effect of X_2 from both X_1 and *Y*, we no longer have 100% of *Y* variance—that is, the whole circle representing 100% of *Y* variance is no longer available. What remains is a crescent or "Pacman" shaped outline that consists of regions *a* and *e*, where region *a* is sr_1^2 and region *e* is unexplained variance. Because pr_1^2 represents the amount of *Y* variance explained by X_1 "after the effect of X_2 has been removed from both X_1 and *Y*," when we express this as the ratio of explained variance to total variance, the base is no longer 100%. In other words, the total variance of *Y*, which is represented by the entire *Y* circle and consists of regions ($a + b + c + e$), is now changed to a subset of this circle that is represented by regions ($a + e$).

When examined mathematically, this base is equal to $1 - r_{Y2}^2$, where "1" represents 100% of *Y* variance, and r_{Y2}^2 is the amount of *Y* variance being explained by X_2, which is the zero-ordered squared correlation coefficient. Thus, pr_1^2 is the ratio of sr_1^2 to $(1 - r_{Y2}^2)$. (*Note:* For the squared semipartial correlation coefficient sr_1^2, the total variance was 100%, or 1, because we did not remove the effect of X_2 from *Y* and therefore there was no change in the base.)

For completeness, the two panels of Figure 3.7(b) depict the concept of pr_2^2. This is exactly the same as Figure 3.7(a) except now the focus is on removing the effect of X_1 from both X_2 and *Y*. Mathematically, the respective equations for pr_1^2 and pr_2^2 are:

$$pr_1^2 = \frac{a}{a+e} = \frac{R_{Y\cdot12}^2 - r_{Y2}^2}{1 - r_{Y2}^2} = \frac{sr_1^2}{1 - r_{Y2}^2} = \frac{.009}{1-.273} = \frac{.009}{.727} = .012$$

$$pr_2^2 = \frac{b}{b+e} = \frac{R_{Y\cdot12}^2 - r_{Y1}^2}{1 - r_{Y1}^2} = \frac{sr_1^2}{1 - r_{Y1}^2} = \frac{.211}{1-.071} = \frac{.211}{.929} = .227$$

Note from Figure 3.7(a) that $a + e = [1 - (b + c)]$ and from Figure 3.7(b) that $b + e = [1 - (a + c)]$. Further note that sr_1^2 and pr_1^2 differ only in regard to their denominators. We will discuss the partial correlation coefficient and its square further in Chapter 9 when we introduce analysis of covariance (ANCOVA) via a multiple regression strategy. For the moment, though, recognize that with respect to pr_1^2, it answers the question, "Which variable accounts for more of the *XY* relationship?" For our running example, comparing $pr_1^2 = .012$ to $pr_2^2 = .227$, X_2 = Age accounts for more of the *XY* relationship than X_1 = Malfunction History.

The (Full) Partial Correlation Coefficient. The partial correlation coefficient, denoted *pr*, is the square root of pr^2 and is analogous to the Pearson *r* in that it describes the direction and strength of a relationship between two variables. For the general two IVs case, the partial correlation coefficient for X_1, denoted pr_1, represents the relationship between *Y* and X_1 where the effects of X_2 have been completely removed from both X_1 and *Y*. Similarly, the partial correlation coefficient for

X_2, denoted pr_2, represents the relationship between Y and X_2 where the effects of X_1 have been completely removed from both X_2 and Y. When examined from the perspective of our running example where Y = ASI scores, X_1 = Malfunction History, and X_2 = Age, the partial correlation coefficients rounded to three decimal places obtained from our statistics software are as follows: (*Note:* When taking the square root, the sign of the result follows the sign of the corresponding B value. For example, if B_1 is negative, then pr_1 must be negative.)

- $pr_1 = \sqrt{pr_1^2} = \sqrt{.012} = .110$ means that when the effect of age is removed *completely* from malfunction history *and* ASI scores, the relationship between malfunction history and ASI scores is positive but relatively small. In other words, when we statistically control for the effect of age, then the relationship between malfunction history and ASI scores is .110.
- $pr_2 = \sqrt{pr_2^2} = \sqrt{.227} = .476$ means that when the effect of malfunction history is removed *completely* from age *and* ASI scores, the relationship between age and ASI scores is positive and somewhat moderate. In other words, when we statistically control for the effect of malfunction history, then the relationship between age and ASI scores is .476.

Table 3.5 contains a summary of the results of all our analyses involving the "partials." From a generic perspective and without qualification, every entry except for the zero-order correlation coefficients is considered a "partial." For example, B_i values are partial regression coefficients, sr_i and sr_i^2 values are semipartial correlation coefficients and their squares, and pr_i and pr_i^2 values are (full) partial correlation coefficients and their squares. Furthermore, the significance values correspond to all the partials in the same row. So, if one partial that corresponds to an IV is significant, then all the partials associated with that IV are also significant.

A Word on Notation

Although we use sr_i and sr_i^2 to represent the semipartial correlation coefficients and their squares, and pr_i and pr_i^2 to represent the partial correlation coefficients and their squares, it is instructional to present alternative notation for these coefficients that is more descriptive of the concepts they represent. For example, with respect to the two IVs case, if the IV under discussion is X_1, then sr_1 represents the relationship X_1 has with Y when the effects of X_2 have been removed from X_1 but not from Y. A more descriptive notation that reflects this concept is $r_{Y(1\bullet2)}$. In this notation, the subscript to the left of the "dot" represents the variable under discussion (X_1) and indicates that the relationship being considered is between Y and X_1. The subscript

TABLE 3.5 *Summary of Multiple Regression Results of the Partials for the ASI Data*

		Zero-Order Correlations		**Semipartial Correlations**		**Full Partial Correlations**		**Significance**	
Factor [a]	B_i	r_{Yi}	r_{Yi}^2	sr_i	sr_i^2	pr_i	pr_i^2	t_i	p_i
Intercept	8.494								
X_1	0.089	.267	.071	.094	.009	.110	.012	0.506	.618
X_2	0.152	.522	.273	.459	.211	.476	.227	2.480	.022

Note. N = 24. All entries rounded to three decimals. $R_{Y\bullet12}^2 = .282$, $F(2, 21) = 4.118$, $p = .031$.
[a] X_1 = Malfunction History, X_2 = Age, Y = ASI scores.

to the right of the "dot" represents the "other" variable that is part of the regression model (X_2) whose effects are being removed. We place the subscripts in parentheses to convey that we are removing the effects of X_2 from X_1, but *not* from *Y*. Similarly, if the IV under discussion is X_2, then $sr_2 = r_{Y(2 \cdot 1)}$.

Now consider the partial correlation coefficient. For the two IVs case, if the IV under discussion is X_1, then pr_1 represents the relationship X_1 has with *Y* when the effects of X_2 have been removed *completely* from both X_1 *and Y*. A more descriptive notation that reflects this concept is $r_{Y1 \cdot 2}$. Note that the absence of the parentheses in the subscript indicates that the relationship being considered is between *Y* and X_1, and that X_2 is being partialled from *both Y* and X_1. Similarly, if the IV under discussion is X_2, then $pr_2 = r_{Y2 \cdot 1}$. Lastly, following this logic and pattern:

- The squared semipartial correlation coefficients for the two IVs case may be denoted more descriptively as $sr_1^2 = r^2_{Y(1 \cdot 2)}$ and $sr_2^2 = r^2_{Y(2 \cdot 1)}$.
- The squared (full) partial correlation coefficients for the two IVs case may be denoted more descriptively as $pr_1^2 = r^2_{Y1 \cdot 2}$ and $pr_2^2 = r^2_{Y2 \cdot 1}$.

Example 3.3 *Semipartial Correlation Coefficients and Their Squares*

In Example 3.2, we conducted an MRC analysis for the reaction times data from the file "Ch_3 Example 3.2 Data." In the current example, we examine the various correlation coefficients and their squares for Example 3.2.

a. Prepare a summary table like Table 3.5. Round three decimal places.
b. Interpret sr_1^2 and sr_2^2 in the context of the given research setting.
c. Prepare a Venn diagram like Figure 3.6 for the given data set.

Solution

a. See Table 3.6.
b. $sr_1^2 = .146$. When the effect of total flight hours is removed from age, but not from reaction times, pilots' age uniquely explain an additional 14.6% of the variance in reaction times. This amount of explained variance is significant, $t(66) = 4.12$, $p = .0001$.

$sr_2^2 = .019$. When the effect of age is removed from total flight hours, but not from reaction times, pilots' total flight hours uniquely explain an additional 1.9% of the variance in reaction times. This amount of explained variance is not significant, $t(66) = 1.52$, $p = .1329$.

c. See Figure 3.8.

TABLE 3.6 *Summary of Multiple Regression Results of the Partials for Example 3.3a*

		Zero-Order Correlations		Semipartial Correlations		Full Partial Correlations		Significance	
Factor [a]	B_i	r_{Yi}	r^2_{Yi}	sr_i	sr_i^2	pr_i	pr_i^2	$t_i(66)$	p_i
Intercept	−16.625							−2.54	.0133
X_1	0.272	.643	.413	.382	.146	.452	.204	4.12	.0001
X_2	0.003	.536	.287	.138	.019	.182	.033	1.52	.1329

Note. $N = 69$. $R^2_{Y \cdot 12} = .433$, $F(2, 66) = 25.244$, $p < .0001$.
[a] X_1 = Age, X_2 = Total Flight Hours, and Y = Reaction Time (in seconds).

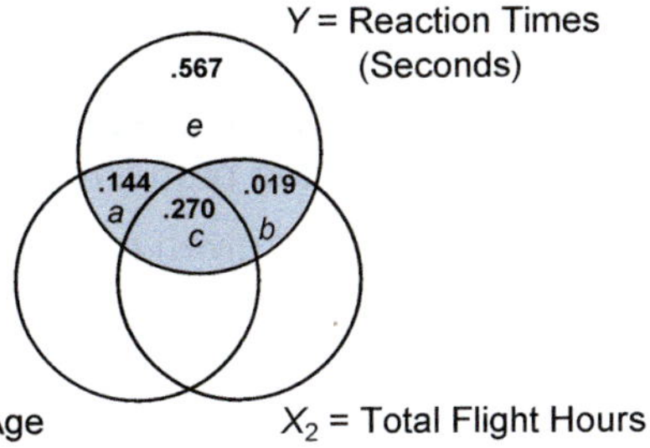

Regions (a + b + c): $R^2_{Y\cdot 12} = .144 + .270 + .019 = .433$. Age and total flight hours collectively explain 43.3% of the variance in reaction times.

Region a: $sr_1^2 = .144$. When the effect of total flight hours is removed from age but not from reaction times, age uniquely explains 14.4% of the variance in reaction times.

Region b: $sr_2^2 = .019$. When the effect of age is removed from total flight hours but not from reaction times, total flight hours uniquely explains 1.9% of the variance in reaction times.

Region c: Age and total flight hours share 27.0% of the variance in reaction times. This is the *redundancy* between X_1 and X_2.

$$
\begin{aligned}
&(a + b + c) - (a + b) \\
&= .433 - (.144 + .019) \\
&= .433 - (.163) \\
&= .270
\end{aligned}
$$

Region e: $1 - R^2_{Y\cdot 12} = 1 - .433 = .567$. 56.7% of the variance in reaction times is unexplained by age and total flight hours.

Regions (a + b + c + e): .144 + .019 + .27 + .567 = 1.00. This is 100% of *Y* variance.

FIGURE 3.8 Summary of complete Venn diagram for the reaction times data for Example 3.3.

3.5 A Guided Example of an MRC Analysis with Two IVs in Research

We now present a guided example of how to conduct an MRC analysis for a research study that involves two IVs. The study we will examine is a slightly modified version of our running example involving pilots' aviation safety index scores. More specifically, X_1 = Total Flight Hours, X_2 = Number of FAA Ratings, and Y = ASI Scores, which can range from 5 to 20, with higher scores reflecting higher levels of aviation safety consciousness.

The data set is provided in the file "Ch_3 Guided Example Data," and the structure of our presentation consists of four parts: Research Questions, Sample Size Planning, Data Analysis, and Interpretation of Results. (*Note:* The reader will recall from the Chapter Overview that we delay the discussion of statistical inference in MRC to Chapter 4, which generalizes from two IVs to *k* IVs. Therefore, we neither present corresponding hypotheses nor do we engage in any post hoc considerations at this time.)

Research Questions

We are primarily interested in the following three questions:

- What is the collective contribution the IVs make in explaining *Y* variance? This question is answered by the overall R^2.

- What is the unique contribution each IV makes in explaining Y variance? This question is answered by the squared semipartial correlation coefficients, sr_i^2.
- Holding all other variables constant, what effect does each IV have on predicting Y? This question is answered by the partial regression coefficients, B_i.

Expressing these RQs in the context of the given research setting:

RQ 1. What proportion of variance in ASI scores is explained collectively by X_1 = Total Flight Hours and X_2 = Number of FAA Ratings?

RQ 2. When holding all other variables constant, what proportion of variance in ASI scores is uniquely explained respectively by X_1 = Total Flight Hours (i.e., sr_1^2) and by X_2 = Number of FAA Ratings (i.e., sr_2^2)?

RQ 3. When holding all other variables constant, what are the respective effects of total flight hours (i.e., B_1) and number of FAA ratings (i.e., B_2) on ASI scores?

Sample Size Planning

As presented in the previous two chapters, sample size planning is an activity researchers conduct a priori to determine the minimum sample size needed (n^*) relative to a minimum power level (commonly set to .8), an alpha level (commonly set to α = .05), and an expected effect size. When applied to MRC, sample size planning also includes the number of IVs (k). In Chapter 4, which generalizes an MRC analysis from two IVs to k IVs, we provide a detailed discussion of the concept of sample size planning applied to MRC, and the interested reader is directed to this chapter advance of our presentation if so inclined. For the current example, though, we will assume that the given data set with N = 66 is sufficiently large.

Data Analysis

We now conduct an MRC analysis for the given data set. A summary of the results is provided in Table 3.7.

Interpretation of Results

The overall regression equation is $\hat{Y} = -0.0004X_1 + 1.2144X_2 + 10.9106$. We now interpret the results relative to the research questions.

RQ 1. This research question relates to the overall R^2. The total proportion of variance in ASI scores explained collectively by X_1 = Total Flight Hours and X_2 = Number of FAA Ratings is $R^2_{Y\bullet 12}$ = .181, with an adjusted R^2 = .155. This means that the two IVs collectively explain 18.1% of the variance in ASI scores and when adjusted for bias, these IVs collectively explain 15.5% of the variance in ASI scores. This result is significant, $F(2, 63) = 6.94$, $p = .0019$.

RQ 2. This research question relates to the squared semipartial correlation coefficients (sr_i^2). The interpretation of each coefficient follows.

• sr_1^2 = .068. X_1 = Total Flight Hours uniquely explains 6.8% of the variance in ASI scores in the presence of X_2. Alternatively, total flight hours explain an additional 6.8% of the variance in ASI scores after considering the amount of variance explained by the number of FAA ratings. This is significant, $t(63) = -2.28$, $p = .0262$.

• sr_2^2 = .136. X_2 = Number of FAA Ratings uniquely explains 13.6% of the variance in ASI scores in the presence of X_1. Alternatively, the number of FAA ratings

TABLE 3.7 *Summary of MRC Analysis for the Data Set Given in the File "Ch_3 Guided Example Data"*

Model Summary

R	.424910
R^2	.180549
Adjusted R^2	.154535
Standard Error of the Estimate	4.847333
Mean of Response	12.80
Sample Size	66

Coefficients (Parameter Estimates)

Term	*B*	*SE*	β	*t*(63)	*p*
Constant	10.9106	1.618291	—	6.74	< .0001
X_1 = Total Flight Hours	–0.0004	0.000173	–0.26228	–2.28	.0262
X_2 = Number of FAA Ratings	1.21438	0.375421	0.37249	3.23	.0019

ANOVA

Source	*SS*	*df*	*MS*	*F*	*p*
Regression	326.1510	2	163.075	6.94	.0019
Residual (Error)	1480.2884	63	23.497		
Total	1806.4394	65			

Partial Correlation Coefficients and Their Squares

	Zero-Order Correlations		Semipartial Correlations		Full Partial Correlations	
Factor [a]	r_{Yi}	r_{Yi}^2	sr_i	sr_i^2	pr_i	pr_i^2
X_1 = Total Flight Hours	–.211	.045	–.259	.068	–.276	.076
X_2 = Number of FAA Ratings	.336	.113	.369	.136	.377	.142

explain an additional 13.6% of the variance in ASI scores after considering the amount of variance explained by total flight hours. This is significant, $t(63) = 3.23$, $p = .0019$.

RQ 3. This research question relates to the partial regression coefficients (B_i). The interpretation of each coefficient follows.

• $B_1 = -0.0004$. Holding all other variables in the regression model constant, for every 1-hour increase in total flight hours, ASI scores *decrease*, on average, by 0.0004 points. From a more practical perspective, for every 10,000-hour increase in total flight hours, ASI scores decrease, on average, by 4.0 points. This is significant, $t(63) = -2.28$, $p = .0262$.

• $B_2 = 1.21438$. Holding all other variables in the regression model constant, for every 1 additional FAA rating, ASI scores increase on average by 1.2 points, which is significant, $t(63) = 3.23$, $p = .0019$.

Standard Error of the Estimate. Recall that this is the root mean square error (*RMSE*), which is a metric that indicates approximately how far "off" a prediction of *Y* will be when using the regression equation. Although this was not part of our RQs, it is informative to report it. For the current study, *RMSE* = 4.85, which means that if we were to predict a pilot's ASI score based on the pilot's total flight hours and number of FAA ratings, then this predicted score will be "off" on average by 4.85 points.

Chapter 3 Summary

3.2 Bivariate vs. Multiple Regression: A Conceptual Approach

- When the bivariate correlation between two IVs, X_1 and X_2, is not zero (i.e., $r_{12} \neq 0$), then part of their respective relationship with Y is redundant and therefore it is not appropriate to examine their relationship with Y separately.
- Venn diagrams are used to visually illustrate the concept of disjoint IVs where $r_{12} = 0$ vs. the concept of redundancy (i.e., $r_{12} \neq 0$).
- Given two IVs, X_1 and X_2, and the DV, Y, if the correlation coefficient between any two variables is calculated in the absence of the third variable, then the coefficient is called a zero-order correlation coefficient. Thus, r_{Y1}, r_{Y2}, and r_{12} are zero-order correlation coefficients.

3.3 Bivariate vs. Multiple Regression: An Empirical Approach

- A multiple regression equation that examines the effect of two IVs on a DV is of the form $\hat{Y} = B_1X_1 + B_2X_2 + B_0$ where B_i is a partial regression coefficient, and B_0 is the regression constant. In this context, B_1 is the effect X_1 has on Y in the presence of X_2, and B_2 is the effect X_2 has on Y in the presence of X_1.
- Unstandardized regression coefficients (B_i) are based on raw data, and standardized regression coefficients (β_i) are the result of standardizing the data such as using z scores.
- The interpretation of a partial regression coefficient is prefaced with "Holding all other variables in the model constant ..." or "When examined in the presence of the other variables ..."
- The regression constant, B_0, is not always interpretable for a given research context.
- In addition to the regression equation, the results of an MRC analysis also include the multiple R and R^2, the adjusted R^2, *RMSE*, and the corresponding ANOVA table, which indicates if the overall regression model is significant. If the overall model is not significant, then we do not interpret the regression coefficients because of the possibility of inflated alpha levels.

3.4 Semipartial and Partial Correlation Coefficients and Their Squares

- The semipartial correlation coefficient (sr_i) represents the unique relationship between an IV and DV when in the presence of all other IVs.
- The squared semipartial correlation coefficient (sr_i^2) represents the amount of Y variance uniquely explained by one variable when the effects of all the other IVs have been removed from this variable but not from Y. As shown in Figures 3.5 and 3.6, this is reflected in regions *a* and *b* for the two IVs case.
- The sum of the sr_i^2s does not equal the overall R^2. This means that in the two IVs case, $R^2_{Y \cdot 12} \neq sr_1^2 + sr_2^2$. This is because the redundancy between X_1 and X_2 is never accounted for.
- In an MRC analysis, sr_i^2 answers the question, "How much additional information about the variance in Y is gained when an IV enters the analysis in the presence of all the other IVs?"

- The partial correlation coefficient (pr_i) and its square (pr_i^2) represent the relationship X_i has with Y when the effects of all the other IVs have been removed from both X_i and Y. This statistically controls for the effect the other IVs have on X_i and Y.
- From a Venn diagram perspective, the circle that represents the full partial correlation coefficient and its square is completely removed from the diagram as shown in Figure 3.7. This results in a change in base when calculating these coefficients because the circle representing Y is no longer whole (i.e., it is not equal to 1).
- In an MRC analysis, pr_i^2 answers the question, "When all the other variables in a regression model are statistically controlled, how much of Y variance does a targeted variable account for?"
- In an MRC analysis, if a regression coefficient is significant, then all the corresponding correlation coefficients and their squares associated with this regression coefficient are also significant.

Key Terms and Concepts

Adjusted R^2
Beta coefficients
Big R squared
Correlation matrix
Disjoint
MRC
Multiple correlation coefficient
Multiple R
Multiple R square (R^2)
Partial correlation coefficient (pr_i)
Partial regression coefficient (B_i)
Redundancy
Regression constant
Residual analysis
Root mean square error ($RMSE$)
$R^2_{Y \cdot 12}$
Semipartial correlation coefficient (sr_i)
Squared partial correlation coefficient (pr^2)
Squared semipartial correlation coefficient (sr^2)
Standard error of the estimate
Standardized regression coefficients
Unexplained variance
Unique contribution
Unstandardized regression coefficients
Venn diagrams
Zero-order correlation coefficients

Review Exercises

A. Checking Conceptual Knowledge

In 1–10, choose the best answer among the choices provided.

1. In the Venn diagram given in Figure 3.9, the circle labeled Y represents 100% of Y variance. Which regions of the diagram are "removed" when pr_1^2 is calculated?
   a. *a*, *b*, *d*
   b. *b*, *c*, *d*
   c. *a*, *b*, *c*
   d. *d*, *b*

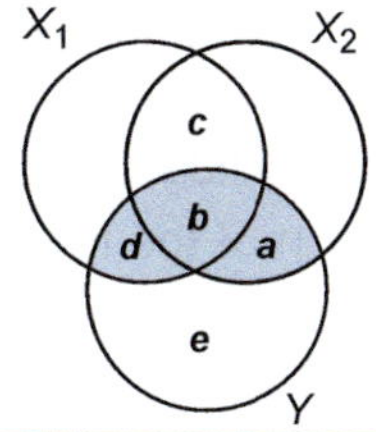

FIGURE 3.9 Venn diagram for Review Exercises, Part A, Item 1.

2. In an MRC analysis, as N increases the difference between R^2 and adjusted R^2
   a. remains the same.
   b. increases slightly.
   c. decreases slightly.
   d. approaches zero.
3. Which statistic do we try to minimize in an MRC analysis?
   a. R
   b. R^2
   c. $1 - R^2$
   d. η^2
4. From a conceptual perspective, the *RMSE* may be considered an estimator of
   a. s
   b. R^2
   c. s^2
   d. Y
5. Below are three prediction equations that resulted from an MRC analysis from all 50 U.S. states. The equations were used to predict aviation safety index (ASI) scores of airline transport pilots from X_1 = Total Flight Hours and X_2 = Number of Ratings held for all aircraft classifications. What is the best interpretation of the intercept term in Model 3?

   Model 1: $\hat{Y} = 0.025X_1 + 5$
   Model 2: $\hat{Y} = 0.5X_2 + 5.5$
   Model 3: $\hat{Y} = 0.05X_1 + 0.25X_2 + 7$

   a. ATPs with 0 flight hours and at least one rating have an ASI score of 7, on average.
   b. In those U.S. states that have ATPs with 0 flight hours but at least one rating, ATPs have an ASI score of 7, on average.
   c. The intercept term is meaningless because it is not possible to have an ATP with 0 flight hours and 0 ratings.
   d. Both (b) and (c) are equally good interpretations.

Use the following research description for Items 6–10.

A research study examined the extent to which flight attendants' income level and education level predicted their level of happiness as flight attendants. The results of the corresponding MRC analysis are given in Table 3.8.

6. Which statement is NOT true for sr_2^2?
   a. Education level is not a significant predictor of flight attendants' happiness scores.
   b. Education level explains 4.3% of the variance in happiness scores.
   c. When statistically controlling for the effect of income level, education level explains 4.3% of the variance in happiness scores.
   d. When examined in the presence of income level, education level contributes an additional 4.3% of the variance in happiness scores.
7. The results reveal that the collective contribution the two variables make in explaining the variance in happiness scores is
   a. significant, $F(2, 181) = 946.779$.
   b. not significant, $F(2, 183) = 30.5439$.
   c. significant, $F(2, 181) = 30.5439$.
   d. not significant, $F(2, 183) = 30.5439$.
8. Which of the following statements is true?
   a. 25.2% of the variance in Happiness scores in the population is explained.
   b. 50.2% of the variance in Happiness scores in the sample is explained.
   c. Education level is the major predictor.
   d. Income level would be a significant predictor if the other IVs were removed.

TABLE 3.8 *Summary of MRC Analysis for Review Exercises-Part A Problems 6–10*

Model Summary

R	.502332
R^2	.252338
Adjusted R^2	.244076
Standard Error of the Estimate	5.567522
Mean of Response	21.0797
Sample Size	184

Coefficients (Parameter Estimates)

Term	*B*	*SE*	β	*t*	*p*
Constant	4.6549601	2.686517	—	1.73	.0849
X_1 = Income Level	0.000448	0.000061	0.47691	7.39	< .0001
X_2 = Education Level	−0.617847	0.190446	−0.20946	−3.24	.0014

ANOVA

Source	*SS*	*df*	*MS*	*F*	*p*
Regression	1893.5584	2	946.779	30.5439	< .0001
Residual (Error)	5610.5108	181	30.997		
Total	7504.0692	183			

Partial Correlation Coefficients and Their Squares

	Zero-Order Correlations		Semipartial Correlations		Full Partial Correlations	
Factor [a]	r_{Yi}	r_{Yi}^2	sr_i	sr_i^2	pr_i	pr_i^2
X_1 = Income Level	.457	.209	.475	.225	.481	.232
X_2 = Education Level	−.164	.027	−.209	.043	−.234	.055

9. The prediction equation is
   a. Predicted Happiness = (0.048)Income Level – (0.617847)Education Level.
   b. Predicted Happiness = 4.65 + (0.000448)Income Level – (0.617847)Education Level.
   c. Predicted Happiness = 4.65 + (.00006)Income Level – (0.190446)Education Level.
   d. Predicted Happiness = (.00006)Income Level – (0.190446)Education Level,
10. The pr_i^2 values are larger in magnitude than their corresponding sr_i^2 values because
    a. they are a fluke of the data.
    b. their calculation relies on total Y variance.
    c. their calculation involves a smaller base even though their numerators are the same.
    d. this is what makes their results "unique."

B. Applying Procedural and Conceptual Knowledge

Use the following research description and the corresponding data set to conduct the activities given in Parts B1–B4. See Section 3.5 for guidance.

> Harwin (2023) examined the relationship between a set of targeted factors and the total flight time students needed to become ready to take the private pilot check ride. For this exercise, we present a research project similar to Harwin's that involves two independent variables: X_1 = Flight Students' Age (in years) and X_2 = Number of Dual Hours reported by students' CFIs. The dependent variable is Y = Flight Time to Proficiency, which is defined as the number of flight hours needed before a CFI determines a student is check-ride proficient. A set of hypothetical data is provided in the file "Ch_3 Review Exercises Data-Part B."

B1. Research Questions. Write the research questions that pertain to the overall regression model, the squared semipartial correlation coefficients, and the partial regression coefficients.

B2. Sample Size Planning. Assume that the given data set is of sufficient size.

B3. Data Analysis. Conduct an MRC analysis of the given data set and summarize the results in a table like Table 3.7.

B4. Interpretation of Results. Interpret the results relative to the research questions stated in Part B1. Also include the regression equation and *RSME*.

References

Barkhuizen, W., Schepers, J., & Coetzee, J. (2002). Rate of information processing and reaction time of aircraft pilots and non-pilots. *Journal of Industrial Psychology, 28*(2), 67–76.

Bell, B. D., Robertson, C. L., & Wagner, G. S. (1995). Aviation safety as a function of pilot experience: Rationale or rationalization? *The Journal of Aviation/Aerospace Education & Research, 5*(3), 27–38.

Harwin, M. S. (2023). *The private pilot check ride: Applying the spacing effect theory to predict time to proficiency for the practical test.* [Doctoral dissertation, Florida Institute of Technology]. Scholarship Repository. https://repository.fit.edu/cgi/viewcontent.cgi?article=2353&context=etd

Schepers, J. M. (2001). *Manual of the computerised information processing test battery.* Johannesburg: R.A.U.

Chapter 4
MRC with More Than Two Predictors

Student Learning Outcomes

After studying this chapter, you will be able to do the following:

1. Interpret the results reported in the parameter estimates table, model summary table, and ANOVA summary table of an MRC analysis with more than two IVs.
2. Prepare a table that summarizes the results of an MRC analysis with more than two IVs and interpret the results of this table, including the various correlation coefficients and their squares.
3. Determine and interpret factors of statistical inference—parameter estimation, hypothesis testing, and effect size—of an MRC analysis with more than two IVs.
4. Conduct an MRC analysis for a research study with more than two IVs by engaging in the activities presented in Section 4.5.

4.1 Chapter Overview

In Chapter 3, we presented key concepts related to a multiple regression/correlation analysis involving two IVs. These concepts included the partial regression coefficients (B_i) and the various correlation coefficients and their squares (R, R^2, sr_i, sr_i^2, pr_i, pr_i^2). In this chapter, we broaden Chapter 3's discussion to research settings that involve more than two IVs. We begin conceptually by presenting a Venn diagram comprised of three IVs and then generalize this concept to k IVs. As part of this discussion, we conduct an MRC analysis involving three IVs and interpret the corresponding results. We then discuss the concept of statistical inference, including confidence intervals and hypothesis testing, relative to an MRC analysis with k IVs. We conclude the chapter with a guided example of how to conduct an MRC analysis for a research study that involves four IVs and a single, continuous DV.

4.2 Generalizing from Two IVs to *k* IVs

Recall from Chapter 3 that we used Venn diagrams to represent the relationships among two IVs and a single DV. We now extend this presentation by introducing a third IV as illustrated in Figure 4.1. Observe from Figure 4.1(a) that the circles representing the IVs do not overlap—that is, they are disjoint—which means their respective zero-order correlations are zero: $r_{12} = 0$, $r_{13} = 0$, and $r_{23} = 0$. As a result, there is no redundancy among the IVs and the relationships X_1, X_2, and X_3 have with Y are reflected in regions a, b, and c, respectively.

Now look at Figure 4.1(b), which has overlapping circles.

- X_1 and X_2 have regions b and f in common.
- X_1 and X_3 are disjoint.
- X_2 and X_3 have regions d and g in common.

DOI: 10.4324/9781003506935-6

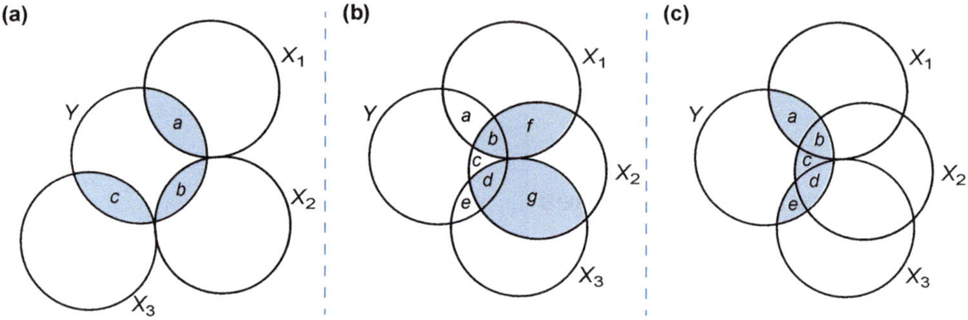

FIGURE 4.1 Graphical representation of the various relationships involving three IVs and a DV.

If we focus on only those regions related to the DV as shown in Figure 4.1(c):

- X_1's relationship with Y is represented by regions a and b.
- X_2's relationship with Y is represented by regions b, c, and d.
- X_3's relationship with Y is represented by regions d and e.

Combining these two sets of observations from Figures 4.2(b) and (c):

- X_1 and X_2 share redundant information about Y via region b.
- X_1 and X_3 do not share any redundant information about Y.
- X_2 and X_3 share redundant information about Y via region d.

As a result, to capture these relationships and account for redundancies, we need to extend our multiple regression model for two IVs so that it includes the third IV. This model is given by the following general equation, which uses the "partialling" notation subscripts introduced in Chapter 3:

$$\hat{Y} = B_{Y1 \cdot 23}X_1 + B_{Y2 \cdot 13}X_2 + B_{Y3 \cdot 12}X_3 + B_0$$

where

- $B_{Y1 \cdot 23}$ is the partial regression coefficient corresponding to X_1's effect on Y in the presence of X_2 and X_3.
- $B_{Y2 \cdot 13}$ is the partial regression coefficient corresponding to X_2's effect on Y in the presence of X_1 and X_3.
- $B_{Y3 \cdot 12}$ is the partial regression coefficient corresponding to X_3's effect on Y in the presence of X_1 and X_2.

This general regression equation is written more simply as:

$$\hat{Y} = B_1X_1 + B_2X_2 + B_3X_3 + B_0$$

For a concrete example, let's revisit the data set from Chapter 3, which was given in the file "Ch_3 ASI Data." Recall that this data set consisted of X_1 = Malfunction History, X_2 = Age, and Y = Aviation Safety Index (ASI) scores, which could range from 5 to 20 with higher scores reflecting higher levels of safety consciousness. For the current example, we add a third variable, namely, X_3 = Total Flight Hours. This revised data set is given in the file "Ch_4 ASI Data with 3 IVs," and Table 4.1 contains a summary of the results of this MRC analysis. (*Note*. Table 4.1 also includes a column for the 95% confidence intervals, which we will discuss in Section 4.4.)

TABLE 4.1 *Summary of Multiple Regression Analysis Results for the Data Set Given in the File "Ch_4 ASI Data with 3 IVs"*

Model Summary

R	.646409
R^2	.417845
Adjusted R^2	.330522
Standard Error of the Estimate	2.22775
Mean of Response	13.75
Sample Size	24

Coefficients (Parameter Estimates)

Term	*B*	*SE*	β	*t*(20)	*p*	95% CI
Constant	8.097214	2.014938	—	4.02	.0007	[3.8941, 12.3003]
X_1 = Malfunction History	−0.009871	0.167672	−0.01111	−0.06	. 9536	[−0.3596, 0.3399]
X_2 = Age	0.126892	0.057629	0.40813	2.20	.0396	[0.0067, 0.2471]
X_3 = Total Flight Hours	0.000354	0.000164	0.40115	2.16	.0429	[3.8941, 12.3003]

ANOVA

Source	*SS*	*df*	*MS*	*F*	*p*
Regression	71.24262	3	23.7475	4.7850	.0113
Residual (Error)	99.25738	20	4.9629		
Total	170.50000	23			

Note from the Coefficients subtable of Table 4.1 that the regression equation is:

$$\hat{Y} = -0.009871X_1 + 0.126892X_2 + 0.000354X_3 + 8.097214$$

Also note from Table 4.1's Model Summary subtable:

- $R = .646409$, which means that the correlation between the observed ASI scores and their respective predicted scores (i.e., $r_{Y\hat{Y}}$) is moderately strong and positive.
- $R^2_{Y \cdot 123} = .417845$, which means that 41.78% of the variance in ASI scores is being explained collectively by the three IVs. This also means that if we know a pilot's malfunction history, age, and total flight hours, then we have 41.78% of the information needed to perfectly predict this pilot's ASI score.
- The adjusted $R^2 = .330522$, which means that when the model is adjusted for bias, the three IVs collectively explain 33% of the variance in ASI scores in the population.
- $RMSE = 2.22775$, which means that if we use this regression equation to predict a pilot's ASI score from the pilot's malfunction history, age, and total flight hours, then the predicted ASI score will be "off" on average by about 2.23 points.
- The mean ASI score ($M_Y = 13.75$) and sample size ($N = 24$) are the same as those reported in Chapter 3 because we are working with exactly the same data set. We just included a third IV.

Lastly, observe from the ANOVA subtable of Table 4.1 that the overall regression model is statistically significant, $F(3, 20) = 4.7850$, $p = .0113$, for a preset alpha level of $\alpha = .05$. Let's now interpret the partial regression coefficients.

- $B_1 = -0.009871$. Holding all other IVs constant (i.e., X_2 and X_3), for every 1 additional inflight malfunction reported by pilots, ASI scores decrease on average by 0.009871 points. More practically, there is a 0-point increase on average in ASI scores. This finding is not significant, $t(20) = -0.06$, $p = .9536$.
- $B_2 = 0.126892$. Holding all other IVs constant (i.e., X_1 and X_3), for every 1-year increase in pilots' age, ASI scores increase on average by 0.126892 points. More practically, for every 10-year increase in age, ASI scores increase on average by approximately 1.3 points. Thus, older pilots are more safety conscious than younger pilots. This finding is significant, $t(20) = 2.20$, $p = .0396$.
- $B_3 = 0.000354$. Holding all other IVs constant (i.e., X_1 and X_2), for every 1-hour increase in total flight hours, ASI scores increase on average by 0.000354 points. More practically, for every 1,000-hour increase in total flight time, ASI scores increase on average by approximately 0.35 points, or for every 10,000-hour increase in total flight time, ASI scores increased on average by approximately 3.5 points. This finding is significant, $t(20) = 2.16$, $p = .0429$.
- $B_0 = 8.097214$. A pilot who has experienced 0 inflight malfunctions, is 0 years old, and has 0 hours of total flight time, will have, on average, an ASI score of approximately 8 points. Although this finding is significant, $t(20) = 4.02$, $p = .0007$, the result has no meaningful interpretation relative to the given research context.

Example 4.1 *Conducting An MRC Analysis with Three IVs*

For this example, we will use the data set from Chapter 3's Guided Example, which was given in the file "Ch_3 Guided Example Data," but now include a third IV:

Y = ASI scores, which can range from 5 to 20, with higher the scores reflecting higher levels of aviation safety consciousness.

X_1 = Total Flight Hours.

X_2 = Number of FAA Ratings

X_3 = Flight Training Program, which is a nominal IV that compares Part 61 vs. Part 141 programs. Because X_3 represents a dichotomy between two variables, we will code the groups 0 = Part 61 and 1 = Part 141.

The data set is provided in the file "Ch_4 Example 4.1 Data." Conduct an MRC analysis using this data set and then respond to Items a–d. (Round calculations to four decimal places. Assume $\alpha = .05$.)

a. Determine the overall regression equation.

b. Determine if the overall model is significant.

c. Interpret the regression coefficients and regression constant in the context of the given research setting and determine if they are significant. Include corresponding *t*, *df*, and *p* values.

d. Determine the *R*, R^2, adjusted R^2, and *RMSE* values and interpret them in the context of the given research setting.

Solution

A summary of the results for Example 4.1 is provided in Table 4.2.

Continued on next page ...

TABLE 4.2 *Summary of Multiple Regression Analysis Results for Example 4.1*

Model Summary

R	.4861
R^2	.2363
Adjusted R^2	.1993
Standard Error of the Estimate	4.7172
Mean of Response	12.80
Sample Size	66

Coefficients (Parameter Estimates)

Term	*B*	*SE*	β	*t*(62)	*p*	95% CI
Constant	9.7660	1.664	—	5.87	< .0001	[6.44, 13.09]
X_1 = Total Flight Hours	−0.0004	0.0002	−0.2612	−2.33	.0231	[−0.0007, −0.0001]
X_2 = Num. of FAA Ratings	1.2265	0.3654	0.3762	3.36	.0014	[0.50, 1.96]
X_3 = Flight Training Program	2.4887	1.1701	0.2361	2.13	.0374	[0.15, 4.83]

ANOVA

Source	*SS*	*df*	*MS*	*F*	*p*
Regression	426.8179	3	142.273	6.3937	.0008
Residual (Error)	1379.6215	62	22.252		
Total	1806.4394	65			

a. The regression equation is $\hat{Y} = -0.0004X_1 + 1.2265X_2 + 2.4887X_3 + 9.7660$.

b. The overall model is significant, $F(3, 62) = 6.3937$, $p = .0008$.

c. $B_1 = -0.0004$. Holding the number of FAA ratings and flight training program constant, for every 1-hour increase in total flight hours, ASI scores *decrease,* on average, by 0.0004 points. More concretely, for every 1,000-hour increase in total flight hours, ASI scores decrease, on average, by four tenths of a point, or equivalently, there is a 4-point decrease, on average, for every 10,000-hour increase in total flight hours. Thus, higher flight time pilots have a lower level of aviation safety consciousness than lower flight time pilots. This result is significant, $t(62) = -2.33$, $p = .0231$.

$B_2 = 1.2265$. Holding total flight hours and flight training program constant, for every 1 additional FAA rating, ASI scores increase, on average, by 1.23 points. Thus, pilots with more FAA ratings have a higher level of aviation safety consciousness than pilots with fewer FAA ratings. This result is significant, $t(62) = 3.36$, $p = .0014$.

$B_3 = 2.4887$. X_3 is a nominal IV representing group membership. Thus, B_3 represents a comparison between the two flight training programs and is aligned to the group coded 1. So, holding total flight hours and number of FAA ratings constant, pilots who receive flight training via a Part 141 program average approximately 2.5 points higher on the ASI than pilots who receive flight training via a Part 61 program. This infers that Part 141-trained pilots have a higher level of aviation safety consciousness than Part 61-trained pilots. This result is significant, $t(62) = 2.13$, $p = .0374$.

Continued on next page ...

B_0 = 9.766: Zeroing out the regression equation ($X_1 = X_2 = X_3 = 0$) reflects a pilot with 0 total flight hours, 0 FAA ratings, and trained under a Part 61 program. Although not practical, this could refer to a new flight student who is about to begin flight training. Thus, such individuals, on average, will have an ASI score of 9.766 points. This result is significant, $t(62) = 5.87$, $p < .0001$.

d. R = .4861. The correlation between the observed ASI scores and the corresponding predicted ASI scores is .4861. This reflects a positive but moderate relationship between the observed and predicted scores.

R^2 = .2363. The coefficient of determination indicates that 23.63% of the variance in ASI scores is being explained collectively by pilots' total flight hours, number of FAA ratings, and type of flight training program. We also can interpret this from a prediction perspective: If we know a pilot's total flight hours, number of FAA ratings, and type of flight training program, then we have 23.63% of the information needed to perfectly predict the pilot's ASI score.

Adjusted R^2 = .1993. When the overall R^2 has been adjusted for bias, 19.93% of the variance in ASI scores is being explained collectively by pilots' total flight hours, number of FAA ratings, and type of flight training program.

RMSE = 4.7172. If we use pilots' total flight hours, number of FAA ratings, and type of flight training program to predict their ASI scores using the corresponding regression equation, then the prediction would be "off," on average, by approximately 4.7 points.

As demonstrated in Example 4.1, conducting an MRC analysis with three IVs is done exactly as we did in Chapter 3 when we conducted an MRC analysis with two IVs. The only difference is for the current situation we included a term in the regression equation to represent the third IV. All the analyses and interpretations, though, are the same. To further emphasize this extension, we provide an example of an MRC analysis with five IVs (see Example 4.2). We also prepared Table 4.3, which shows the general form of a regression equation with two IVs to five IVs, and then generalizes to k IVs using both descriptive subscript notation and simple subscripts.

TABLE 4.3 *General Forms of Regression Equations for 2 IVs, 3 IVs, 4 IVs, 5 IVs, and k IVs*

# of IVs	*Form*	**Regression Equation**
$k = 2$	*Descriptive*	$\hat{Y} = B_{Y1 \cdot 2}X_1 + B_{Y2 \cdot 1}X_2 + B_0$
	Simple	$\hat{Y} = B_1X_1 + B_2X_2 + B_0$
$k = 3$	*Descriptive*	$\hat{Y} = B_{Y1 \cdot 23}X_1 + B_{Y2 \cdot 13}X_2 + B_{Y3 \cdot 12}X_3 + B_0$
	Simple	$\hat{Y} = B_1X_1 + B_2X_2 + B_3X_3 + B_0$
$k = 4$	*Descriptive*	$\hat{Y} = B_{Y1 \cdot 234}X_1 + B_{Y2 \cdot 134}X_2 + B_{Y3 \cdot 124}X_3 + B_{Y4 \cdot 123}X_4 + B_0$
	Simple	$\hat{Y} = B_1X_1 + B_2X_2 + B_3X_3 + B_4X_4 + B_0$
$k = 5$	*Descriptive*	$\hat{Y} = B_{Y1 \cdot 2345}X_1 + B_{Y2 \cdot 1345}X_2 + B_{Y3 \cdot 1245}X_3 + B_{Y4 \cdot 1235}X_4 + B_{Y5 \cdot 1234}X_5 + B_0$
	Simple	$\hat{Y} = B_1X_1 + B_2X_2 + B_3X_3 + B_4X_4 + B_5X_5 + B_0$
Generalizing to k IVs		
k	*Descriptive*	$\hat{Y} = B_{Y1 \cdot 23 \ldots k}X_1 + B_{Y2 \cdot 13 \ldots k}X_2 + B_{Y3 \cdot 12 \ldots k}X_3 + \ldots + B_{Yk \cdot 123 \ldots k-1}X_k + B_0$
	Simple	$\hat{Y} = B_1X_1 + B_2X_2 + B_3X_3 + \ldots + B_kX_k + B_0$

Note. In the descriptive form, the respective partialling is given in the subscripts for the partial regression coefficients (B_i). In the simple form, the subscripts for each B_i correspond to the IV and the respective partialling is understood relative to the number of IVs in the equation.

Example 4.2 *Conducting An MRC Analysis with Five IVs*

A study investigated the effect five factors had on pilots' visual reaction times. The targeted factors are:

X_1 = Age is a self-reported continuous variable measured in years.

X_2 = Corrected Vision is a self-reported dichotomous nominal variable that represents a comparison between pilots with corrected vision (i.e., those wear glasses/contact lenses or had Lasik surgery) vs. pilots who did not have corrected vision. This variable was coded 1 = corrected vision (yes) and 0 = no corrected vision.

X_3 = Visual Acuity Angle is a measure of visual angle on the retina equivalent to the more conventional Snellen measure (e.g., 20/20) except it is specified as a single number that is calculated by dividing the denominator by the numerator of a Snellen measure. For example, someone with 20/10 vision would have a visual angle of 0.5, and someone with 20/40 vision would have a visual angle of 2.0. The smaller the visual angle, the better a person's eyesight.

X_4 = Flight Hours is a self-reported continuous variable that represents pilots' total number of accrued flight time measured in whole hours.

X_5 = Number of Ratings is a self-reported discrete variable that represents the total number of FAA pilot certificate ratings such as Private Pilot License (PPL), Instrument Rated Pilot (IR), Commercial Pilot License (CPL), Airline Transport Pilot (ATP), Certified Flight Instructor (CFI), Certified Flight Instructor–Instrument (CFII), and Multiengine–Instrument (MEI).

The dependent variable, Y = Visual Reaction Time, is measured in milliseconds (ms) by having participants press a joystick immediately at the appearance of a suprathreshold spot target. Note that 1 second = 1,000 ms, so a reaction time of 279 ms is equal to 0.279 seconds, or approximately three tenths of a second. The data set corresponding to this study is given in the file "Ch_4 Example 4.2 Data." (Round calculations to four decimal places. Assume $\alpha = .05$.)

a. Determine the overall regression equation.

b. Determine if the overall model is significant.

c. Interpret the regression coefficients and regression constant in the context of the given research setting and determine if they are significant. Include corresponding *t*, *df*, and *p* values.

d. Determine the *R*, R^2, adjusted R^2, and *RMSE* values and interpret them in the context of the given research setting.

Solution

A summary of the results for Example 4.2 is provided in Table 4.4.

a. The regression equation is $\hat{Y} = 6.3571X_1 - 71.0436X_2 - 4.5468X_3 + 0.0089X_4 + 2.0391X_5 + 45.1123$

b. The overall model is significant, $F(5, 39) = 4.2883$, $p = .0033$.

Continued on next page ...

TABLE 4.4 *Summary of Multiple Regression Analysis Results for Example 4.2*

Model Summary

R	.5956
R^2	.3548
Adjusted R^2	.2720
Standard Error of the Estimate	79.1941
Mean of Response	274.24
Sample Size	45

Coefficients (Parameter Estimates)

Term	B	SE	β	$t(39)$	p	95% CI
Constant	45.1123	69.7120	—	0.65	.5213	[–95.89, 186.12]
X_1 = Age	6.3571	1.8442	0.5125	3.45	.0014	[2.63, 10.09]
X_2 = Corrected Vision	–71.0436	28.2971	–0.3846	–2.51	.0163	[–128.28, –13.81]
X_3 = Visual Acuity Angle	–4.5468	15.2690	–0.0385	–0.30	.7675	[–35.43, 26.34]
X_4 = Flight Hours	0.0089	0.0034	0.3580	2.61	.0128	[0.002, 0.016]
X_5 = Num. of Ratings	2.0391	7.8245	0.0352	0.26	.7958	[–13.79, 17.87]

ANOVA

Source	SS	df	MS	F	p
Regression	134475.82	5	26895.2	4.2883	.0033
Residual (Error)	244596.49	39	6271.7		
Total	379072.31	44			

c. B_1 = 6.3571. Holding all other variables constant, for every 1-year increase in age, a pilot's visual reaction time increases, on average, by approximately 6.4 ms. More concretely, as pilots get older, their visual reaction times *increase*, which means they react *slower*. This result is significant, $t(39)$ = 3.45, p = .00014.

B_2 = –71.0436. Holding all other variables constant, pilots with corrected vision on average have a visual reaction time that is 71.04 ms *lower* than pilots without corrected vision. More concretely, pilots who wear glasses or contact lenses or had Lasik surgery have *faster* visual reaction times than pilots whose vision is not corrected. This result is significant, $t(39)$ = –2.51, p =.0163.

B_3 = –4.5468. Holding all other variables constant, for every 1-point increase in visual acuity angle (poorer eyesight), visual reaction times *decrease*, on average, by 4.55 ms. More concretely, as pilots' vision becomes worse, they have faster reaction times. This result is not significant, $t(39)$ = –0.30, p =.7675.

B_4 = 0.0089. Holding all other variables constant, for every 1-hour increase in flight time, pilots' visual reaction times increase, on average, by about 0.01 ms. More concretely, for every 1,000-hour increase in flight hours, pilots' visual reaction times increase, on average, by 8.9 ms (visual reaction times become slightly slower). This result is significant, $t(39)$ = 2.61, p =.0128.

Continued on next page ...

B_5 = 2.0391. Holding all other variables constant, for every 1 additional rating, visual reaction times increase, on average, by 2.04 ms. More concretely, as pilots accrue more ratings, their visual reaction times also increase, which means it takes longer for them to react. This result is not significant, $t(39) = 0.26$, $p = .7958$.

B_0 = 45.1123. This is not interpretable because it would represent the average visual reaction time of a pilot who is 0 years old, does not have corrected vision, has 20/0 vision, 0 flight hours, and 0 ratings.

d. $R = .5956$. The correlation between the observed visual reaction times and the corresponding predicted visual reaction times is .5956. This reflects a positive and relatively strong relationship between the observed and predicted reaction times.

$R^2 = .3548$. The coefficient of determination indicates that 35.48% of the variance in visual reaction times is being explained collectively by the targeted five factors. It also means that if we know a pilot's age, whether or not the pilot has corrected vision, the pilot's visual acuity angle, total flight hours, and number of ratings, then we have 35.48% of the information needed to perfectly predict the pilot's visual reaction time.

Adjusted $R^2 = .2720$. When the overall R^2 is adjusted for bias, 27.2% of the variance in visual reaction times is being explained collectively by the five targeted factors.

$RMSE$ = 79.1941. If we use the targeted five factors and apply them to the corresponding regression equation to predict pilots' visual reaction times, then the prediction would be "off," on average, by approximately 79 ms.

4.3 Determining and Interpreting the Partials for *k* IVs

In this section, we discuss the semipartial and partial correlation coefficients and their squares for MRC analyses with more than two IVs. To motivate the conceptual development of these coefficients, we will restrict our discussion to three IVs as configured in Figure 4.2(a), and we will let the circles represent the total variance in a variable. We also will focus first on the squared semipartial correlation coefficient, sr_i^2, followed by the squared partial correlation coefficient, pr_i^2. (*Note:* Recall from Chapter 3 that sr_i and pr_i may be determined by taking the square root of their respective squared coefficients and that the sign of these square roots follows the sign of their corresponding B_i values.)

The Semipartial Correlation Coefficient and Its Square (sr_i and sr_i^2)

As shown in Figure 4.2(a), the regions of the three IVs that are common with the dependent variable (Y) are a, b, c, d, and e. As a result, the overall R^2 is represented by these regions and from a conceptual perspective the total amount of variance explained collectively by the three IVs is $R^2_{Y \cdot 123} = a + b + c + d + e$. Let's now examine the unique contribution to Y variance made by each IV.

To determine the squared semipartial correlation coefficient for X_1 we begin with a model that currently contains X_2 and X_3 and then let X_1 enter the model in the presence of these two IVs. As illustrated in Figure 4.2(b), when X_1 enters this model, the

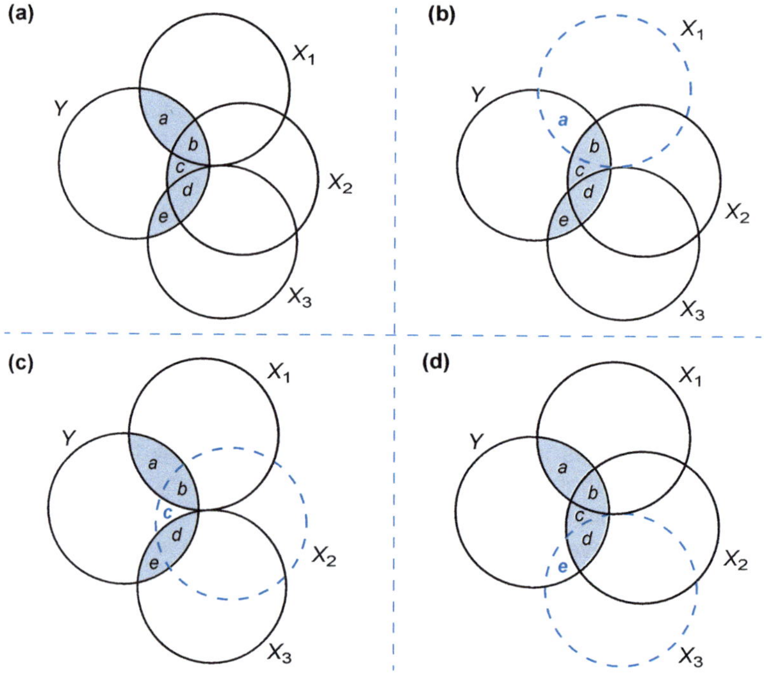

FIGURE 4.2 Graphical representation for developing the concept of sr_i^2 with three IVs.

only region that is not occupied by either X_2 or X_3 is region *a*, which represents the *unique* contribution X_1 makes to explaining *Y* variance in the presence of X_2 and X_3. Another way of looking at this is as follows:

- The amount of variance explained by X_2 is represented by regions *b*, *c*, and *d*.
- The amount of variance explained by X_3 is represented by regions *d* and *e*.
- When taken collectively, these regions represent the regression of *Y* on X_2 and X_3. In other words, and from a conceptual perspective, $R^2_{Y \bullet 23} = b + c + d + e$.
- Therefore, region *a*, which represents sr_1^2, is determined by subtracting the amount of variance explained by the IVs currently in the model from the total amount of overall variance explained by all three IVs:

$$\begin{aligned} sr_1^2 &= R^2_{Y \bullet 123} - R^2_{Y \bullet 23} \\ &= (a + b + c + d + e) - (b + c + d + e) \\ &= a \end{aligned}$$

Now focus on Figure 4.2(c). This Venn diagram illustrates the conceptual development of sr_2^2. We begin with a model that currently contains X_1 and X_3, and then let X_2 enter this model in the presence of these two IVs. The only region that is not claimed by X_1 or X_3 is region *c*, which represents X_2's *unique* contribution to explaining *Y* variance in the presence of X_1 and X_3. Thus:

$$\begin{aligned} sr_2^2 &= R^2_{Y \bullet 123} - R^2_{Y \bullet 13} \\ &= (a + b + c + d + e) - (a + b + d + e) \\ &= c \end{aligned}$$

Lastly, Figure 4.2(d) illustrates the conceptual development of sr_3^2, which is X_3's unique contribution to explaining Y variance in the presence of X_1 and X_2. Following the emerging pattern demonstrated above, we have:

$$\begin{aligned} sr_3^2 &= R^2_{Y\bullet 123} - R^2_{Y\bullet 12} \\ &= (a + b + c + d + e) - (a + b + c + d) \\ &= e \end{aligned}$$

As a result of this development, to determine each sr_i^2 for an MRC model with three IVs requires two multiple regression analyses. The first involves the overall model, namely, $R^2_{Y\bullet 123}$, and the second involves regressing Y on all the IVs *except* the one being targeted (X_i). From a conceptual perspective, this makes sense because a squared semipartial correlation coefficient represents the unique contribution a targeted IV makes in the presence of all the other IVs, which means that the collective contributions of these other IVs have already been accounted for. Generalizing to k IVs:

$$sr_i^2 = R^2_{Y\bullet 123\ldots i\ldots k} - R^2_{Y\bullet 123\ldots(i)\ldots k}$$

where the i in the subscript of the second term is placed in parentheses to signify that the targeted IV (i.e., X_i) is being omitted in the regression. For example, if an MRC model contains six IVs and we want to calculate sr_4^2, then the targeted IV is X_4 and $sr_4^2 = R^2_{Y\bullet 123456} - R^2_{Y\bullet 12356}$.

Let's apply this discussion to the data set from the file "Ch_4 ASI Data with 3 IVs," which consists of Y = ASI scores, X_1 = Malfunction History, X_2 = Age, and X_3 = Total Flight Hours. As summarized earlier in Table 4.1, $R^2_{Y\bullet 123} = .4178$. We now run three additional multiple regression analyses, $R^2_{Y\bullet 23}$, $R^2_{Y\bullet 13}$, and $R^2_{Y\bullet 12}$, to calculate the squared semipartial correlation coefficients for the respective IVs: (*Note:* Although some statistics software packages include these results, it is both instructional and informative to demonstrate their calculations.)

$$\begin{aligned} sr_1^2 &= R^2_{Y\bullet 123} - R^2_{Y\bullet 23} & sr_2^2 &= R^2_{Y\bullet 123} - R^2_{Y\bullet 13} & sr_3^2 &= R^2_{Y\bullet 123} - R^2_{Y\bullet 12} \\ &= .4178 - .4177 & &= .4178 - .2767 & &= .4178 - .2817 \\ &= .0001 & &= .1411 & &= .1361 \end{aligned}$$

Interpreting these results:

- $sr_1^2 = .0001$. When X_1 = Malfunction History enters an analysis that has already considered the impact of age and total flight hours on ASI scores (i.e., *in the presence of* X_2 and X_3), malfunction history contributes an additional 0% to explaining the variance in ASI scores. Thus, its unique contribution is 0%, which is not significant, $t(20) = -0.06$, $p = .9536$. Recall that the significance of the correlation coefficients and their squares "follows" the corresponding regression coefficients. From Table 4.1, because B_1 is not significant, sr_1^2 also is not significant.
- $sr_2^2 = .1411$. When X_2 = Age enters an analysis that has already considered the impact of malfunction history and total flight hours on ASI scores (i.e., *in the presence of* X_1 and X_3), then its unique contribution to explaining the variance in ASI scores is 14.11%, which is significant, $t(20) = 2.20$, $p = .0396$.
- $sr_3^2 = .1361$. When X_3 = Total Flight Hours enters an analysis that has already considered the impact of malfunction history and age on ASI scores (i.e., *in the presence of* X_1 and X_2), then its unique contribution to explaining the variance in ASI scores is 13.61%, which is significant, $t(20) = 2.16$, $p = .0429$.

TABLE 4.5 *Summary of the Correlation Coefficients and Their Squares for the Data Set Given in the File "Ch_4 ASI Data with 3 IVs"*

		Zero-Order Correlations		Semipartial Correlations		Full Partial Correlations		Significance	
Factor [a]	B_i	r_{Yi}	r^2_{Yi}	sr_i	sr^2_i	pr_i	pr^2_i	$t_i(20)$	p_i
Constant	8.09721							4.02	.0007
X_1	–0.00987	.2671	.0713	.01	.0001	.0130	.0002	–0.06	.9536
X_2	0.12689	.5224	.2729	.3756	.1411	.4417	.1951	2.20	.0396
X_3	0.00035	.5175	.2678	.3689	.1361	.4353	.1895	2.16	.0429

Note. $N = 24$. $R^2_{Y \cdot 123} = .417845$, $F(3, 20) = 4.785$, $p = .0113$. Readers' results might differ due to rounding.

[a] X_1 = Malfunction History, X_2 = Age, X_3 = Total Flight Hours, and Y = ASI scores.

A summary of the squared semipartial correlation coefficients is provided in Table 4.5. Based on these findings, X_2 = Age is the strongest predictor of ASI scores among the three IVs.

As for the corresponding semipartial correlation coefficients sr_1, sr_2, and sr_3, their interpretations follow those of the two IVs case presented in Chapter 3. For example, sr_2 represents the correlation between X_2 = Age and Y = ASI scores *after* X_2 is free of any relationship it has with X_1 and X_3. To determine sr_2, we simply take the square root of sr^2_2. Thus, when age is liberated from the influences of malfunction history and total flight hours, the correlation between this remaining portion of X_2 and Y is $sr_2 = \sqrt{.1411} = .3756$. The reader is informed that because our primary focus is on explained variance, we will report semipartial correlation coefficients for completeness, but we will not interpret them henceforth.

The Partial Correlation Coefficient and Its Square (pr and pr²)

Recall from the two IVs case in Chapter 3 that the partial correlation coefficient, pr_i, and its square, pr^2, represent the relationship between a targeted IV and the DV after the effects of all the other IVs have been removed from both the targeted IV *and* the DV. To illustrate this concept using Venn diagrams, we blocked out the entire circle that represented the "other IVs." This approach may be extended to a regression model with more than two IVs. To demonstrate this, we provide in Figure 4.3 the conceptual development of the squared partial correlation coefficients for three IVs: pr^2_1, pr^2_2, and pr^2_3.

In Figure 4.3(a), we replicate the Venn diagram configuration for the three IVs case introduced earlier in Figure 4.2(a) to highlight the regions of the IVs common with the dependent variable. From a conceptual perspective, this configuration indicates that the total amount of variance explained collectively by the three IVs is $R^2_{Y \cdot 123} = a + b + c + d + e$. Focusing on Figure 4.3(b), to determine the squared partial correlation coefficient for X_1 (i.e., pr^2_1), we must remove the effects of X_2 and X_3 completely from both X_1 and Y. To do this, we completely block out the entire circles representing X_2 and X_3. These removed effects can be observed by placing your thumb over the X_2 and X_3 circles. Notice that this action removes regions *b*, *c*, *d*, *e* from the model. More specifically:

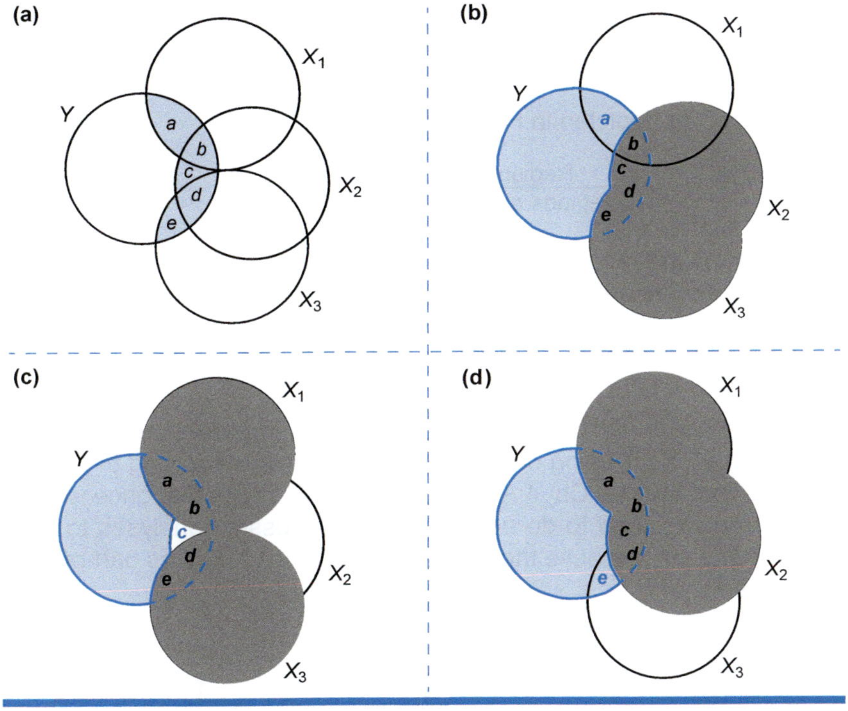

FIGURE 4.3 Graphical representation for developing the concept of pr_i^2 with three IVs.

- X_2's effect is removed from both X_1 and Y. This consists of regions b, c, and d, where region b is redundant with X_1.
- X_3's effect is removed from Y. This consists of regions d and e, where region d is redundant with X_2.

Recall from Chapter 3 that as a result of these actions, the only remaining region associated with any of the IVs is region a, which represents the squared semipartial correlation coefficient for X_1 (i.e., sr_1^2). You also might recall from Chapter 3 that by completing removing the effects of "all other variables" from the model results in a "fractured" circle that represents the DV. In other words, Y is no longer whole because parts of it—regions b, c, d, ~~e~~—were removed. Therefore, pr_1^2 is represented conceptually in Figure 4.3(b) as the ratio between region a and the remaining part of Y. The part of Y that was removed, though, is actually the amount of Y variance explained collectively by X_2 and X_3, which is conceptually, $R^2_{Y \cdot 23} = b + c + d + e$. Therefore, the part of Y that remains is equal to the "whole Y" minus $R^2_{Y \cdot 23}$, that is, $1 - R^2_{Y \cdot 23}$, and the corresponding ratio that represents pr_1^2 is

$$pr_1^2 = \frac{\text{Region } a}{1 - (\text{Regions } b + c + d + e)} = \frac{sr_1^2}{1 - R^2_{Y \cdot 23}}$$

Now focus on Figure 4.3(c), which represents pr_2^2, where the effects of X_1 and X_3 are completely removed from both X_2 and Y. Following the discussion above, we get the following ratio:

$$pr_2^2 = \frac{\text{Region } c}{1-(\text{Regions } a+b+d+e)} = \frac{sr_2^2}{1-R_{Y\bullet 13}^2}$$

Lastly, the ratio for pr_3^2 as depicted in Figure 4.3(d) is:

$$pr_3^2 = \frac{\text{Region } e}{1-(\text{Regions } a+b+c+d)} = \frac{sr_3^2}{1-R_{Y\bullet 12}^2}$$

Generalizing, pr_i^2 is the ratio of the amount of Y variance uniquely explained by X_i to the amount of Y variance that is independent of all the other IVs in the model.

$$pr_i^2 = \frac{sr_i^2}{1-R_{Y\bullet 12\ldots(i)\ldots k}^2}$$

where the i in the subscript of the second term is placed in parentheses to signify that the targeted IV (i.e., X_i) is being omitted in the regression. Applying this discussion to the data set from the file "Ch_4 ASI Data with 3 IVs," note that once we have calculated sr_i^2, we do not need to do any additional regression analyses to determine the corresponding pr_i^2 because they were previously acquired as part of determining sr_i^2. As a result:

$$pr_1^2 = \frac{sr_1^2}{1-R_{Y\bullet 23}^2} = \frac{.0001}{1-.4177} = \frac{.0001}{.5823} = .000172$$

$$pr_2^2 = \frac{sr_2^2}{1-R_{Y\bullet 13}^2} = \frac{.1411}{1-.2767} = \frac{.1411}{.7233} = .1951$$

$$pr_3^2 = \frac{sr_3^2}{1-R_{Y\bullet 12}^2} = \frac{.1361}{1-.2817} = \frac{.1361}{.7183} = .1895$$

These are summarized in Table 4.5. The reader also might notice from Table 4.5 that the pr_i^2 values are never smaller than their corresponding sr_i^2 values. This is because of the difference in their respective bases. The squared semipartial correlation coefficient represents the unique contribution of an IV to Y variance that has a base of 1; that is, it is expressed relative to *total* Y variance. On the other hand, the base of the squared partial correlation coefficient represents only part of Y variance. It is the part of Y variance that is not accounted for by the other IVs in the model.

With respect to their interpretations, recall from Chapter 3 that we preface all such interpretations with phrases such as "when the (IVs) have been statistically controlled" or "when the effects of (IVs) have been completely removed from both X_i and Y." As a result:

- $pr_1^2 = .000172$. When the effects of age and total flight hours are completely removed from both malfunction history and ASI scores, malfunction history explains 0% of the variance in ASI scores. This is not significant, $t(20) = -0.06$, $p = .9536$.
- $pr_2^2 = .1951$. When the effects of malfunction history and total flight hours are completely removed from both age and ASI scores, age explains 19.51% of the variance in ASI scores. This is significant, $t(20) = 2.20$, $p = .0396$.

- $pr_3^2 = .1895$. When the effects of malfunction history and age are completely removed from both total flight hours and ASI scores, total flight hours explain 18.95% of the variance in ASI scores. This is significant, $t(20) = 2.16$, $p = .0429$.

The interpretations of the corresponding partial correlation coefficients, pr_1, pr_2, and pr_3, follow those of the two IVs case presented in Chapter 3 and reflect correlations between X_i and Y after the effects of all the other IVs are removed from both X_i and Y. As noted earlier, our primary focus is on explained variance and therefore we will report partial correlation coefficients for completeness, but we will not interpret them.

Example 4.3 *Examining the Squared Semipartial and Partial Correlation Coefficients*

Reference the data set given in the file "Ch_4 Example 4.2 Data," which was presented in Example 4.2 and involved five IVs:

a. Prepare a table that summarizes the partials like Table 3.6 from Example 3.3 in Chapter 3.

b. Interpret the squared semipartial correlation coefficients in the context of the given research setting.

c. Interpret the squared partial correlation coefficients in the context of the given research setting.

Solution

a. See Table 4.6

b. $sr_1^2 = .1966$. When examined in the presence of the other IVs, X_1 = Age uniquely accounts for an additional 19.66% of the variance in visual reaction times. This result is significant, $t(39) = 3.45$, $p = .00014$.

$sr_2^2 = .1043$. When examined in the presence of the other IVs, the comparison between pilots who have corrected vision vs. those who do not uniquely accounts for an additional 10.43% of the variance in visual reaction times. This result is significant, $t(39) = -2.51$, $p = .0163$.

$sr_3^2 = .0015$. When examined in the presence of the other IVs, X_3 = Vision Acuity Angle uniquely accounts for an additional 1.5% of the variance in visual reaction times. This result is not significant, $t(39) = -0.30$, $p = .7675$.

Continued on next page ...

TABLE 4.6 *Summary of the Various Correlation Coefficients and Their Squares for Example 4.3a*

		Zero-Order Correlations		Semipartial Correlations		Full Partial Correlations		Significance	
Factor [a]	B_i	r_{Yi}	r_{Yi}^2	sr_i	sr_i^2	pr_i	pr_i^2	$t_i(39)$	p_i
Constant	45.11							0.65	.5213
X_1	6.357	.4090	.1673	.4434	.1966	.4832	.2335	3.45	.0014
X_2	–71.043	–.0245	.0006	–.3229	.1043	–.3730	.1391	–2.51	.0163
X_3	–4.547	.0105	.0001	–.0387	.0015	–.0479	.0023	–0.30	.7675
X_4	0.009	.3718	.1382	.3357	.1127	.3856	.1487	2.61	.0128
X_5	2.039	.0906	.0082	.0332	.0011	.0412	.0017	0.26	.7958

Note. $N = 45$. $R^2_{Y \cdot 12345} = .35$, $F(5, 39) = 4.2883$, $p = .0033$.

[a] X_1 = Age, X_2 = Corrected Vision (yes/no), X_3 = Vision Acuity Angle, X_4 = Total Flight Hours, X_5 = Number of Ratings, and Y = Visual Reaction Time (msec).

sr_4^2 = .1127. When examined in the presence of the other IVs, X_4 = Total Flight Hours uniquely accounts for an additional 11.27% of the variance in visual reaction times. This result is significant, $t(39) = 2.61$, $p = .0128$.

sr_5^2 = .0011. When examined in the presence of the other IVs, X_5 = Number of Ratings uniquely accounts for an additional 0.11% of the variance in visual reaction times. This result is not significant, $t(39) = 0.26$, $p = .7958$.

Among the five IVs, the strongest predictor of visual reaction times is X_1 = Age.

c. pr_1^2 = .2335. When all the other IVs in the model are statistically controlled, which means their effects are completely removed from both X_1 = Age and Y = ASI scores, age explains 23.35% of the variance in ASI scores. This result is significant, $t(39) = 3.45$, $p = .0014$.

pr_2^2 = .1391. When all the other IVs in the model are statistically controlled, which means their effects are completely removed from both X_2 = Corrected Vision (yes/no) and Y = ASI scores, the comparison between pilots with vs. without corrected vision explains 13.91% of the variance in ASI scores. This result is significant, $t(39) = -2.51$, $p = .0163$.

pr_3^2 = .0023. When all the other IVs in the model are statistically controlled, which means their effects are completely removed from both X_3 = Vision Acuity Angle and Y = ASI scores, visual acuity angle explains 0.23% of the variance in ASI scores. This result is not significant, $t(39) = -0.30$, $p = .7675$.

pr_4^2 = .1487. When all the other IVs in the model are statistically controlled, which means their effects are completely removed from both X_4 = Total Flight Hours and Y = ASI scores, total flight hours explain 14.87% of the variance in ASI scores. This result is significant, $t(39) = 2.61$, $p = .0128$.

pr_5^2 = .0017. When all the other IVs in the model are statistically controlled, which means their effects are completely removed from both X_5 = Number of Ratings and Y = ASI scores, number of ratings explains 0.17% of the variance in ASI scores. This result is not significant, $t(39) = 0.26$, $p = .7958$.

Therefore, X_1 = Age accounts for more of the relationship with Y = ASI scores than the other four IVs.

4.4 Statistical Inference for MRC

Thus far in our presentation, when interpreting the results of an MRC analysis, we have been including a statement about the statistical significance of the overall MRC model, the partial regression coefficients, and the semipartial and partial correlation coefficients and their squares. Within these statements of statistical significance, we reported the corresponding *t* or *F*, *df*, and *p* values without explanation because we presume that the reader understands the concept of statistical significance as well as the meaning of these statistical tests. In this section, we do not dwell on these concepts, which are more appropriate for a first-level statistics book, but instead focus on the concept of statistical inference and how it applies to MRC.

Recall that the concept of statistical inference is grounded in inferential statistics, which consists of various statistical procedures whose results help inform any conclusions or decisions we make about a population parameter based on sample data. For our purposes, the statistical procedure is MRC, and the parameters of interest, which are estimated by their corresponding sample statistics, include: (a) the squared multiple correlation coefficient, R^2; (b) the partial regression coefficients, B_i; and (c) the squared semipartial and partial correlation coefficients, sr_i^2 and pr_i^2. Recall that R^2 represents the collective contribution the IVs make to explaining Y variance, sr_i^2 and pr_i^2 represent the contribution each IV makes to explaining Y variance, and B_i represents the weight IVs are given when making predictions of y from x via the regression equation. As a result, when examining the concept of statistical inference with respect to MRC, the underlying question is, "To what extent are these explained variances and predictions representative of the population from which the sample was selected?" Two general approaches that can be used to answer this question are parameter estimation and hypothesis testing.

Parameter Estimation

Parameter estimation is considered a direct approach to statistical inference because it uses sample statistics as estimates of the corresponding population parameters. Two such estimations are point estimates and confidence intervals.

Point Estimates. A point estimate uses a sample statistic that resulted from an analysis as a direct estimate of the corresponding population parameter. For example, consider the results of the MRC analysis we conducted for the data set given in the file "Ch_4 ASI Data with 3 IVs." As summarized in Table 4.1, $R^2_{Y \cdot 123} = .4178$, which is a sample statistic. We can use this result to estimate the corresponding population parameter and claim that the collective contribution the three IVs make to explaining the variance in Y is $\rho^2 = .4178$. (Recall that we use Greek letters to represent population parameters. The symbol for "r" is the Greek letter "rho" denoted ρ.)

As another example, notice also from Table 4.1 that $B_2 = 0.13$, rounded to two decimal places, which indicates that when holding the other two IVs in the model constant, for every 1 year older a pilot is, ASI scores increase, on average, by 0.13 points. More concretely, for every 10-year increase in age, ASI scores increase, on average, by 1.3 points. Using this as a point estimate, we would report that in the population, $\beta_2 = 0.13$ and make the same interpretation. Thus, we can use any of the sample statistics of an MRC analysis as point estimates for the corresponding population parameters.

Confidence Intervals and AIPE. In addition to point estimates, we could use confidence intervals (CIs) to give us an idea of how close a sample statistic is to the true population parameter. For example, as reported in Table 4.1, the 95% CI for B_2 is [0.0067, 0.2471]. Recall from Chapter 2 that one way to interpret this is, "If we were to randomly select 100 samples of the same size from the same population, then 95 of these samples would have a regression coefficient for B_2 that is between 0.0067 and 0.2471." An alternative interpretation is, "In the population, 95% of the time we expect the partial regression coefficient for B_2 to range anywhere between 0.0067 and 0.2471. Interpreting this CI in the context of the given research setting: "95% of the time, we expect that in the population, for every 1-year increase in age, ASI scores will increase anywhere between 0.0067 points and 0.2471 points."

Also recall from the earlier chapters the concept of accuracy in parameter estimation (AIPE), which is a judgment call based on the width of the confidence interval. For example, with respect to B_2, the width of the corresponding 95% CI is 0.2471 – 0.0067 = 0.2404, which is very small compared to the DV where ASI scores can range between 5 and 20. Therefore, we would judge this width as being very narrow and report that the AIPE for this 95% CI is very high. (*Note*: Recall from Chapter 2 that the width of a CI refers to its precision and narrow CIs have higher precision than wide CIs.)

Let's now examine the CIs for each of the three parameters of interest: the squared multiple correlation coefficient, the partial regression coefficients, and the squared semipartial correlation coefficients.

Confidence Intervals for R^2. Recall from Chapter 2 the general form of a CI:

$$\text{Population Parameter} = \text{Point Estimate} \pm (SE)(t_{\text{Critical}})$$

where SE is the standard error of the targeted parameter, and t_{Critical} is the t value that corresponds to the desired level of significance. When applied to multiple regression, the degrees of freedom for t are $df = (n - k - 1)$ and we use a two-tailed test, which means that for the 95% CI, $\alpha = .05$ with $\alpha = .025$ in each tail. Applying this general CI form to the overall R^2, the form is:

$$\rho^2 = R^2 \pm SE_{R^2}(t_{\text{Critical}})$$

So, to determine the 95% CI for the squared multiple correlation coefficient, we need the corresponding standard error and t_{Critical}. Unfortunately, most statistics programs do not report the CI for the overall R^2, and they do not report the corresponding SE, which means we need to calculate these values. Olkin and Finn (1995, as cited in Cohen et al., 2003, p. 89) provided a formula that can be used to approximate SE_{R^2}. Following is the application of this formula relative to the data set given in the file "Ch_4 ASI Data with 3 IVs," where $R^2 = .42$ (rounded to two decimal places), $n = 24$, and $k = 3$ IVs.

$$SE_{R^2} = \sqrt{\frac{4R^2(1-R^2)^2(n-k-1)^2}{(n^2-1)(n+3)}}$$

$$= \sqrt{\frac{4(.42)(1-.42)^2(24-3-1)^2}{(24^2-1)(24+3)}}$$

$$= \sqrt{\frac{1.68(.58)^2(20)^2}{(576-1)(27)}}$$

$$= \sqrt{\frac{1.68(.3364)(400)}{(575)(27)}}$$

$$= \sqrt{\frac{226.0608}{15525}} = \sqrt{.01456} = 0.120669$$

Therefore, the standard error for R^2 is approximately 0.12. We next acquire t_{Critical} from the t table given in Appendix B / Table B1. For the given sample size of $n = 24$, $df = (n - k - 1) = 20$ and the corresponding t_{Critical} for $\alpha = .025$ in each tail is $t(20) = 2.086$. We now calculate the 95% CI for R^2:

$$\begin{aligned}
95\% \text{ CI for } \rho^2 &= R^2 \pm SE_{R^2}\,(t_{\text{Critical}}) \\
&= .42 \pm (.12)(2.086) \\
&= .42 \pm .25032 \\
&= .42 \pm .25 \\
&= .42 - .25 \text{ and } .42 + .25 \\
&= [.17, .67]
\end{aligned}$$

As a result, the 95% CI for the overall R^2 of the ASI data with three IVs is [.17, .67], which means that 95% of the time we expect that the true R^2 in the population will be somewhere between .17 and .67. In other words, *95% of the time we expect that in the population the three IVs of X_1 = Malfunction History, X_2 = Age, and X_3 = Total Flight Hours will collectively explain anywhere between 17% and 67% of the variance in ASI scores.* The width of this interval is .50, which is large given that R^2 can range between 0 and 1. Therefore, the corresponding AIPE is not very good.

As an alternative to performing these manual calculations, Soper (2025a) provides an online calculator for determining the 90%, 95%, and 99% CI for the overall R^2. Although this certainly makes things easier, the calculator does not provide the corresponding standard error.

Confidence Intervals for B_i. Unlike R^2, most statistics software packages provide CIs and respective standard errors for the partial regression coefficients, eliminating the need to calculate them manually. Nevertheless, for completeness and instructional purposes, we provide an example of how to calculate the 95% CI for these coefficients. First, recognize that the general form of the CI for B_i is:

$$\beta_i = B_i \pm SE_{B_i}\,(t_{\text{Critical}})$$

Let's apply this formula to the ASI data set with 3 IVs for X_2 = Age. As reported earlier, as well as in Table 4.1, B_2 = 0.126892 and the corresponding standard error is SE_{B_2} = 0.057629. We will round these values to four decimal places. Because we are using the same data set as we did for R^2, the corresponding t_{Critical} is $t(20)$ = 2.086. Therefore:

$$\begin{aligned}
95\% \text{ CI for } \beta_2 &= B_2 \pm SE_{B_2}\,(t_{\text{Critical}}) \\
&= 0.126892 \pm (0.057629)(2.086) \\
&= 0.126892 \pm .0.120214 \\
&= (0.126892 - 0.120214) \text{ and } (0.126892 + 0.120214) \\
&= [0.006678, 0.247106]
\end{aligned}$$

Note that this result matches what is reported in Table 4.1.

In the event that neither the 95% CIs nor the standard errors for the partial regression coefficients is reported, then the following formula can be used to calculate the standard error, which then may be applied to the CI formula given above:

$$SE_{B_i} = \frac{sd_Y}{sd_i}\sqrt{\frac{1}{1-R_i^2}}\sqrt{\frac{1-R_Y^2}{n-k-1}}$$

where sd_Y = the standard deviation of Y, sd_i = the standard deviation of the targeted partial regression coefficient, R_Y^2 = the overall R^2, and R_i^2 = the R^2 that results from regressing X_i (as the DV) on all the other IVs. For example, focusing on B_2 from our running example, observe that: $sd_Y = 2.72$, $sd_{B_2} = 8.76$, $R_Y^2 = .42$, and $R_2^2 = .15$. When these values are substituted into the standard error formula we get:

$$SE_{B_2} = \frac{2.72}{8.76}\sqrt{\frac{1}{1-.15}}\sqrt{\frac{1-.42}{24-3-1}}$$

$$= (0.3105)\sqrt{\frac{1}{.85}}\sqrt{\frac{.58}{20}}$$

$$= (0.3105)\sqrt{1.1765}\sqrt{0.029}$$

$$= (0.3105)(1.08466)(0.17029)$$

$$= 0.05735$$

Note that this result matches what is reported in Table 4.1 (to three decimal places).

Confidence Intervals for sr_i^2. According to Cohen et al. (2003, p. 93), the CIs for the squared semipartial correlation coefficient are "asymmetrical and complex" and therefore we will not address them.

Hypothesis Testing

Hypothesis testing is considered an *indirect* approach to statistical significance because researchers first make a hypothesis about a population parameter and then use sample statistics to either support or refute this hypothesis. The hypothesis a researcher makes is often referred to as a *statistical hypothesis* and is presented in two parts: null and alternative.

The null hypothesis, denoted H_0, is a statistical statement about the targeted population parameter. This statement conveys the message that whatever effect we observe in a sample is nothing more than a "chance effect" and any similar effect that might occur in the population will also be random. In other words, the null hypothesis asserts "there will be no statistically significant effect" relative to the targeted parameter.

The alternative hypothesis, denoted H_1, is a statistical statement about the same targeted population parameter as the null hypothesis. This statement, however, reflects the researcher's prediction about the relationship between/among the variables, which is that the population parameter will be different than what is stated in the null hypothesis. In other words, the alternative hypothesis asserts "there will be a statistically significant effect" relative to the targeted parameter.

The decision to *reject* H_0 (and accept H_1) or *fail to reject* H_0 is governed by the results of data analysis. The decision surrounding H_0 is then used to answer the corresponding research question. To illustrate the concept of hypothesis testing for MRC we will use the results from the chapter's running example of ASI scores with three IVs as summarized in Table 4.1. Because the results of hypothesis testing answer the corresponding research questions, we will include the RQs as well as the corresponding research hypotheses as part of this discussion.

Research Questions and Hypotheses

As presented earlier, there are three primary RQs and corresponding hypotheses: The first is related to the overall R^2, the second is related to the squared semipartial correlation coefficients (sr_i^2), and the third is related to the regression coefficients (B_i).

RQ 1. What proportion of the variance in ASI scores is explained collectively by X_1 = Malfunction History, X_2 = Age, and X_3 = Total Flight Hours?

Hyp. 1. The focus is on the collective contribution of the IVs to Y variance.

- H_0: $\rho^2_{Y \cdot 123} = 0$ (The three IVs collectively will not explain a significant proportion of Y variance in the population.)
- H_1: $\rho^2_{Y \cdot 123} > 0$ (The three IVs collectively will explain a significant proportion of Y variance in the population.)

To determine whether to reject or fail to reject H_0, we examine the results from Table 4.1. Note from the Model Summary and ANOVA subtables that $R^2 = .42$, $F(3, 20) = 4.785$, $p = .0113$. Consulting the F table in Appendix B/Table B4, the corresponding $F_{\text{Critical}} = 3.10$. Thus, R^2 is significant for the preset alpha level of $\alpha = .05$ and correspondingly, we reject H_0 and conclude that the three IVs collectively explain a significant amount of the variance in ASI scores.

For completeness, the F formula for determining the significance of R^2 is

$$F = \frac{R^2(N-k-1)}{(1-R^2)k}$$

where N = total sample size, k = number of IVs, $df_{\text{Num}} = k$ and $df_{\text{Den}} = N - k - 1$. With the respect to the current situation where $R^2 = .42$, $k = 3$, and $N = 24$:

$$F = \frac{.42(24-3-1)}{(1-.42)3} = \frac{.42(20)}{(.58)3} = \frac{8.4}{1.74} = 4.8276$$

This F value is larger than what our statistics program reported due to rounding.

RQ 2. How much additional knowledge does each IV make in explaining the variance in ASI scores when it enters the model in the presence of the other two IVs? Because there are three IVs, the corresponding hypothesis has three parts.

Hyp. 2a. The focus is on X_1's unique contribution to explaining Y variance (i.e., sr_1^2) in the population when X_1 enters the model in the presence of X_2 and X_3.

- H_0: $\rho^2_{Y \cdot 123} - \rho^2_{Y \cdot 23} = 0$ (X_1's unique contribution to Y variance is not significant.)
- H_1: $\rho^2_{Y \cdot 123} - \rho^2_{Y \cdot 23} > 0$ (X_1's unique contribution to Y variance is significant.)

Hyp. 2b. The focus is on X_2's unique contribution to explaining Y variance (i.e., sr_2^2) in the population when X_2 enters the model in the presence of X_1 and X_3.

- H_0: $\rho^2_{Y \cdot 123} - \rho^2_{Y \cdot 13} = 0$ (X_2's unique contribution to Y variance is not significant.)
- H_1: $\rho^2_{Y \cdot 123} - \rho^2_{Y \cdot 13} > 0$ (X_2's unique contribution to Y variance is significant.)

Hyp. 2c. The focus is on X_3's unique contribution to explaining Y variance (i.e., sr_3^2) in the population when X_3 enters the model in the presence of X_1 and X_2.

- H_0: $\rho^2_{Y \cdot 123} - \rho^2_{Y \cdot 12} = 0$ (X_3's unique contribution to Y variance is not significant.)
- H_1: $\rho^2_{Y \cdot 123} - \rho^2_{Y \cdot 12} > 0$ (X_3's unique contribution to Y variance is significant.)

To determine whether to reject or fail to reject each H_0, we examine the results from Table 4.5.

- sr_1^2 = .0001, $t(20) = -0.06$, $p = .9536$. Therefore, we *fail to reject* Null Hypothesis 2a and conclude that X_1 = Malfunction History does not provide any significant amount of additional knowledge to the explained variance of ASI scores when it enters the analysis in the presence of X_2 and X_3.
- sr_2^2 = .1411, $t(20) = 2.20$, $p = .0396$. Therefore, we *reject* Null Hypothesis 2b and conclude that X_2 = Age provides a significant amount of additional knowledge to the explained variance of ASI scores when it enters the analysis in the presence of X_1 and X_3. Its unique contribution to explaining Y variance is 14.11%.
- sr_3^2 = .1361, $t(20) = 2.16$, $p = .0429$. Therefore, we *reject* Null Hypothesis 2c and conclude that X_3 = Total Flight Hours provides a significant amount of additional knowledge to the explained variance of ASI scores when it enters the analysis in the presence of X_1 and X_2. Its unique contribution to explaining Y variance is 13.61%.

For completeness, there are two formulas for determining the significance of sr_i^2:

$$F = \frac{sr_i^2(N-k-1)}{(1-R^2)} \quad \text{and} \quad t_i = sr_i\sqrt{\frac{N-k-1}{1-R^2}}$$

Most statistics software packages provide the t test results, which is what we reported in Table 4.5. As an illustration, we apply the t formula for sr_2^2 = .1411:

$$\begin{aligned} t_2 &= (.3756)\sqrt{\frac{24-3-1}{1-.42}} \\ &= (.3756)\sqrt{\frac{20}{.58}} \\ &= (.3756)\sqrt{34.4827} \\ &= (.3756)(5.8722) \\ &= 2.2056 \end{aligned}$$

This matches what is reported in Table 4.5 for sr_2^2. Also note that this formula can be used to determine the respective t values for any of the partials listed in Table 4.5.

RQ 3. When holding all other variables constant, what are the respective effects of X_1 = Malfunction History, X_2 = Age, and X_3 = Total Flight Hours on Y = ASI scores? Because there are three IVs, the corresponding hypothesis has three parts.

Hyp. 3a. The focus is on the effect X_1 has on Y holding all other IVs constant.

- H_0: $\beta_1 = 0$ (X_1 will not have a significant effect on Y.)
- H_1: $\beta_1 \neq 0$ (X_1 will have a significant nonzero effect on Y.)

Hyp. 3b. The focus is on the effect X_2 has on Y holding all other IVs constant.

- H_0: $\beta_2 = 0$ (X_2 will not have a significant effect on Y.)
- H_1: $\beta_2 \neq 0$ (X_2 will have a significant nonzero effect on Y.)

Hyp. 3c. The focus is on the effect X_3 has on Y holding all other IVs constant.

- H_0: $\beta_3 = 0$ (X_3 will not have a significant effect on Y.)
- H_1: $\beta_3 \neq 0$ (X_3 will have a significant nonzero effect on Y.)

To determine whether to reject or fail to reject each H_0, we examine the results from either Table 4.1 or Table 4.5.

- $B_1 = -0.00987$, $t(20) = -0.06$, $p = .9536$. Therefore, we *fail to reject* Null Hypothesis 3a and conclude that when holding all other IVs constant, pilots' malfunction history does not have a significant effect on their ASI scores.
- $B_2 = 0.12689$, $t(20) = 2.20$, $p = .0396$. Therefore, we *reject* Null Hypothesis 3b and conclude that when holding all other IVs constant, pilots' age has a significant effect on their ASI scores: For every 1-year increase in age, ASI scores increase on average by 0.12689 points. More concretely, for every 10-year increase in age, ASI scores increase on average by approximately 1.3 points.
- $B_3 = 0.00035$, $t(20) = 2.16$, $p = .0429$. Therefore, we *reject* Null Hypothesis 3c and conclude that when holding all other IVs constant, pilots' total flight hours have a significant effect on ASI scores: For every 1-hour increase in flight time, ASI scores increase on average by 0.00035 points. More concretely, for every 10,000-hour increase in flight time, ASI scores increase on average by 3.5 points.

We do not provide a *t* or *F* formula for B_i because the formulas for sr_i^2 may be used.

Parameter Estimation vs. Hypothesis Testing

Although both parameter estimation and hypothesis testing make inferences about the parent population from sample data, their objectives are different. The purpose of parameter estimation is to use sample statistics to estimate the corresponding unknown population parameter, whereas the purpose of hypothesis testing is to determine how likely an effect observed in a sample also exists in the parent population. Nevertheless, as summarized in Table 4.7, parameter estimation is more informative and is often preferable because of its direct approach. Also note from Table 4.7, that parameter estimation can be used as an alternative to hypothesis testing for statistical inference.

As an example, consider Hypothesis 3a given above for B_1 where H_0: $B_1 = 0$ and H_1: $B_1 \neq 0$. Referencing Table 4.1, observe that the corresponding 95% CI for $B_1 = [-0.3596, 0.3399]$, which is interpreted as, "95% of the time we can expect that, in the population, for every additional inflight malfunction a pilot reports, ASI scores on average will range between a *decrease* of 0.36 points to an *increase* of .34 points." Note that 0 is within this interval. Thus, it is possible for B_1 to equal 0, which means we would *fail to reject* H_0. On the other hand, look at Hypothesis 3b given above for B_2. As reported in Table 4.1, its corresponding 95% CI = [0.0067, 0.2471]. Because 0 is not in this interval, it is not possible for $B_2 = 0$ (95% of the time) and therefore we *reject* its null hypothesis.

TABLE 4.7 *Parameter Estimation vs. Hypothesis Testing*

Considerations	Parameter Estimation	Hypothesis Test
Approach	Direct	Indirect
Provides a point estimate of the population parameter.	Yes	No
Provides an interval of probable values for the population parameter.	Yes	No
Provides a degree of confidence with respect to knowing the population parameter	Yes	No
Determines if an effect is statistically significant.	Yes	Yes

Sample Size Planning

A key question when undertaking any research study is, "How large of a sample do I need?" This question is answered by engaging in an activity known as sample size planning, which is done prior to conducting a study. Sample size planning helps determine how large a sample should be so that the results of data analysis will lead to either (a) a highly precise confidence interval and strong AIPE or (b) a significant effect that provides strong support to reject H_0. The strategy used for sample size planning differs between parameter estimation and hypothesis testing. If the objective of a study is parameter estimation, then the focus of sample size planning is with respect to precision, whereas if the objective is hypothesis testing, then the focus of sample size planning is with respect to power analysis.

Precision. Recall that precision refers to the width of a CI. For example, consider the effect of X_2 = Age on ASI scores. From Table 4.1, the corresponding 95% CI is [0.0067, 0.2471], which has a width of 0.2471 – 0.0067 = 0.1801, or 0.18 rounded to two decimal places. One-half of this width is 0.18/2 = .09, which is the margin of error. (Recall that when calculating a CI it is always "± *SE*.") Thus, the 95% CI for B_2, which is based on sample data, provides an estimate of the true effect of age on ASI scores within 0.09 points. This is a very precise estimate and the corresponding accuracy in parameter estimation also is very high.

This finding, though, is post hoc (i.e., after the fact) and is left to chance. For example, it is possible that the 95% CI for B_2 from a different sample might be [1.5, 4.5], which means that 95% of the time, in the population, for every 1-year increase in age, ASI scores will increase on average between 1.5 and 4.5 points. Note here that the width is 3 points, and the margin of error is 1.5 points, which is much less precise—and has poorer AIPE—than what was initially found.

So, if a study is focused on parameter estimation, then sample size planning requires determining an acceptable probability level as well as the amount of error that will be tolerated so the resulting CI has high precision (narrow interval) and high AIPE. Our focus, though, is not on parameter estimation but instead on hypothesis testing, which is more commonly done in practice. For readers interested in sample size planning for precision, we direct you to Kelley and Maxwell (2003).

Power Analysis. Recall that power is the probability of making a strong correct decision relative to a hypothesis test. In this context, sample size planning is often referred to as power analysis, and making a strong correct decision means that an effect observed in the sample truly exists in the population and the correct decision is to reject H_0.

The elements of a power analysis for MRC include the desired level of significance (α), the expected effect size (*ES*), the desired level of power, the number of IVs, and sample size (*n*). If we know any four of these components, we can then determine the fifth. In practice, to determine sample size, the level of significance is commonly set to α = .05, the expected effect size is "medium," and the minimum power is .80, which means there will be an 80% chance of correctly rejecting a false H_0. We also denote the minimum sample size acquired from a power analysis as n^* to distinguish it from the actual sample size, *N*.

When applied to an MRC analysis, we must consider power analysis from two perspectives: the minimum sample size needed for the overall model (R^2), which involves all the IVs collectively, and the minimum size needed for testing the individual IVs for significance. For the overall model, Cohen (1992) recommended at least 15 to 30 participants for every IV, Green (1991) recommended a minimum sample size of $8k + 50$, where k = the number of IVs, and Khamis and Kepler (2010) recommended a minimum sample size of $5k + 20$. For testing the individual IVs, Green also recommended a minimum sample size of $k + 104$. The larger of the two sample sizes (collective vs. individual) is the sample size that should be acquired. In addition to these cited guidelines, some statistics software programs provide power analysis routines, and there are several online sample size calculators available. The reader should be aware that these different strategies for conducting an a priori power analysis will not necessarily yield the same n^*.

The reader also is informed that the sample size guidelines cited above are relative to a *medium effect size*. Although Cohen et al. (2003, p. 93) prefer researchers determine the expected effect size via theory or prior research, in the event that this information is lacking, they suggest values of 0.02, 0.13, and 0.26 be used to respectively represent "small," "medium," and "large" effect sizes for the overall R^2. Similarly (p. 95), they suggest values of 0.02, 0.15, and 0.35 be used to respectively represent "small," "medium," and "large effect sizes for individual IVs, which represent the proportion of Y variance being explained by X_i that is not being explained by the other IVs in the model.

As an example of conducting an a priori power analysis, let's pretend that we are planning to conduct the ASI study with three IVs and would like to know what the respective minimum sample sizes should be for the overall R^2 and the individual IVs. Table 4.8 contains a summary of the minimum sample sizes obtained by the cited references as well as from Soper's (2025b, 2025c) online sample size calculators. Note the differences in the results among the approaches. This is why it is important for researchers to specify the approach they use for power analysis. Also remember that these are *minimum* sample sizes. When compared to the actual sample size of our running example of ASI scores with three IVs, our sample size of $N = 24$ was inadequate relative to a medium effect size. [*Note:* With respect to a large effect of 0.35, Soper (2025b) returns a minimum sample size of $n^* = 36$.)]

In addition to conducting an a priori power analysis to determine the minimum sample size, we also can conduct a post hoc power analysis to determine the power of our study after it has been conducted. For example, with respect to our running example, we now know that $N = 24$ and the significance level was $\alpha = .05$. If we know the effect size, then we can determine the power of the study. To determine the effect

TABLE 4.8 *Power Analysis Results for the ASI Study with Three IVs*

Method [a]	Minimum N^* for R^2	Minimum N^* for sR^2
15 to 30 participants per IV [a]	Between 15(3) = 45 and 30(3) = 90	Indirect
$8k + 50$ (R^2) and $k + 104$ (sr^2) [b]	8(3) + 50 = 74	3 + 104 = 107
$5k + 20$ [c]	5(3) + 20 = 35	No
Soper (2025b and 2025c)	88 [d]	56 [e]

Note. All calculations based on Cohen et al.'s (2003) suggested minimum guidelines for a medium effect size.

[a] Cohen (1992). [b] Green (1991). [c] Khamis and Kepler (2010). [d] $f^2 = 0.13$, $\alpha = .05$, power = .8 and $k = 3$. [e] $f^2 = 0.15$, $\alpha = .05$, power = .8, $k_A = 2$, and $k_B = 1$.

size, we can use an online effect size calculator such as Soper (2025d) or we can calculate the effect sizes directly using the following formulas.

- The effect size for the overall R^2 is $f^2 = \frac{R^2}{1-R^2}$
- The effect size for the squared semipartial correlation coefficient is $f^2 = \frac{sr_i^2}{1-R^2}$

Once we know the effect size, we can then determine power. For our running example involving ASI scores with three IVs, recall from Table 4.1 that $R^2 = .42$ and $1 - R^2 = .58$. Therefore, the effect size is $f^2 = .42/.58 = 0.724$. To determine the power of the study, we consult Soper (2025e). This returns a power of .99999781, which means that the power of the MRC analysis for the overall R^2 is greater than 99.99%. It also means there is a greater than 99.99% probability that the decision to reject the corresponding null hypothesis was correct and there is a greater than 99.99% chance that the effect found in the sample also exists in the population. A similar procedure is followed to determine for the respective squared semipartial correlation coefficients, except we use Soper (2025f) to determine power.

Example 4.4 *Conducting An A Priori Power Analysis for An MRC Study with Four IVs*

A researcher is planning a study that will consist of four IVs. Based on a review of the literature, she expects the overall population R^2 to be around .40, and she believes the unique contributions the IVs will make to explaining Y variance are 5% each for X_1 and X_2, 7% for X_3, and 11% for X_4. Conduct a power analysis to determine the minimum sample size needed for this study based on an alpha level of $\alpha = .05$ and power = .8.

Solution

Overall R^2. If the overall population R^2 is expected to be .40, then the corresponding effect size is (.4) / (1 – .4) = (.4) / (.6) = 0.67 (rounded to two decimal places). Consulting Soper (2025b), the minimum sample size is $n^* = 24$.

X_1. Given four IVs, we presume that three of the IVs (X_2, X_3, and X_4) are currently in the model and we are interested in the sample size needed for the unique contribution X_1 makes when it enters the model in the presence of the other three IVs. Consulting Soper (2025c), "set B" refers to X_1 because it is the IV entering the model presently, and "set A" refers to X_2, X_3, and X_4 because they are the IVs currently in the model. Therefore, given an anticipated effect size of 0.05 for X_1, the minimum sample size is $n^* = 159$.

X_2. "Set B" refers to X_2 and "set A" refers to the other three IVs, and the anticipated effect size for "set B" is 0.05. Thus, the minimum sample size is $n^* = 159$.

X_3. "Set B" refers to X_3, "set A" refers to the other IVs, and the anticipated effect size for "set B" is 0.07. Thus, the minimum sample size is $n^* = 115$.

X_4. "Set B" refers to X_4, "set A" refers to the other IVs, and the anticipated effect size is 0.11. Thus, the minimum sample size is $n^* = 75$.

Based on these results, the minimum sample sizes vary from $n^* = 24$ for the overall R^2 to $n^* = 159$ for sr_1^2 and sr_2^2. As a result, the researcher should strive for a sample size of at least $n^* = 159$.

4.5 Conducting an MRC Analysis with *k* IVs in Research: A Guided Example

We now present a guided example of how to conduct an MRC analysis for a research study that involves $k = 4$ IVs. Our study is a facsimile of Baran (2018) who, in part, examined the relationship between various factors and the average air fare an airline assigns to a city-pair market, which is a specific market in which at least one commercial airline is operating. Examples of city-pair markets include New York–Orlando, Miami–Atlanta, and Boston–Houston. Baran restricted his study to passenger airline routes that have both origin and destination airports within the borders of the United States and was for the 10-year time period 2006–2015, inclusive. For this Guided Example, the targeted factors are:

- X_1 = Route Length (in miles).
- X_2 = Per Capita Income in U.S. dollars of City 1, which represents one end-point of the city-pairs that form the route.
- X_3 = Per Capita Income in U.S. dollars of City 2, which represents the other end-point of the city-pairs that form the route.
- X_4 = Business Model of the carrier having the largest average air fare in a route. The business model is defined as either a full-service carrier (FSC), which is coded 0, or a low-cost carrier (LCC), which is coded 1.

The dependent variable is Y = Average Air Fares measured in U.S. dollars for a city-pair market. The data set is provided in the file "Ch_4 Guided Example Data," and the structure of our presentation consists of five parts: Research Questions and Statistical Hypotheses, Sample Size Planning, Data Analysis, Interpretation of Results, and Post Hoc Considerations.

Research Questions and Hypotheses

There are three research questions and corresponding hypotheses. The first is for the overall R^2, the second is for the unique contribution each IV makes in explaining Y variance (sr_i^2), and the third is for the regression coefficients' (B_i) effect on Y.

RQ 1. What proportion of variance in average air fares is explained collectively by the four targeted IVs?

Hyp. 1. The focus is on the collective contribution of the IVs to Y variance.

- H_0: $\rho^2_{Y \cdot 1234} = 0$ (The four IVs collectively will not explain a significant proportion of Y variance in the population.)
- H_1: $\rho^2_{Y \cdot 1234} > 0$ (The four IVs collectively will explain a significant proportion of Y variance in the population.)

RQ 2. How much additional knowledge does each IV make in explaining the variance in average air fares when entered into an analysis that contains the other three IVs? Because there are four IVs, the corresponding hypothesis has four parts.

Hyp. 2a. The focus is on X_1's unique contribution to explaining Y variance (sr_1^2) in the population when X_1 enters the model in the presence of X_2, X_3, and X_4.

- H_0: $\rho^2_{Y \cdot 1234} - \rho^2_{Y \cdot 234} = 0$ (X_1's unique contribution to Y variance is not significant.)
- H_1: $\rho^2_{Y \cdot 1234} - \rho^2_{Y \cdot 234} > 0$ (X_1's unique contribution to Y variance is significant.)

Hyp. 2b. The focus is on X_2's unique contribution to explaining Y variance (sr_2^2) in the population when X_2 enters the model in the presence of X_1, X_3, and X_4.

- H_0: $\rho^2_{Y\bullet1234} - \rho^2_{Y\bullet134} = 0$ (X_2's unique contribution to Y variance is not significant.)
- H_1: $\rho^2_{Y\bullet1234} - \rho^2_{Y\bullet134} > 0$ (X_2's unique contribution to Y variance is significant.)

Hyp. 2c. The focus is on X_3's unique contribution to explaining Y variance (sr_3^2) in the population when X_3 enters the model in the presence of X_1, X_2, and X_4.

- H_0: $\rho^2_{Y\bullet1234} - \rho^2_{Y\bullet124} = 0$ (X_3's unique contribution to Y variance is not significant.)
- H_1: $\rho^2_{Y\bullet1234} - \rho^2_{Y\bullet124} > 0$ (X_3's unique contribution to Y variance is significant.)

Hyp. 2d. The focus is on X_4's unique contribution to explaining Y variance (sr_4^2) in the population when X_4 enters the model in the presence of X_1, X_2, and X_3.

- H_0: $\rho^2_{Y\bullet1234} - \rho^2_{Y\bullet123} = 0$ (X_4's unique contribution to Y variance is not significant.)
- H_1: $\rho^2_{Y\bullet1234} - \rho^2_{Y\bullet123} > 0$ (X_4's unique contribution to Y variance is significant.)

RQ 3. When holding all other variables constant, what are the respective effects (i.e., slopes) of each IV on average air fares? Because there are four IVs, the corresponding hypothesis has four parts.

Hyp. 3a. The focus is X_1's effect on Y when holding all other IVs constant.

- H_0: $\beta_1 = 0$ (X_1 does not have a significant effect on Y.)
- H_1: $\beta_1 \neq 0$ (X_1 has a significant nonzero effect on Y.)

Hyp. 3b. The focus is X_2's effect on Y when holding all other IVs constant.

- H_0: $\beta_2 = 0$ (X_2 does not have a significant effect on Y.)
- H_1: $\beta_2 \neq 0$ (X_2 has a significant nonzero effect on Y.)

Hyp. 3c. The focus is X_3's effect on Y when holding all other IVs constant.

- H_0: $\beta_3 = 0$ (X_3 does not have a significant effect on Y.)
- H_1: $\beta_3 \neq 0$ (X_3 has a significant nonzero effect on Y.)

Hyp. 3d. The focus is X_4's effect on Y when holding all other IVs constant.

- H_0: $\beta_4 = 0$ (X_4 does not have a significant effect on Y.)
- H_1: $\beta_4 \neq 0$ (X_4 has a significant nonzero effect on Y.)

Sample Size Planning

We now conduct a power analysis to see if the data set is sufficiently large. Of the guidelines given earlier (see Table 4.8), we will use Soper (2025b) for the overall R^2, Soper (2025c) for the sr_i^2 values, and Cohen et al.'s (2003) definitions for a medium effect for R^2 (p. 93) and a medium effect for sr_i^2 (p. 95).

- Overall R^2. If $ES = f^2 = 0.13$, power = .8, IVs = 4, and $\alpha = .05$, then $n^* = 96$.
- sr_i^2. If $ES = f^2 = 0.15$, power = .8, $k_{\text{Set A}} = 3$, $k_{\text{Set B}} = 1$, and $\alpha = .05$, then $n^* = 57$.

Therefore, we will need at least $n^* = 96$ cases to have an 80% probability of finding a significant medium effect for the overall R^2 and each sr_i^2. The given data set has $N = 233$, so we have met this minimum sample size.

TABLE 4.9 *Summary of MRC Analysis Results for Chapter 4's Guided Example*

Model Summary

R	.771
R^2	.595
Adjusted R^2	.588
Standard Error of the Estimate	36.89
Mean of Response	198.24
Sample Size	233

Coefficients (Parameter Estimates)

Term	B	SE	β	t(228)	p	95% CI
Constant	33.1227	21.9124	—	1.51	.1320	[−10.05, 76.30]
X_1 = Route Length	0.0553	0.0044	0.5469	12.60	< .0001	[0.05, 0.06]
X_2 = Per Capita Income City 1	0.0009	0.0004	0.1035	2.40	.0172	[0.00017, 0.00172]
X_3 = Per Capita Income City 2	0.0019	0.00036	0.2380	5.26	< .0001	[0.0012, 0.0026]
X_4 = Business Model	−29.7863	5.1739	−0.2592	−5.76	< .0001	[−39.98, −19.59]

Note. $N = 233$. $R^2_{Y\bullet1234} = .595$, $F(4, 228) = 83.8461$, $p < .0001$. Y = Average Air Fare ($).

ANOVA

Source	SS	df	MS	F	p
Regression	456511.71	4	114127.93	83.8461	< .0001
Residual (Error)	310344.80	228	1361.16		
Total	766856.51	232			

Summary of the Correlation Coefficients and Their Squares

	Zero-Order Correlations		Semipartial Correlations		Full Partial Correlations	
Factor	r_{Yi}	r^2_{Yi}	sr_i	sr^2_i	pr_i	pr^2_i
X_1 = Route Length	.646	.417	.531	.282	.641	.411
X_2 = Per Capita Income City 1	.206	.042	.101	.010	.157	.025
X_3 = Per Capita Income City 2	.423	.179	.222	.049	.329	.108
X_4 = Business Model	−.462	.213	−.243	.059	−.356	.127

Data Analysis

We now conduct an MRC analysis for the given data set. The results are summarized in Table 4.9.

Interpretation of Results

The interpretations of the results of this analysis as reported in Table 4.9 are presented here with respect to the RQs. Prior to doing so, first note that the overall regression equation is $\hat{Y} = 0.0553X_1 + 0.0009X_2 + 0.0019X_3 - 29.7863X_4 + 33.1227$.

RQ 1. This research question corresponds to the overall R^2. The total proportion of variance in average air fares explained collectively by the four targeted IVs is $R^2_{Y\bullet1234} = .595$, with an adjusted $R^2 = .588$. Thus, the four IVs collectively explain 59.5% of the variance in average air fares and when adjusted for bias, these IVs collectively explain 58.8% of the variance in average air fares. This result is significant,

$F(4, 228) = 83.8461$, $p < .0001$ and therefore we reject the corresponding null hypothesis and conclude that the four IVs collectively explain a significant proportion of the variance in average air fares.

RQ 2. This research question relates to the squared semipartial correlation coefficients (sr_i^2) and has four corresponding null hypotheses.

Hyp. 2a. X_1 = Route Length uniquely contributes an additional 28.2% to the knowledge in Y variance and is significant, $sr_1^2 = .282$, $t(228) = 12.60$, $p < .0001$. Therefore, we reject Null Hypothesis 2a and conclude that route length explains a significant proportion of the variance in average air fares when examined in the presence of X_2, X_3, and X_4.

Hyp. 2b. X_2 = Per Capita Income of City 1 uniquely contributes an additional 1.0% to the knowledge in Y variance and is significant, $sr_2^2 = .010$, $t(228) = 2.40$, $p = .0172$. Therefore, we reject Null Hypothesis 2b and conclude that the per capita income of the first city in a city-pair that makes up a route explains a significant proportion of the variance in average air fares when examined in the presence of X_1, X_3, and X_4.

Hyp. 2c. X_3 = Per Capita Income of City 2 uniquely contributes an additional 4.9% of the knowledge in Y variance and is significant, $sr_3^2 = .049$, $t(228) = 5.26$, $p < .0001$. Therefore, we reject Null Hypothesis 2c and conclude that the per capita income of the second city in a city-pair that makes up a route explains a significant proportion of the variance in average air fares when examined in the presence of X_1, X_2, and X_4.

Hyp. 2d. X_4 = Business Model Comparison (LCC vs. FSC) uniquely contributes an additional 5.9% of the knowledge in Y variance and is significant, $sr_4^2 = .059$, $t(228) = -5.76$, $p < .0001$. Therefore, we reject Null Hypothesis 2d and conclude that the comparison between the two business models explains a significant proportion of the variance in average air fares when examined in the presence of X_1, X_2, and X_3.

Based on these results of the squared semipartial correlation coefficients, the strongest predictor of average air fares is X_1 = Route Length (in miles).

RQ 3. This research question relates to the partial regression coefficients (B_i) and has four corresponding null hypotheses.

Hyp. 3a. Holding all other IVs in the regression model constant, $B_1 = 0.0553$, which means that for every 1-mile increase in route length, average air fares increase, on average, by $0.0553. More concretely, for every 100-mile increase in route length, average air fares increase, on average, by $5.53. This result is significant, $t(228) = 12.60$, $p < .0001$. Therefore, we reject Null Hypothesis 3a and conclude that route length has a significant effect on average air fares.

Hyp. 3b. Holding all other IVs in the regression equation constant, $B_2 = 0.0009$, which means that for every $1.00 increase in the per capita income of City 1, average air fares increase, on average, by $0.0009. More concretely, for every $10,000 increase in the per capita income of City 1, average air fares increase, on average, by $9.00. This result is significant, $t(228) = 2.40$, $p = .0172$. Therefore, we reject Null Hypothesis 3b and conclude that the per capita income of the first city in a city-pair route has a significant effect on average air fares.

Hyp. 3c. Holding all other IVs in the regression model constant, $B_3 = 0.0019$, which means that for every \$1.00 increase in the per capita income of City 2, average air fares increase, on average, by \$0.0019. More concretely, for every \$10,000 increase in the per capita income of City 1, average air fares increase, on average, by \$19.00. This result is significant, $t(228) = 5.26$, $p < .0001$. Therefore, we reject Null Hypothesis 3c and conclude that the per capita income of the second city in a city-pair route has a significant effect on average air fares.

Hyp. 3d. As given in the research description, X_4 is a nominal group membership variable that represents a comparison between low-cost carriers vs. full-service carriers where the IV is coded 1 for LCC. Therefore, holding all other IVs in the regression model constant, $B_4 = -29.7863$ indicates that the average air fare of LCC is on average \$29.79 *less* than the average air fare of FSC. This result is significant, $t(228) = -5.76$, $p < .0001$. Therefore, we reject Null Hypothesis 3d and conclude that the comparison between the business models of FSC and LCC has a significant effect on average air fares.

Standard Error of the Estimate. As reported in Table 4.9, $RMSE = 36.89$, which means that if we were to predict the average air fare for a city-pair market based on its route length, the respective per capita incomes of the two cities that comprise the city-pair, and whether the airline's business model is either FSC or LCC, then the prediction would be "off" on average by \$36.89.

Post Hoc Considerations

We now comment on the effect sizes, 95% CIs, and power for the regression model.

Effect Size. Following are the effect sizes for the overall model and the squared semipartial correlation coefficients.

ES for $R^2_{Y \cdot 1234}$. $f^2 = \dfrac{R^2}{1-R^2} = \dfrac{.595}{1-.595} = \dfrac{.595}{.405} = 1.47$ (large effect)

ES for sr_1^2. $f^2 = \dfrac{sr_1^2}{1-R^2} = \dfrac{.282}{1-.595} = \dfrac{.282}{.405} = 0.696$ (large effect)

ES for sr_2^2. $f^2 = \dfrac{sr_2^2}{1-R^2} = \dfrac{.010}{1-.595} = \dfrac{.010}{.405} = 0.02$ (small effect)

ES for sr_3^2. $f^2 = \dfrac{sr_3^2}{1-R^2} = \dfrac{.049}{1-.595} = \dfrac{.049}{.405} = 0.121$ (medium effect)

ES for sr_4^2. $f^2 = \dfrac{sr_4^2}{1-R^2} = \dfrac{.059}{1-.595} = \dfrac{.059}{.405} = 0.146$ (medium effect)

95% CIs. The 95% confidence intervals are reported in Table 4.9.

Power. Consulting Soper (2025e) for the overall model and Soper (2025f) for the squared semipartial correlation coefficients, the respective powers are as follows:

Power for $R^2_{Y \cdot 1234}$ **= 1.0.** There is a greater than 99% probability that the collective effect of the four IVs found in the sample also exists in the population, and there is a greater than 99% probability that the decision to reject the corresponding null hypothesis is correct.

Power for sr_1^2 = 1.0. There is a greater than 99% probability that the effect found in the sample relative to X_1's unique contribution also exists in the population, and there is a greater than 99% probability that the decision to reject the null hypothesis is correct.

Power for sr_2^2 = .58. There is a 58% probability that the effect found in the sample relative to X_2's unique contribution to explaining *Y* variance also exists in the population, and there is a 58% probability that the decision to reject the null hypothesis is correct.

Power for sr_3^2 = .9996. There is a 99.96% probability that the effect found in the sample relative to X_3's unique contribution to explaining *Y* variance also exists in the population, and there is a 99.96% probability that the decision to reject the null hypothesis is correct.

Power for sr_4^2 = .9999. There is a 99.99% probability that the effect found in the sample relative to X_4's unique contribution to explaining *Y* variance also exists in the population, and there is a 99.99% probability that the decision to reject the null hypothesis is correct.

Chapter 4 Summary

4.2 Generalizing from Two IVs to k IVs

- The general form of a regression equation with three or more IVs is of the same form as that with two IVs with a new term "+ B_iX_i" being included in the equation to represent the new IV. This pattern was illustrated in Table 4.3.
- The interpretation of each B_i follows that of regression equations with two IVs, with each interpretation being prefaced with "Holding all other IVs in the model constant."
- The interpretation of the overall results of an MRC analysis with three or more IVs, including *R*, R^2, adjusted R^2, *RMSE*, are done exactly as an MRC analysis with two IVs.
- Examples 4.1 and 4.2 contain demonstrations of an MRC analysis with three IVs and five IVs, respectively.

4.3 Determining and Interpreting the Partials for k IVs

- The squared semipartial correlation coefficient sr_i^2 for more than two IVs is equal to the difference between two squared multiple regression coefficients: the overall R^2 that involves all the IVs and one that involves all the IVs except the one under discussion: $R^2_{Y \cdot 123 \ldots i \ldots k} - R^2_{Y \cdot 123 \ldots (i) \ldots k}$.
- Some statistics software programs will provide either sr_i or sr_i^2. If one is given, the second can be acquired by either squaring or taking the square root of the one provided.
- The interpretation of sr_i^2 follows a regression analyses with two IVs except now the targeted IV is being analyzed in the presence of two or more IVs.
- The squared partial correlation coefficient pr_i^2 for more than two IVs is equal to the ratio of the corresponding sr_i^2 and the amount of *Y* variance that remains

after the effects of all the other IVs except the one under discussion have been partialled out from the IV under discussion as well as from Y. In general:

$$pr_i^2 = \frac{sr_i^2}{R^2_{Y\cdot 123\ldots(i)\ldots k}}.$$

- Some statistics software programs will provide either pr_i or pr_i^2. If one is given, the second can be acquired by either squaring or taking the square root of the one provided.
- The interpretation of pr_i^2 follows a regression analyses with two IVs except now we preface the interpretation with, "when the effects of all other IVs have been completely removed from both the targeted IV (X_i) and Y." This also is known as "statistically controlling for the effects of all the other IVs."
- pr_i^2 will never be smaller than sr_i^2 and is sometimes thought of as the "true" relationship between X and Y.

4.4 Statistical Inference for MRC

- Statistical inference refers to making judgments, conclusions, or decisions about a population parameter based on a sample selected from the given population. The statistical strategies used for this assessment is known as inferential statistics.
- Statistical inference is generally approached either directly via parameter estimation or indirectly via hypothesis testing.
- Parameter estimation consists of using sample statistics as point estimates of the corresponding population parameters or confidence intervals, which provide an interval of values that the population parameter can attain relative to a probability level. The width of a CI reflects the precision of the estimate with narrower widths being more precise and having higher accuracy in parameter estimation (AIPE). Point estimates and CIs may be determined for the overall R^2, the squared semipartial and partial correlation coefficients, sr_i^2 and pr_i^2, and the partial regression coefficients, B_i.
- Hypothesis testing involves establishing null and alternative hypotheses about a targeted parameter. The null hypothesis (H_0) states the parameter has no effect in the population, whereas the alternative hypothesis (H_1) reflects the researcher's prediction about the parameter's effect in the population. Based on the results of data analysis, a decision is made to either fail to reject H_0 and claim the parameter has no significant effect or reject H_0 and claim the parameter has a significant effect. Hypotheses generally are formed relative to three RQs: One relates to the overall R^2, the second relates to the squared semipartial correlation coefficients, sr_i^2, and the third relates to the partial regression coefficients, B_i.
- Regardless of the approach (parameter estimation or hypothesis testing), statistical inference is governed by three factors—sample size, effect size, and significance level—and all three must be considered for sample size planning, which involves determining the minimum sample size needed for a study.
- If a study's objective is parameter estimation, then the focus of sample size planning is on precision. The larger the sample size, the smaller the standard

error, and the more precise the CI. The higher the probability level of confidence (e.g., 99% vs. 95%), the less precise the estimate.

- If a study's objective is hypothesis testing, then the focus of sample size planning is on power analysis, which involves the study's effect size, sample size, number of IVs, and power level. As sample size increases, the probability of finding statistical significance also increases. A large sample size is not required to find a large effect, but a large sample size is needed to find a small effect. As the number of IVs increases, power decreases. More stringent alpha levels (e.g., .01 vs. .05) make it more challenging to find a significant effect.
- Soper (2025a–f) provides online calculators that can be used for sample size planning, including those for determining minimum sample sizes, effect sizes, and power for the various parameters associated with an MRC analysis.

Key Terms and Concepts

Accuracy in parameter estimation (AIPE)
Alternative hypothesis (H_1)
Confidence interval (CI)
Hypothesis testing
Inferential statistics
Margin of error
Null hypothesis (H_0)
Parameter estimation
Point estimate
Power analysis
Precision
Sample size planning

Review Exercises

A. Checking Conceptual Knowledge

In 1–10, choose the best answer among the choices provided.

1. In an MRC analysis with *k* predictors, when constructing a confidence interval for any of the partial regression coefficients, the degrees of freedom for the table value of *t* is

   a. $n - 1$ b. $n - 2$ c. $n - k - 1$ d. $k - 1$

2. Let's assume that the 95% CI for B_3 of an MRC analysis is [-5.65, 2.61]. What would a test for H_0: $B_3 = 0$ vs. H_1: $B_3 \neq 0$ conclude given a preset alpha level of $\alpha = .05$?

   a. Reject the null hypothesis at $\alpha = .05$ and all smaller α.
   b. Fail to reject the null hypothesis at $\alpha = .05$ and all smaller α.
   c. Reject the null hypothesis at $\alpha = .05$ and all larger α.
   d. Fail to reject the null hypothesis at $\alpha = .05$ and all larger α.

3. The reason we preface the interpretation of a partial regression coefficient B_i with "holding all other IVs in the model constant" is

   a. to inform readers that B_i is being analyzed in the presence of the other IVs.
   b. because B_i represents the unique contribution to explaining *Y* variance.
   c. to single out B_i so that it is not being confused with any of the other IVs in the model.
   d. because we are zeroing out all the other IVs.

4. Given an MRC analysis with nine IVs, which of the following denotes sr_6^2?
   a. $R^2_{Y\cdot123456789} - R^2_{Y6\cdot12345789}$
   b. $R^2_{Y\cdot123456789} - R^2_{Y(6\cdot12345789)}$
   c. $R^2_{Y\cdot123456789} - R^2_{Y(6)\cdot12345789}$
   d. $R^2_{Y\cdot123456789} - R^2_{Y\cdot123\ldots(6)\ldots9}$
5. In an MRC analysis with six IVs, if $B_2 = -2.3$ and $sr_2^2 = .36$, then $sr_2 =$
   a. .6
   b. .06
   c. –.6
   d. ± .6

Use the given research description and corresponding results to answer Items 6–10.

Let's assume a study was conducted that involved examining the effect of five independent variables on Y = Attitudes toward Aviation Safety as measured by the Aviation Safety Attitude Scale (ASAS), which consists of 25 items measured on a 6-point Likert-type response scale. Thus, scores on the ASAS could range from 25 to 150 points, with higher scores reflecting a more positive attitude toward aviation safety. The targeted variables are: X_1 = Age, X_2 = Years' Experience in Current Job, X_3 = Group membership that represents Pilots (coded 1) vs. Air Traffic Controllers (ATCs, coded 0), X_4 = Level of Education (in years), and X_5 = Annual Income. Let's further assume that $B_3 = 2.0$, $sr_3^2 = .03$, and the 95% CI for B_3 is [0.211, 7.05].

6. Which statement is a correct interpretation of B_3?
   a. The difference in ASAS scores between pilots and ATCs is significant.
   b. There is no significant difference in ASAS scores between pilots and ATCs.
   c. ATCs scored on average 2 more points on the ASAS than pilots.
   d. Pilots scored on average 2 more points on the ASAS than ATCs.
7. Which statement is correct with respect to B_3?
   a. B_3 is a continuous variable.
   b. $B_3 = 2.0$ reflects the difference on ASAS scores between pilots and ATCs.
   c. B_3 represents the amount of explained variance group membership has on ASAS scores.
   d. $B_3 = 2.0$ means that two groups are being compared.
8. Which statement is a correct interpretation of the 95% CI for B_3?
   a. It represents a point estimate for the mean ASAS score in the population.
   b. It shows that X_3 is significant because $B_3 = 2.0$ is in the interval.
   c. It shows that X_3 is significant because $B_3 = 0$ is not in the interval.
   d. It suggests that the AIPE for B_3 is extremely low because precision is low.
9. Which statement is correct with respect to sr_3^2?
   a. sr_3^2 is a group membership variable.
   b. sr_3^2 represents the amount of variance in ASAS scores being explained by X_3 when the other IVs are statistically controlled.
   c. $sr_3^2 = .03$ infers that pilots average 0.03 points higher on the ASAS than ATCs.
   d. sr_3^2 represents the amount of additional variance in ASAS scores being explained by group membership over and above what was previously explained by the other IVs.
10. Which null hypothesis is NOT correct for X_3?
    a. H_0: $\rho^2_{Y\cdot1234} - \rho^2_{Y\cdot124}$
    b. H_0: $X_3 = 0$
    c. H_0: $\rho^2_{Y(3\cdot1245)} = 0$
    d. H_0: $\beta_3 = 0$

B. Applying Procedural and Conceptual Knowledge

Use the following research description and the corresponding data set to conduct the activities given in Parts B1–B5. See Section 4.5 for guidance.

Motivated by Torres (2023), a faculty member at a local College of Aeronautics examined the relationship between a set of targeted factors and students' college grade-point average (GPA). The factors were X_1 = Students' Age (in years), X_2 = students' High School GPA, X_3 = Flight vs. Nonflight Students (coded 1 for flight students), and X_4 = Total Credit Hours earned to date. A set of hypothetical data is provided in the file "Ch_4 Review Exercises Data-Part B."

B1. Research Questions and Hypotheses. Write the research questions and corresponding hypotheses that pertain to the overall regression model, the squared semipartial correlation coefficients, and the partial regression coefficients.

B2. Sample Size Planning. Conduct an a priori power analysis to determine if the given data set is sufficiently large.

B3. Data Analysis. Conduct an MRC analysis of the given data set and summarize the results in a table like Table 4.9.

B4. Interpretation of Results. Interpret the results relative to the research questions and corresponding null hypotheses stated in Part B1.

B5. Post Hoc Considerations. Calculate and comment on the effect sizes, 95% CIs, and power for the overall R^2 and sr_i^2.

References

Baran, S. (2018). *Survival in the U.S. domestic airline market: Strategies for entry, exit, and air fare competition.* [Doctoral dissertation, Florida Institute of Technology]. Scholarship Repository. https://repository.fit.edu/etd/3/

Cohen, J. (1988). *Statistical power analysis for the behavioral sciences* (2nd ed.). Lawrence Erlbaum Associates.

Cohen, J. (1992). A power primer. *Psychological Bulletin*, *112*, 155–159. https://doi.org/10.1037/0033-2909.112.1.155

Cohen, J., Cohen, C., West, S. G., & Aiken., L. S. (2003). *Applied multiple regression/correlation analysis for the behavioral sciences* (3rd ed.). Lawrence Erlbaum Associates.

Green, S. B. (1991). How many subjects does it take do a regression analysis? *Multivariate Behavioral Research*, *26*, 449–510.

Khamis, H., & Kepler, M. (2010). Sample size in multiple regression: 20 + 5k. *Journal of Applied Statistical Science*, *17*, 505–517.

Kelley, K., & Maxwell, S. (2003). Sample size for multiple regression: Obtaining regression coefficients that are accurate, not simply significant. *Psychological Methods, 8(3), 305*–321. DOI: 10.1037/1082-989X.8.3.305. https://www3.nd.edu/~kkelley/publications/articles/Kelley_Maxwell_2003.pdf

Soper, D. (2025a). *R-square confidence interval calculator.* https://www.danielsoper.com/statcalc/calculator.aspx?id=28

Soper, D. (2025b). *A-priori sample size calculator for multiple regression*. https://www.danielsoper.com/statcalc/calculator.aspx?id=1

Soper, D. (2025c). *A-priori sample size calculator for hierarchical multiple regression* https://www.danielsoper.com/statcalc/calculator.aspx?id=16

Soper, D. (2025d). *Effect size calculator for multiple regression*. https://www.danielsoper.com/statcalc/calculator.aspx?id=5

Soper, D. (2025e). *Post-hoc statistical power calculator for multiple regression*. https://www.danielsoper.com/statcalc/calculator.aspx?id=9

Soper, D. (2025f). *Post-hoc statistical power calculator for hierarchical multiple regression*. https://www.danielsoper.com/statcalc/calculator.aspx?id=17

Torres, J. (2023). *How students learn: Examining the differences in learning strategies between flight and nonflight students*. [Doctoral dissertation, Florida Institute of Technology]. Scholarship Repository. https://repository.fit.edu/etd/1387

Chapter 5
Preliminary Data Screening: Regression Assumptions and Diagnostics

Student Learning Outcomes

After studying this chapter, you will be able to do the following:

1. Describe the underlying assumptions of multiple regression.
2. Determine if a data set is compliant with the assumptions of multiple regression and resolve noncompliance issues.
3. Perform regression diagnostics on a data set in preparation for an MRC analysis.
4. Assess the predictive performance of a regression model.
5. Conduct an MRC analysis for a research study with more than two IVs by engaging in the activities presented in Section 5.5.

5.1 Chapter Overview

In the previous two chapters, we developed the conceptual foundation for multiple regression/correlation and provided examples on how to conduct and interpret the results of an MRC analysis with two or more independent variables. As part of this discussion, we also examined statistical inference via parameter estimation and hypothesis testing. In this chapter, we turn our attention to the concept of preliminary data screening, which consists of a set of activities researchers perform to ensure they are analyzing a "clean" data set, which means it is free from certain problems that could emerge from the analysis and could lead to spurious results. We begin by introducing the underlying *assumptions of multiple regression*. As part of this discussion, we demonstrate how to determine if a data set is compliant with these assumptions and how to resolve any violations. We then consider several practical issues that can emerge from an MRC analysis, including *outliers*, *multicollinearity*, *missing data*, *measurement error*, and *IV relevance*. Consideration of these issues is tantamount to performing *regression diagnostics* because we are examining the data set for potential "problems" that could impact the "health" (i.e., confidence) of the results and any conclusions or recommendations made from these results. We also introduce the concept of *cross-validation*, which addresses the generalizability of a regression model with respect to its predictive performance when applied to a new set of data. We conclude the chapter with a detailed example of an MRC analysis, similar to the guided examples of Chapters 3 and 4, but now include a section on *preliminary data screening*.

5.2 Underlying Assumptions of Multiple Regression

The underlying assumptions of multiple regression are the same as those for bivariate linear regression: *linearity, equal variances, independence, and normality*. The difference, though, is we must consider them from a multivariate perspective, which

DOI: 10.4324/9781003506935-7

means we must examine all the variables in the data set collectively, not individually. One of the best ways to test for these assumptions is to conduct a residual analysis, which involves examining scatter plots that contain the residuals. Recall from Chapter 2 that residuals represent the errors between observed scores and predicted scores. Most statistics software packages provide an option that will save the residuals to the given data set. A brief discussion of the four assumptions and their respective residual plots follows.

Linearity

The linearity assumption specifies that the *form* of the relationship between the DV and IVs is linear. Violations of this assumption may result in regression coefficients and standard errors that are biased, which means that the corresponding results acquired from confidence intervals or hypothesis testing could be spurious. One way to test for this assumption is to examine a scatter plot where the residuals are plotted on the vertical axis and the DV is plotted on the horizontal axis. If the resulting pattern of points reflects a linear relationship, then this assumption is satisfied.

As an illustration, consider Figure 5.1, which shows the residual plot for the data set given in the file "Ch_4 Example 4.2 Data," which was used for Example 4.2. Recall that this data set consists of five factors: X_1 = Age; X_2 = Corrected Vision (1 = Yes and 0 = No); X_3 = Visual Acuity Angle (smaller angles = better eyesight); X_4 = Total Flight Hours; and X_5 = Number of FAA Ratings. The dependent variable is Y = Visual Reaction Time (measured in milliseconds), and the objective of the study is to examine the effect the five IVs have on pilots' reaction times. For our discussion of this data set in the current chapter, we relabel this file as "Ch_5 Visual Reaction Times Data." Observe from Figure 5.1 that the residual plot shows a definite linear pattern and hence the visual reaction times data set is compliant with the linearity assumption.

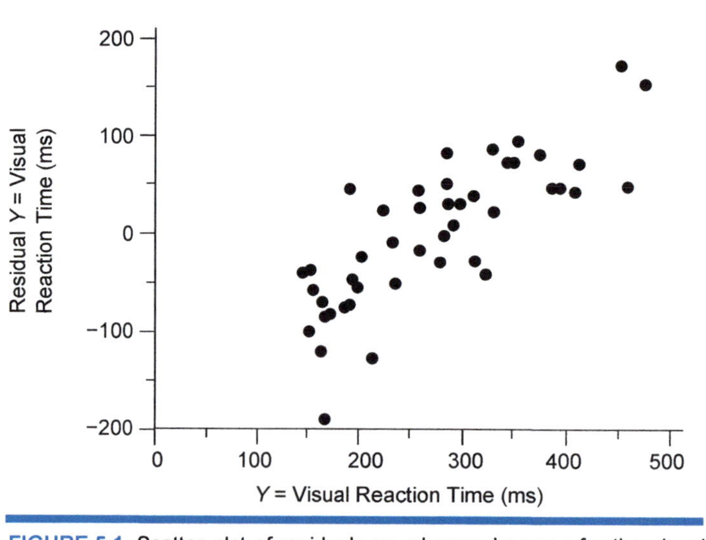

FIGURE 5.1 Scatter plot of residuals vs. observed scores for the visual reaction times data set. This is used to confirm that the data set is compliant with the linearity assumption.

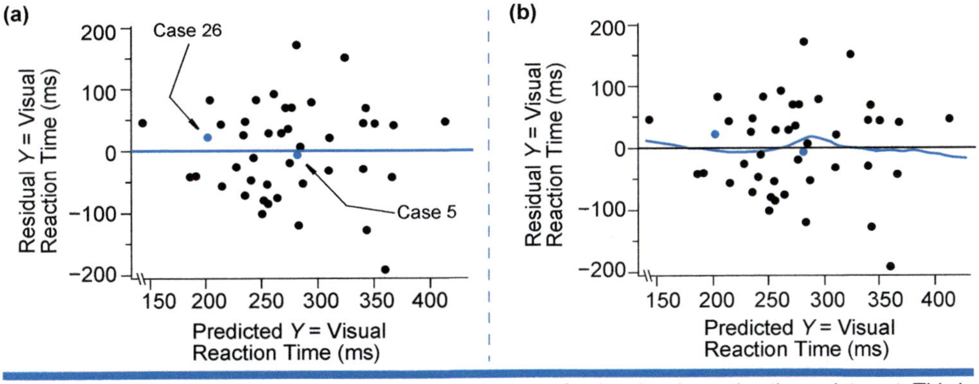

FIGURE 5.2 Scatter plot of residuals vs. predicted scores for the visual reaction times data set. This is used to confirm that the data set is compliant with the linearity and equal variances assumptions.

Another residual plot that can be used to test for compliance with the linearity assumption involves plotting the residuals on the vertical axis and the predicted y values on the horizontal axis. This residual plot for the visual reaction times data set is shown in Figure 5.2(a). We included with this plot a horizontal line at $y = 0$. This line is called the 0-line because any point where the predicted value is equal to the observed value will have a residual of 0 and fall on this line. For example, the data point that corresponds to Case 26 in the data set has an observed score of $y = 225$ ms and the corresponding predicted score is $\hat{y} = 201.65$ ms. Thus, the residual y value is $225 - 201.65 = 23.35$, and this point is 23.35 units above the 0-line. Now look at the point that corresponds to Case 5. The observed y value is 282 ms, the corresponding predicted value is $\hat{y} = 283.65$, and the residual is −1.65, which is just slightly below the 0-line. Thus, the 0-line serves as a reference line for the residuals. If the residuals appear to be randomly scattered around the 0-line—as opposed to depicting a pronounced or systematic pattern—then this suggests compliance with the linearity assumption. To help with the visual inspection of the residuals, in Figure 5.2(b) we superimpose a kernel smoother line (highlighted in blue), which follows the trend of the data. Notice how the kernel smoother line "hugs" the 0-line, which confirms the absence of any pronounced or systematic pattern.

Equal Variances

The equal variances assumption states that for any value of the independent variable X, the variance of the residuals around the regression line in the population is assumed to be constant. This idea of "equal or constant variance of the residuals" is known as homoscedasticity of the residuals. In the multiple IVs case, the variance of the residuals should not be related to any of the IVs or to the predicted Y values. If the variance changes as X changes, then we have a condition known as *heteroscedasticity*, which is a violation of this assumption.

To confirm the equal variances assumption in MRC, we examine the residuals vs. predicted plot we used to confirm linearity (see Figure 5.2). If the residuals appear to be approximately equally scattered above and below the 0-line without any pronounced pattern—for example, the points might fan out in a triangular fashion—then this assumption is satisfied, which is the case for the visual reaction times data set.

Independence

In addition to the equal variances assumption, the residuals of the observations must also be independent of one another. What this means is the respective errors that result from predicting *Y* using the regression equation for each x_i value—that is, the residuals $(y - \hat{y})$—must be independent of each other. If a data set is not compliant with this assumption, the regression coefficients are not affected but the standard errors are, and this will culminate in incorrect parameter estimations (i.e., confidence intervals) and spurious hypothesis test results.

One way to do check for compliance with the independence assumption is to examine a scatter plot of the residuals plotted on the vertical axis vs. the case numbers, which are the row numbers, plotted on the horizontal axis as shown in Figure 5.3 for the visual reaction times. We then examine the distribution of the points to confirm there is no systematic pattern. Given the absence of any systematic pattern as confirmed by the kernel smoother line in Figure 5.3, the visual reaction times data set appears to satisfy the independence assumption.

A more precise approach to examining a data set for compliance with the independence assumption is the Durbin-Watson statistic, which is a measure of autocorrelation relative to the residuals. Without providing any explanation—the reader is directed to Cohen et al. (2003, pp. 136–137)—autocorrelation of the residuals involves correlating the errors of prediction associated with adjacent rows in a data set (e.g., correlating Row 1's residual with Row 2's residual). If the Durbin-Watson statistic is not significant, then there is no relationship among the residuals and the data set is compliant with the independence assumption. Furthermore, if the corresponding autocorrelation metric is positive, then the overall MRC model is subject to inflated Type I error, and if it is negative, then the power of the overall model is

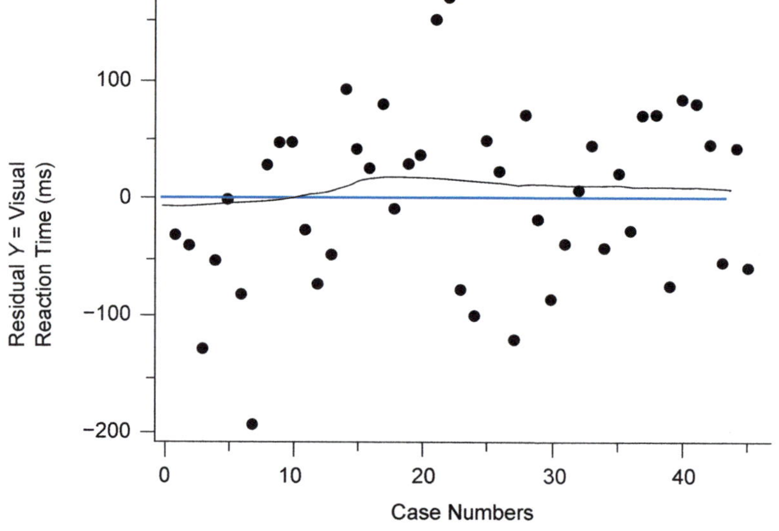

FIGURE 5.3 Scatter plot of residuals vs. case numbers for the visual reaction times data set. This is used to confirm that the data set is compliant with the independence assumption.

reduced. When applied to the visual reaction times data set, the Durbin-Watson statistic for the residuals of the visual reactions provided by our statistics program is 1.654, $p = .1239$, which indicates the independence assumption is met. However, the corresponding autocorrelation is 0.1638, which is not 0 and therefore suggests that the overall model's power might be reduced.

Normality

The last underlying assumption for MRC states that for any value of the IV, the residuals are assumed to be normally distributed. This assumption can be confirmed by visually examining a *normal probability plot* called a *quantile-quantile* plot, which is abbreviated as *q-q* plot. The reader might recall that quantiles are simply specific points that partition a data set into equal parts. For example, common quantiles are quartiles (Q_1, Q_2, and Q_3) and percentiles (P_1, P_2, P_3, ..., P_{97}, P_{98}, P_{99}). A normal *q-q* plot compares two probability distributions: the one being tested for normality—in our current example, this is the set of residuals—and a theoretical one such as the *z* distribution. Observed scores are placed on the horizontal axis, expected *z* scores are placed on the vertical axis, and the resulting plot is a straight line placed at a 45-degree angle. Some statistics software packages also include a 95% confidence band. If the sample data are from a population that is normally distributed, then the corresponding points will "hug" the straight line with the majority (if not all) appearing within the 95% confidence band. If, however, the points are scattered considerably from the straight line with many falling outside the 95% confidence band, then the sample data are presumed not to be from a normally distributed population. As illustrated in Figure 5.4, the normal *q-q* plot for the residuals associated with the visual reaction times confirms the normality assumption: All the points fall within the 95% confidence band with most "hugging" the straight line.

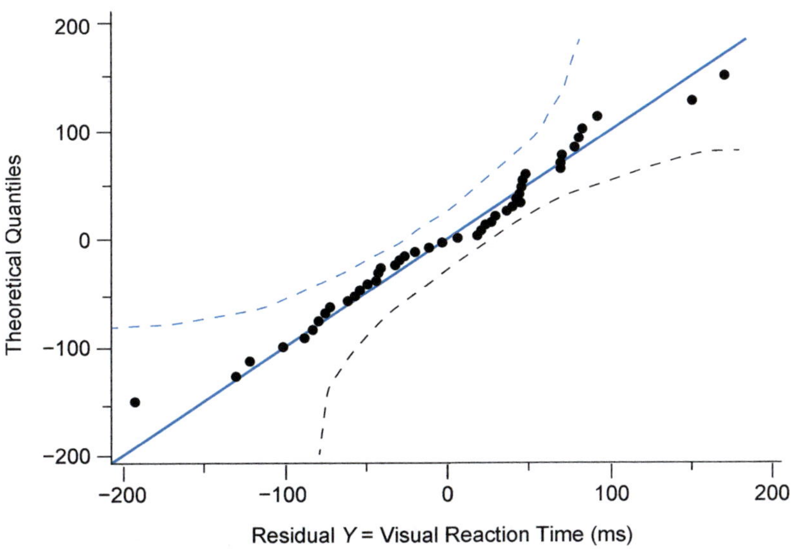

FIGURE 5.4 A normal *q-q* plot is used to confirm compliance with the normality assumption for the visual reaction times data set.

A complement to normal *q-q* plots is the Shapiro-Wilk Goodness of Fit test, which tests the null hypothesis that the data are normally distributed. If the decision is "fail to reject H_0," then the normality assumption is met; otherwise, it is not. With respect to the visual reaction times data set, the results of the Shapiro-Wilk Goodness of Fit test for the residuals are $W = .984$, $p = .4728$, which means we fail to reject H_0 and therefore this assumption is satisfied.

Based on the results of the residual analyses, we affirm that the visual reaction times data set is compliant with the four underlying assumptions of multiple regression.

5.3 Regression Diagnostics

In addition to checking if a data set is compliant with the underlying assumptions of multiple regression, we also need to perform regression diagnostics by examining the data set for potential problems that can emerge from an MRC analysis. These "problems," which can impact the results of an analysis, as well as any conclusions or recommendations made from these results, include *outliers*, *multicollinearity*, *missing data*, *measurement error*, and *IV relevance*. For this discussion, we will examine these regression diagnostics with respect to the visual reaction times data set given in the "Ch_5 Visual Reaction Times Data" file.

Outliers

Recall that outliers are extreme scores relative to the other scores within a data set and may be classified generally as either contaminants or rare cases. An example of the former is a data entry error or a flawed data collection instrument, and an example of the latter is an ATP with 40,000 hours. Of the various approaches to detecting *multivariate* outliers, we use Jackknife distances. This approach generates a centroid, which is the mean of a multivariate set of variables, and then calculates the distance each observed score is from the centroid. Scores with distances that exceed a threshold, which is determined by the Jackknife algorithm, are flagged as outliers. Some statistics software packages do not support Jackknife distances but instead provide a similar routine called Mahalanobis distances. We prefer the former.

If outliers are identified, the first course of action is to determine if the outliers are contaminants or rare cases. If they are contaminants, then the next step is to correct them, replace them, or delete them, whichever is appropriate or feasible. If the outliers are rare cases, though, then this presents a conundrum: Should we keep the outliers or delete them? Although the easier solution is to delete them, this action raises two additional questions: (a) Will the remaining sample still be representative of the parent population? and (b) Will the new data set now have new outliers? Although the first question is important from a research methodology perspective, we will focus on the second question.

Because outliers are relative to a given data set, if the outliers in the initial data set (i.e., DS-1) are deleted, it is possible that the new data set (i.e., DS-2)—which is absent of the original outliers—will now have a set of new outliers. If these outliers are deleted, which will result in DS-3, it is possible that DS-3 will now have a new set of outliers. This cycle of having new outliers present in each succeeding data set, which resulted from deleting outliers in the previous data set, can continue ad infinitum because, once again, outliers are relative to a given data set.

TABLE 5.1 *Summary of Outlier Analysis Iterations for the Data Set from the File "Ch_3 Review Exercises Data–Part B"*

		MRC Model				
Run #	***N***	$R^2_{Y \cdot 12}$	**Adj. R^2**	***RMSE***	***F***	**Outlier Cases Flagged and Deleted Prior to Next Run**
Initial	136	.54	.53	4.3	$F(2, 133) = 77.37$	51, 53, 80, 96, 103, 111, 128
1st	129	.56	.55	3.9	$F(2, 126) = 79.35$	5, 10, 79, 91
2nd	125	.59	.58	3.7	$F(2, 122) = 87.63$	8, 32, 47, 66, 81, 95, 97
3rd	118	.68	.67	3.4	$F(2, 115) = 123.58$	23, 24, 57, 70, 78, 94, 118, 130
4th	110	.67	.66	3.1	$F(2, 107) = 108.64$	22, 34, 90, 129, 134
5th	105	.63	.62	3.1	$F(2, 102) = 85.72$	41, 102, 115
6th	102	.61	.60	3.1	$F(2, 99) = 76.89$	58, 77
7th	100	.59	.58	3.1	$F(2, 97) = 70.50$	117
8th	99	.58	.57	3.1	$F(2, 96) = 66.51$	46
9th	98	.56	.55	3.1	$F(2, 95) = 60.75$	11, 73
10th	96	.53	.52	3.1	$F(2, 93) = 52.65$	6, 42, 107, 123
11th	92	.50	.49	3.1	$F(2, 89) = 45.13$	48, 60, 119
12th	89	.48	.47	3.1	$F(2, 86) = 40.17$	35, 61, 65, 122
13th	85	.46	.45	3.1	$F(2, 82) = 35.70$	25, 59, 125
14th	82	.47	.46	3.1	$F(2, 79) = 35.48$	43, 62, 69, 85, 114
15th	77	.49	.48	3.1	$F(2, 74) = 35.53$	29
16th	76	.48	.47	3.1	$F(2, 73) = 34.18$	<No Outliers>

Note: Outlier analyses performed using Jackknife distances.

As an example, consider the data set given in the file "Ch_3 Review Exercises Data–Part B," which contains hypothetical data relative to Harwin (2023) who examined the relationship between a set of targeted factors and the total flight time students need to become ready to take the private pilot check ride. This data set consists of two IVs, X_1 = Flight Student's Age and X_2 = Number of Dual Hours reported by students' CFIs, and the DV is Y = Number of Flight Hours to Check-Ride Proficiency. As summarized in Table 5.1, when examined for outliers via Jackknife distances, the initial data set of $N = 136$ contains seven outliers. After deleting these outliers and examining the data set a second time, there are another four outliers. Continuing this process, it takes 16 outlier analyses and a 44% reduction in the sample (from $N = 136$ to $N = 76$) before the data set has no outliers.

Although it is tempting to improve a regression model's fit by deleting outliers, note from Table 5.1 that the fit of the regression model did not improve after the third iteration when the overall R^2 maxed out at .68. What also is noteworthy is there are no contaminants in this data set. Every outlier flagged is a rare case. So, by deleting 60 cases in this march to removing all the outliers, not only do we lose information about real-world phenomena relating to flight instruction, but the remaining sample is probably biased and not representative of the parent population. The reader is also informed that once a change is made to a data set, such as deleting outliers, then this "new" data set must be examined for compliance with the underlying regression assumptions.

When confronted with outliers, it is important to not lose sight of the primary purpose for examining outliers: It is to alert us that we need to examine these extreme cases more closely, not to delete them to simply improve model fit. Furthermore, as observed by Kleinbaum et al. (1998, p. 232), "Scientific judgment is more important here than statistical tests, once influential observations have been flagged." As a result, to address the issue of how to handle outliers, we provide the following guidance, which is modeled after Judd et al. (2017):

- If the outliers are contaminants, then try to rectify the problem by correcting them, replacing them, or deleting them, whichever is appropriate or feasible.
- If the outliers are rare cases, then conduct two separate analyses, one in the presence and one in the absence of the outliers, and compare the results. If the results are similar and there is no difference in how the two data sets affect the assumptions, then keep the outliers because they reflect real-world phenomena; otherwise choose the data set that is compliant with the assumptions.
- If the data set with outliers is significant, but the data set without outliers is not significant, then the outliers are inflating significance. If the situation is reversed, then the outliers are suppressing significance. In either scenario, the outliers *probably* should be deleted, but the decision to keep or delete them will depend on how influential they are, and which data set (with or without outliers) is compliant with the assumptions.

Regardless of the situation or path taken, it is imperative that we (a) inform the reader what we decided to do and why, and (b) report the results of the separate analyses (with vs. without outliers) so readers can make their own decision about the final model's appropriateness. It also is important to provide the reason(s) for keeping or deleting outliers to help inform readers about the generalizability of the results. Lastly, the method used to detect outliers and how they were handled are to be reported as delimitations to the study.

With respect to the situation illustrated in Table 5.1, because all the outliers are rare cases, we follow the second guideline given above. More specifically, we would delete the initial seven outliers and work with the MRC model with $N = 129$. This is because of the seven outliers, five cases (51, 53, 103, 111, 128) involved CFIs with dual hours between 3,700 and 7,500, which far exceeded the rest of the group, and two cases (80, 96) involved CFIs with the fewest number of dual hours (< 33). Thus, these cases reflect either highly experienced CFIs or those just starting their career.

Example 5.1 *Conducting An Outlier Analysis*

Conduct an outlier analysis using Jackknife distances for the visual reaction times data set given in the file "Ch_5 Visual Reaction Times Data."

Solution

As summarized in Table 5.2, the initial data set contained two outliers: Cases 7 and 10. Inspecting these cases, it appears that Case 7 has a small visual angle acuity whereas Case 10 has a large visual angle acuity. After checking these entries, we confirm they are rare cases but will delete them because the sample size is relatively small and outliers, regardless of their nature, have a greater impact on smaller samples than larger samples.

Continued on next page ...

TABLE 5.2 *Summary of Outlier Analysis for Example 5.1*

		MRC Model				Outlier Cases Flagged and Deleted Prior to Next Run
Run #	N	$R^2_{Y \cdot 12}$	Adj. R^2	*RMSE*	F	
Initial	45	.35	.27	79.2	$F(5, 39) = 4.29$ [a]	7, 10
1st	43	.40	.32	73.4	$F(5, 37) = 4.95$ [b]	3, 22
2nd	41	.54	.48	62.1	$F(5, 35) = 8.35$ [c]	<No Outliers>

Note: The data set is from the file, "Ch_5 Visual Reaction Times Data." Outlier analyses performed using Jackknife distances.
[a] $p = .0033$. [b] $p = .0014$. [c] $p < .0001$.

After deleting Cases 7 and 10, the resulting MRC model is an improvement to the original. It has a higher R^2 (.40 vs. .35) and a smaller *RMSE* (73.4 vs. 79.2). However, this data set also contains two outliers: Cases 3 and 22. Case 3 has a high number of flight hours coupled with a low visual acuity angle, and Case 22 has a very high visual acuity angle. After inspecting these cases, we confirm that they too are rare cases. Before accepting this model, though, we run another outlier analysis for the same reason as given above.

After deleting Cases 3 and 22, the resulting MRC model is a considerable improvement to the previous model. The overall R^2 increases by .14 and the *RMSE* decreases by 11.3 points. Furthermore, this data set does not contain any additional outliers. Therefore, we choose to use this model instead of either of the previous two models because it appears that the four outliers were influential to the point that they were suppressing significance and impacting model fit.

Based on our earlier discussion, we know that the initial model with $N = 45$ is compliant with the regression assumptions, but we do not know if this final model with the four outliers deleted ($N = 41$) is also compliant. Examining this new data set with respect to the four underlying assumptions, we confirm that it is indeed compliant with the linearity and equal variances assumptions (see Figure 5.5a); the independence assumption (see Figure 5.5b), which has a Durbin-Watson statistic of 2.03, $p = .5374$, and autocorrelation = –0.33; and the normality assumption (See Figure 5.5c), with Shapiro-Wilk $W = .9792$, $p = .6464$. We will now use this data set *without* outliers for the remaining regression diagnostics.

Multicollinearity

Because an MRC analysis involves two or more IVs, it is possible for some of the IVs to be highly correlated with each other, which makes it difficult to interpret the findings. Furthermore, the stronger the correlation, the less unique contribution an IV will make to explaining Y variance. This concept of high correlation among a set of IVs is known as multicollinearity. An exaggerated illustration of the concept of multicollinearity is given in Figure 5.6, which shows the partitioning of Y variance between two IVs. First note that X_1 and X_2 *collectively* account for a large proportion of Y variance (regions *a*, *b*, and *c*). However, further note that of this large proportion of explained variance, a considerable part of it is *shared variance* (region *c*). Recall that this region of the Venn diagram represents where X_1 and X_2 are carrying redundant information about Y. Because of this redundancy, there is very little *unclaimed* variance in Y that can be accounted *uniquely* by either X_1 (region *a*) or X_2 (region *b*).

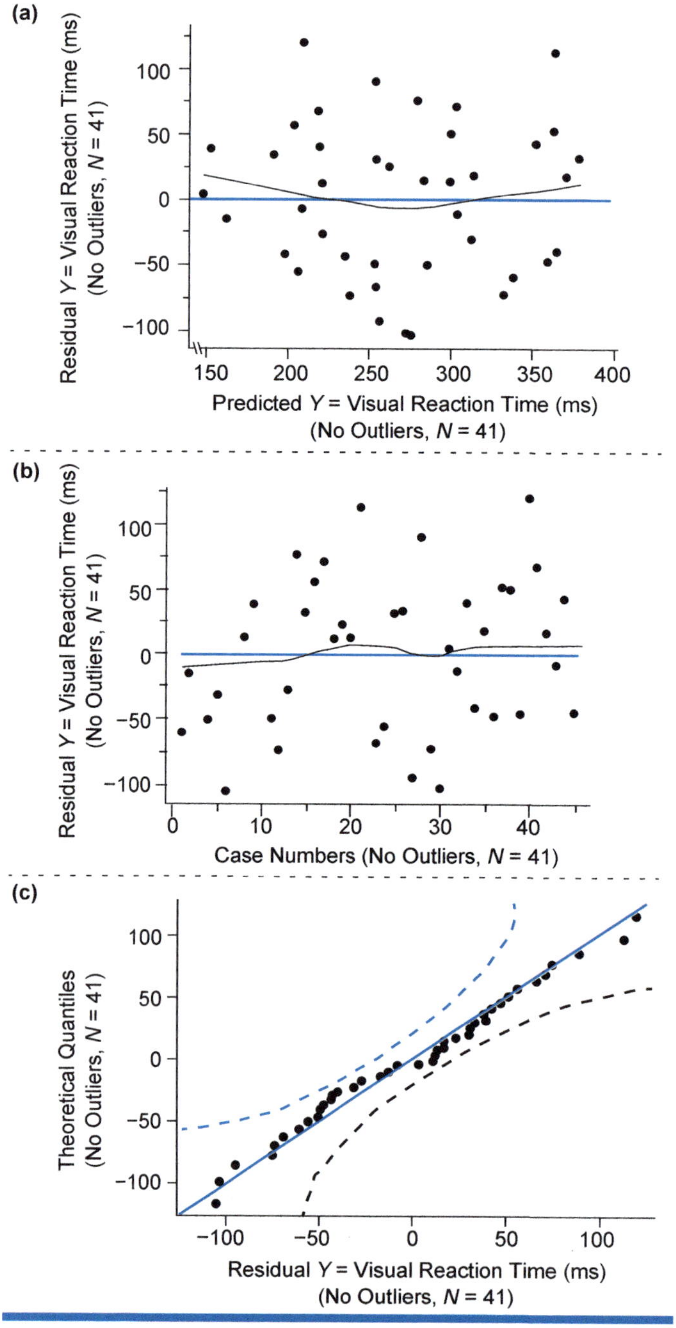

FIGURE 5.5 Scatter plots for checking the regression assumptions of the revised data set for Example 5.1 after deleting the outliers. Figure 5.5(a) confirms the linearity and equal variances assumption, Figure 5.5(b) confirms the independence assumption, and Figure 5.5(c) confirms the normality assumption.

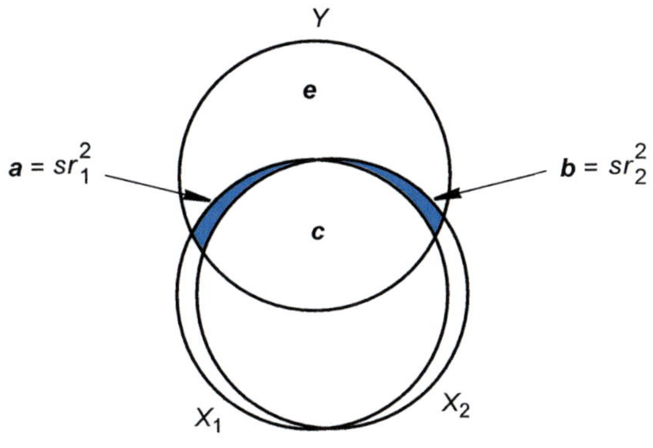

FIGURE 5.6 Illustration of multicollinearity.

To determine the presence of multicollinearity in a data set, we rely on the variance inflation factor (*VIF*), which is a metric that is calculated for each regression coefficient in the regression equation. According to Cohen et al. (2003, p. 423), "any *VIF* of 10 or more provides evidence of serious multicollinearity involving the corresponding IV." Furthermore, the square root of the *VIF* indicates how much larger the standard error associated with a regression coefficient will increase compared to what it would be if the IV was not correlated with any other IV. For example, if B_1's $VIF = 10$, then the corresponding square root, which is approximately 3.16, indicates that we can expect B_1's standard error to be about three times larger than it would be if B_1 was not correlated with any of the IVs in the data set. For our running example of visual reaction times based on the data set without outliers, the *VIF*s for all the IVs are between 1.0 and 1.3. Thus, multicollinearity is not a concern.

Because multicollinearity involves two or more highly correlated IVs, a common approach to dealing with multicollinearity is to combine the IVs. For example, the two variables "age" and "years' experience" can provide different information, but they often are highly correlated. As such, it is reasonable to drop one of the IVs from the final data set. For additional information on the pros and cons of doing so, the reader is directed to Cohen et al. (2003, pp. 422–430).

Missing Data

The issue of missing data is always a concern in every research study. For example, when completing a questionnaire, some participants might neglect to respond to certain items, and during an intervention study, some participants might withdraw before the study ends, resulting in not only a mortality threat to internal validity but also could lead to incomplete data. Missing data can occur randomly or systematically. An example of the former is skipping a questionnaire item with the intent to return to it but forgetting to do so, and an example of the latter is intentionally not responding to a questionnaire item because it is too personal (e.g., reporting annual salary or illegal drug use). The strategy a researcher uses to manage missing data can impact both a study's findings and its generalizability. Therefore, it is important to minimize the amount of missing data.

Of the various strategies available for handling missing data, the single best strategy is to not have missing data to begin with, which suggests that researchers carefully plan their data collection with sufficient time and opportunity to acquire any missing information. It also suggests that their data collection instrument should be free of any items that might be ambiguous or have the potential to elicit no response from participants. In the event that the former is not feasible—for example, it is not possible to acquire missing data in a cross-sectional observational study that uses an electronic questionnaire to collect data anonymously via a web hosting site—then any of the following strategies may be used. Be forewarned, though: Each strategy has its own set of advantages and disadvantages, and each strategy has its share of supporters and critics. Therefore, readers are encouraged to consult the references cited prior to choosing a particular strategy to implement.

Delete the IVs with Missing Data. Deleting variables with missing data is acceptable if the IV is not critical to the research, but if this is the case, then why was the IV included as part of the data set initially? Deleting a variable also means losing the information that was being represented by the IV. Furthermore, if *all* the IVs of a study have missing data, then there would not be any factors to consider if they all are deleted. As a result, unless an IV has an inordinate number of missing entries, which would leave no choice but to delete the variable from the final model, this approach is not recommended.

Delete the Cases with Missing Data. An alternative to deleting IVs with missing data is to delete from the study those participants who have missing data with respect to an IV. This is a reasonable strategy if (a) there is no systematic pattern to the missing data, (b) there are not many cases with missing data, and (c) the data are missing on different IVs. This strategy also can be performed in one of two ways: *listwise deletion* or *pairwise deletion*. With listwise deletion, the participants with missing data are simply deleted from the data set, which means they are not part of the final data analysis. This is generally the default strategy for many statistics programs. One advantage to listwise deletion is that once the cases are dropped, the sample size is fixed for all correlations. The disadvantage, though, is that the sample size will be smaller than the initial sample, which means that power is reduced, and the sample might be biased.

With pairwise deletion, all cases with missing data remain in the data set, but these cases are now examined on an IV-by-IV basis when calculating correlations. For example, given a data set with $N = 50$ and three predictors, let's assume that X_1 has data missing on five cases: 7, 19, 31, 33, 43; X_2 has data missing on 11 cases: 9, 17, 18, 23, 29, 34, 36, 37, 41, 44, 49; and X_3 has data missing on two cases: 2, 7. When considering the zero-order correlations, r_{Y1} will be based on $n = 45$, r_{Y2} will be based on $n = 39$, r_{Y3} will be based on $n = 48$, r_{12} will be based on $n = 34$, r_{13} will be based on $n = 43$, and r_{23} will be based on $n = 37$. Thus, pairwise deletion maintains the largest possible sample size for each correlation, but the sample sizes across these correlations are different, which could impact the integrity of the correlations.

Substitute Missing Data with a Reasonable Alternative. Instead of deleting IVs or cases, we could replace the missing values with a reasonable estimate and then include these estimates in the final model. This concept of replacing missing data with reasonable estimates is referred to as imputation. Many of today's statistics programs now provide various imputation routines that can be used to make this approach for handling missing data simple and feasible.

To illustrate this using our statistics program, consider the data set given in the file, "Ch_5 Missing Data Example," which consists of four predictors: X_1 = Years as Pilot-in-Command (PIC), X_2 = Total Flight Hours, X_3 = Years in Aviation, and X_4 = Biological Sex (coded 1 = Female and 0 = Male). The DV is Y = Annual Salary ($), and the data were acquired from a random sample of N = 87 airline transport pilots (ATPs) working at a regional airline in the U.S. Three IVs have missing data: X_1 has 8 cases, X_2 has 17 cases, and X_3 has 16 cases. Of these 41 cases, 30 are unique. The first two subtables of Table 5.3 contain a summary of the results from regressing Y on the four IVs from two perspectives. The results in the first subtable are based on leaving the missing values in the analysis as blanks, and the results in the second subtable are from our statistics program's missing data imputation routine. This routine uses a regression analysis to predict the missing data from nonmissing values in the same row and then replaces the missing data with these estimates.

TABLE 5.3 *Summary of Missing Data Analyses for "Ch_5 Missing Data Example"*

Summary of Results When Missing Data Are Discarded

		Factor	B_i	SE_i	t_i	p_i
R^2	.493	Constant	38348.76	2740.71	13.99	< .0001
Adj. R^2	.454	X_1 = Years as PIC	1052.16	294.09	3.58	.0008
RMSE	6674.84	X_2 = Total Flight Hours	2.12	0.61	3.47	.0011
M	55774.02	X_3 = Years in Aviation	5.18	102.78	0.05	.9600
N	57	X_4 = Biological Sex	–142.12	1840.45	0.08	.9387

Note. $F(4, 52) = 12.6306$, $p < .0001$.

Summary of Results Using Our Statistics Program's Missing Data Routine [a]

		Factor	B_i	SE_i	t_i	p_i
R^2	.471	Constant	39657.44	2337.65	16.96	< .0001
Adj. R^2	.445	X_1 = Years as PIC	1100.76	255.73	4.30	< .0001
RMSE	6898.92	X_2 = Total Flight Hours	1.83	0.53	3.47	.0008
M	57699.55	X_3 = Years in Aviation	28.86	98.78	0.29	.7709
N	87	X_4 = Biological Sex	1218.54	1540.51	0.79	.4312

Note. $F(4, 82) = 15.6892$, $p < .0001$.
[a] The results are from JMP's Multivariate Normal Imputation procedure (with shrinkage), which estimates missing values using regression to predict missing values from nonmissing data,

Summary of Results From Substituting with the Mean

		Factor	B_i	SE_i	t_i	p_i
R^2	.451	Constant	40118.66	2492.94	16.09	< .0001
Adj. R^2	.417	X_1 = Years as PIC	1159.43	233.41	4.97	< .0001
RMSE	7071.55	X_2 = Total Flight Hours	1.50	0.54	2.80	.0063
M	57699.55	X_3 = Years in Aviation	26.03	93.12	0.28	.7805
N	87	$X_{3\text{-Missing}}$	3398.55	1987.032	1.71	.0910
		X_4 = Biological Sex	871.50	1589.29	0.55	.5850

Note. $F(5, 81) = 13.2998$, $p < .0001$.

Focusing on the first two subtables, note that the sample sizes are different. This is because the analysis where the missing data are discarded is performed using listwise deletion and hence the 30 unique cases with missing data are deleted. In the second subtable, though, the missing data are replaced (*imputed*) with their respective estimates and therefore all 87 cases are included. Further notice that the overall results are similar, but the R^2 values and standard errors (SE_js) are smaller for the analysis based on the imputation procedure. Another noteworthy difference is the result for X_4 = Biological Sex. With missing data discarded, $B_4 = -142.12$, which means that female ATPs' annual salary is, on average, \$142.12 *less* than male ATPs. However, when estimated values are substituted for the missing data, $B_4 = 1218.54$, which means that female ATPs' annual salary is, on average, \$1,218.54 *more* than their male counterparts. Both results are not significant. These differences are the result of sample size and bias on the part of the imputation procedure.

The regression imputation approach presented here is one of many different substitution strategies for handling missing data. Other substitution strategies include making an educated guess, imputing the mean (see below), and hot deck imputation. Regardless of the imputation strategy, all have a certain level of inherent bias. For more information about these and other imputation routines, including corresponding advantages and disadvantages, see Cohen et al. (2003, pp. 444–446) and Tabachnick and Fidell (2013, pp. 67–68).

Plug with Means. A simple and popular substitution routine used for missing data is mean substitution, which involves calculating the mean of an IV with missing data with respect to the IV's nonmissing values and then replacing the missing values with this mean. To mitigate some of the criticism of this approach, Cohen et al. (2003, pp. 442–444) recommended the following:

(a) For each X_j with missing data, create a *missing data variable*—for example, X_j-Missing—and enter 0 in this variable for those cases where data are present in X_j, and 1 where data are missing in X_j.

(b) Regress Y on X_j-Missing. If the result is not significant, then the data are missing randomly; otherwise, the data are missing systematically.

(c) Substitute the missing entries in X_j with the mean of the nonmissing data in X_j and include X_j-Missing in the final model if the data are missing systematically but not if the data are missing randomly.

As an example, consider X_1 = Years as PIC in the "Ch_5 Missing Data Example" data set. For this IV, we first create X_1-Missing and enter 1 for rows 7, 8, 25, 26, 35, 53, 70, 71, which have missing data in X_1, and 0 for all the other rows. When we regress Y on X_1-Missing, we get $R^2 = .013$, $F(1, 85) = 1.1494$, $p = .2867$. Because this result is not significant, we substitute the missing values in X_1 with the mean of the nonmissing values in X_1, namely, $M_1 = 8.77$, but do not include X_1-Missing in the final model. Repeating this procedure for X_2 and X_3:

- The relationship between Y and X_2-Missing is not significant, $R^2 = .013$, $F(1, 85) = 1.1407$, $p = .2885$. So, we replace X_2's missing data with the mean of the nonmissing data, $M = 3874.29$, but do not include X_2-Missing in the final model.
- The relationship between Y and X_3-Missing is significant, $R^2 = .046$, $F(1, 85) = 4.1163$, $p = .0456$. So, we replace X_3's missing data with the mean of the nonmissing data, $M = 16.54$, but we *keep* X_3-Missing in the final regression model.

Based on these analyses, it appears that the missing data in X_1 and X_2 are missing randomly, but the missing data in X_3 are missing systematically. This latter finding should be a flag to review X_3 = Years in Aviation and try to understand why 16 of 87 participants (nearly 20%) did not respond to this item. The last subtable of Table 5.3 contains a summary of the MRC analysis where Y is regressed on the four IVs with their "plugged means" along with X_3-Missing. Note the following:

- The overall model is significant, $R^2 = .451$, Adj. $R^2 = .417$, $RMSE = 7071.55$, $F(5, 81) = 13.3$, $p < .0001$, and these results are not substantially different from the previous two analyses.
- The sample size is $N = 87$ because all missing data are "plugged" with means.
- Most of the results are similar to those from the analysis based on the computer routine for missing data given in the second subtable.
- The inclusion of the X_3-Missing variable provides additional information with respect to the missing data for X_3 = Years in Aviation: $B_{3\text{-Missing}} = 3398.55$ indicates that the annual salary of participants who did not specify the number of years they have worked in the aviation profession is, on average, $3,398.55 *more* than the annual salary of participants who responded to this item. Although this finding is not significant, $t(81) = 1.71$, $p = .0910$, it nevertheless provides added-value to the analysis.

As noted earlier, mean substitution has its critics and readers are advised to consult Cohen et al. (2003, pp. 442–444 and 447–450) for additional information. For readers who opt to use this strategy for missing data, it is advisable *not* to include the missing data IV in the final model when either (a) a small proportion (e.g., < 5%) or a large proportion (e.g., > 30%) of cases have missing data, or (b) when many IVs have nonrandomly missing data. In the latter case, if multiple missing data IVs are used, it is possible that they could be highly correlated with each other, which would then introduce multicollinearity.

Overall Guidance. Given the various approaches to handling missing data, the obvious question that emerges is, "Which approach should I use?" and the answer to this question is, "It depends on the context of the study." For example, it might be appropriate to use listwise deletion or mean substitution for a one-time thesis or dissertation research study, but inappropriate for large-scale funded studies or those in which the data set will be used for secondary analyses. In such situations, the missing data routine provided by a highly regarded statistics program might be the better choice. It also might be beneficial to review the imputation routines various U.S. government organizations such as the U.S. Census Bureau or Centers for Disease Control and Prevention (CDC) use for handling missing data. The good news is that, as demonstrated with the Ch_5 Missing Data Example data set, regardless of the approach—with the possible exception of pairwise deletion and deleting IVs—the overall results and corresponding implications/conclusions should be similar.

Measurement Error

Measurement error is related to instrumentation reliability. According to Cohen et al. (2003), if an IV is measuring psychological constructs such as attitudes, locus of control, achievement, aptitude, personality traits, and cognition, then these measures are to have perfect reliability. This means there is zero measurement error, and the corresponding reliability coefficient is $r = 1.0$. If this is not the case, then it is possible that an IV's regression coefficient and standard error could be biased, which in turn could result in spurious confidence intervals and significance tests. Given that it is impractical to have zero measurement error, Cohen et al. (p. 123) recommend a reliability coefficient of $r \geq .70$. For our running example involving visual reaction times, the issue of measurement error is not applicable to the IVs.

Relevant IVs

When planning a study, researchers determine which factors to analyze by reviewing the literature, consulting theory, conducting a preliminary study, or relying on personal experience or expertise. These approaches, which are both acceptable and appropriate, do not mean that all the targeted IVs will be correctly specified and relevant to the study. This raises the question, "What happens if I include an IV in the final regression model that has little or no relationship with the DV?" Although this will not bias the regression coefficients of the other IVs, the presence of an irrelevant IV in the model could inflate standard errors, which in turn would result in wider confidence intervals and less precision, and a decrease in overall power. To check for the relevance of the IVs, we examine *leverage plots*, and one specific type of leverage plot is the added variable plot or AVP. The concept of an AVP is illustrated in Figure 5.7, which examines the relevancy of X_1 with respect to a regression model that contains Y, X_2, and X_3.

As shown in Figure 5.7(a), we first consider a regression model with Y, X_2, and X_3, and then partial out the effects of X_2 and X_3 from Y. What remains is the amount of Y variance not explained by X_2 and X_3. In Figure 5.7(b), we examine a separate regression model with X_1, X_2, and X_3, where X_1 is the dependent variable, and then partial out the effects of X_2 and X_3 from X_1. What remains is the amount of X_1 variance not explained by X_2 and X_3. In Figure 5.7(c), we examine the relationship between the unexplained Y variance with the unexplained X_1 variance both of which are free of any effects from X_2 and X_3. If this relationship is significant, then X_1 is relevant to the regression model.

Although some statistics software packages provide AVPs, it is instructional to demonstrate how they are constructed. To illustrate this, we will use the visual reaction times data set and focus on X_3 = Visual Acuity Angle. To determine if X_3 is relevant, we begin with a regression model that contains all the other IVs but not X_3. For our example, this will include X_1 = Age, X_2 = Corrected Vision, X_4 = Flight Hours, and X_5 = Number of Ratings:

$$\hat{Y} = B_1X_1 + B_2X_2 + B_4X_4 + B_5X_5 + B_0$$

We regress Y on these four IVs and save the residuals. These residuals represent that part of Y variance that is not accounted for by X_1, X_2, X_4, and X_5. In other words, it's that part of the DV that has not been claimed by these other IVs. This is the error variance of Y (e_Y), which is conceptually illustrated in Figure 5.7(a).

We next regress the targeted IV, in this case, X_3, which becomes the DV in this situation, on all the other IVs. This results in the following model:

$$\hat{X}_3 = B_1X_1 + B_2X_2 + B_4X_4 + B_5X_5 + B_0$$

As we did for the previous regression analysis, we save the residuals from the current model. These residuals represent that part of X_3's variance that is not accounted

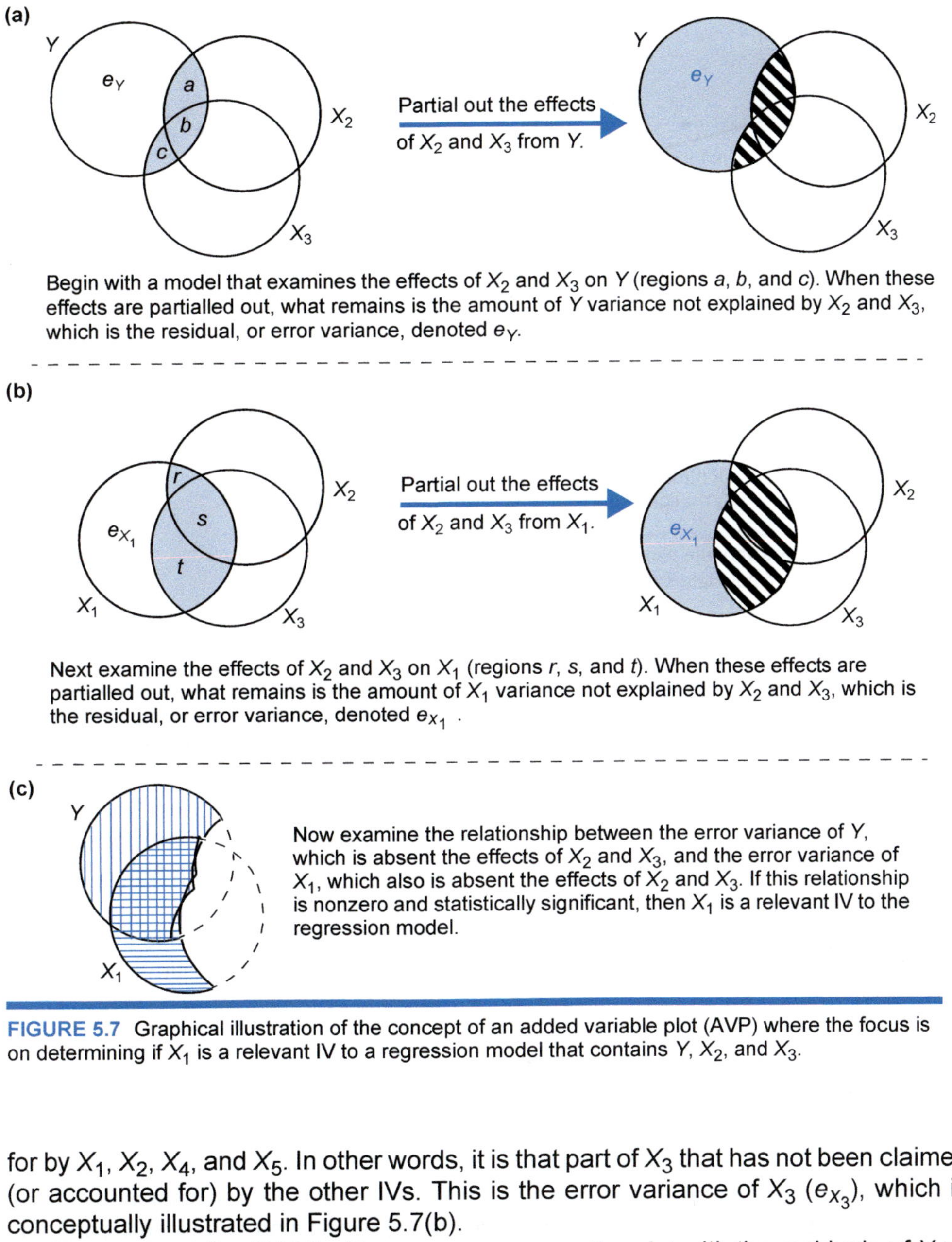

FIGURE 5.7 Graphical illustration of the concept of an added variable plot (AVP) where the focus is on determining if X_1 is a relevant IV to a regression model that contains Y, X_2, and X_3.

for by X_1, X_2, X_4, and X_5. In other words, it is that part of X_3 that has not been claimed (or accounted for) by the other IVs. This is the error variance of X_3 (e_{X_3}), which is conceptually illustrated in Figure 5.7(b).

To construct the AVP for X_3, we prepare a scatter plot with the residuals of Y on the vertical axis and the residuals of X_3 on the horizontal axis. This is shown in Figure 5.8. We also insert the regression line that results from this analysis along with a kernel smoother line. Observe from Figure 5.8 that the regression line shows there is a negative linear relationship, and the kernel smoother line confirms this relationship because it follows the trend of the data: it does not stray that much from the regression line. Therefore, the AVP in Figure 5.8 illustrates there is a linear relationship

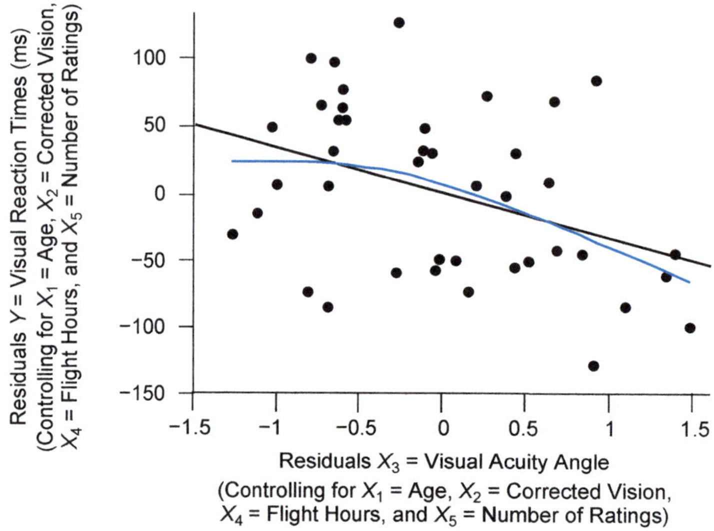

FIGURE 5.8 Added variable plot for X_3 = Visual Acuity Angle with respect to the visual reaction times data set given in the file "Ch_5 Visual Reaction Times Data."

between Y = Visual Reaction Time (ms) and X_3 = Visual Acuity Angle when controlling for the effects of X_1 = Age, X_2 = Corrected Vision, X_4 = Flight Hours, and X_5 = Number of Ratings. Because the slope of the regression line is not zero, X_3 is a relevant IV: It has a nonzero relationship with Y when both Y and X_1 have been freed of any influences from the other IVs. This relationship also is significant, $R^2 = .15$, $F(1, 39) = 7.19$, $p = .0107$. and hence X_3 will remain in the final regression model. Figure 5.9 contains AVPs for the other IVs of the visual reaction times data set. A brief discussion of each AVP follows.

- Figure 5.9(a) contains the AVP for X_1 = Age. The regression line is positive, and the kernel smoother line follows the regression line closely confirming the linear relationship, which is significant, $R^2 = .34$, $F(1, 39) = 20.36$, $p < .0001$. Therefore, X_1 will remain in the final regression model.
- Figure 5.9(b) contains the AVP for X_2 = Corrected Vision. The regression line is negative, and the kernel smoother line follows the regression line closely confirming the linear relationship, which is significant, $R^2 = .24$, $F(1, 39) = 12.20$, $p = .0012$. Therefore, X_2 will remain in the final regression model.
- Figure 5.9(c) contains the AVP for X_4 = Flight Hours. The regression line is positive, and the kernel smoother line follows the regression line closely confirming the linear relationship, which is significant, $R^2 = .33$, $F(1, 39) = 19.64$, $p < .0001$. Therefore, X_4 will remain in the final regression model.
- Figure 5.9(d) contains the AVP for X_5 = Number of Ratings. The regression line is slightly positive, and the kernel smoother line follows the regression line closely confirming the linear relationship, which is not significant, $R^2 = .008$, $F(1, 39) = 0.33$, $p = .5690$. Therefore, X_5 will *not* remain in the final regression model.

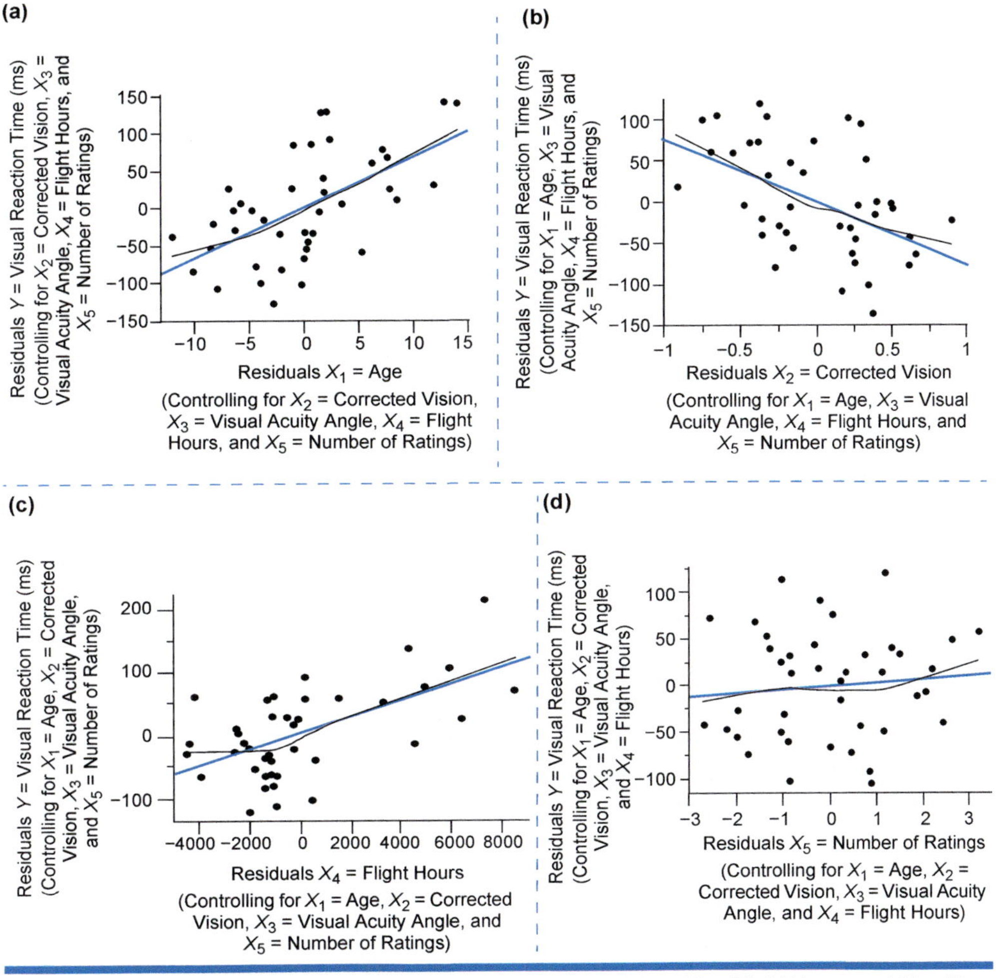

FIGURE 5.9 Added variable plots for X_1 = Age, X_2 = Corrected Vision, X_4 = Flight Hours, and X_5 = Number of Ratings with respect to the visual reaction times data set given in the file "Ch_5 Visual Reaction Times Data."

As a result of these leverage plots, we will remove X_5 = Number of Ratings from the final model. Because the data set was modified, we once again must examine this most recently modified data set for compliance with regression assumptions. We will not do so here, but instead leave it to the reader to confirm. (*Hint:* The new data set is indeed compliant.)

Summary of Regression Diagnostics Applied to the Running Example

With respect to our running example, we confirmed in the previous section that the original visual reaction times data set is compliant with the underlying regression assumptions. In the current section, though, we discovered that this data set had four outliers, which were rare cases, but we deleted them because they were influential and doing this resulted in a better model fit (see Table 5.2). This reduced the sample size from N = 45 to N = 41. Because the initial data set was modified as a

TABLE 5.4 *Summary of MRC Analysis Results for the Visual Reaction Times Data Post Preliminary Data Screening*

Model Summary

R^2	.540
Adjusted R^2	.489
Standard Error of the Estimate	61.51
Mean of Response	269.34
Sample Size	41

ANOVA

Source	*SS*	*df*	*MS*	*F*	*p*
Regression	159885.31	4	39971.3	10.5640	< .0001
Residual (Error)	136213.91	36	3783.7		
Total	269099.22	40			

Coefficients (Parameter Estimates)

Term	*B*	*SE*	β	*t*(36)	*p*	95% CI	*VIF*
Constant	67.7637	56.4793		1.20	.2381	[–46.78, 182.31]	
X_1 = Age	6.8822	1.5603	0.5542	4.41	< .0001	[3.72, 10.05]	1.24
X_2 = Corrected Vision	–71.8342	21.7906	–0.4225	–3.30	.0022	[–116.03, –27.64]	1.29
X_3 = Visual Acuity Angle	–33.4409	13.0751	–0.2915	-2.56	.0149	[–59.96, –6.92]	1.02
X_4 = Flight Hours	0.0133	0.0029	0.5305	4.55	< .0001	[0.0074, 0.192]	1.06

Note. N = 41.

Summary of the Correlation Coefficients and Their Squares

Factor	Zero-Order		Semipartial		Full Partial	
	r_{Yi}	r^2_{Yi}	sr_i	sr^2_i	pr_i	pr^2_i
X_1 = Age	.3804	.1447	.4986	.2486	.5923	.3508
X_2 = Corrected Vision	–.0845	.0071	–.3727	.1389	–.4816	.2319
X_3 = Visual Acuity Angle	–.2157	.0465	–.2891	.0836	–.3922	.1538
X_4 = Flight Hours	.4346	.1889	.5149	.2651	.6046	.3656

result of deleting the outliers, we rechecked the assumptions with respect to this modified data set and found that it is compliant. We then discovered from the added variable plots that X_5 = Number of Ratings is not a correctly specified or relevant IV and removed it from the final model. We also confirmed that removing this IV from the data set did not affect the data set's compliance with the regression assumptions. As a result, the final data set that will be examined via an MRC analysis consists of N = 41 cases and includes four IVs: X_1, X_2, X_3, and X_4. Note that the results, which are provided in Table 5.4, are not that much different from the last row of Table 5.2, which is in the absence of outliers but still has X_5 in the model. This is further confirmation that X_5 has little relationship with Y.

5.4 Assessing a Regression Model's Reliability: Cross-Validation

Up to this point in our discussion, we have focused on the overall R^2, adjusted R^2, and *RMSE* as indicators of how well a regression model fits the sample data from which the regression model was derived. Although these metrics are appropriate and informative, they are *relative to the sample* and do not provide us with any information about the *generalizability* of the model. Given that one of the primary uses of multiple regression is for prediction, it is helpful to know how well a regression model might work if we were to apply it to the parent population or to other samples from the same population. If the predicted scores are very "close" to the true scores in the population (or to those observed in a new sample), then the regression model would be a strong predictor and considered to be a reliable model; otherwise, the model would not be a strong predictor and it would be considered unreliable.

To measure a regression model's predictive performance, and hence its reliability, we conduct a cross-validation, which can be performed via three different approaches. We can: (a) conduct a *cross-validation study*, (b) use a formula to manually calculate the *cross-validated* R^2, or (c) use a built-in routine of a statistics program to perform a *quasi-cross-validation study*. These latter two approaches are more commonly used because they do not require a second sample. A brief discussion of each approach follows. (*Note:* When assessing the predictive performance of a regression model, we presume that the model is compliant with the regression assumptions and was examined with respect to regression diagnostics.)

Cross-Validation Study

A cross-validation study involves collecting two independent samples from the same population. The first sample is called the screening sample, and it is the sample we use to derive the regression model. The second sample is called the calibration sample, and it is the sample we use to test the regression model. The terms "screening sample" and "calibration sample" are from Lord and Novick (1968, p. 285), but for convenience and ease of notation, we will denote the screening sample, S_1, and the calibration sample, S_2. Because the regression model of one sample is being applied to a second sample, a cross-validation study is sometime referred to as a *follow-up study*.

To demonstrate a cross-validation study, we provide two data sets, "Ch_5 S1=Screening Sample Data" and "Ch_5 S2=Calibration Sample Data," which were independently acquired from a target population of CFIs. The predictors are X_1 = Number of Dual Hours, X_2 = Complacency Scores (higher scores equate to higher levels of complacency), and X_3 = Biological Sex (coded 1 = Female and 0 = Male). The dependent variable is Y = Career Satisfaction Scores (higher scores equate to higher career satisfaction). We now follow the procedure as outlined by Kleinbaum et al. (1998, pp. 402–403).

1. Conduct a regression analysis on the screening sample, S_1. This yields the regression equation (rounded to two decimal places):

$$\hat{Y}_1 = 0.01X_1 + 1.35X_2 - 3.88X_3 + 0.48 \text{ with } R^2(1) = .734$$

   where $\hat{Y}_1$ indicates that the regression equation is from Sample 1, and $R^2(1)$ denotes that the overall R^2 is from Sample 1.

2. Use the regression equation from the screening sample, S_1, to predict the *Y* values of the calibration sample, S_2. We denote the predicted *Y* values as $\hat{Y}^*_{2i}$, where the subscript 2 denotes Sample 2, *i* reflects the case number of Sample 2, and * indicates that the predicted values are from Sample 1's regression equation. For example, in the Ch_5 S2=Calibration Sample Data file, the regression weights for Case 1 are: X_1 = 2900, X_2 = 19, and X_3 = 0. When we substitute these weights into the regression equation obtained from S_1 we get:

$$\hat{Y}^*_{21} = 0.01(2900) + 1.35(19) - 3.88(0) + 0.48$$
$$= 29 + 25.65 - 0 + 0.48$$
$$= 55.13$$

   Therefore, the predicted career satisfaction score for Sample 2 / Case 1, which is acquired by using the regression equation from Sample 1, is 55.13. Table 5.5 contains a summary of the $\hat{Y}^*_{2i}$ values for S_2.

3. Calculate the Pearson *r* between Sample 2's observed and predicted *Y* values, denoted $r(Y_2, \hat{Y}^*_2)$, and then square this correlation to get the corresponding coefficient of determination. For our example:

$$r(Y_2, \hat{Y}^*_2) = .8232$$

$$\text{and } r^2(Y_2, \hat{Y}^*_2) = (.8232)^2 = .6777$$

   Recognize that this r^2 value indicates that the three IVs, based on Sample 1's regression equation, are predicted to explain 67.77% of the *Y* variance in Sample 2. As a result, we denote this $R^2_*(2)$, where * indicates that R^2 was predicted using Sample 1's regression equation, and the parenthetical (2) indicates that R^2 is for Sample 2. This metric is called the cross-validation correlation (Kleinbaum et al., 1998, p. 402).

4. Calculate the difference between $R^2(1)$ and $R^2_*(2)$.

$$R^2(1) - R^2_*(2)$$
$$= .734 - .6777$$
$$= .0563$$

   This difference, called the shrinkage on cross-validation (Kleinbaum et al., 1998, p. 403), is a measure of Sample 1's regression model's predictive performance. According to Kleinbaum et al. (1998), if the shrinkage is less than .10, then this is an indication of a reliable model. Kleinbaum et al. also indicated that in such cases it would be reasonable to combine the two samples (i.e., pool the data) and use the regression equation for the combined samples for future predictions.

As a result of this demonstration, the corresponding shrinkage of .0563 indicates that the regression equation from Sample 1 is a strong predictor and therefore a reliable model. Furthermore, although the two samples were collected independently from the same population, it would be reasonable to combine the two samples if a larger sample size were needed to increase statistical power.

TABLE 5.5 *Sample 2's (S_2) Data with Predicted Y Values Acquired from Sample 1's (S_1) Regression Equation* [a]

Case #	Y	X_1	X_2	X_3	$\hat{Y}^*_{2i}$
1	56	2900	19	0	55.13
2	52	1600	18	0	40.78
3	40	2400	13	0	42.03
4	39	1400	8	0	25.28
5	23	1200	5	0	19.23
6	36	2200	10	0	35.98
7	16	1500	5	0	22.23
8	53	3200	15	0	52.73
9	62	2100	21	0	49.83
10	45	2100	16	0	43.08
11	41	2700	22	0	57.18
12	46	2600	19	1	48.25
13	52	3000	23	1	57.65
14	51	2900	21	1	53.95
15	57	3400	18	1	54.90
16	62	2500	26	1	56.70
17	38	1600	10	1	26.10

Note. [a] $\hat{Y} = 0.01X_1 + 1.35X_2 - 3.88X_3 + 0.48$, where Y = Career Satisfaction Scores, X_1 = Number of Dual Hours, X_2 = Complacency Scores, and X_3 = Biological Sex (coded 1 = Female and 0 = Male).

Cross-Validated R^2

In most realistic research settings, collecting two independent samples can be expensive with respect to time and financial resources, and in many cases is not always practical. An alternative to conducting a cross-validation study, is to calculate the cross-validated R^2, which provides an estimate of the amount of variance a sample's regression model will explain in the population or to a different sample selected from the same population. Denoted $\hat{R}^2$ (Rozeboom, 1979, as cited in Cohen et al., 2003, p. 98), the "hat" symbol indicates that R^2 is a predicted value, which is similar to the way we use $\hat{Y}$. The formula for the cross-validated R^2 is:

$$\hat{R}^2 = 1 - \left[(1 - R^2)\frac{(N + k)}{(N - k)}\right]$$

The cross-validated R^2 answers the question, "If I were to apply the *sample* regression weights to the population, or to another sample from the same population, what proportion of the Y variance would my thus-predicted Y values explain?" (Cohen et al., 2003, p. 98). Applying the cross-validated R^2 formula to the visual reaction times data where $N = 41$, $k = 4$, and $R^2_{Y \cdot 1234} = .54$:

$$\begin{aligned} \hat{R}^2 &= 1 - \left[(1 - .54)\frac{(41 + 4)}{(41 - 4)}\right] = 1 - \left[(1 - .54)\frac{(45)}{(37)}\right] \\ &= 1 - [(.46)(1.22)] \\ &= 1 - (.5612) \\ &= .4388 \end{aligned}$$

Thus, if we were to apply the regression model from the visual reaction times sample data to the population, or to a new sample from the same population, the predicted amount of variance in visual reaction times the four IVs will collectively explain is estimated to be $\hat{R}^2$ = .4388, or approximately 44%. This is a relatively large cross-validated R^2. Furthermore, because it is not that much different from S_1's R^2 = .54, this suggests that the regression model is reliable.

Quasi-Cross-Validation Study

Instead of conducting a cross-validation study or manually calculating the cross-validated R^2, we can use our statistics program to perform a quasi-cross-validation study, which splits a study's sample into two subsamples. The first subsample, called the training sample, which we will denote S_1, is used to derive the regression model. The second subsample, called the test sample, which we will denote S_2, is used to test the regression model. Thus, the regression model is being "trained" via the training sample, and the generalizability of this model is then being tested via the test sample.

The simplest way to do this is called the Holdout method and involves invoking an X–Y split of the initial data set, with X% being used for training and Y% being used for testing. For example, for our running example involving visual reaction times (N = 41), if we opt for a 70–30 split, which is common, we randomly select 70% of the cases for the training sample (n_{Training} = 29) and 30% for the test sample (n_{Test} = 12). We then use the regression equation from the Training sample to predict the Y values of the Test sample relative to the IVs. To measure the predictive performance of the regression equation, we use the mean absolute error (MAE) of the Test sample. MAE is the average absolute error between the observed Y values from the Test sample and the corresponding predicted Y values acquired from the Training sample's regression equation. As the differences between the predicted and actual scores become smaller, MAE also becomes smaller. Applying this to the visual reaction times data, the MAE is 1641 ms, which indicates that when applying the regression model to the population or to a new data set from the same population, the predicted visual reaction times will have an average error of 1,641 milliseconds. (Readers will get a different MAE because of the random selection process.)

An extension of the Holdout method, and a more thorough strategy, is called K-fold cross-validation, which generates K separate Holdout models or "folds." In each fold, the initial data set is split into K subsamples of approximately the same size with $1/K$ cases designated as the Test sample and $(K - 1)/K$ cases designated as the Training sample. For example, if K = 10, then 10 folds are generated and in each fold there are 10 equal subsamples: 1/10 = 10% of the cases serve as the Test sample and the remaining 9/10 = 90% of the cases collectively serve as the Training sample. This is illustrated in Figure 5.10. Thus, if an initial data set has a sample size of N = 150 and K = 10, then n = 15 cases are randomly selected for the Test sample in each fold, and the remaining n = 135 cases are used for the Training sample in each fold.

Observe from Figure 5.10 how the K-fold cross validation approach is an improvement to the Holdout approach. Instead of working with one overall model that splits the initial data set into Training and Tests samples, we have 10 separate submodels (or more generally, K folds), and each fold has a different randomly selected Test sample and Training sample.

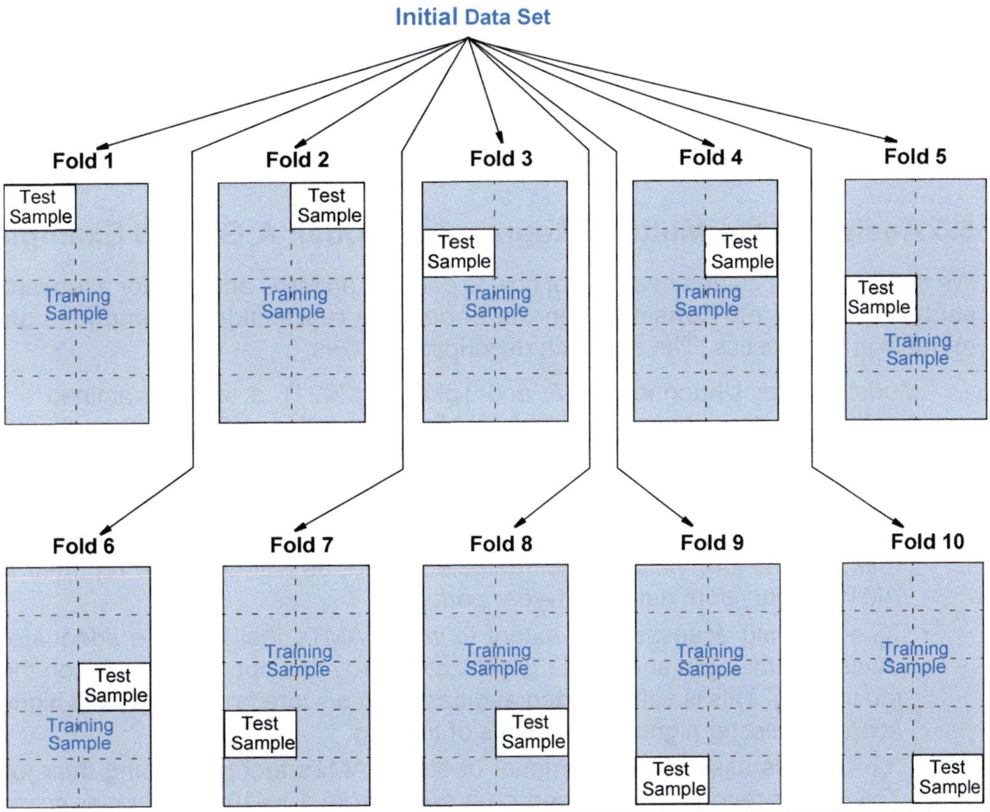

FIGURE 5.10 Graphical illustration of the K-fold cross-validation approach for $K = 10$.

To measure the predictive performance of each fold, the statistics program basically repeats the Holdout method K times, and then averages each fold's error to get the error for the overall model. The program randomly selects the Test and Training samples for each fold and ensures that every score is included in a Test sample exactly one time and is included in a Training sample $K - 1$ times.

As part of its output for K-fold cross-validation, our statistics program also generates a K-fold R^2, which is similar to the cross-validated R^2 (i.e., $\hat{R}^2$). Both are estimates of the explained Y variance in the population that is being predicted by the sample's regression model. For example, for the visual reaction times data where the overall R^2 is .54, when $K = 5$, K-fold $R^2 = .3261$, and when $K = 10$, K-fold $R^2 = .3970$. Thus, based on the K-fold cross-validation approach, for $K = 5$ folds, the estimated amount of explained variance in the population is 32.61%, and for $K = 10$ folds, it is 39.70%.

When conducting a K-fold cross-validation, although $K = 5$ and $K = 10$ are commonly used, there are no general guidelines governing the number of folds to use. Keep in mind, though, that small K values generally lead to lower variance and higher error, whereas the opposite occurs for larger K values. Given these relationships, a common strategy called leave-one-out cross-validation is to set K equal to the sample size, N. For this strategy, the Training sample consists of all the observations except one ($N - 1$), and the Test sample consists of the score that was left out. The regression model derived from the Training sample is then used to predict the observation that was left out. This process is repeated N times where a different observation is left out for each iteration. In the visual reaction times cross-validation analysis, when we let $K = N = 41$, K-fold $R^2 = .4060$. (*Note:* This strategy is equivalent to the PRESS statistic, which stands for Predicted Residual Sum of Squares.)

5.5 Analyzing the Multiple Regression Model: A Guided Example

We now present a guided example of how to conduct an MRC analysis for a research study that takes into consideration the underlying regression assumptions and regression diagnostics. The research description follows.

> Modeled after Uhuegho (2017) and Tokarski (2021), a study examined aviation maintenance technicians' (AMTs') perceptions of the safety climate at their aviation maintenance, repair, and overhaul (MRO) organization. The factors are:

- X_1 = Maintenance Errors represent the number of maintenance errors AMTs make on the job, including self-detected errors as well as those reported by AMTs' supervisors during a 1-year period.
- X_2 = Training assesses the extent to which AMTs feel they are adequately trained for their job and if their organization provides opportunities for on-the-job training. This is self-reported and based on a score from 1 to 5, with higher scores reflecting higher perceptions of training.
- X_3 = Stress assesses the amount of stress AMTs feel from doing their job. This is self-reported and based on a range from 1 to 10, with higher scores reflecting higher perceived levels of on-the-job stress.
- X_4 = Years employed as an AMT.

The dependent variable, Y = Safety Climate, is operationally defined as "the individual's perceptions of the organization's policies, procedures, and rewards relevant to safety in the organization" (Fogarty, 2005, p. 75) and measured using Fogarty's (2005) 48-item Maintenance Environment Survey (MES). The MES is scored on a traditional Likert response scale ranging from 1 = Strongly Disagree to 5 = Strongly Agree. Thus, scores could range from 48 to 240, with higher scores reflecting a more positive perception of an MRO's safety climate. The data set is provided in the file "Ch_5 Guided Example Data," and the structure of our presentation consists of six parts: Research Questions and Hypotheses, Sample Size Planning, Preliminary Data Screening, Data Analysis, Interpretation of Results, and Post Hoc Considerations.

Research Questions/Hypotheses

There are three research questions and corresponding hypotheses. The first RQ is for the overall R^2, the second RQ is with respect to the unique contribution each IV makes in explaining Y variance (sr_i^2), and the third RQ is with respect to the effect the regression coefficients (B_i) have on Y.

RQ 1. What proportion of variance in safety climate scores is explained collectively by the four targeted IVs?

Hyp. 1. The focus is on the collective contribution of the IVs to *Y* variance.

- H_0: $\rho^2_{Y \cdot 1234} = 0$ (The four IVs collectively do not explain a significant proportion of *Y* variance in the population.)
- H_1: $\rho^2_{Y \cdot 1234} > 0$ (The four IVs collectively explain a significant proportion of *Y* variance in the population.)

RQ 2. How much additional knowledge does each IV make in explaining the variance in safety climate scores when entered into an analysis that contains the other three IVs? Because there are four IVs, the corresponding hypothesis has four parts.

Hyp. 2a. The focus is on X_1's unique contribution to explaining *Y* variance (sr_1^2) in the population when X_1 enters the analysis in the presence of X_2, X_3, and X_4.

- H_0: $\rho^2_{Y \cdot 1234} - \rho^2_{Y \cdot 234} = 0$ (X_1's unique contribution is not significant.)
- H_1: $\rho^2_{Y \cdot 1234} - \rho^2_{Y \cdot 234} > 0$ (X_1's unique contribution is significant.)

Hyp. 2b. The focus is on X_2's unique contribution to explaining *Y* variance (sr_2^2) in the population when X_2 enters the analysis in the presence of X_1, X_3, and X_4.

- H_0: $\rho^2_{Y \cdot 1234} - \rho^2_{Y \cdot 134} = 0$ (X_2's unique contribution is not significant.)
- H_1: $\rho^2_{Y \cdot 1234} - \rho^2_{Y \cdot 134} > 0$ (X_2's unique contribution is significant.)

Hyp. 2c. The focus is on X_3's unique contribution to explaining *Y* variance (sr_3^2) in the population when X_3 enters the analysis in the presence of X_1, X_2, and X_4.

- H_0: $\rho^2_{Y \cdot 1234} - \rho^2_{Y \cdot 124} = 0$ (X_3's unique contribution is not significant.)
- H_1: $\rho^2_{Y \cdot 1234} - \rho^2_{Y \cdot 124} > 0$ (X_3's unique contribution is significant.)

Hyp. 2d. The focus is on X_4's unique contribution to explaining *Y* variance (sr_4^2) in the population when X_4 enters the analysis in the presence of X_1, X_2, and X_3.

- H_0: $\rho^2_{Y \cdot 1234} - \rho^2_{Y \cdot 123} = 0$ (X_4's unique contribution is not significant.)
- H_1: $\rho^2_{Y \cdot 1234} - \rho^2_{Y \cdot 123} > 0$ (X_4's unique contribution is significant.)

RQ 3. When holding all other variables constant, what are the respective effects (i.e., slopes) of each IV on safety climate scores? Because there are four IVs, the corresponding hypothesis has four parts.

Hyp. 3a. The focus is on X_1's effect on *Y* when holding all other IVs constant.

- H_0: $\beta_1 = 0$ (X_1 does not have a significant effect on *Y*.)
- H_1: $\beta_1 \neq 0$ (X_1 has a significant nonzero effect on *Y*.)

Hyp. 3b. The focus is on X_2's effect on *Y* when holding all other IVs constant.

- H_0: $\beta_2 = 0$ (X_2 does not have a significant effect on *Y*.)
- H_1: $\beta_2 \neq 0$ (X_2 has a significant nonzero effect on *Y*.)

Hyp. 3c. The focus is on X_3's effect on *Y* when holding all other IVs constant.

- H_0: $\beta_3 = 0$ (X_3 does not have a significant effect on *Y*.)
- H_1: $\beta_3 \neq 0$ (X_3 has a significant nonzero effect on *Y*.)

Hyp. 3d. The focus is on X_4's effect on *Y* when holding all other IVs constant.

- H_0: $\beta_4 = 0$ (X_4 does not have a significant effect on *Y*.)
- H_1: $\beta_4 \neq 0$ (X_4 has a significant nonzero effect on *Y*.)

Sample Size Planning

To determine the minimum sample size needed, we use Soper (2025a) for the overall R^2, Soper (2025b) for sr_i^2, and Cohen et al.'s (2003) definitions for a medium effect for R^2 (p. 93) and a medium effect for sr_i^2 (p. 95).

- Overall R^2. Given *ES* of $f^2 = 0.13$, power = .8, IVs = 4, and $\alpha = .05$: $n^* = 96$.
- sr_i^2. Given *ES* of $f^2 = 0.15$, power = .8, $k_{\text{Set A}} = 3$, $k_{\text{Set B}} = 1$, and $\alpha = .05$: $n^* = 57$.

Therefore, our sample should consist of at least $N = 96$ cases to have an 80% probability of finding a significant medium effect for the overall R^2 and each sr_i^2. The given data set has $N = 90$, which will reduce power from .8 to approximately .77. This reduction in power can be determined by calculating R^2 based on an effect size of $f^2 = 0.13$, and then using Soper (2025c). Because the overall power is still acceptable, we continue with the analysis.

Preliminary Data Screening

Preliminary data screening involves confirming that the data set is compliant with the underlying assumptions of multiple regression and performing regression diagnostics. Recall from Section 5.4 that we checked for compliance first followed by diagnostics. Because we made changes to the model as a result of the diagnostics we performed, we then had to recheck the modified model for compliance again. As a result, in this example we will do diagnostics first, get the model of best fit, and then confirm it is compliant with the assumptions.

Outlier Analysis. Examining the initial data set for outliers via Jackknife distances yielded four outliers: Cases 8, 34, 59, and 89. A brief analysis of each case follows.

- Case 8 has the second lowest safety climate score ($Y = 130$) and the second lowest stress level ($X_3 = 3$).
- Case 34 has the highest safety climate score ($Y = 220$) and the highest number of maintenance errors ($X_1 = 22$).
- Case 59 has the fourth lowest safety climate score ($Y = 113$), fewest number of maintenance errors ($X_1 = 5$), and the second highest stress level ($X_3 = 9$).
- Case 89 has the lowest safety climate score ($Y = 109$), the fewest number of maintenance errors ($X_1 = 5$), and the lowest stress level ($X_3 = 2$).

As summarized in Table 5.6, with outliers present, $R^2 = .613$, adjusted $R^2 = .595$, *RMSE* = 11.45, and $F(4, 85) = 33.7138$, $p < .0001$. With outliers absent, $R^2 = .547$, adjusted $R^2 = .524$, *RMSE* = 10.74, and $F(4, 81) = 24.4271$, $p < .0001$. These results suggest that the outliers are inflating significance. Unfortunately, we do not know if they are contaminants or rare cases. For example, Case 89 could be suspect. This AMT reported the lowest perception of the MRO's safety climate, but also reported the fewest number of maintenance errors and the lowest stress level. On one hand, this is plausible, but on the other hand, an argument could be made that challenges these data. As a result of this uncertainty, we will delete the outliers from the data set. In doing so, we will lose .066 of explained *Y* variance, but we do get a slightly better *RMSE*. When we examine this new data set ($N = 86$) for outliers via Jackknife distances, there are no new outliers.

TABLE 5.6 *Summary of MRC Analysis Results for Chapter 5's Guided Example Data (with vs. without Outliers)*

Model Summary

	With Outliers	Without Outliers
R^2	.613	.547
Adjusted R^2	.595	.524
Standard Error of the Estimate	11.45	10.74
Mean of Response	146.43	146.59
Sample Size	90	86

ANOVA (with Outliers)

Source	*SS*	*df*	*MS*	*F*	*p*
Regression	17683.856	4	4420.96	33.7138	< .0001
Residual (Error)	11146.244	85	131.13		
Total	28830.100	89			

Coefficients (with Outliers)

Term	*B*	*SE*	*t*(85)	*p*	95% CI	*VIF*
Constant	26.135	11.495	2.27	.0255	[3.28, 48.99]	
X_1 = Maintenance Errors	4.976	0.535	9.29	< .0001	[3.91, 6.04]	1.30
X_2 = Training	13.095	1.774	7.38	< .0001	[9.57, 16.62]	1.21
X_3 = Stress	2.046	0.829	2.47	.0156	[0.40, 3.69]	1.27
X_4 = Years as an AMT	0.011	0.118	0.10	.9224	[–0.22, 0.25]	1.06

Note. *N* = 90.

ANOVA (without Outliers)

Source	*SS*	*df*	*MS*	*F*	*p*
Regression	11277.641	4	2819.41	24.4271	< .0001
Residual (Error)	9349.115	81	115.42		
Total	20626.756	85			

Coefficients (without Outliers)

Term	*B*	*SE*	*t*(81)	*p*	95% CI	*VIF*
Constant	38.574	11.479	3.36	.0012	[15.74, 61.41]	
X_1 = Maintenance Errors	3.985	0.573	6.95	< .0001	[2.84, 5.12]	1.17
X_2 = Training	13.333	1.680	7.94	< .0001	[9.99, 16.68]	1.16
X_3 = Stress	1.587	0.893	1.78	.0794	[–0.19, 3.36]	1.27
X_4 = Years as an AMT	0.085	0.117	0.73	.4657	[–0.15, 0.32]	1.10

Note. *N* = 86.

Multicollinearity. Examining the revised data set without outliers for multicollinearity, we focus on the *VIFs* for each IV. As summarized in the Coefficients (without Outliers) subtable in Table 5.6, the *VIF*s are all less than 1.27 and therefore multicollinearity is not a concern.

Missing Data. The revised data set without outliers does not have any missing data.

Measurement Error. All four IVs reflect factually-based self-reported data and therefore measurement error is not a concern. (*Note:* The reader will recognize that the dependent variable is an attitude questionnaire and therefore its reliability coefficient should be reported. However, because this instrument is measuring the DV and not an IV, it does not get reported here as part of regression diagnostics.)

Relevant IVs. To determine if the IVs are relevant (or correctly specified) to the current model, we examine each IV's added variable plot. As presented in Figure 5.11(a) and Figure 5.11(b), X_1 = Maintenance Errors and X_2 = Training both have a nonzero linear relationship with Y = Safety Climate Scores and these respective relationships are significant. Therefore, both X_1 and X_2 are relevant and will remain in the final model.

With respect to X_3 = Stress and X_4 = Years as AMT, each has a nonzero linear relationship with Y as depicted in Figure 5.11(c) and Figure 5.11(d), respectively, but neither relationship is statistically significant with respect to the preset alpha level of $\alpha = .05$. Furthermore, X_3 accounts for nearly 4% of the variance in Y whereas X_4

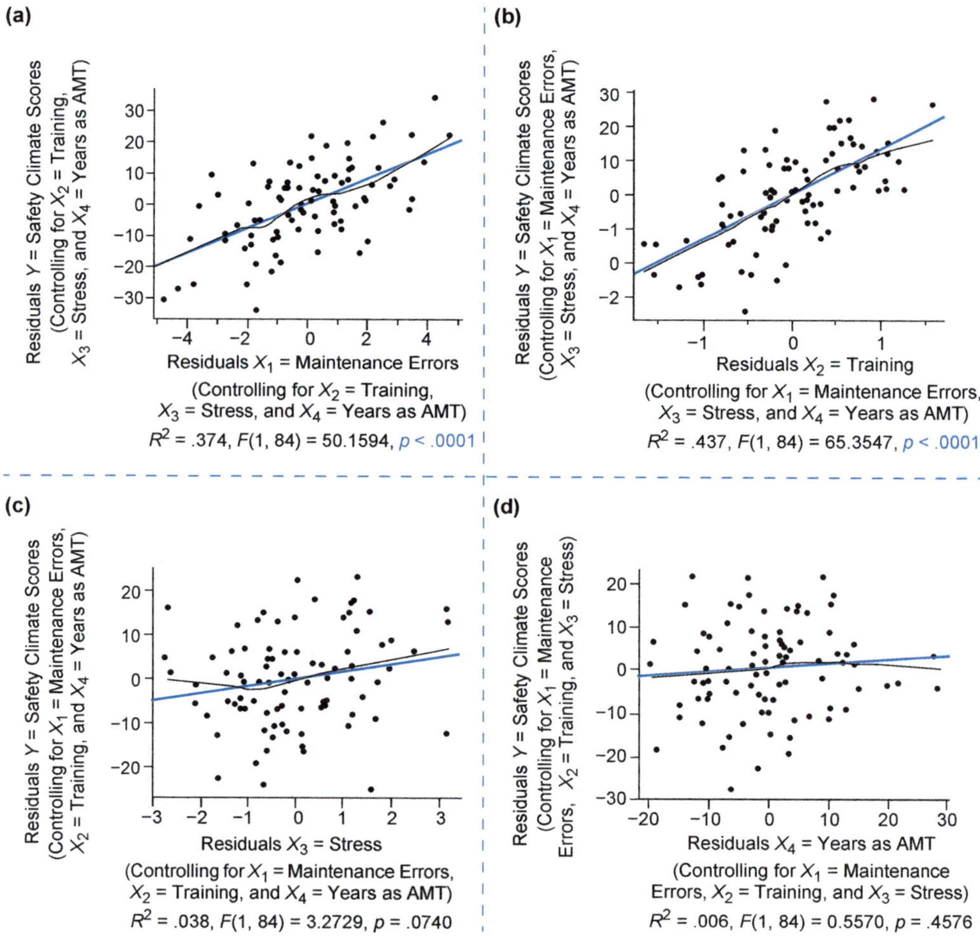

FIGURE 5.11 Added variable plots for each IV in Chapter 5's Guided Example.

accounts for 0.6%. Based on these results, there is no doubt that X_4 is not relevant and will not be included in the final model. The decision to keep or not keep X_3, though, is a bit problematic. Because stress on the job is generally accepted as a factor that could influence employees' perceptions of their work environment, we will keep X_3 in the final model. (This is a judgment call and is considered a delimitation.)

As a result of regression diagnostics, our final model was modified from its original state as follows: Instead of $N = 90$, the final model consists of $N = 86$ cases, and instead of the four targeted IVs, the final model has three IVs: X_1 = Maintenance Errors, X_2 = Training, and X_3 = Stress. We now check to see if this model is compliant with the assumptions of multiple regression. The corresponding residual plots are provided in Figure 5.12.

Linearity. As shown in Figure 5.12(a), the residuals appear to be randomly scattered around the 0-line without exhibiting any pronounced or systematic pattern and the corresponding kernel smoother line confirms this trend. As a result, the final model is compliant with the linearity assumption.

Equal Variances (Homoscedasticity of the Residuals). Figure 5.12(a) also confirms the equal variances assumption because the residuals appear to be approximately equally scattered above and below the 0-line without any pronounced pattern, which is confirmed by the kernel smoother line. Therefore, the final model is compliant with the equal variances assumption.

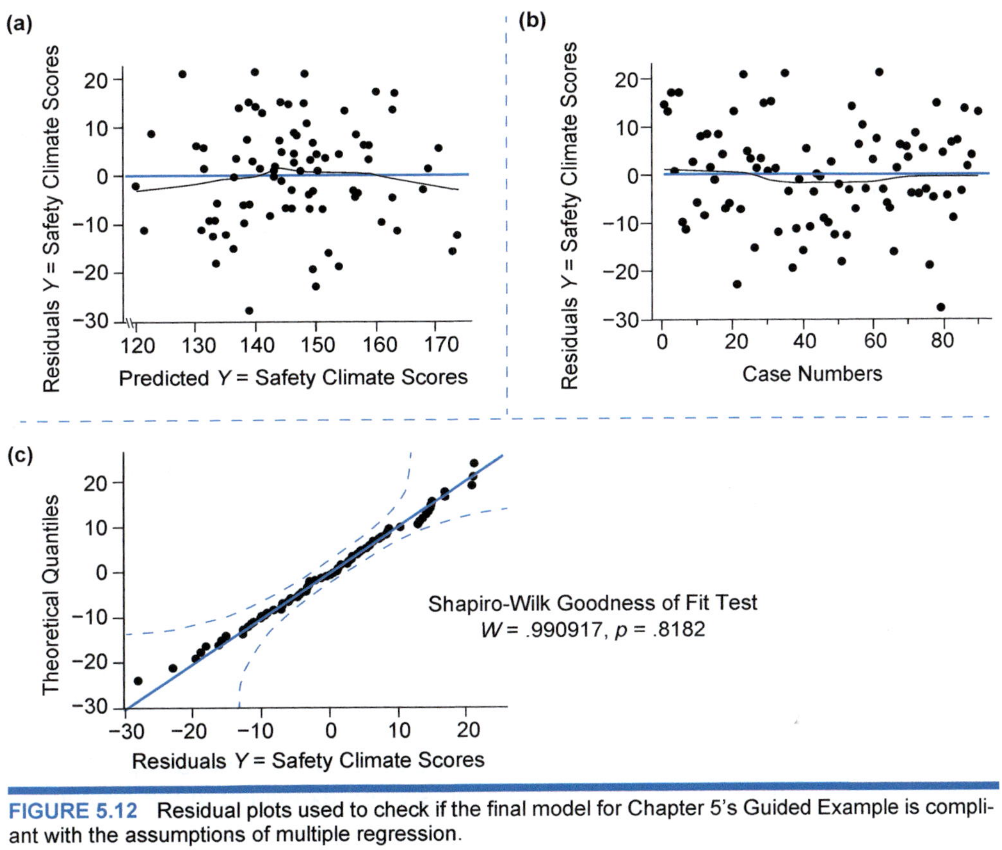

FIGURE 5.12 Residual plots used to check if the final model for Chapter 5's Guided Example is compliant with the assumptions of multiple regression.

Independence of the Residuals. As shown in Figure 5.12(b), when the residuals are plotted against the case numbers, there is no systematic pattern, which is confirmed by the kernel smoother line. Therefore, the final model is compliant with the independence assumption.

Normality of the Residuals. As shown in Figure 5.12(c), the *q-q* plot of the residuals shows that the residuals closely "hug" the normal line and all the points are within the 95% confidence band. The corresponding Shapiro-Wilk Goodness of Fit test also is not significant. Therefore, the final model is compliant with the normality assumption.

Data Analysis

We now conduct an MRC analysis for the final data set. The results are summarized in Table 5.7.

Interpretation of Results

The overall regression equation (with coefficients rounded to two decimal places) is $\hat{Y} = 3.95X_1 + 13.20X_2 + 1.76X_3 + 40.05$. We now interpret the results from Table 5.7 relative to the research questions and corresponding hypotheses.

RQ 1. This research question relates to the overall R^2. The total proportion of variance in safety climate scores explained collectively by the three IVs, X_1 = Maintenance Errors, X_2 = Training, and X_3 = Stress, is $R^2_{Y \cdot 123} = .544$, with an adjusted $R^2 = .527$. Thus, the three IVs collectively explain 54.4% of the variance in safety climate scores and when adjusted for bias, these IVs collectively explain 52.7% of the variance in safety climate scores. This result is significant, $F(3, 82) = 32.57$, $p < .0001$ and therefore we reject the corresponding null hypothesis. The three IVs collectively explain a significant proportion of Y variance.

RQ 2. This research question relates to the squared semipartial correlation coefficients (sr_i^2) and has four parts to it with respect to the corresponding hypothesis.

Hyp. 2a. When X_1 = Maintenance Errors enters the model in the presence of the other IVs, $sr_1^2 = .2678$, which means that it explains an additional 26.78% of Y variance. This result is significant, $t(82) = 6.94$, $p < .0001$, and therefore we reject Null Hypothesis 2a and conclude that maintenance errors explain a significant proportion of the variance in safety climate scores when examined in the presence of X_2 and X_3.

Hyp. 2b. When X_2 = Training enters the model in the presence of the other IVs, $sr_2^2 = .3498$, which means that it explains an additional 34.98% of Y variance. This result is significant, $t(82) = 7.93$, $p < .0001$, and therefore we reject Null Hypothesis 2b and conclude that participants' perceptions of their training explain a significant proportion of the variance in safety climate scores when examined in the presence of X_1 and X_3.

Hyp. 2c. When X_3 = Stress enters the analysis in the presence of the other IVs, $sr_3^2 = .0231$, which means it explains an additional 2.31% of Y variance. This result is significant, $t(82) = 2.04$, $p = .0447$, and therefore we reject Null Hypothesis 2c and conclude that participants' perceptions of their stress level explain a significant proportion of the variance in safety climate scores when examined in the presence of X_1 and X_2.

TABLE 5.7 *Summary of MRC Analysis Results for the Final Model of Chapter 5's Guided Example*

Model Summary

R^2	.544
Adjusted R^2	.527
RMSE	10.71
Mean of Response	146.59
Sample Size	86

ANOVA

Source	*SS*	*df*	*MS*	*F*	*p*
Regression	11215.647	3	3738.55	32.5744	< .0001
Residual (Error)	9411.109	82	114.77		
Total	20626.756	85			

Coefficients (Parameter Estimates)

Term	*B*	*SE*	β	*t*(82)	*p*	95% CI	*VIF*
Constant	40.049	11.270	—	3.55	.0006	[17.63, 62.47]	
X_1 = Maintenance Errors	3.952	0.570	.559	6.94	< .0001	[2.82, 5.08]	1.17
X_2 = Training	13.197	1.664	.632	7.93	< .0001	[9.89, 16.51]	1.14
X_3 = Stress	1.755	0.861	.165	2.04	.0447	[0.04, 3.47]	1.18

Note. *N* = 86.

Summary of the Correlation Coefficients and Their Squares

	Zero-Order		Semipartial		Full Partial	
Factor	r_{Yi}	r_{Yi}^2	sr_i	sr_i^2	pr_i	pr_i^2
X_1 = Maintenance Errors	.4401	.1937	.5175	.2678	.6082	.3699
X_2 = Training	.4288	.1839	.5914	.3498	.6587	.4339
X_3 = Stress	.1616	.0261	.1520	.0231	.2195	.0482

Hyp. 2d. This hypothesis was for X_4 = Years as AMT, which was deleted from the final model because it was not correctly specified (i.e., it was not relevant) based on its AVP plot. Therefore, we fail to reject Null Hypothesis 2d and conclude that years as an AMT do not explain a significant proportion of the variance in safety climate scores when examined in the presence of X_1, X_2, and X_3.

Based on the results of the squared semipartial correlation coefficients, it appears that the strongest predictor of safety climate scores is X_2 = Training.

RQ 3. This research question relates to the partial regression coefficients (B_i) and has four parts to it with respect to the corresponding hypothesis.

Hyp. 3a. Holding the other IVs in the regression model constant, B_1 = 3.952, which means that for every 1 additional maintenance error reported by AMTs, safety climate scores increase, on average, by approximately 4 points. More concretely, AMTs' perceptions of their MRO's safety climate become more positive the more maintenance errors they report. This result is significant, $t(82) = 6.94$, $p < .0001$, and therefore we reject Null Hypothesis 3a and conclude that maintenance errors have a significant effect on safety climate.

Hyp. 3b. Holding all other IVs in the regression model constant, $B_2 = 13.197$, which means that for every 1-unit increase in training scores, safety climate scores increase, on average, by approximately 13 points. More concretely, the more AMTs believe that they are adequately trained for their job and that their organization provides opportunities for on-the-job training, the more positive their perceptions are of their MRO's safety climate. This result is significant, $t(82) = 7.93$, $p < .0001$, and therefore we reject Null Hypothesis 3b and conclude that participants' perceptions of the training they receive have a significant effect on safety climate.

Hyp. 3c. Holding all other IVs in the regression model constant, $B_3 = 1.755$, which means that for every 1-unit increase in stress scores, safety climate scores increase, on average, by 1.75 points. More concretely, as AMTs' perceived on-the-job stress levels increase, their perceptions of their MRO's safety climate also increase. This result is significant, $t(82) = 2.04$, $p = .0447$, and therefore we reject Null Hypothesis 3c and conclude that participants' perceived stress levels have a significant effect on safety climate.

Hyp. 3d. This hypothesis was for X_4 = Years as AMT, which was deleted from the final model because it was not correctly specified (i.e., it was not relevant) based on its AVP plot. Therefore, we fail to reject the corresponding null hypothesis and conclude that years as an AMT do not have a significant effect on safety climate.

Standard Error of the Estimate. As reported in the Model Summary subtable in Table 5.7, *RMSE* = 10.71, which means that if we were to predict the safety climate score of an MRO relative to the three IVs, then the prediction would be "off," on average, by nearly 11 points where the scale can vary between 48 and 240 points.

Post Hoc Considerations

We now present the effect sizes, 95% CIs, power for the overall regression model, and the model's generalizability for the reader's benefit and interpretation. Note that the descriptors for small, medium, and large effect sizes are from Cohen et al. (2003, pp. 93–95).

Effect Size. Following are the effect sizes for the overall model and the squared semipartial correlation coefficients.

ES for $R^2_{Y\cdot123}$. $f^2 = \dfrac{R^2}{1-R^2} = \dfrac{.544}{1-.544} = \dfrac{.544}{.456} = 1.19$ (large effect)

ES for sr^2_1. $f^2 = \dfrac{sr^2_1}{1-R^2} = \dfrac{.268}{1-.544} = \dfrac{.268}{.456} = 0.59$ (large effect)

ES for sr^2_2. $f^2 = \dfrac{sr^2_2}{1-R^2} = \dfrac{.35}{1-.544} = \dfrac{.35}{.456} = 0.77$ (large effect)

ES for sr^2_3. $f^2 = \dfrac{sr^2_3}{1-R^2} = \dfrac{.023}{1-.544} = \dfrac{.023}{.456} = 0.05$ (small effect)

95% CIs. The 95% confidence intervals are reported in the Coefficients subtable in Table 5.7. Note that the widths of the intervals for X_1, X_2, and X_3 are 2.26, 6.62, and 3.43, respectively. Given that the possible range of the dependent measure, Y = Safety Climate Scores, is between 48 and 240, these confidence intervals are very narrow, which implies relatively high accuracy in parameter estimation with respect to each IV. The interpretation of these CIs is left for the reader.

Power. Consulting Soper (2025c) for the overall model and Soper (2025d) for the squared semipartial correlation coefficients, the respective post hoc powers are as follows:

Power for $R^2_{Y \cdot 123}$ = 1.0. There is a greater than 99% probability that the effect found in the sample relative to the three IVs also exists in the population, and there is a greater than 99% probability that the decision to reject the corresponding null hypothesis was correct.

Power for sr_1^2 = .9999. There is a 99.99% probability that the effect found in the sample relative to X_1's unique contribution to explaining Y variance also exists in the population, and there is a greater than 99% probability that the decision to reject the corresponding null hypothesis was correct.

Power for sr_2^2 = 1.0. There is a greater than 99% probability that the effect found in the sample relative to X_2's unique contribution to explaining Y variance also exists in the population, and there is a greater than 99% probability that the decision to reject the corresponding null hypothesis was correct.

Power for sr_3^2 = .537. There is a 53.7% probability that the effect found in the sample relative to X_3's unique contribution to explaining Y variance also exists in the population, and there is a 53.7% probability that the decision to reject the corresponding null hypothesis was correct. Recall that we opted to keep X_3 = Stress in the final model. Although its effect is statistically significant, its post hoc power is slightly better than a coin toss.

Generalizability of the Regression Model. Because we did not collect a separate sample from the population, we will assess the generalizability of the regression model using both the cross-validated R^2 and the K-fold cross-validation method.

Cross-Validated R^2.

$$\begin{aligned}
\hat{R}^2 &= 1 - \left[(1 - R^2)\frac{(N+k)}{(N-k)}\right] \\
&= 1 - \left[(1 - .544)\frac{(86+3)}{(86-3)}\right] \\
&= 1 - \left[(1 - .54)\frac{(89)}{(83)}\right] \\
&= 1 - [(.456)(1.07)] \\
&= 1 - (.48792) \\
&= .51208
\end{aligned}$$

Therefore, if we were to use the regression model derived from the sample data, the estimated amount of variance in safety climate scores in the population that will be explained collectively by the three IVs is predicted to be .512 or about 51.2%.

K-Fold Cross-Validation. We let $K = N = 86$ so that our statistics program will run a leave-one-out cross-validation. The result is R^2 *K*-fold = .4962. So, if we were to apply the overall regression model to the parent population of AMTs working at MROs, or to a new sample selected from the same parent population, the estimated amount of explained variance in safety climate scores that will be predicted by the three IVs is 49.62% or approximately 50%.

Based on the results of the cross-validated R^2 and the *K*-fold cross-validation with $K = 86$, it appears that the final regression model is a good fit to the sample data (overall $R^2 = .544$) and has relatively strong generalizability ($\hat{R}^2 = .5121$ and R^2 *K*-fold = .4962). With respect to the latter, we conclude that the analysis resulted in a strong predictor and reliable model.

Chapter 5 Summary

5.2 Underlying Assumptions

- The underlying assumptions of regression are linearity, equal variances, independence, and normality.
- The linearity assumption is examined from a multivariate perspective and requires that the relationship between the DV and the IVs taken collectively is linear. This is confirmed via a scatter plot where the *Y* residuals are plotted on the vertical axis and the *Y* values are plotted on the horizontal axis. If the pattern reflects a linear relationship, then the data set is compliant with this assumption. An alternative is to plot the *Y* residuals on the vertical axis and the predicted $\hat{Y}$ values on the horizontal axis, and then superimpose the 0- and kernel smoother lines. If the kernel smoother line closely follows the 0-line without any pronounced or systematic pattern, then the data set is compliant.
- The equal variances assumption, also known as homoscedasticity of the residuals, requires that the variance of the residuals is not related to any of the IVs or to the $\hat{Y}$ values. This is confirmed via a scatter plot of the *Y* residuals vs. $\hat{Y}$ predicted values.
- The independence of residuals assumption requires that the residuals of the observations are independent of one another. This is confirmed by a scatter plot of the *Y* residuals vs. the case numbers and includes the 0-line and kernel smoother line to ensure there is no pronounced or systematic pattern. This assumption also can be confirmed by a nonsignificant Durban-Watson statistic.
- The normality assumption requires that the residuals form a normal distribution. This is confirmed graphically by a *q-q* plot where the residuals fall closely along the normal line of the plot and are contained within the corresponding 95% confidence band. This also can be confirmed by a nonsignificant Shapiro-Wilk Goodness of Fit test.

5.3 Regression Diagnostics

- Regression diagnostics involve examining the data set for potential problems that could arise from outliers, multicollinearity, missing data, measurement error, and IV relevance.

- Outliers are extreme scores relative to a given data set that can inflate or suppress statistical significance and must be examined from a multivariate perspective. One strategy for doing this is Jackknife distances, which calculates a centroid and then determines how far each score is from the centroid. Scores with distances that exceed a threshold are outliers, which are then examined to determine if they are contaminants or rare cases. Contaminants are to be corrected, replaced, or deleted, whichever is appropriate, whereas rare cases may be retained or deleted. Regardless, a data set is to be examined with outliers present and outliers absent to help determine the course of action and to allow readers to assess the extent to which the final model is appropriate.
- Multicollinearity involves two (or more) IVs that are so closely correlated that they carry redundant information about *Y*. To check for multicollinearity, we examine each IV's variable inflation factor (*VIF*). A $VIF > 10$ requires attention.
- Missing data can occur due to a variety of reasons and the best strategy for dealing with this issue is to not have missing data. If this is not possible, strategies for managing missing data include deleting the IVs with missing data, deleting cases with missing data via listwise deletion or pairwise deletion, using a statistic program's imputation routine, and substituting the missing values with means. Although no single strategy is better than another, some are more appropriate depending on the context of the study, and all have their biases, advantages, and disadvantages.
- Measurement error refers to the reliability coefficient of a questionnaire (or similar) that is assigned to an IV to assess a psychological construct such as attitudes or self-efficacy. Ideally, measurement error should be 0, but reliability coefficients of at least .70 are acceptable.
- IV relevance considers whether an IV is correctly specified, that is, does the IV "belong" in the regression model? This can be determined by examining each IV's added variable plot (AVP), which consists of the residuals of *Y* after all the other IVs have been partialled out vs. the residuals of X_i after it too has been freed of the influences from the other IVs. If the plot forms a significant, nonzero linear relationship, then X_i is correctly specified; otherwise, it is not and should be removed from the final regression model.

5.4 Assessing a Regression Model's Reliability: Cross-Validation

- Independent of a regression model's fit to the sample data, the reliability of the model also should be examined to assess the model's predictive performance relative to its generalizability.
- One strategy for assessing the predictive performance of a regression model is to conduct a cross-validation study, which involves collecting two separate samples from the same population and using the regression equation from S_1 to predict the *Y* values of S_2. The predicted *Y* values from S_2 are correlated with the actual *Y* values, and this correlation is then subtracted from the overall R^2 from S_1. If the difference, called "shrinkage," is less than .10, then the regression model from S_1 is reliable and generalizable to the population or a new sample drawn from the same population.

- A second strategy for assessing the predictive performance of a regression model is to calculate the cross-validated R^2 ($\hat{R}^2$) using the formula given in Section 5.4.
- A third strategy for assessing the predictive performance of a regression model is to have a statistics program perform a quasi-cross-validation study by splitting a single sample into two subsamples: a Training sample and a Test sample. The Training sample is used to generate a regression model, which is then tested on the Test sample. This strategy can be accomplished using the "Holdout" method or the "*K*-fold" method.
- The Holdout method splits the data set one time and uses the regression model from the Training sample to predict the *Y* values of Test sample. It then calculates the mean absolute error (MAE) between the predicted and actual *Y* values of the Test sample to determine the model's predictive performance.
- The *K*-fold method splits, or "folds," the data set *K* times and each fold consists of different training and test samples, with $1/K$ cases randomly assigned to the Test sample and $(K - 1)/K$ cases randomly assigned to the Training sample. The Holdout method is then repeated *K* times and the overall model's predictive performance is equal to the overall average of each fold's MAE.
- Some statistics programs provide an R^2 value that results from the *K*-fold process. This R^2 is an estimate of the amount of *Y* variance the IVs in the final model are predicted to explain if the model is used in the population or in new samples from the population.

Key Terms and Concepts

0-line
Added variable plot (AVP)
Autocorrelation
Calibration sample
Centroid
Cross-validated R^2 ($\hat{R}^2$)
Cross-validation
Cross-validation correlation
Cross-validation study
Durbin-Watson statistic
Holdout method
Homoscedasticity of the residuals
Imputation
Jackknife distances
K-fold cross-validation
Kernel smoother line
Leave-one-out cross-validation
Linearity assumption
Listwise deletion
Mahalanobis distances
Mean absolute error (MAE)
Mean substitution
Multicollinearity
Outliers
Pairwise deletion
q-q plot
Quasi-cross-validation study
Regression diagnostics
Residual analysis
Screening sample
Shrinkage on cross-validation
Test sample
Training sample
Variance inflation factor (*VIF*)

Review Exercises

A. Checking Conceptual Knowledge

In 1–10, choose the best answer among the choices provided.

1. When determining if a data set is compliant with regression assumptions, we primarily examine
   a. outliers.
   b. residuals.
   c. multicollinearity.
   d. residual plots.
2. Consider a study that examines the effect of X_1 = Total Flight Hours, X_2 = Number of FAA Ratings, X_3 = Biological Sex, and X_4 = Number of Years as PIC have on Y = Annual Salaries of ATPs at a LCC. With respect to the regression assumptions, which of the following must have a normal distribution and constant variance?
   a. Total Flight Hours
   b. Annual Salaries
   c. Number of Years as PIC
   d. Number of FAA Ratings
3. In a multiple regression model, multicollinearity occurs when
   a. at least two IVs contain redundant information about the variance in the DV.
   b. at least two IVs contain opposite information about the variance in the DV.
   c. at least two IVs are strong predictors of the DV.
   d. at least two IVs are weak predictors of the DV.
4. In a general sense, cross-validation
   a. examines the correlation between the overall R^2 of the regression model in the sample and the corresponding overall R^2 of the regression model in the population.
   b. is the same as *RMSE*.
   c. assesses how strong a sample's regression model is if it were applied to the population.
   d. can only be conducted using a statistics program.
5. Among the choices provided, the best strategy for handling outliers is to
   a. delete them from the data set because they can have an unjustified effect on the results.
   b. keep them in the data set because their discovery can lead to something completely unexpected or they can provide new insights relative to the targeted research.
   c. conduct two separate analyses—with and without outliers—and decide whether to keep or delete the outliers based on the respective results.
   d. ignore them by not checking for them.
6. Which statement is NOT true about the residual plots used to check regression assumptions?
   a. The 0-line is where observed and predicted values of the DV are equal to each other.
   b. The 0-line is always horizontal.
   c. The kernel smoother line will always follow the 0-line.
   d. If a data set is compliant with the linearity assumption, then the mean of the residuals should be 0.

7. Given a regression model with $R^2_{Y\bullet1234} = .65$. If $\hat{R}^2_{Y\bullet1234} = .005$, then
   a. the shrinkage on cross-validation is less then .10 and therefore, the regression model is a strong predictor.
   b. the regression model is not reliable and therefore has poor generalizability.
   c. the four IVs collectively explain .65 + .05 = .70 = 70% of *Y* variance in the population.
   d. there must be an error in the calculation of $\hat{R}^2_{Y\bullet1234}$.
8. Let's assume that the original data set prior to performing an MRC analysis consists of six IVs and that as part of her findings, the researcher reported $R^2_{Y\bullet135} = .47$, $F(3, 96) = 6.21$, $p = .0027$. What is a plausible conclusion from this information?
   a. The researcher conducted two separate analyses: one with the even numbered IVs and one with the odd numbered IVs.
   b. The original data set had 99 cases, but three were eliminated from the final model.
   c. Three IVs were eliminated from the final model because their AVPs were not statistically significant.
   d. There is not enough information given in the problem to make any type of conclusion.
9. Which statement is true about the cross-validated R^2 (i.e., $\hat{R}^2$)?
   a. The cross-validated R^2 and the adjusted R^2 are both estimates of the population R^2.
   b. The cross-validated R^2 and *RMSE* are both estimates of the population R^2.
   c. The cross-validated R^2 and *K*-fold R^2 are both estimates of the population R^2.
   d. The cross-validated R^2 and the overall R^2 must be equal.
10. Based on the scatter plot given in Figure 5.13, which statement is true?
    a. The corresponding data set is compliant with the independence of residuals assumption.
    b. The corresponding data set is compliant with the homoscedasticity of residuals assumption.
    c. The corresponding data set is complaint with the linearity assumption.
    d. The corresponding data set is not compliant with the linearity assumption.

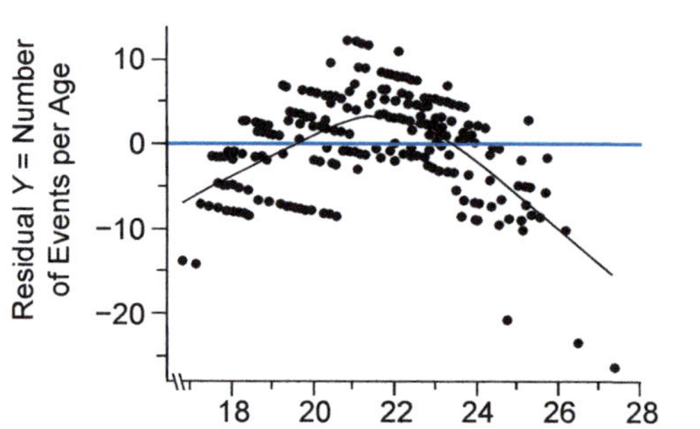

FIGURE 5.13 Scatter plot for Chapter 5 Review Exercises Part A, Number 10.

B. Applying Procedural and Conceptual Knowledge

Use the following research description and the corresponding data set to conduct the activities given in Parts B1–B6. See Section 5.5 for guidance.

Consider a study designed to investigate the effect various factors have on CFIs' risk perception. The targeted factors are:

- X_1 = Total Dual Hours Given.
- X_2 = Years Held CFI Certificate.
- X_3 = HES Score, defined as participants' score on Hunter's (2002) 10-item Hazardous Events Scale, which asks participants to reflect on the number of times they were involved in various aviation-related hazardous events in the past 24 months. Possible responses for each item are 0, 1, 2, 3, and 4 or more, with higher scores indicating a higher level of actions by pilots that could lead or contribute to an unplanned or undesired event such as an accident.
- X_4 = *Self-Efficacy*. A continuous variable that is measured using Chen et al.'s (2001) New General Self-Efficacy Scale, an 8-item instrument scored on a traditional 5-point Likert response scale ranging from 1 = Strongly Disagree to 5 = Strongly Agree. Higher scores reflect a higher level of self-efficacy. (*Note:* Assume this is a reliable instrument by referencing the related citation.)

The dependent variable, Y = Risk Perception, is measured using Hunter's (2006) 17-item Risk Perception–Other instrument, which is scored on a 100-point response scale (1 to 100). Each item presents a scenario and participants assess the amount of risk they perceive in the given scenario. Higher scores indicate a higher degree of risk perception, which suggests that participants are more risk aversive than risk prone. We will assume it is the researcher's belief that: (a) the IVs will collectively explain a significant amount of the variance in risk perception scores, (b) each IV will uniquely explain a significant amount of the variance in risk perception scores, and (c) X_1, X_2, and X_3 will have a positive relationship with Y, but X_4 will have a negative relationship with Y. A set of hypothetical data is provided in the file "Ch_5 Review Exercises Data-Part B."

B1. Research Questions and Hypotheses. Write the research questions and corresponding hypotheses that pertain to the overall regression model, the squared semipartial correlation coefficients, and the partial regression coefficients.

B2. Sample Size Planning. Conduct an a priori power analysis to see if the given data set is sufficiently large.

B3. Preliminary Data Screening. Perform regression diagnostics on the data set and then confirm that the final model is compliant with the underlying regression assumptions.

B4. Data Analysis. Conduct an MRC analysis of the final data set and summarize the results in a table like Table 5.7.

B5. Interpretation of Results. Interpret the results relative to the research questions and corresponding null hypotheses stated in Part B1.

B6. Post Hoc Considerations. Determine and comment on the effect sizes, 95% CIs, power for the overall R^2 and sr_i^2, and the reliability of the regression model.

References

Chen, G., Gully, S. M., & Eden, D. (2001). Validation of a new general self- efficacy scale. *Organizational Research Methods*, *4*(1), 62–83.

Cohen, J., Cohen, C., West, S. G., & Aiken., L. S. (2003). *Applied multiple regression/correlation analysis for the behavioral sciences* (3rd ed.). Lawrence Erlbaum Associates.

Fogarty, G. J. (2005). Psychological strain mediates the impact of safety climate on maintenance errors. *International Journal of Applied Aviation Studies*, *5*(1), 53–63.

Harwin, M. S. (2023). *The private pilot check ride: Applying the spacing effect theory to predict time to proficiency for the practical test.* [Doctoral dissertation, Florida Institute of Technology]. Scholarship Repository. https://repository.fit.edu/cgi/viewcontent.cgi?article=2353&context=etd

Hunter, D. R. (2002). *Risk perception and risk tolerance in aircraft pilots* (Rep. No. DOT/FAA/AM–02/17). Washington, DC: Federal Aviation Administration.

Hunter, D. R. (2006). Risk perception among general aviation pilots. *The International Journal of Aviation Psychology, 16*(2), 135–144. https://doi.org/10.1207/s15327108ijap1602_1

Judd, C. M., McClelland, G. H. & Ryan, S.C. (2017). *Data analysis: A model-comparison approach to regression, ANOVA, and beyond* (3rd ed.). Routledge. https://doi.org/10.4324/9781315744131

Kleinbaum, D. G., Kupper, L. L., Muller, K. E., & Nizam, A. (1998). *Applied regression analysis and other multivariate methods* (3rd ed.). Duxbury Press.

Lord, F. M., & Novick, M. R. (1968). *Statistical theories of mental test scores.* Addison-Wesley.

Soper, D. (2025a). *A-priori sample size calculator for multiple regression.* https://www.danielsoper.com/statcalc/calculator.aspx?id=1

Soper, D. (2025b). *A-priori sample size calculator for hierarchical multiple regression* https://www.danielsoper.com/statcalc/calculator.aspx?id=16

Soper, D. (2025c). *Post-hoc statistical power calculator for multiple regression.* https://www.danielsoper.com/statcalc/calculator.aspx?id=9

Soper, D. (2025d). *Post-hoc statistical power calculator for hierarchical multiple regression.* https://www.danielsoper.com/statcalc/calculator.aspx?id=17

Tabachnick, B. G., & Fidell, L. S. (2013). *Using multivariate statistics* (6th ed.). Pearson.

Tokarski, R. V. (2021). *Examining the safety climate of a general aviation maintenance, repair, and overhaul (MRO) organization: A replication study and application of Bandura's reciprocal causation model.* [Doctoral dissertation, Florida Institute of Technology]. Scholarship Repository. https://repository.fit.edu/etd/41/

Uhuegho, K. O. (2017). *Examining the safety climate of U.S. based aviation maintenance, repair, and overhaul (MRO) organizations.* [Doctoral dissertation, Florida Institute of Technology]. Scholarship Repository. https://repository.fit.edu/etd/37/

Part C
Data Analytic Strategies for MRC

Chapter 6
Multiple Regression Strategies for Variable Selection

Student Learning Outcomes

After studying this chapter, you will be able to do the following:

1. Contrast among the three different multiple regression variable selection strategies: simultaneous, hierarchical, and statistical.
2. Perform and interpret the results from a hierarchical regression analysis.
3. Perform and interpret the results from a statistical regression analysis.
4. Conduct a hierarchical regression analysis for a research study by engaging in the activities presented in Section 6.5 and contrast these results to those from simultaneous and statistical regression.

6.1 Chapter Overview

In Chapters 3–5, we performed all the MRC analyses by entering the IVs into the regression model simultaneously without any concern for a specific variable entry order or any consideration of their relationship with the DV, logic, theory, past or preliminary studies, or professional experience. In this chapter, we present two alternative multiple regression strategies to the *simultaneous* approach we have used thus far: *hierarchical* and *statistical*. We begin the chapter with a review of the simultaneous approach. We then introduce hierarchical regression, which involves entering the IVs into the regression model via a specific variable entry order. We follow this presentation by introducing statistical regression, which uses various statistical criteria to determine the best fitting model. We conclude the chapter with an example of an MRC analysis of a research study based on a hierarchical regression strategy and compare the results to those from simultaneous and statistical regression.

6.2 Simultaneous Regression

A Conceptual Review

As noted in the Chapter Overview, a simultaneous regression strategy involves entering all the IVs into a regression model in no logical order. We simply "collect" all the IVs and place them into the analysis together at the same time, which is what we did in all the MRC analyses presented in Chapters 3–5. Focusing specifically on Chapter 4, recall that the semipartial correlation coefficient and its square (sr_i, sr_i^2), as well as the full partial correlation coefficient and its square (pr_i, pr_i^2), reflect the correlation of each X_i with Y in the presence of all other IVs (i.e., all the IVs are partialled from each other). Further recall that because we partial out the effects of each IV from all the other IVs, the *shared contribution* the IVs make in explaining Y variance is never fully accounted for, which means that the sum of the sr_i^2s will not equal the overall R^2. This is graphically illustrated in Figure 6.1 for three IVs. The reader will note that Figure 6.1 is a composite of Figures 4.2(b, c, d) from Chapter 4.

DOI: 10.4324/9781003506935-9

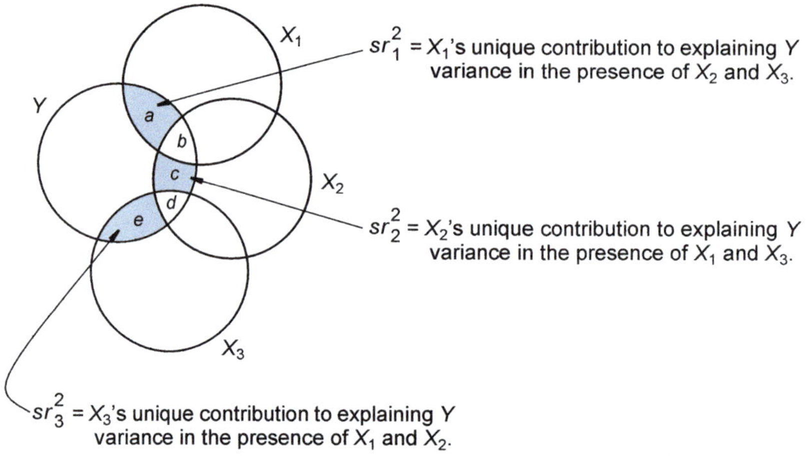

FIGURE 6.1 Venn diagram depiction of a simultaneous regression analysis involving three IVs.

Observe from Figure 6.1 that the collective contribution the three IVs make to explaining Y variance consists of regions, a, b, c, d, and e; that is, conceptually, $R^2_{Y \cdot 123} = a + b + c + d + e$. However, note that although X_1 and X_2's collective contribution to explaining Y variance is represented by regions a, b, c, and d, to determine X_1's *unique* contribution to Y variance in the presence of X_2 and X_3, we partial out X_2 and X_3's effect on X_1, which leaves region a. Similarly, X_2's unique contribution to explaining Y variance in the presence of X_1 and X_3 is represented by region c, and X_3's unique contribution in the presence of X_1 and X_2 is represented by region e. As a result, $sr_1^2 = a$, $sr_2^2 = c$, and $sr_3^2 = e$. Because of this partialling, though, we do not get a full accountability of the three IVs' collective contribution to explaining Y variance. In other words: $R^2_{Y \cdot 123} \neq sr_1^2 + sr_2^2 + sr_3^2$.

A Concrete Example

To refresh the reader's memory from Chapter 4's presentation of MRC with k IVs, consider the data set given in the file "Ch_6 Multiple Regression Strategies Example Data." This data set consists of manipulated data for a study that examines the effect flight students' characteristics have on the ratings they assign to their CFIs. The factors are X_1 = Age, X_2 = Biological Sex (coded 1 = Female and 0 = Male), and X_3 = Total Flight Hours. The dependent variable is Y = CFI Ratings, which are measured using a 16-item instrument scored on a traditional Likert response scale that ranges from 1 = Strongly Disagree to 5 = Strongly Agree. Thus, scores could range from 16 to 80, with higher scores reflecting a more positive perception flight students have of their CFI. Table 6.1 contains a summary of the corresponding simultaneous regression analysis.

As reported in Table 6.1, $R^2_{Y \cdot 123} = .1435$. Thus, the three factors of age, biological sex, and total flight hours collectively explain 14.35% of the variance in CFI ratings. Further note from Table 6.1 that $R^2_{Y \cdot 12} = .1261$, $R^2_{Y \cdot 13} = .0938$, and $R^2_{Y \cdot 23} = .0901$. Therefore:

TABLE 6.1 *Summary of Simultaneous Regression Analysis Results for Chapter 6's Regression Strategies Example Data*

Model Summary

$R^2_{Y \cdot 123}$	.1435
Adjusted R^2	.1063
Standard Error of the Estimate	17.683
Mean of Response	61.700
Sample Size	73

$R^2_{Y \cdot 12}$ = .1261

$R^2_{Y \cdot 13}$ = .0938

$R^2_{Y \cdot 23}$ = .0901

ANOVA

Source	*SS*	*df*	*MS*	*F*	*p*
Regression	3615.042	3	1205.01	3.8539	.0131
Residual (Error)	21574.327	69	312.67		
Total	25189.370	72			

Coefficients (Parameter Estimates)

Term	*B*	*SE*	sr_i^2	*t*(69)	p_i	95% CI
Constant	45.310	7.715		5.87	< .0001	[29.92, 60.70]
X_1 = Age	0.448	0.216	.0534	2.07	.0418	[0.02, 0.88]
X_2 = Biological Sex	8.576	4.286	.0497	2.00	.0493	[0.03, 17.13]
X_3 = Total Flight Hours	− 0.154	0.130	.0174	−1.19	.2400	[−0.41, 0.11]

- $sr_1^2 = (R^2_{Y \cdot 123} - R^2_{Y \cdot 23}) = (.1435 - .0901) = .0534$
- $sr_2^2 = (R^2_{Y \cdot 123} - R^2_{Y \cdot 13}) = (.1435 - .0938) = .0497$
- $sr_3^2 = (R^2_{Y \cdot 123} - R^2_{Y \cdot 12}) = (.1435 - .1261) = .0174$

Observe that when we add each IV's unique contribution to explaining *Y* variance in the presence of the other IVs, the sum of the respective squared semipartial correlation coefficients does *not* equal the total collective contribution made by the three IVs: $sr_1^2 + sr_2^2 + sr_3^2 = .0534 + .0497 + .0174 = .1205$, but $R^2_{Y \cdot 123} = .1435$, a difference of .1435 − .1205 = .0227. Thus, when analyzing this data set via a simultaneous regression strategy, 2.27% of the total variance explained collectively by the three IVs is unaccounted for. Once again, this is because when we partial out the effects of all the other IVs from the IV under discussion via a simultaneous regression strategy, various "regions" (conceptually speaking) will be left out of the analysis, and therefore we will not get a full accountability of the explained variance.

It is important to note that this concept of "unaccountable" explained variance via simultaneous regression occurs when the IVs are related to each other as depicted in Figure 6.1. A simultaneous approach, however, will provide a full accountability of the collective contribution a set of IVs makes in explaining *Y* variance if the IVs are not correlated. For example, if $r_{12} = r_{13} = r_{23} = 0$ as depicted in Figure 6.2, then $R^2_{Y \cdot 123} = sr_1^2 + sr_2^2 + sr_3^2$. Notice further that in this situation, the squared semipartial correlation coefficients are equal to their respective zero-order squared correlation coefficients.

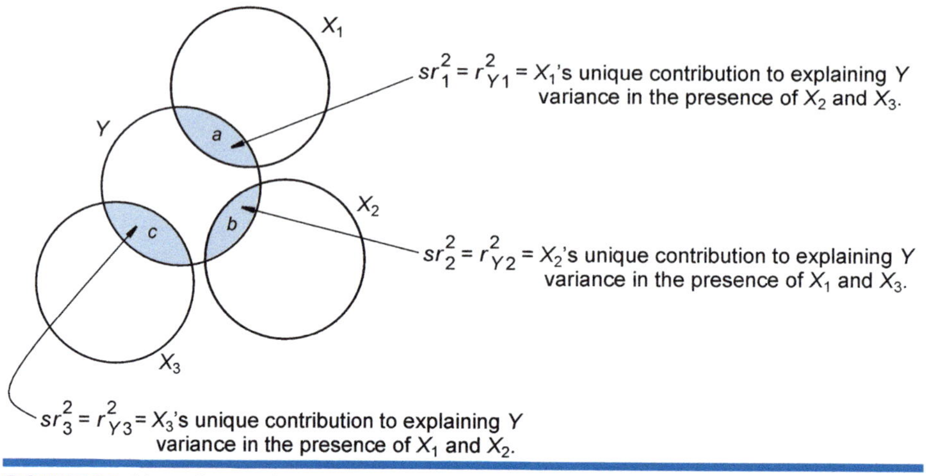

FIGURE 6.2 Venn diagram for a simultaneous regression involving three IVs that are not related.

Continuing with this review, also recall from Chapter 4 that the respective *F* tests for determining the significance of the overall regression model and the partial variances are:

$$F = \frac{R^2(N-k-1)}{(1-R^2)k} \text{ and } F = \frac{sr_i^2(N-k-1)}{(1-R^2)}$$

Let's now apply these formulas to the current example's data set from the file "Ch_6 Multiple Regression Strategies Example Data," where X_1 = Age, X_2 = Biological Sex (coded 1 = Female and 0 = Male), X_3 = Total Flight Hours, and Y = CFI Rating:

- $R^2_{Y \bullet 123}$. $F = \frac{R^2(N-k-1)}{(1-R^2)k} = \frac{.1435(73-3-1)}{(1-.1435)3} = \frac{.1435(69)}{(.8565)3} = \frac{9.9015}{2.5695} = 3.85$
- sr_1^2. $F = \frac{sr_1^2(N-k-1)}{(1-R^2)} = \frac{.0534(73-3-1)}{(1-.1435)} = \frac{.0534(69)}{(.8565)} = \frac{3.6846}{.8565} = 4.30$
- sr_2^2. $F = \frac{sr_2^2(N-k-1)}{(1-R^2)} = \frac{.0497(73-3-1)}{(1-.1435)} = \frac{.0497(69)}{(.8565)} = \frac{3.4293}{.8565} = 4.00$
- sr_3^2. $F = \frac{sr_3^2(N-k-1)}{(1-R^2)} = \frac{.0174(73-3-1)}{(1-.1435)} = \frac{.0174(69)}{(.8565)} = \frac{1.2006}{.8565} = 1.40$

From these calculations, observe the following. [*Note*: F_{Critical} (1, 69) = 3.9798 is from Soper (2025a), and the respective *p* values for sr_i^2 are from Soper (2025c). *F* values rounded to two decimal places.]

- The collective contribution the three IVs make to explaining the variance in CFI ratings is significant, $R^2_{Y \bullet 123}$ = .1435, $F(3, 69) = 3.85$, $p = .0131$. (This matches what we reported in Table 6.1.)
- X_1's unique contribution to explaining the variance in CFI ratings (in the presence of X_2 and X_3) is significant, sr_1^2 = .0534, $F(1, 69) = 4.30$, $p = .0418$.

- X_2's unique contribution to explaining the variance in CFI ratings (in the presence of X_1 and X_3) is significant, $sr_2^2 = .0497$, $F(1, 69) = 4.00$, $p = .0493$.
- X_3's unique contribution to explaining the variance in CFI ratings (in the presence of X_1 and X_2) is not significant, $sr_3^2 = .0174$, $F(1, 69) = 1.40$, $p = .2400$.

Given this review of simultaneous regression, we now present another approach to variance partitioning: *hierarchical regression*.

6.3 Hierarchical (Sequential) Regression

Concept of Unique Partitioning

An alternative to a simultaneous regression approach—and one that will provide a full accountability of the overall R^2 when the IVs are related to each other—is hierarchical regression. Because the term "hierarchical" is also used for a statistical strategy called *hierarchical linear model* (*HLM*), hierarchical regression is commonly referred to as sequential regression to avoid confusion with *HLM* and is more descriptive of the way in which the regression analysis is performed.

In a hierarchical regression analysis, the IVs are entered into the model one at a time via a predetermined sequence or variable entry order. As each IV enters the model, a regression analysis is performed that consists of the IV currently entering the model plus all the IVs that were previously entered into the model. For example, in the three IVs case, if the variable entry order is X_1–X_2–X_3, then there would be three separate analyses, one each for examining $R^2_{Y \cdot 1}$, $R^2_{Y \cdot 12}$, and $R^2_{Y \cdot 123}$. Thus, this strategy results in successive regression analyses where each step consists of one more IV than the preceding step. This is graphically illustrated in Figure 6.3 involving three IVs.

Observe from Figure 6.3 how a hierarchical regression strategy provides a "unique" partitioning of *Y* variance. For example, at Step 1, X_1 enters the model first. Because there are no other IVs in the model at this step, regions *a* and *b represent* the amount of *Y* variance X_1 explains. Recognize that this is r^2_{Y1}, which is equal to sr_1^2. At Step 2, X_2 enters the model in the presence of X_1. Note that X_2's relationship with *Y* is represented by regions *b*, *c*, and *d*, but region *b* was previously claimed by

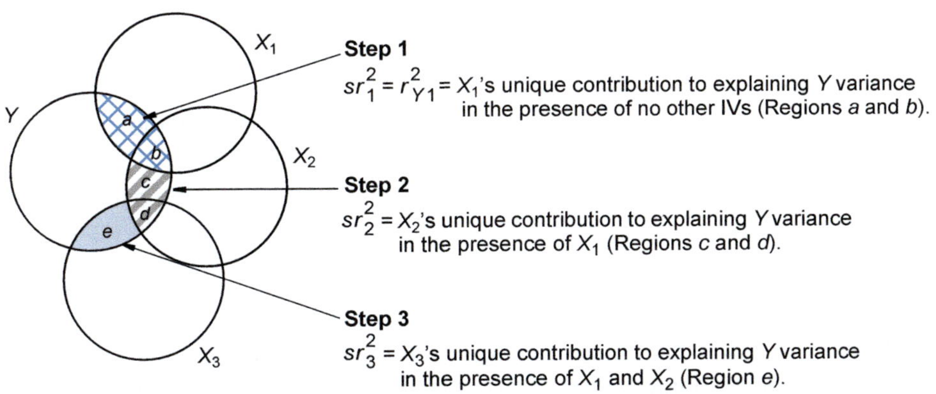

FIGURE 6.3 Venn diagram depiction of a hierarchical regression analysis involving three IVs with variable entry order X_1–X_2–X_3.

X_1. This leaves regions *c* and *d*, which is sr_2^2, and represents X_2's unique contribution to explaining *Y* variance. At Step 3, X_3 enters the model in the presence of X_1 and X_2. Although X_3's relationship with *Y* is represented by regions *d* and *e*, region *d* was previously claimed by X_2. This leaves region *e*, which is sr_3^2, and represents X_3's unique contribution to explaining *Y* variance. As a result, the area representing the amount of explained *Y* variance has been uniquely partitioned among the three IVs, which leads to a full accountability of *Y* variance explained collectively by the three IVs:

$$\begin{aligned} R^2_{Y\bullet 123} &= a + b + c + d + e \\ &= (a + b) + (c + d) + (e) \\ &= sr_1^2 + sr_2^2 + sr_3^2 \end{aligned}$$

Examining Statistical Significance at Each Step of the Analysis

One of the primary reasons for performing a hierarchical regression analysis is to answer the research question, "What is the incremental effect of each IV relative to the DV?" In other words, when an IV enters the model at a specific stage of the analysis, how much additional knowledge does this IV provide in explaining *Y* variance and is this increment in explained *Y* variance statistically significant? To answer this RQ, we examine the significance of sr_i^2 at each step of the analysis.

To illustrate this, let's reexamine the data set from the "Ch_6 Multiple Regression Strategies Example Data" file using a hierarchical regression perspective with the variable entry order of X_1–X_2–X_3, where X_1 = Age, X_2 = Biological Sex, X_3 = Total Flight Hours, and Y = CFI Ratings.

Step 1. X_1 enters the model in the absence of any other variables and the cumulative R^2 at this step is $R^2_{Y\bullet 1} = r^2_{Y1} = .0850$, $F(1, 71) = 6.59$, $p = .0124$. Because there are no other IVs in the model prior to this step, X_1's *incremental effect* is .0850 because it increased the amount of explained *Y* variance from 0 to .0850. Thus, $R^2_{Y\bullet 1} = r^2_{Y1} = sr_1^2$ at this step of the analysis, and when interpreted in the context of the given research setting, X_1 = Age explains 8.5% of the variance in CFI ratings when examined with no other IVs in the model.

Step 2. X_2 enters the model in the presence of X_1 and the cumulative R^2 at this step is $R^2_{Y\bullet 12} = .1261$, $F(2, 70) = 5.05$, $p = .0089$. Thus, X_2's *incremental effect* is $sr_2^2 = R^2_{Y(2\bullet 1)} = (R^2_{Y\bullet 12} - R^2_{Y\bullet 1}) = (.1261 - .0850) = .0411$. More concretely, X_2 = Biological Sex, which compares female vs. male flight students, explains an additional 4.11% of the variance in CFI ratings when examined in the presence of X_1. Note that we included in the string of "=" terms for sr_2^2 the expression $R^2_{Y(2\bullet 1)}$. Recall this denotes that we are examining the relationship between *Y* and X_2 in the presence of X_1, and hence are partialling the effects of X_1 from X_2 but not from *Y*.

Step 3. X_3 enters the model in the presence of X_1 and X_2 and the cumulative R^2 at this step is $R^2_{Y\bullet 123} = .1435$, $F(3, 69) = 3.85$, $p = .0131$. Thus, X_3's *incremental effect* is $sr_3^2 = R^2_{Y(3\bullet 12)} = (R^2_{Y\bullet 123} - R^2_{Y\bullet 12}) = (.1435 - .1261) = .0174$. More concretely, X_3 = Total Flight Hours explains an additional 1.74% of the variance in CFI ratings when examined in the presence of X_1 and X_2.

Note that we performed three separate regression analyses, one for each IV, with each succeeding analysis building on the immediately preceding one similar to the way we built on the regions of explained *Y* variance depicted conceptually in Figure 6.3. Further note that at each step we reported the cumulative R^2 and its corresponding F and p values, and then determined the entering IV's incremental effect by subtracting the current model's cumulative R^2 by the preceding model's cumulative R^2. The question now is, "Are these incremental effects statistically significant?" For the model at Step 1, we can answer this question via the reported F and p values because $R^2_{Y \bullet 1} = sr^2_1$. For all subsequent steps, though, the reader will recognize by the degrees of freedom that the F and p values represent the *overall* model, which contains *all* the IVs present at each step, whereas the increment is attributed to *one* IV at each step.

To determine the F and p values of the increments, denoted F_I and p_I, respectively, there are three options: (a) use a statistics program, assuming it provides this information; (b) manually calculate F_I and then compare this F value to the corresponding $F_{Critical}$ to see if F_I is significant at the preset alpha level; (c) consult an online calculator. We demonstrate the latter two.

Manually Calculating F_I. The F formula for the increment is an extension of the F formula we provided in Chapter 4 for determining if sr^2_i is significant. This extension involves making accommodations for the multiple steps of a hierarchical regression analysis. Recall that at each step of the analysis there are always two models to consider: the model that is being examined at the current step and the model that was examined at the previous step. We also must keep track of the number of IVs in each model. To do this we let set A = the number of IVs in the previous model and denote this k_A, and we let set B = the number of IVs entering the current model and denote this k_B. The reader is cautioned not to attribute any meaning to the set designations of A and B. They are simply being used in this context as labels. In addition, because we are examining the incremental effect of each IV when it enters the current model, this effect is relative to the cumulative R^2 for the *current* model, denoted $R^2_{Y \bullet AB}$, and not the cumulative R^2 for the *full* model with all the targeted IVs present. (*Note:* In Chapter 7 we will introduce another version of this F test where these last two points are not considered.) With this in mind, here is the formula for the F test of the increment:

$$F_I = \frac{sr^2_i(N - k_A - k_B - 1)}{(1 - R^2_{AB})k_B}$$

where sr^2_i is the increment for X_i, N = the overall sample size, k_A = the number of IVs in the *previous* model, k_B = the number of IVs entering the *current* model, and $R^2_{Y \bullet AB}$ = the cumulative R^2 for the *current* model, which contains the IVs from the previous model *and* the IVs entering the model at the current step. Applying this formula to our running example:

- Step 1: $F(sr^2_1) = \dfrac{.0850(73 - 0 - 1 - 1)}{(1 - .0850)(1)} = \dfrac{.0850(71)}{.9150} = \dfrac{6.035}{.9150} = 6.59$
- Step 2: $F(sr^2_2) = \dfrac{.0411(73 - 1 - 1 - 1)}{(1 - .1261)(1)} = \dfrac{.0411(70)}{.8739} = \dfrac{2.877}{.8739} = 3.29$
- Step 3: $F(sr^2_3) = \dfrac{.0174(73 - 2 - 1 - 1)}{(1 - .1435)(1)} = \dfrac{.0174(69)}{.8565} = \dfrac{1.2006}{.8565} = 1.40$

To determine if the respective increments are significant, we consult Soper (2025a) for the respective $F_{Critical}$ values, and we consult Soper (2025c) for the respective *p* values. Based on the results given above, and assuming a preset alpha level of α = .05, we conclude the following:

Step 1. The calculated *F* for sr_1^2 is $F(1, 71) = 6.59$ vs. $F_{Critical} = 3.9758$. Therefore, X_1's incremental effect where $sr_1^2 = .0850$ is significant, $p = .0124$.

Step 2. The calculated *F* for sr_2^2 is $F(1, 70) = 3.29$ vs. $F_{Critical} = 3.9778$. Therefore, X_2's incremental effect where $sr_2^2 = .0411$, is *not* significant, $p = .0739$.

Step 3. The calculated *F* for sr_3^2 is $F(1, 69) = 1.40$ vs. $F_{Critical} = 3.9798$. Therefore, X_3's incremental effect where $sr_3^2 = .0174$, is *not* significant, $p = .2408$.

Using an Online Calculator to Determine F_I and p_I. Soper (2025b) provides an online calculator for determining the *F* values for a hierarchical regression analysis, and as noted above, Soper (2025c) provides *p* values for *F*. When using Soper (2025b), the user must enter: (a) Number of predictors in set A, which is k_A; (b) Number of predictors in set B, which is k_B; (c) R^2 for set A ($R^2_{Y \bullet A}$), which is the cumulative R^2 of the previous model; (d) R^2 for sets A and B ($R^2_{Y \bullet AB}$), which is the cumulative R^2 for the current model; and (e) the overall sample size, *N*.

Step 1. Using Soper (2025b) at Step 1, the reader is cautioned that an error will occur because $k_A = 0$ and Soper requires k_A to be between 1 and 99. This is not a concern, though, because Step 1 does not actually reflect a hierarchical regression, per se, but instead may loosely be considered a simultaneous regression where "all the IVs in the model" are only those being examined at Step 1. Thus, the *F* value of the increment may be obtained directly from our statistics program or from Soper (2025g). Regardless, the *F* value for sr_1^2 is $F(1, 71) = 6.59$ and $p = .0124$.

Step 2. Using Soper (2025b) at Step 2: $k_A = 1$, $k_B = 1$, $R^2_{Y \bullet A} = .0850$, $R^2_{Y \bullet AB} = .1261$, and $N = 73$. This yields $F(1, 70) = 3.29$, and from Soper (2025c), $p = .0740$.

Step 3. Using Soper (2025b) at Step 3: $k_A = 2$, $k_B = 1$, $R^2_{Y \bullet A} = .1261$, $R^2_{Y \bullet A} = .1435$, $N = 73$. This yields $F(1, 69) = 1.40$, and from Soper (2025c), $p = .2408$.

These results match those we calculated manually.

Reporting the Results of a Hierarchical Regression Analysis

There are three ways to report the results of a hierarchical regression analysis. One is to follow the structure of the style manual that is used for a particular discipline. For aviation research, the style manual we use is from the American Psychological Association (APA, 2020). A second approach is to follow the guidelines of the targeted audience. This would include a thesis/dissertation committee, a college or university's graduate school or office, and the journal where the research will be published. A third approach is to present the results in a manner that you feel is more informative and easier for the reader to follow. Our focus here is on the first and third approaches.

APA Compliant Table. Table 6.2 contains a summary of the results of the analysis involving our running example and is structured to be compliant with the sample hierarchical multiple regression table given in APA (2020, p. 121). Note how Table 6.2 is organized:

TABLE 6.2 *APA (2020) Compliant Table for Summarizing the Results of the Hierarchical Regression Analysis for Chapter 6's Regression Strategies Example Data with Variable Entry Order X_1– X_2– X_3*

		95% CI for B_i [a]					
Variables in Model	B_i	*LL*	*UL*	$SE\ B_i$	β_i	R^2 [b]	ΔR^2 [c]
Step 1						0850*	.0850*
Constant	44.02***	29.65	58.38	7.20			
X_1 = Age	0.55*	0.12	0.98	0.21	0.29*		
Step 2						.1261**	.0411
Constant	41.91***	27.58	56.24	7.18			
X_1 = Age	0.48*	0.06	0.91	0.21	0.26*		
X_2 = Biological Sex	7.68	−0.76	16.12	4.23	0.21		
Step 3						.1435*	.0174
Constant	45.31***	29.92	60.70	7.72			
X_1 = Age	0.45*	0.02	0.88	0.22	0.24*		
X_2 = Biological Sex	8.58*	0.03	17.13	4.29	0.23*		
X_3 = Total Flight Hours	−0.15	−0.41	0.11	0.13	−0.13		

Note. $N = 73$. All entries rounded to two decimal places except R^2 and ΔR^2, which are rounded to four decimal places. The DV is Y = CFI Rating.
[a] CI = confidence interval; *LL* = lower limit; *UL* = upper limit. [b] R^2 = the overall R^2 for the current model at each step. [c] $\Delta R^2 = sR^2$, which is the increment in explained Y variance for each X_i at a given step. It is the difference between the R^2 at a current stage and the R^2 of the previous stage.
$^*p < .05$. $^{**}p < .01$. $^{***}p < .001$.

- The results of the analysis are reported separately for each step. This enables the reader to clearly see the variable entry order and the respective regression models. For example: at Step 1, X_1 enters the model in the presence of no other IVs; at Step 2, X_2 enters the model in the presence of X_1; and at Step 3, X_3 enters the model in the presence of X_1 and X_2. Thus, the variable entry order is X_1–X_2–X_3.
- R^2 is the *cumulative* R^2 at each step of the analysis. Note that this column reflects the collective contribution to explained Y variance made by the IVs that are in the current model. For example, at Step 2, the R^2 for the current model is $R^2_{Y\bullet12} = .1261$, which is the amount of Y variance collectively explained by X_1 and X_2. The two asterisks indicate that this model is statistically significant ($p < .01$). Recognize that at the final step of the analysis, the cumulative R^2 is equal to the overall R^2 from a simultaneous regression analysis.
- ΔR^2 is the *increment* (I) in explained Y variance that is attributed to the variable entering the model at each step. Note that this increment is the unique contribution to explained Y variance a variable makes when it enters the current model in the presence of those IVs that were in the previous model. Thus, $\Delta R^2 = I = sr_i^2$. For example, when X_2 enters the model at Step 2, it is entering the model in the presence of X_1. Therefore, X_2's increment is $I_2 = sr_2^2 = .1261 - .0850 = .0411$. The absence of an asterisk indicates that the increment is not statistically significant ($p > .05$).

- The sum of the increments is equal to the overall R^2 given in the final model, which occurs at the last step of the analysis: $R^2_{Y \cdot 123}$ = .0850 + .0411 + .0174 = .1435. Once again, note how the results of a hierarchical regression analysis provide a full accountability of the explained *Y* variance among the IVs, which is something that a simultaneous regression approach does not do.

Researcher-Derived Table. Recall that one of the primary reasons for performing a hierarchical regression analysis is to answer the research question, "What is the incremental effect of each IV relative to the DV?" Although Table 6.2 is APA (2020) compliant, its structure does not lend itself to an easy examination of the increment. For example, it lacks specific information about the increment's *df*, *F*, and *p* values. As an alternative to Table 6.2, we prepared Table 6.3 to reconcile our perceived deficiencies of Table 6.2. Note that Table 6.3 includes the regression coefficients, overall R^2, and increment ($I_i = sr_i^2$) for each model similar to Table 6.2, but it also has a greater emphasis on the increment by including each increment's corresponding degrees of freedom, *F* value, and exact *p* value. This enables the reader to more completely answer the corresponding RQ. Table 6.3, however, does not include the regression constants at each step of the analysis. Although the regression constant is not the primary focus, it can easily be inserted to make the table more complete with respect to the regression equations at each step. Nevertheless, the reader is once again cautioned that because Table 6.3 is not APA compliant, it might not be acceptable for publication purposes such as for a thesis, dissertation, or research journal.

Determining the Variable Entry Order for a Hierarchical Regression Analysis

A key consideration for hierarchical regression is the variable entry order. Because each model's cumulative R^2 and sr_i^2 are determined at each step of the analysis, the cumulative R^2 and the increment an IV makes in explaining *Y* variance can change relative to the variable entry order. To illustrate this, we prepared Table 6.4, which contains a summary of the results from six hierarchical regression analyses for our running example's data set that consists of all possible variable entry order permutations. With respect to Table 6.4, observe the following:

- At Step 1 of Analysis 1, when X_1 enters the model first, the cumulative R^2 = .0850 and p = .012 ($p < .05$). However, at Step 1 of Analysis 5, when X_3 enters the analysis first, the cumulative R^2 = .0163 and p = .282 ($p > .05$).

TABLE 6.3 *Alternative Table for Summarizing the Results of the Hierarchical Regression Analysis for Chapter 6's Regression Strategies Example Data with Variable Entry Order X_1– X_2– X_3*

Variables in Model [a]	B_i	Cum. R^2 [b]	$I_i = sr_i^2$ [c]	df_I [d]	F_I [d]	p_I [d]
X_1	B_1 = 0.55	.0850	.0850	1, 71	6.59	.0124
X_1, X_2	B_1 = 0.48, B_2 = 7.68	.1261	.0411	1, 70	3.29	.0740
X_1, X_2, X_3	B_1 = 0.45, B_2 = 8.58, B_3 = –0.15	.1435	.0174	1, 69	1.40	.2408

Note. N = 73. The DV is Y = CFI Rating.
[a] X_1 = Age, X_2 = Biological Sex, X_3 = Total Flight Hours. [b] Cum. R^2 = the R^2 for the current model at each step of the analysis. [c] I = the increment in explained *Y* variance by the variable entering the model at each step. [d] df_I, F_I, p_I = the respective *df*, *F*, and *p* values for the increment.

TABLE 6.4 *Summary of the Results of Six Hierarchical Regression Analyses for Chapter 6's Regression Strategies Example Data for All Possible Variable Entry Order Permutations*

Analysis	Variables in Model [a]	B_i	Cum. R^2 [b]	$I_i = sr_i^2$ [c]	df_I	F_I	p_i
1	X_1	$B_1 = 0.55$	.0850	.0850	1, 71	6.59	.012
	X_1, X_2	$B_1 = 0.48, B_2 = 7.68$	.1261	.0411	1, 70	3.29	.074
	X_1, X_2, X_3	$B_1 = 0.45, B_2 = 8.58, B_3 = -0.15$	.1435	.0174	1, 69	1.40	.241
2	X_1	$B_1 = 0.55$	.0850	.0850	1, 71	6.59	.012
	X_1, X_3	$B_1 = 0.53, B_3 = -0.11$	.0938	.0088	1, 70	0.68	.412
	X_1, X_3, X_2	$B_1 = 0.45, B_3 = -0.15, B_2 = 8.58$	.1435	.0497	1, 69	4.00	.049
3	X_2	$B_2 = 9.29$	.0620	.0620	1, 71	4.69	.034
	X_2, X_1	$B_2 = 7.68, B_1 = 0.48$	.1261	.0641	1, 70	5.13	.027
	X_2, X_1, X_3	$B_2 = 8.58, B_1 = 0.45, B_3 = -0.15$	.1435	.0174	1, 69	1.40	.241
4	X_2	$B_2 = 9.29$	.0620	.0620	1, 71	4.69	.034
	X_2, X_3	$B_2 = 10.26, B_3 = -0.19$	.0901	.0281	1, 70	2.16	.146
	X_2, X_3, X_1	$B_2 = 8.58, B_3 = -0.15, B_1 = 0.45$	.1435	.0534	1, 69	4.30	.042
5	X_3	$B_3 = -0.15$	.0163	.0163	1, 71	1.17	.282
	X_3, X_1	$B_3 = -0.11, B_1 = 0.53$	.0938	.0775	1, 70	5.99	.017
	X_3, X_1, X_2	$B_3 = -0.15, B_1 = 0.45, B_2 = 8.58$	.1435	.0497	1, 69	4.00	.049
6	X_3	$B_3 = -0.15$	.0163	.0163	1, 71	1.17	.282
	X_3, X_2	$B_3 = -0.19, B_2 = 10.26$	.0901	.0738	1, 70	5.68	.020
	X_3, X_2, X_1	$B_3 = -0.15, B_2 = 8.58, B_1 = 0.45$	.1435	.0534	1, 69	4.30	.042

Note. $N = 73$. The DV is Y = CFI Rating.
[a] X_1 = Age, X_2 = Biological Sex, X_3 = Total Flight Hours. [b] Cum. R^2 = the R^2 for the current model at each step of the analysis. [c] I = the increment in explained variance by the X_i entering the model at each step.

- At Step 2 of Analysis 2, where the variable entry order at this step is X_1–X_3, when X_3 enters the model in the presence of X_1, its increment is $sr_3^2 = .0088$, which is less than 1%, and the corresponding p value is $p = .412$. However, at Step 2 of Analysis 4, where the variable entry order at this step is X_2–X_3, when X_3 enters the model in the presence of X_2, its increment is $sr_2^2 = .0281$, which is 2.8%, and the corresponding p value is $p = .146$.

Notice also from Table 6.4 that regardless of the variable entry order, the results are exactly the same for the last step of each analysis. This should not be surprising because each analysis is based on the same data set and all the IVs are now in the model at this step.

Given the differences in results that can occur based on the variable entry order, a key question is, "What criteria do we use to determine the order?" Prior to answering this question, the reader is reminded that the choice of variable entry order is to be made a priori. The reader also should recognize that early-entry variables will get most of the shared Y variance, whereas later-entry variables likely will have little-to-no shared variance. With this in mind, one approach to determining the variable entry order is to consider the variables' presumed causal priority: *We should not let any variable X_a enter the model before a variable X_b if we believe (i.e., presume) that X_b is the cause of X_a.* For example, given the model in Figure 6.4, it is presumed that X_2

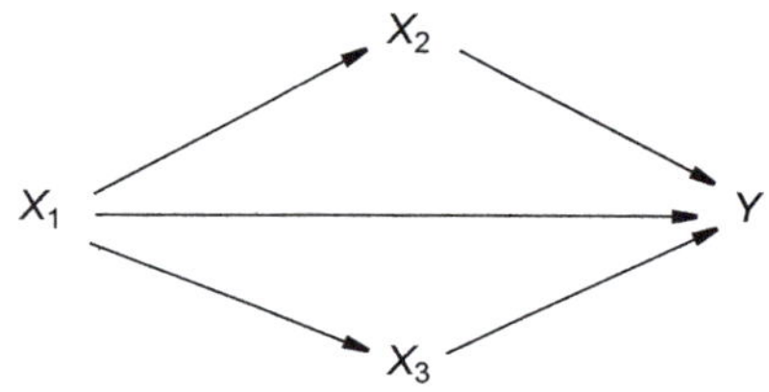

FIGURE 6.4 Illustration of the concept of presumed causal priority.

and X_3 are caused by X_1 and hence X_1 has presumed causal priority. Alternatively, we could say that X_2 and X_3 are *presumptive causes* of X_1. As a result, neither X_2 nor X_3 should precede X_1 in a hierarchical regression analysis, which means the corresponding variable entry order would be X_1–X_2–X_3 or X_1–X_3–X_2.

Applying this to our current example, where X_1 = Age, X_2 = Biological Sex, and X_3 = Total Flight Hours, it is plausible to consider age a presumptive cause of flight hours if we presume that the older you are the more flight hours you are likely to have. Thus, an appropriate variable entry order relative to this presumption would be X_1–X_2–X_3 or X_1–X_3–X_2. Given the male-female ratio among pilots, it also might be plausible to consider biological sex a presumptive cause of flight hours and therefore we would not let X_3 precede X_2.

Another consideration for determining the variable entry order for a hierarchical regression analysis is the relevance of the IVs with respect to the research study: *Primary IVs are to enter the model before secondary IVs.* As an example, consider the data set from the file "Ch_4 Example 4.2 Data," which was used for the study presented in Example 4.2 (Chapter 4). Recall that this study examined the effect of five factors on pilots' visual reaction times: X_1 = Age, X_2 = Corrected Vision (yes/no), X_3 = Visual Acuity Angle, X_4 = Flight Hours, and X_5 = Number of FAA Ratings. (*Note*: We also examined this data set in Chapter 5 for outliers in Example 5.1, and then again for IV relevance later in the chapter when discussing regression diagnostics. See the file "Ch_5 Visual Reaction Times Data.")

Given the DV, Y = Visual Reaction Times, we might declare pilots' age, visual acuity angle, and whether or not their vision was corrected (i.e., they wore glasses/contact lenses or had Lasik surgery) as primary IVs, but the number of flight hours and FAA ratings as secondary IVs. As a result, X_1, X_2, and X_3 would be early-entry variables and X_4 and X_5 would be later-entry variables. If we also apply the concept of presumed causal priority, then pilots' visual acuity angle and whether or not they have corrected vision would be considered presumptive causes of age. Similarly, because whether or not pilots have corrected vision is based on their visual acuity angle, corrected vision would be a presumptive cause of visual acuity angle. When all these considerations are taken collectively, an appropriate variable entry order would be X_1–X_3–X_2–X_4–X_5 or X_1–X_3–X_2–X_5–X_4.

6.4 Statistical (Stepwise) Regression

An alternative to simultaneous and hierarchical regression is a third variable entry strategy called statistical regression. This strategy is sometimes referred to as stepwise regression because IVs are entered into or removed from the regression model one step at a time based on various variable selection criteria. Although "stepwise regression" aptly describes how a statistical regression is conducted, the term also refers to a specific statistical regression option (discussed later) within the SPSS data analysis community. Thus, the reader is cautioned to be sensitive to the context in which "stepwise regression" is used. To avoid any confusion, we will use the term "statistical regression."

As summarized in Figure 6.5, there are four statistical regression options—*forward addition*, *backward elimination*, *mixed (stepwise)*, and *best model*—and three different selection criteria that determine if an IV should be included in or excluded from the regression model—*p values*, *Akaike Information Criteria (AIC)*, and *Bayesian Information Criteria (BIC)*. Before we examine the four statistical regression options, we discuss the three variable selection criteria because they may be applied to all four options.

Variable Selection Criteria

***p* Value Approach.** The *p* value approach—also referred to as the *p* value stopping rule—compares the *p* values that are associated with the *F* tests of each IV's increment (sr_i^2) to a preset significance level, α. This preset alpha level is called α-to-enter for the forward addition option and α-to-leave for the backward elimination option. In the forward addition option, if $p <$ α-to-enter, then the IV enters (or is "added to") the model, whereas in the backward elimination option, if $p >$ α-to-leave, then the IV is removed (or is "eliminated") from the model. Thus, α-to-enter is a threshold or "cut-off" value for determining if an IV enters or does not enter the model, and α-to-leave is a cut-off value for determining if an IV remains in or is removed from the model. The order in which an IV is entered or removed from a regression model is based on the IV's increment to R^2. In forward addition, the IV with the highest increment and $p <$ α-to-enter at each stage is added. In backward elimination, the IV with the smallest increment and $p >$ α-to-leave at each stage is removed. The reader is cautioned that if alpha is set to the traditional $\alpha = .05$ or $\alpha = .01$ level, it is possible that "important" variables could be left out, which would result in an underspecified model with biased regression coefficients and spurious predicted *Y* values. As a result, Bendel and Afifi (1977) and Hosmer et al. (2013) recommend setting α between $\alpha = .15$ and $\alpha = .20$, whereas Harrell (2001) proposes an even more liberal alpha level of $\alpha = .50$. Thus, different regression models could result by using different alpha levels.

Akaike Information Criterion and Bayesian Information Criterion. Without going into too much detail—the interested reader is directed to Aho et al. (2014), Burnham and Anderson (2004), and Chowdhury and Turin (2020)—both AIC and BIC seek a final model that is analogous to the "Three Little Bears" fairy tale: The final model is neither underspecified (too few IVs) nor over specified (too many IVs), but "just right." With respect to AIC, given a set of targeted variables, the AIC algorithm generates and compares different regression models with different IVs and ranks these models via an AIC metric. The model that the algorithm determines to fit the data the "best" will have the smallest *RMSE*, which will yield the smallest AIC metric (i.e., "minimum AIC"). However, if the sample size, *N*, is relatively "small," then there is a tendency for the final model to be over specified. To remedy this tendency, a modified version of AIC is used. This modified version is designated AICc,

Forward Addition

Begin with no IVs in the model.

Step 1. Examine each IV's zero-order squared correlation coefficient (r^2) one at a time. The IV with the largest statistically significant r^2 enters the model first.

Step 2. Examine the remaining IVs one at a time. The IV that has the largest significant increment (sr^2) in the presence of the IV(s) currently in the model enters the model next.

Step 3. Continue with Step 2 until the designated variable selection criterion is met.

Variable Selection Criteria

***p* values.** Set the significance level (α-to-enter) and compare the *p* value of each IV's increment to α-to-enter. The process stops when $p > \alpha$ for the increment of the remaining IVs.

Akaike Information Criterion (AIC). Select the model with minimum AIC.

Bayesian Information Criterion (BIC). Select the model with minimum BIC.

Backward Elimination

Begin with all the IVs in the model.

Step 1. Examine the IVs in the current model one at a time. The IV that makes the least significant contribution to R^2 leaves the model.

Step 2. Refit the model after the IV from Step 1 has been removed.

Step 3. Continue this process until the designated variable selection criterion is met.

Variable Selection Criteria

***p* values.** Set the significance level (α-to-leave) and compare the *p* value of each IV's increment to α-to-leave. The process stops when $p < \alpha$ for the increment of the remaining IVs in the model.

Akaike Information Criterion (AIC). Select the model with minimum AIC.

Bayesian Information Criterion (BIC). Select the model with minimum BIC.

Mixed (*Stepwise*)

Begin with either the Forward Addition or Backward Elimination option.

If the analysis begins with Forward Addition:

- Add the IVs into the model one at a time as described in the Forward Addition option.
- After an IV is added to the model, refit this new model and apply Backward Elimination to determine if any of the IVs in the current model need to be removed.
- Continue this process until no more IVs can be added to or removed from the model, which implies that the increments of the IVs in the model are significant and the increments of the IVs not in the model are not significant.

If the analysis begins with Backward Elimination:

- Remove IVs from the full model one at a time as described in the Backward Elimination option.
- After the Backward Elimination stopping rule takes effect, reexamine the model with the IVs that were dropped to see if they can now reenter the model.

Note. For the *p* value stopping rule, both α-to-enter and α-to-leave are needed, where α-to-enter < α-to-leave.

Best Model

The computer's Best Model algorithm will examine *all possible models* relative to the subsets associated with the number of IVs in a given data set. In general, a data set with *k* IVs will have 2^k subsets. For example, with $k = 4$, there are 16 subsets:

- 1 subset with 0 IVs, which is the empty set: { }
- 4 subsets with 1 IV: $\{X_1\}, \{X_2\}, \{X_3\}, \{X_4\}$
- 6 subsets with 2 IVs: $\{X_1, X_2\}, \{X_1, X_3\}, \{X_1, X_4\}, \{X_2, X_3\}, \{X_2, X_4\}, \{X_3, X_4\}$
- 4 subsets with 3 IVs: $\{X_1, X_2, X_3\}, \{X_1, X_2, X_4\}, \{X_1, X_3, X_4\}, \{X_2, X_3, X_4\}$
- 1 subset with 4 IVs: $\{X_1, X_2, X_3, X_4\}$

The computer will display the various models associated with the subsets and the researcher will then determine which model to accept relative to *RMSE*, AIC, and BIC.

Note. Because the subset with zero IVs will result in $R^2 = 0$, the Best Model algorithm examines $2^k - 1$ models.

FIGURE 6.5 Summary of statistical regression options and corresponding IV selection criteria.

where the lowercase "c" denotes that the corresponding AIC metric was "corrected" for small sample sizes. Because AICc approaches AIC as *N* gets larger, many statistics programs report AICc.

BIC is similar to AIC in that a BIC metric is calculated for each model and the "best" model is the one with minimum BIC. The primary difference between AIC and BIC, though, is that BIC has a *consistency property* incorporated into its algorithm, which essentially means, "As *N* grows large, the BIC selects the true model with probability approaching 1" (Vrieze, 2012, p. 233). According to Vrieze (2012), this property enables researchers "to conduct what amounts to data-driven theory building. One can collect data, fit models, and compare models with BIC, knowing that as *N* grows large the true model will be selected" (p. 233). Comparing AIC and BIC,

Chowdhury and Turin (2020), who referenced Aho et al. (2014), provided the following summary (p. 6):

> If our objective is to select the best model that will provide maximum predictive accuracy, then AIC is superior (because there is no true model, and the best model is selected to maximise the predictive accuracy and represent an approximate true relation). However, if the goal is to select a correct model that is consistent, then BIC is superior (because BIC consistently selects the correct model from among the candidate models that best represent the true model).

For our purposes, we will report both AICc and BIC values. We now present a discussion of each of the statistical regression options.

Statistical Regression Options

To demonstrate the four statistical regression options as well as the different variable selection criteria, we will use the data set given in the file "Ch_6 Statistical Regression Example Data." This data set is a replication of "Ch_4 Review Exercises Data-Part B," which was used for Part B of Chapter 4's Review Exercises but relabeled for the current discussion. Recall that this review exercise examines the relationship between four student related factors, X_1 = Age, X_2 = High School GPA, X_3 = Flight vs. Nonflight Students (coded 1 for flight students), and X_4 = Total Credit Hours earned to date. The DV is Y = College GPA.

Forward Addition. In the forward addition option, the computer is presented with a group of IVs where none of the IVs are in the model. It then performs a series of regression analyses to determine which IV yields the highest statistically significant increment to explaining Y variance at each step of the analysis. The computer first examines each IV's zero-order squared correlation coefficient (r_i^2), which is also equal to (sr_i^2) at the initial stage of the analysis, and the IV with the largest significant r^2 enters the model first. The computer then examines the remaining IVs, one at a time, to determine which IV makes the largest statistically significant increment to R^2 in the presence of the IV(s) currently in the model and places it into the model. This process continues until all the IVs have been added into the model or when the remaining IVs do not make a significant incremental contribution to R^2. Once an IV enters the model it remains in the model and is not removed. Table 6.5 contains a summary of the results of the forward addition option for the data set given in the file "Ch_6 Statistical Regression Example Data." The reader will note that the p value stopping rule is α-to-enter = .20. Observe the following from Table 6.5:

Step 1. No IVs are in the model. Each IV is examined separately for its contribution to explaining Y variance. All IVs meet the α-to-enter threshold (their respective p values are less than .20), but X_2 makes the largest significant contribution, $R^2_{Y\bullet 2} = sr^2_2 = .4291$, $F(1, 295) = 221.73$, $p < .0001$. Therefore, it enters the model first.

Step 2. X_2 is in the model. X_1 and X_4 both meet the α-to-enter threshold, but X_4 has the larger significant increment, $sr^2_4 = R^2_{Y\bullet 24} - R^2_{Y\bullet 2} = .4822 - .4291 = .0531$, $F(1, 294) = 30.15$, $p < .0001$. Therefore, it enters the model next.

Step 3. X_2 and X_4 are in the model. X_3 meets the α-to-enter threshold, $sr^2_3 = R^2_{Y\bullet 243} - R^2_{Y\bullet 24} = .4879 - .4822 = .0057$, $F(1, 293) = 3.26$, $p = .0720$. Therefore, it enters the model at this step.

TABLE 6.5 *Summary of Results of the Forward Addition Option for Statistical Regression Relative to the Data Set Given in the File "Ch_6 Statistical Regression Example Data"*

Step-by-Step Analysis Results

Analysis	IVs in Model	IV Being Examined [a]	Cum. R^2 [b]	$I_i = sr_i^2$ [c]	df_I	F_I	p_I	AICc	BIC
Step 1		X_1	.0242	.0242	1, 295	7.33	.0072		
		X_2	.4291	.4291	1, 295	221.73	< .0001	607.16	618.16
		X_3	.1526	.1526	1, 295	53.10	< .0001		
		X_4	.1946	.1946	1, 295	71.30	< .0001		
Step 2	X_2	X_1	.4323	.0032	1, 294	1.66	.1986		
		X_3	.4306	.0015	1, 294	0.77	.3809		
		X_4	.4822	.0531	1, 294	30.15	< .0001	580.24	594.87
Step 3	X_2, X_4	X_1	.4830	.0008	1, 293	0.45	.5029		
		X_3	.4879	.0057	1, 293	3.26	.0720	579.00	597.26
Step 4	X_2, X_4, X_3	X_1	.4884	.0005	1, 292	0.28	.5971		

Note. Final Model is $R^2_{Y\cdot243}$ = .4879, $F(3, 293) = 93.05$, $p < .0001$. The second color reflects the "best" IV at each step relative to the other IVs currently present.
[a] X_1 = Age, X_2 = High School GPA, X_3 = Flight vs. Nonflight Student, X_4 = Total Credit Hours. [b] Cum. R^2 = the R^2 for the current model at each step of the analysis. [c] I_i = the increment in explained variance by the X_i entering the model at each step.

Coefficients (Parameter Estimates) of Final Model

Term	B_i	SE_i	$t(81)$	p_i	95% CI
Constant	−1.15	0.311	−3.70	.0003	[−1.76, −0.54]
X_2 = High School GPA	1.09	0.114	9.61	< .0001	[0.87, 1.31]
X_4 = Total Credit Hours	0.03	0.004	5.73	< .0001	[0.02, 0.04]
X_3 = Flight vs. Nonflight Student	0.16	0.09	1.81	.0712	[−0.01, 0.34]

Note. N = 297 (three outliers were removed from the original data set of N = 300). Results based on α-to-enter = .20.

Step 4. X_2, X_4, and X_3 are in the model. X_1 is now examined but it does not meet the α-to-enter threshold, $sr_1^2 = R^2_{Y\cdot2431} - R^2_{Y\cdot243} = .4884 - .4879 = .0005$, $F(1, 292) = 0.28$, $p = .5971$. Therefore, it does not get added to the model.

Final Model. As a result, the final model from the forward addition option is X_2–X_4–X_3, and $R^2_{Y\cdot243} = .4879$, $F(3, 293) = 93.05$, $p < .0001$. The corresponding Coefficients table is included in Table 6.5.

Recall that the results of the final model are based on α-to-enter = .20. Looking at the p column for the increment (p_I) in Table 6.5, note that all the p values are less than .20 except for X_3 at Step 2 and X_1 at Step 3 and Step 4. To add X_3 to the model at Step 2, α-to-enter would have to be set to .40 at the start of the analysis. This change in the significance level, though, would not have altered the results because at Step 2, $sr_4^2 > sr_3^2$. However, if α-to-enter were set at .60, then all four IVs would be placed in the final model because at Step 4, X_1's increment has a p value that is less than the α-to-enter = .60. Continuing this discussion of changing α-to-enter, look at what happens if α-to-enter = .05. Focusing once again on the p_I column in Table 6.5, X_2 would enter the model first followed by X_4 at Step 2. However, neither X_3 nor X_4 would be in the final model because their respective p values are greater than α-to-enter = .05 at the remaining steps of the analysis.

Now look at the AICc and BIC values that are included in Table 6.5. Note that minimum AICc yields the same model as the p value stopping rule. However, minimum BIC yields the model X_2–X_4 with R^2 = .4822, $F(2, 294)$ = 136.88, p < .0001. So, which model should be reported as the final model? The answer is "it depends." Referencing our previous discussion on AICc and BIC, we opt to report the final model as X_2–X_4–X_3 because our focus is on "maximum predictive accuracy."

Backward Elimination

The backward elimination option is the opposite of forward addition. The analysis begins with all k IVs in the model and then removes the IV that makes the *smallest* contribution to R^2 where p > α-to-leave. In other words, when this IV is removed, its absence results in the largest R^2 among the remaining $(k - 1)$ IVs in the model. This process of removing one IV at a time is repeated until having fewer IVs in the model do not result in a statistically better model fit. With respect to the p value stopping rule, the IVs in this final model also will have p values less than the α-to-leave threshold. Table 6.6 contains a summary of the results of the backward elimination option for our running example involving the college GPA data set where α-to-leave = .20. Observe the following from Table 6.6:

Step 1. With all the IVs in the model, X_1 has the largest p value (p = .5912) that is greater than α-to-leave. X_1 also makes the least significant contribution to R^2 and hence, when X_1 is removed from the model, $R^2_{Y \cdot 1234}$ is reduced by the smallest amount, which leads to the highest overall R^2 among the remaining IVs. Therefore, the decision is to remove X_1.

TABLE 6.6 *Summary of Results of the Backward Elimination Option for Statistical Regression Relative to the Data Set Given in the File "Ch_6 Statistical Regression Example Data"*

Analysis	Overall Model	X_i [a]	p_i	Remove X_i	Loss in R^2	Comments [b]
Step 1	$R^2_{Y \cdot 1234}$ = .4884	X_1	.5912	$R^2_{Y \cdot 234}$ = .4879	.0005	• X_1's absence has smallest loss in R^2.
		X_2	< .0001	$R^2_{Y \cdot 134}$ = .3282	.1602	• X_1's p > α-to-leave.
		X_3	.0792	$R^2_{Y \cdot 124}$ = .4830	.0054	• Decision: Remove X_1.
		X_4	< .0001	$R^2_{Y \cdot 123}$ = .4335	.0549	
Step 2	$R^2_{Y \cdot 234}$ = .4879	X_2	< .0001	$R^2_{Y \cdot 34}$ = .3267	.1612	• X_3's absence has smallest loss in R^2.
		X_3	.0712	$R^2_{Y \cdot 24}$ = .4822	.0057	• All p values are less than α-to-leave.
		X_4	< .0001	$R^2_{Y \cdot 23}$ = .4306	.0573	• Stop the analysis.

Note. N = 297 (three outliers were removed from the original data set of N = 300). Results based on α-to-leave = .20.

[a] X_1 = Age, X_2 = High School GPA, X_3 = Flight vs. Nonflight Student, X_4 = Total Credit Hours. [b] For the initial model with all four IVs, AICc = 580.79 and BIC = 602.66. With X_1 removed, AICc = 579.00 and BIC = 597.26.

Step 2. X_1 is removed and the model is refitted for X_2, X_3, and X_4. This leads to an overall R^2 of $R^2_{Y \cdot 234}$ = .4879. Although X_3 makes the least significant contribution to the overall R^2, the *p* values of *all* the IVs in the model are *less than* α-to-leave. Therefore, the analysis stops, and the final model contains X_2, X_3, and X_4. Note also that minimum AICc and BIC occur when X_1 is removed: AICc = 579.00 and BIC = 597.26 (see Table 6.6's table note).

Although not displayed in Table 6.6, it is instructional to observe what happens if α-to-leave = .05 instead of .20. Based on the respective *p* values given in Table 6.6 at Step 2 for X_2, X_3, and X_4, the *p* value for X_3 is p = .0712, which is greater than α-to-leave = .05. Therefore, the backward elimination method would execute one more step (Step 3) and remove X_3, which results in a final model containing X_2 and X_4. Furthermore, the corresponding AICc = 580.24 and BIC = 594.88. Although BIC at Step 3 is the minimum when compared to Step 1 and Step 2, AICc at Step 3 is *not* the minimum when compared to Step 1 and Step 2. Thus, two different final models emerge: one with X_2 and X_4 if the selection criterion is either α-to-leave = .05 or minimum BIC, and one with X_2, X_3, and X_4 if the selection criterion is minimum AICc.

Mixed (Stepwise)

In the mixed option, the forward addition and backward elimination procedures are combined. This option is also called *Stepwise Selection* in SPSS or *Bidirectional Elimination* in other contexts. Regardless of its name, the analysis involves moving in both directions and hence requires α-to-enter *and* α-to-leave values for the stopping rule. If the analysis begins with forward addition, then at each step, after an IV is added to the model, this new model is examined from a backward elimination approach to see if any IV is to be eliminated. This process continues until all the IVs in the model make a significant contribution to R^2, and all the IVs that were removed are not statistically significant. If the analysis begins with backward elimination, then the least significant IV is removed from the model one at a time, but any previously removed IV is also reexamined for possible reentry. For our running example involving the college GPA data set, the mixed option results in the same final model that was rendered from the forward addition option summarized in Table 6.5.

Best Model

In the previous three options, the respective statistical regression analyses examined one IV at a time and generated one final model. In the best model option, the computer's algorithm examines all the different possible combinations of IVs and presents the "best" models based on various statistical parameters. As a result, this option is often referred to as *All Possible Models*. To explain the concept of "all the different possible combinations of IVs," we invoke the language of set theory: *Given a universal set with k elements, where each element is an IV, there are* 2^k *possible subsets and each subset is considered a regression model that gets examined.* For example, with respect to our running example, the universal set consists of four IVs, $\{X_1, X_2, X_3, X_4\}$, and hence there are 2^4 = 16 *subsets* or different possible combinations of IVs:

- 1 subset has all four elements: $\{X_1, X_2, X_3, X_4\}$
- 4 subsets have three elements each:

  $\{X_1, X_2, X_3\}$, $\{X_1, X_2, X_4\}$, $\{X_1, X_3, X_4\}$, $\{X_2, X_3, X_4\}$

- 6 subsets have two elements each:
  $\{X_1, X_2\}, \{X_1, X_3\}, \{X_1, X_4\}, \{X_2, X_3\}, \{X_2, X_4\}, \{X_3, X_4\}$
- 4 subsets have one element each: $\{X_1\}, \{X_2\}, \{X_3\}, \{X_4\}$
- 1 subset has zero elements (called the *empty set*): { }

Because the model with zero elements (i.e., zero IVs) will result in $R^2 = 0$, in practice, $2^k - 1$ models are examined. So, for our running example, 15 models are examined and the "best" ones are those with the largest cumulative R^2, smallest *RMSE*, minimum AICc, and minimum BIC. Table 6.7 contains a summary of the results of this analysis organzied from the "best" to worst model based on number of IVs.

Observe from Table 6.7 there is no "empty set" model and hence $2^4 - 1 = 15$ models are reported. Further observe the different "best" models based on the number of IVs being considered:

- Of the four models with one IV, the best model is $\{X_2\}$.
- Of the six models with two IVs, the best model is $\{X_2, X_4\}$.
- Of the four models with three IVs, the best model is $\{X_2, X_3, X_4\}$.
- The single model with all four IVs is $\{X_1, X_2, X_3, X_4\}$.

So, which is the overall "best" model? Once again, it depends on the researcher. For example, a case can be made that the three IVs model of $\{X_2, X_3, X_4\}$ is the "best" because it has the second largest R^2, the smallest *RSME*, minimum AICc, and the second smallest BIC. On the other hand, a researcher might place higher credence on minimum BIC and decide that the two IVs model of $\{X_2, X_4\}$ is the "best."

TABLE 6.7 *Summary of Results of the Best Model Option for Statistical Regression Relative to the Data Set Given in the File "Ch_6 Statistical Regression Example Data"*

Number of IVs in Model	Model [a]	Cum. R^2 [b]	*RMSE*	AICc	BIC
1	X_2	.4291	0.6679	607.159	618.159
	X_4	.1947	0.7933	709.347	720.346
	X_3	.1525	0.8137	724.482	735.481
	X_1	.0242	0.8732	766.353	777.352
2	X_2, X_4	.4822	0.6372	580.238	594.876
	X_1, X_2	.4323	0.6671	607.549	622.187
	X_2, X_3	.4306	0.6682	608.451	623.089
	X_3, X_4	.3267	0.7266	658.233	672.871
	X_1, X_4	.2016	0.7912	708.832	723.470
	X_1, X_3	.1637	0.8097	722.593	737.231
3	X_2, X_3, X_4	.4879	0.6347	579.002	597.265
	X_1, X_2, X_4	.4830	0.6378	581.847	600.109
	X_1, X_2, X_3	.4335	0.6676	608.972	627.234
	X_1, X_3, X_4	.3282	0.7270	659.616	677.879
4	X_1, X_2, X_3, X_4	.4884	0.6355	580.792	602.665

Note. $N = 297$ (three outliers were removed from the original data set of $N = 300$).
[a] X_1 = Age, X_2 = High School GPA, X_3 = Flight vs. Nonflight Student, X_4 = Total Credit Hours.
[b] Cum. R^2 = the R^2 for the current model at each step of the analysis.

The Concerns of Using Statistical Regression as a Variable Selection Strategy

In practice, we may consider statistical regression as a variable selection strategy from a "good news–bad news" perspective. The good news is it is an easy way to determine which IVs should be selected for the final model. The bad news is it is an easy way to determine which IVs should be selected for the final model. Although this might seem paradoxical, it is not. This is because the final models from a statistical regression strategy are based on *statistical* parameters such as R^2, sr^2, F, p, *RMSE*, AICc, and BIC and hence are void of any logical considerations, relationships to theory, findings from past or preliminary studies, IV relevance, or presumed causal priority.

For example, consider the results from the forward addition option for our running example. Based on α-to-enter = .20, which is an arbitrary threshold, the final model's variable entry order is X_2–X_4–X_3. When examined from a presumed causal priority perspective, this model suggests that X_4 = Total Credit Hours earned to date is a presumptive cause of X_2 = HS GPA, and that X_3 = Type of Student (flight vs. nonflight) is a presumptive cause of X_4 = Total Credit Hours. Although it is reasonable for high school GPA to have the greatest relevance to the DV (Y = College GPA) and therefore should enter the model first, an argument could be made that the type of student (flight vs. nonflight) dictates students' total credit hours. For example, because of differences in curriculum, flight students might require taking more courses than nonflight students. If this were the case, then this would make total credit hours a presumptive cause of type of student and a more logical variable entry order would be X_2–X_3–X_4. We also should not dismiss X_1 = Age, which was not included in the final model. Because older students generally accumulate more credit hours than younger students (e.g., think first-year college students' mean age vs. that of fourth-year college students), total credit hours would be a presumptive cause of age. So, if a hierarchical regression strategy were used, a more appropriate variable entry order would be X_2–X_1–X_3–X_4, which would result in a larger R^2 than what resulted from the forward addition option of statistical regression.

Let's now consider the backward elimination option. Because this option begins with a model in which all k IVs are entered simultaneously, the final model that results from eliminating one IV at a time relative to an arbitrary cut-off has no specific variable entry order. The same also can be said for the mixed option given that it involves adding and removing IVs from the model one at time. Lastly, with the best option, the researcher is presented with $2^k - 1$ models and selects which model to report as the final one. Once again, in all cases there is no concern for logic, theory, findings from past or preliminary studies, IV relevance, or causal priority.

In short, because statistical regression uses statistical parameters for selecting variables for the final regression model, it can: (a) exploit chance relationships; (b) lead to different variable entry orders for different samples selected from the same population; and (c) as observed by Cohen et al. (2003) and Warner (2008), encourage "lazy" research that effectively becomes a data fishing expedition. These issues notwithstanding, statistical regression can play an important role as a variable selection strategy in certain situations. For example, statistical regression can be very beneficial when used for exploratory purposes or seminal research where the objective is model-building (as opposed to model-testing) to acquire an understanding of the relationship among a targeted set of IVs. It also can help identify IVs that are highly collinear to free a data set from multicollinearity, and the best model option can help researchers determine the "best" linear combination of IVs that leads to the strongest prediction of the DV. For readers who are considering using statistical regression as a variable selection strategy, it would be prudent to heed Cohen et al.'s (p. 162) guidelines for conditions that support such use: (a) restrict the strategy to preliminary research where the relationships among variables are largely unknown, (b) ensure that the k/N ratio, where k = the number of original IVs and N = sample size, is at least 1/40, and (c) perform a cross-validation of the final results. Readers are also cautioned not to overstate the results of statistical regression.

6.5 Applying Hierarchical Regression in Research: A Guided Example

In this chapter, we presented three multiple regression strategies for variable selection: simultaneous, hierarchical, and statistical. For our guided example, though, we will focus on a hierarchical strategy for two reasons. First, and with respect to simultaneous regression, the guided examples in Chapters 4 and 5 were performed using a simultaneous approach and there is no reason to continue with this approach here. Furthermore, unlike hierarchical regression, a simultaneous approach does not provide a full accountability of the explained Y variance and does not require an a priori variable entry order involving the targeted IVs. The reader also is reminded that the final model obtained from the last step of hierarchical regression will be the same as that acquired from simultaneous regression. Second, and with respect to statistical regression, the results can be manipulated by the variable selection criteria, which can lead to chance relationships. As a result, this strategy is better suited for preliminary data analyses and not as a primary strategy. For completeness and instructional purposes, though, we provide an end-of-section commentary that highlights the salient differences between examining the data set for the current Guided Example using a hierarchical regression strategy vs. using a simultaneous or statistical regression strategy. We now present a guided example of how to conduct a hierarchical regression analysis for a research study. A description of the research, which was motivated by Burian (2002), follows.

> An aviation researcher is conducting a study to determine the effect various factors have on general aviation (GA) pilots' knowledge of basic weather concepts. A random sample of N = 85 GA pilots is asked to self-report the following information, which represents the targeted factors:

- X_1 = Total Flight Hours since pilots began flying.
- X_2 = Formal Education Background given as the total number of years of formal schooling.
- X_3 = Number of Formal Weather-Related or Meteorological Courses completed.
- X_4 = Age.
- X_5 = Aviation Safety Scores assessed via a 20-item instrument scored on a traditional 5-point Likert response scale where 1 = Strongly Disagree and 5 = Strongly Agree. Thus, scores could range from 20 to 100, with higher scores reflecting a more positive attitude toward aviation safety.

The dependent variable is Y = Weather Test Scores, which are measured using a researcher-developed instrument that consists of 100 dichotomously scored multiple-choice items that assess pilots' knowledge of basic weather concepts. Scores could range from 0 to 100, with higher scores reflecting greater understanding of basic weather concepts. Following are our thoughts on what the variable entry order will be for the hierarchical regression analysis:

- Based on the given DV, it is plausible for X_3 = Number of Formal Weather Related or Meteorological Courses pilots completed to have the strongest relationship to Y compared to the other IVs, so X_3 will enter the analysis first.
- It is equally plausible for older pilots to have more weather related experiences than younger pilots and therefore X_4 = Age will enter the model second.
- Because X_1 = Total Flight Hours and X_2 = Formal Education Background may be considered presumptive causes of age—for example, older pilots most likely will have more flight hours and more years of formal education than younger pilots—these two IVs will enter the model third and fourth, respectively.
- Finally, it is our belief that X_5 = Aviation Safety Scores will have the least relationship with the DV because this IV measures attitudes toward safety, which is a mind-set or psychological construct, and stands in stark contrast to the other IVs, which are more concrete and practical.

As a result, we will use the variable entry order X_3–X_4–X_1–X_2–X_5. The data set is provided in the file "Ch_6 Guided Example Data." The structure of our presentation follows Chapter 5's Guided Example and consists of six parts: Research Questions and Hypotheses, Sample Size Planning, Preliminary Data Screening, Data Analysis, Interpretation of Results, and Post Hoc Considerations.

Research Questions and Hypotheses

For hierarchical regression, there are two primary research questions and corresponding hypotheses. The first is related to the overall R^2, the second is related to the increments in R^2.

RQ 1. What is the total proportion of variance in weather test scores explained collectively by the IVs?

Hyp. 1. The focus is on the collective contribution of the IVs to Y variance.

- H_0: $\rho^2_{Y \cdot 12345} = 0$ (The five IVs collectively do not explain a significant proportion of Y variance in the population.)
- H_1: $\rho^2_{Y \cdot 12345} > 0$ (The five IVs collectively explain a significant proportion of Y variance in the population.)

RQ 2. When examined from a hierarchical perspective with the given variable entry order, what is the unique contribution each IV makes in explaining Y variance at each step of the analysis? Because there are five IVs, there are five hypotheses.

Hyp. 2a. The focus is on X_3's increment to explaining Y variance in the population when X_3 enters the model first and in the presence of no other IVs.

- H_0: $\rho^2_{Y3} = 0$ (There is no significant predictive gain from 0% in explained Y variance with respect to X_3's increment.)
- H_1: $\rho^2_{Y3} > 0$ (There is a significant predictive gain from 0% in explained Y variance with respect to X_3's increment.)

Hyp. 2b. The focus is on X_4's increment to explaining *Y* variance in the population when X_4 enters the model second and in the presence of X_3.

- H_0: $\rho^2_{Y\bullet 34} - \rho^2_{Y3} = 0$ (There is no significant predictive gain in explained *Y* variance with respect to X_4's increment.)
- H_1: $\rho^2_{Y\bullet 34} - \rho^2_{Y3} > 0$ (There is a significant predictive gain in explained *Y* variance with respect to X_4's increment.)

Hyp. 2c. The focus is on X_1's increment to explaining *Y* variance in the population when X_1 enters the model third and in the presence of X_3 and X_4.

- H_0: $\rho^2_{Y\bullet 341} - \rho^2_{Y\bullet 34} = 0$ (There is no significant predictive gain in explained *Y* variance with respect to X_1's increment.)
- H_1: $\rho^2_{Y\bullet 341} - \rho^2_{Y\bullet 34} > 0$ (There is a significant predictive gain in explained *Y* variance with respect to X_1's increment.)

Hyp. 2d. The focus is on X_2's increment to explaining *Y* variance in the population when X_2 enters the model fourth and in the presence of X_3, X_4, and X_1.

- H_0: $\rho^2_{Y\bullet 3412} - \rho^2_{Y\bullet 341} = 0$ (There is no significant predictive gain in explained *Y* variance with respect to X_2's increment.)
- H_1: $\rho^2_{Y\bullet 3412} - \rho^2_{Y\bullet 341} > 0$ (There is a significant predictive gain in explained *Y* variance with respect to X_2's increment.)

Hyp. 2e. The focus is on X_5's increment to explaining *Y* variance in the population when X_5 enters the model last and in the presence of X_3, X_4, X_1, and X_2.

- H_0: $\rho^2_{Y\bullet 34125} - \rho^2_{Y\bullet 3412} = 0$ (There is no significant predictive gain in explained *Y* variance with respect to X_5's increment.)
- H_1: $\rho^2_{Y\bullet 34125} - \rho^2_{Y\bullet 3412} > 0$ (There is a significant predictive gain in explained *Y* variance with respect to X_5's increment.)

The reader will note that these sr_i^2 values are different from those in simultaneous regression, and we do not pose RQs and corresponding hypotheses for the regression coefficients (B_i). This is because the focus on hierarchical regression is each IV's incremental effect, which reflects the unique partitioning of explained *Y* variance captured by sr_i^2. It also is not appropriate to interpret each B_i value at each step of a hierarchical regression to report the effect the respective X_i has on *Y* when it enters the analysis as we do for sr_i^2. This is because the B_i values collectively reflect an assortment of regression coefficients that were acquired from separate analyses—a "mixed bag" if you will—and therefore cannot be used to make any accurate predictions about the effect each IV has on the DV. Instead, we will report and interpret the regression coefficients of the final model, which will be the same as that acquired from a simultaneous analysis. (*Note:* If the focus is on B_i, then the reader is advised to develop a *causal model* and examine it via *path analysis* or *structural equation modeling*, both of which are beyond the scope of the current text.)

Sample Size Planning

Recall that sample size planning, in general, involves conducting an a priori power analysis to determine the minimum sample size needed based on: (a) an alpha level that corresponds to the probability of committing a Type I error and is typically set to $\alpha = .05$; (b) a minimum power, which is the probability of making a strong, correct decision for rejecting the null hypothesis and is commonly set to .80; and (c) an effect size (*ES*). Thus, an a priori power analysis answers the question, "What size sample is needed to have an 80% chance of finding an effect of size *ES* with respect to a Type I error rate of 5%?" Further recall that for multiple regression we also need to know the number of IVs (k). When applied to hierarchical regression, *ES* is based on the expected overall population R^2 and the expected increments in R^2 at each step of the analysis, (The reader will recall that the expected population parameters for an a priori power analysis are determined by theory, the results from past research or preliminary studies, or professional experience/expertise.)

Overall R^2. Let's assume that we expect the five IVs of the given data set to collectively explain 20% of the variance in the weather test scores. Therefore, $R^2 = .20$, which is equal to an effect size of $ES = f^2 = (.20) / (1 - .20) = (.2) / (.8) = 0.25$, which according to Cohen et al. (2003, p. 93) is a large effect. Consulting Soper (2025d) with $f^2 = 0.25$, power = .8, number of IVs = 5, and $\alpha = .05$, the minimum sample size is $n^* = 57$. Because the given data set has $N = 85$, it is of sufficient size for having an 80% chance of finding an effect size $f^2 = 0.25$ at a Type I error rate of 5%.

The Increments ($I = sr_i^2$). The increments of a hierarchical regression analysis are the sr_i^2 values. The reader is reminded that these sr_i^2 values are different from those obtained in a simultaneous regression. As a refresher, we recommend the reader compare the conceptual diagram of a simultaneous regression (Figure 6.1) to that of a hierarchical regression (Figure 6.3). Remember: In simultaneous regression, each IV is examined in the presence of all the other IVs and hence each IV's contribution to Y variance is partialled from these other IVs. This is why simultaneous regression does not provide a full accountability of explained Y variance. In hierarchical regression, though, the IV that is entering the model at the current stage gets partialled only by those IVs that were in the previous model.

As a result, we need to determine the minimum sample size needed for the increments at each step of the analysis. To do this, we consult Soper (2025e). Although this is the same source we used for simultaneous regression, the difference is with respect to the number of IVs in set A. For example, recall in Chapter 5's Guided Example where the number of IVs = 4, to determine the sample size needed for each sr_i^2, the number of IVs in set A was always 3 and the number of IVs in set B was 1. This reinforces that each IV is being partialled by the other three IVs. This is not the case for the current situation. Because we will be adding one IV into the model at each step of the analysis, the number of IVs in set A will increase by 1 at each step. For example, at Step 1, there will be 0 IVs in the previous model and 1 IV will be entering the current model. Therefore, set A is 0 and set B is 1; that is, $k_A = 0$ and $k_B = 1$. At Step 2, there will be 1 IV in the previous model and 1 IV will be entering the current model and hence $k_A = 1$ and $k_B = 1$. Continuing in this manner, at Step 3, $k_A = 2$, $k_B = 1$; at Step 4, $k_A = 3$, $k_B = 1$; and at Step 5, $k_A = 4$, $k_B = 1$. Table 6.8 contains a summary of the sample size planning for these increments relative to the preset variable entry order X_3–X_4–X_1–X_2–X_5. Focusing on the column headings of Table 6.8, observe the following:

- k_A = the number of IVs in the previous model and k_B = the number of IVs entering the current model.
- $R^2_{Y \cdot AB}$ = the cumulative R^2 of the current model.

TABLE 6.8 *Summary of the A Priori Power Analysis Results Involving the Increments for the Data Set Given in the File "Ch_6 Guided Example Data"*

Step [a]	k_A [b]	k_B [c]	$R^2_{Y \bullet AB}$ [d]	$I_i = sr_i^2$ [e]	f^2 [f]	N^* [g]	Power (N = 85)
				Subtable A			
1 (X_3)	0	1	.08	.08	.087	91	.78
2 (X_4)	1	1	.13	.05	.057	138	.59
3 (X_1)	2	1	.16	.03	.036	217	.49
4 (X_2)	3	1	.18	.02	.024	324	.29
5 (X_5)	4	1	.20	.02	.025	313	.29
				Subtable B			
1 (X_3)	0	1	.10	.10	.111	72	.87
2 (X_4)	1	1	.18	.08	.097	82	.81
3 (X_1)	2	1	.24	.06	.079	101	.73
4 (X_2)	3	1	.30	.06	.086	95	.76
5 (X_5)	4	1	.35	.05	.077	106	.71

Note. N = 85. The dependent variable is Y = Weather Test Scores, which could range from 0 to 100, with higher scores reflecting higher levels of understanding of basic weather concepts. Analyses based on α = .05 and power = .8.
[a] X_1 = Total Flight Hours, X_2 = Formal Education Background (number of years of formal schooling), X_3 = Number of Formal Weather Related or Meteorological Courses Completed, X_4 = Age, X_5 = Aviation Safety Scores (20-item researcher-developed instrument scored on a traditional 5-point Likert response scale; scores could range from 20 to 100; higher scores reflect a more positive attitude toward aviation safety). [b] k_A = number of IVs in previous model. [c] k_B = number of IVs entering current model. [d] $R^2_{Y \bullet AB}$ = cumulative R^2 at this step of the analysis. [e] I_i = expected increment in R^2 for the IV entering the model. [f] $f^2 = (sr_i^2) / (1 - R^2_{Y \bullet AB})$ is the effect size for the increment: [g] Minimum sample size needed for the increment to be statistically significant.

- Each IV is assigned an expected increment (I_i) to the expected population R^2 and the sum of these increments equals the overall expected population R^2, which we assume to be .20.
- The effect sizes of the increments are calculated by $f^2 = (sr_i^2) / (1 - R^2_{Y \bullet AB})$.
- n^* is the minimum sample size needed for each increment relative to the given effect size, power = .80, and α = .05.
- The power for each increment is reported from Soper (2025f) and is based on the sample size of the given data set, namely, N = 85.
- At Step 1 where k_A = 0, Soper (2025e) and Soper (2025f) will give an error because they require k_A to be between 1 and 99. Because there are no other IVs in the model at Step 1 other than the one under discussion, we use Soper (2025d) to determine n^* and Soper (2025g) to determine power.

As an example of how to interpret the contents of Table 6.8, let's focus on Step 3 of Subtable A. At this step, X_1 enters the model in the presence of X_3 and X_4, which means there are two IVs in the previous model (k_A = 2) and 1 IV is entering the current model (k_B = 1). Also at this step, the total amount of Y variance previously explained by X_3 and X_4 is $R^2_{Y \bullet A}$ = .13. When X_1 enters the model, the total amount of explained Y variance with all three IVs present is $R^2_{Y \bullet AB}$ = .16, which means that X_1's increment is $sr_1^2 = (R^2_{Y\bullet AB} - R^2_{Y\bullet A}) = (.16 - .13) = .03$. Another way to denote this is $sr_1^2 = R^2_{Y(1\bullet 34)}$, which indicates that sr_1^2 is equal to the squared bivariate correlation coefficient between Y and X_1 where the effects of X_3 and X_4 are being partialled from X_1 but not from Y. Thus, when examined in the presence of X_3 and X_4, X_1 explains an additional 3% of Y variance and its corresponding effect size is $f^2 = 0.036$. To determine the required minimum sample size for X_1's increment, we consult Soper (2025e) with $f^2 = 0.036$, power = .8, $k_A = 2$, $k_B = 1$, and $\alpha = .05$. This returns a required minimum sample size of $n^* = 217$. In other words, to have an 80% chance of correctly rejecting the null hypothesis associated with X_1's increment at a 5% level of significance requires at least 217 participants, which means that the given data set of $N = 85$ is insufficient.

Focusing on the last column of Step 3, to quantify how insufficient the sample size of the given data set is, we consult Soper (2025f) with $f^2 = 0.036$, $k_A = 2$, $k_B = 1$, $\alpha = .05$, and sample size $N = 85$. This returns a power of .49, which means there is a 49% chance of finding a significant effect size of 0.036 for X_1's increment based on a sample size of 85. This also means there is a 49% chance of correctly rejecting the corresponding H_0. Note that this probability is less than that of a coin flip, which means we would have a better chance of correctly rejecting H_0 by flipping a coin than by conducting this part of the study.

Returning to Subtable A of Table 6.8, the reader will note that with the exception of Step 1, which examines X_3's increment, the respective power values for all the other increments as given in the last column are unacceptable. More specifically, the sample size of the given data set ($N = 85$) is insufficient to detect significant incremental effects for four of the five IVs. So, what do we do? There are three possibilities: (a) reconstitute the study and try to acquire a larger sample size where $N \geq 324$, which is the largest minimum n^*; (b) forget about conducting the study if it is not possible to reconstitute it; or (c) experiment with varying the parameters of the power analysis. Focusing on this latter option, recall that when conducting a power analysis for sample size planning, n^* is a function of the number of IVs (k), desired power, ES, and α. If we were to change one of these parameters, then this would have an impact on n^*. For example, in Step 3 of Subtable A of Table 6.8, if we increase the significance level from $\alpha = .05$ to $\alpha = .10$, then the required minimum sample size decreases from $n^* = 217$ to $n^* = 169$, which in turn increases power from .49 to .54.

To emphasize this point, we prepared Subtable B of Table 6.8 where we assumed the overall expected amount of explained Y variance in the population is $R^2 = .35$. We then altered the respective increments from those given in Subtable A, which in turn increased the corresponding effect sizes. Now look at Step 3 of Subtable B. We have exactly the same situation as before, but here the minimum sample size needed to have an 80% chance of detecting an effect size of $f^2 = 0.079$ at an alpha level of $\alpha = .05$ is $n^* = 101$. Although this is still more than the sample size of the given data set, the corresponding power for this increment based on a sample size of $N = 85$ is .73, which is acceptable. Furthermore, if we made the significance level a bit more liberal by changing it to $\alpha = .08$, then $n^* = 87$ and power is now .795.

This illustrates the strategy researchers use when conducting an a priori power analysis for sample size planning with multiple regression, particularly with hierarchical regression. The reader is cautioned, though, that when altering any of these parameters, the final choice should not be based on some arbitrary number just because it provides acceptable results. Instead, choices should be grounded in theory, prior research, preliminary studies, or professional experience. For example, we might claim that the overall R^2 = .35 is based on a preliminary study we conducted, and the partitioning of explained variance for the increments is based on personal experience. Alternatively, we might report that we used a more liberal significance level because a faculty advisor or journal editor indicated it is acceptable for dissertation research. The bottom line is there needs to be some authentic justification for these decisions. For the current Guided Example, we will use the data reported in Subtable B of Table 6.8 and conclude that the sample size of the given data set is sufficient because all the respective power values are greater than .70.

Preliminary Data Screening

As presented in Chapter 5's Guided Example, preliminary data screening consists of performing regression diagnostics to get the model of best fit, and then checking to see if this model is compliant with the regression assumptions. Unlike Chapter 5, though, we will not provide copies of detailed summary tables, leverage plots, or residual plots but instead simply report the results from our statistics program.

Outlier Analysis. Examining the initial data set for outliers via Jackknife distances yields one outlier, Case 65 (3500), which is with respect to X_1 = Total Flight Hours. As summarized in Table 6.9: (a) with the outlier present, N = 85, R^2 =.838, adjusted R^2 = .828, *RMSE* = 5.92, and $F(5, 79)$ = 81.94, $p < .0001$; and (b) with the outlier absent, N = 84, R^2 = .839, adjusted R^2 = .828, *RMSE* = 5.95, and $F(5, 78)$ = 81.15, $p < .0001$. The difference is negligible. There also is no substantive difference in the mean weather test scores, $M_{\text{with outlier}}$ = 75.88, $M_{\text{without outlier}}$ = 75.9. Therefore, we will keep the outlier. (*Note:* Other outlier procedures might yield different results.)

Multicollinearity. The VIFs for each IV are less than 1.50 and therefore multicollinearity is not a concern.

Missing Data. The data set has no missing data.

Measurement Error. X_5 = Aviation Safety Scores represents a psychological construct and therefore we need to determine its reliability coefficient. However, because we do not have access to the raw scores, we are unable to do this so instead we will assume that $r \geq .70$. (*Note:* The absence of raw scores should be reported as a study limitation, and the assumption that $r \geq .70$ should be reported as a study delimitation.)

TABLE 6.9 *Summary of MRC Analysis Results for Chapter 6's Guided Example Data (with vs. without Outliers)*

	With Outliers	***Without Outliers***
Sample Size (*N*)	85	84
R^2	.838	.839
Adjusted R^2	.828	.828
RMSE	5.92	5.95
M (Weather Test Score)	75.88	75.9
df (num, den)	(5, 79)	(5, 78)
F	81.94	81.15
p	< .0001	< .0001

Relevant IVs. To determine if the IVs are correctly specified, we examine each IV's added variable plot (see Chapter 5). Of these plots, X_1 = Total Flight Hours and X_4 = Age are not relevant to the model. For X_1, the regression line is positive, and the kernel smoother line follows the regression line closely confirming the linear relationship, but it is not significant, $R^2 = .012$, $F(1, 83) = 1.01$, $p = .3174$. For X_4, the regression line is slightly positive, and the kernel smoother line follows the regression line closely confirming the linear relationship, but it also is not significant, $R^2 = .015$, $F(1, 83) = 1.24$, $p = .2680$. (Readers are encouraged to confirm these results themselves.). These results suggest that both X_1 and X_4 be excluded from the final model. However, because we are performing a hierarchical regression and examining the increment each IV makes as it enters the model one at a time in the presence of other IVs, we will keep both IVs in the final model and will report this decision as a study delimitation. (*Note:* See the End-of-Section Commentary.)

We now check to see if the model is compliant with the assumptions of regression.

Linearity. A scatter plot of the residuals vs. the predicted values—in conjunction with the corresponding kernel smoother line—confirms that the scores are randomly scattered around the 0-line and do not show any pronounced or systematic pattern. As a result, the final model is compliant with the linearity assumption.

Equal Variances (Homoscedasticity of the Residuals). The scatter plot of the residuals vs. the predicted values used to check the linearity assumption also confirms the equal variances assumption because the residuals appear to be approximately equally scattered above and below the 0-line without any pronounced pattern. Therefore, the final model is compliant with the equal variances assumption.

Independence of the Residuals. A scatter plot of the residuals vs. the case numbers—in conjunction with the corresponding kernel smoother line—confirms there is no systematic pattern and therefore the model is compliant with the independence assumption.

Normality of the Residuals. The *q-q* plot of the residuals closely "hugs" the normal line and all the data are within the 95% confidence band. The corresponding Shapiro-Wilk Goodness of Fit test also is not significant, $W = .978$, $p = .1635$. Therefore, the model is compliant with the normality assumption.

Data Analysis

We now conduct a hierarchical regression analysis for the final data set with the variable entry order X_3–X_4–X_1–X_2–X_5. The results are summarized in Table 6.10 (APA compliant) and Table 6.11 (non-APA compliant).

Interpretation of Results

We next interpret the results from Tables 6.10 and 6.11 relative to the RQs and corresponding hypotheses.

RQ 1 and Hyp 1. This research question relates to the overall R^2. The total proportion of variance in weather test scores explained collectively by the five IVs is $R^2_{Y\bullet12345} = .838$, with an adjusted $R^2 = .828$. Thus, X_1 = Total Flight Hours, X_2 = Formal Education Background, X_3 = Number of Weather Courses Taken, X_4 = Age, X_5 = Aviation Safety Score, collectively explain 83.8% of the variance in weather test scores and when adjusted for bias, these IVs collectively explain 82.8% of the vari-

TABLE 6.10 *Summary of Results of the Hierarchical Regression Analysis for Chapter 6's Guided Example with Variable Entry Order X_3–X_4–X_1–X_2–X_5 (APA Compliant Table)*

		95% CI for B_i [a]					
Variables in Model	B_i	*LL*	*UL*	*SE* B_i	β_i	R^2 [b]	ΔR^2 [c]
Step 1						.156***	.156***
Constant	71.97***	68.50	75.44	1.74			
X_3 = Num. Weather Courses	1.55***	0.76	2.34	0.39	0.39***		
Step 2						.206***	.050*
Constant	58.75***	46.66	70.85	6.08			
X_3 = Num. Weather Courses	1.72***	0.93	2.50	0.39	0.44***		
X_4 = Age	0.36*	0.04	0.67	0.16	0.23*		
Step 3						.231***	.025
Constant	55.93***	43.47	68.38	6.26			
X_3 = Num. Weather Courses	1.88***	1.08	2.68	0.40	0.48***		
X_4 = Age	0.37*	0.06	0.68	0.16	0.24*		
X_1 = Total Flight Hours	0.003	–0.001	0.006	0.002	0.16		
Step 4						.278***	.047*
Constant	33.78**	10.87	56.69	11.51			
X_3 = Num. Weather Courses	1.45**	0.58	2.31	0.43	0.37**		
X_4 = Age	0.36*	0.05	0.66	0.15	0.23*		
X_1 = Total Flight Hours	0.004*	0.0004	0.008	0.002	0.23*		
X_2 = Formal Education Background	1.50*	0.19	2.82	0.66	0.26*		
Step 5						.838***	.560***
Constant	0.08	–11.55	11.72	5.85			
X_3 = Num. Weather Courses	0.36	–0.07	0.80	0.22	0.09		
X_4 = Age	0.08	–0.07	0.23	0.07	0.05		
X_1 = Total Flight Hours	0.001	–0.001	0.003	0.001	0.05		
X_2 = Formal Education Background	0.83*	0.20	1.47	0.32	0.14*		
X_5 = Aviation Safety Score	0.82***	0.73	0.92	0.05	0.83***		

Note. N = 85. The DV is Y = Weather Test Scores.
[a] CI = confidence interval; *LL* = lower limit; *UL* = upper limit. [b] R^2 = the cumulative R^2 for the current model at each step. [c] $\Delta R^2 = sr_i^2$ = the increment in explained Y variance for each X_i at each step.
*$p < .05$. **$p < .01$. ***$p < .001$.

ance in weather test scores. This result is significant, $F(5, 79) = 81.94$, $p < .0001$. Therefore, we reject Null Hypothesis 1 and conclude that the five IVs collectively explain a significant proportion of the variance in weather test scores.

RQ 2. This research question relates to the increment in explained Y variance each IV makes when it enters the model at each step of the analysis based on the variable entry order X_3–X_4–X_1–X_2–X_5. The corresponding hypothesis has five parts.

Hyp 2a. When X_3 = Number of Weather Courses Taken enters the model at Step 1 without any other IVs present, the amount of explained variance in weather test scores increases from 0% to 15.6%, and this increment is significant, $R^2_{Y \cdot 3} = r^2_{Y3} = sr^2_3 = .156$, $F(1, 83) = 15.37$, $p = .0002$. Therefore, we reject Null Hypothesis 2a and conclude that the number of weather courses pilots take results in a significant gain in the proportion of explained variance in weather test scores.

TABLE 6.11 *Summary of Results of the Hierarchical Regression Analysis for Chapter 6's Guided Example with Variable Entry Order X_3–X_4–X_1–X_2–X_5 (Non-APA Compliant Table)*

Variables in Model [a]	B_i	Cum. R^2 [b]	$I_i = sr_i^2$ [c]	df_I [d]	F_I [d]	p_I [d]
X_3	$B_3 = 1.55$	.156	.156	1, 83	15.37	.0002
X_3, X_4	$B_3 = 1.72$, $B_4 = 0.36$	.206	.050	1, 82	5.16	.0257
X_3, X_4, X_1	$B_3 = 1.88$, $B_4 = 0.37$, $B_1 = 0.003$	.231	.025	1, 81	2.63	.1087
X_3, X_4, X_1, X_2	$B_3 = 1.45$, $B_4 = 0.36$, $B_1 = 0.004$, $B_2 = 1.50$	.278	.047	1, 80	5.21	.0251
X_3, X_4, X_1, X_2, X_5	$B_3 = 0.36$, $B_4 = 0.08$, $B_1 = 0.001$, $B_2 = 0.83$, $B_5 = 0.82$	.838	.560	1, 79	273.09	< .0001

Note. $N = 85$. The DV is Y = Weather Test Scores.
[a] X_3 = Number of Weather Courses Taken, X_4 = Age, X_1 = Total Flight Hours, X_2 = Formal Education Background, X_5 = Aviation Safety Score. [b] Cum. R^2 = the R^2 for the current model at each step. [c] I = the increment in explained Y variance by the X_i entering the model at each step. [d] df_I, F_I, p_I = the respective *df*, *F*, and *p* values for the increment.

Hyp 2b. When X_4 = Age enters the model at Step 2, X_4's increment in explaining Y variance is $sr_4^2 = R^2_{Y(4 \cdot 3)} = (R^2_{Y \cdot 34} - r^2_{Y3}) = (.206 - .156) = .05$, $F(1, 82) = 5.16$, $p = .0257$. Thus, participants' age explains an additional 5% of the variance in weather test scores when examined in the presence of the number of weather courses pilots take and this increment is significant. As a result, we reject Null Hypothesis 2b and conclude that pilots' age provides a significant gain in the proportion of explained variance in weather test scores.

Hyp 2c. When X_1 = Total Flight Hours enters the model at Step 3, X_1's increment in explaining Y variance is $sr_1^2 = R^2_{Y(1 \cdot 34)} = (R^2_{Y \cdot 341} - R^2_{Y \cdot 34}) = (.231 - .206) = .025$, $F(1, 81) = 2.63$, $p = .1087$. Thus, participants' total flight hours explain an additional 2.5% of the variance in weather test scores when examined in the presence of the number of weather courses pilots take and pilots' age but this increment is not significant. As a result, we fail to reject Null Hypothesis 2c and conclude that pilots' age does not provide a significant gain in the proportion of explained variance in weather test scores.

Hyp 2d. When X_2 = Formal Education Background enters the model at Step 4, X_2's increment in explaining Y variance is $sr_2^2 = R^2_{Y(2 \cdot 341)} = (R^2_{Y \cdot 3412} - R^2_{Y \cdot 341}) = (.278 - .231) = .047$, $F(1, 80) = 5.21$, $p = .0251$. Thus, participants' formal education background explains an additional 4.7% of the variance in weather test scores when examined in the presence of the number of weather courses pilots take, pilots' age, and their total flight hours and this increment is significant. Therefore, we reject Null Hypothesis 2d and conclude that pilots' formal education background provides a significant gain in the proportion of explained variance in weather test scores.

Hyp 2e. When X_5 = Aviation Safety Scores enters the model at Step 5, X_5's increment in explaining Y variance is $sr_5^2 = R^2_{Y(5 \cdot 3412)} = (R^2_{Y \cdot 34125} - R^2_{Y \cdot 3412}) = (.838 - .278) = .560$, $F(1, 79) = 273.09$, $p < .0001$. Thus, participants' aviation safety scores explain an additional 56% of the variance in weather test scores when examined in the presence of the number of weather courses pilots take, age, total flight hours, and formal education background, and this increment is significant. As a result, we reject Null Hypothesis 2e and conclude that pilots' aviation safety scores provide a significant gain in the proportion of explained variance in weather test scores.

Summarizing: (a) The increment to R^2 for four of the five IVs is statistically significant when they first enter the model. The exception is X_1 = Total Flight Hours at Step 3. (b) Of the five IVs, X_5 = Aviation Safety Score is, unexpectedly, the strongest predictor even though it did enter the model last. Overall, the final regression model is $\hat{Y} = 0.36X_3 + 0.08X_4 + 0.001X_1 + 0.83X_2 + 0.82X_5 + 0.08$, which is statistically significant, $R^2 = .838$, $F(5, 79) = 81.94$, $p < .0001$, Adj. $R^2 = .828$, $RMSE = 5.9$. Thus, if examined via a simultaneous regression, the five targeted IVs collectively explain 83.8% of the variance in weather test scores, and of these IVs, two are significant: X_2 = Formal Education Background and X_5 = Aviation Safety Score. Note how the results from this final model are different when compared to the findings from a hierarchical perspective.

Post Hoc Considerations

We now present the effect sizes and post hoc power values for the overall model and respective increments for the reader's benefit and interpretation. (*Note:* The descriptors of small, medium, and large effect sizes are relative to Cohen et al., 2003, pp. 93–95). We do not present any confidence intervals because the focus of the analysis is on the increments to R^2 and not the regression coefficients.

Effect Size. Following are the effect sizes for the overall model and the respective increments.

ES for $R^2_{Y \cdot 12345}$. $f^2 = \dfrac{R^2}{1 - R^2} = \dfrac{.838}{1 - .838} = \dfrac{.838}{.162} = 5.17$ (large effect)

ES for sr^2_3. $f^2 = \dfrac{sr_3^2}{1 - R^2_{Y \cdot 3}} = \dfrac{.156}{1 - .156} = \dfrac{.156}{.844} = 0.18$ (medium effect)

ES for sr^2_4. $f^2 = \dfrac{sr_4^2}{1 - R^2_{Y \cdot 34}} = \dfrac{.05}{1 - .206} = \dfrac{.05}{.794} = 0.06$ (small effect)

ES for sr^2_1. $f^2 = \dfrac{sr_1^2}{1 - R^2_{Y \cdot 341}} = \dfrac{.025}{1 - .231} = \dfrac{.025}{.769} = 0.03$ (small effect)

ES for sr^2_2. $f^2 = \dfrac{sr_2^2}{1 - R^2_{Y \cdot 3412}} = \dfrac{.047}{1 - .278} = \dfrac{.047}{.722} = 0.07$ (small effect)

ES for sr^2_5. $f^2 = \dfrac{sr_5^2}{1 - R^2_{Y \cdot 34125}} = \dfrac{.56}{1 - .838} = \dfrac{.56}{.162} = 3.46$ (large effect)

Power. Following are the respective post hoc powers for the overall model and increments for the IVs.

Power for the Overall Model ($R^2_{Y \cdot 12345}$). Consulting Soper (2025g) with $k = 5$, $R^2 = .838$, $p = .05$, and $N = 85$, the power for the overall model is 1.0, which we report as "greater than .99," because we do not make any claim of 100% certainty. Thus, there is a greater than 99% probability that the large effect found in the sample relative to the five targeted IVs also exists in the population, and there is a greater than 99% probability that the decision to reject the corresponding null hypothesis is correct.

Power for X_3's Increment **(sr_3^2).** As noted earlier, because Step 1 of a hierarchical regression does not include any other IVs in the model other than the one under discussion, we consult Soper (2025g) with $k = 1$, $R^2 = .156$, $p = .05$, and $N = 85$. This returns a power of .9767. Thus, there is a 97.67% probability that the medium effect found in the sample relative to X_3's increment to explaining the variance in weather test scores also exists in the population, and there is a 97.67% probability that the decision to reject the corresponding null hypothesis is correct.

Power for X_4's Increment **(sr_4^2).** Consulting Soper (2025f) with $f^2 = 0.06$, $k_A = 1$, $k_B = 1$, $p = .05$, and $N = 85$, the power for X_4's increment is .6136. Thus, there is a 61.36% probability that the small effect found in the sample relative to X_4's increment to explaining the variance in weather test scores also exists in the population, and there is a 61.36% probability that the decision to reject the corresponding null hypothesis is correct.

Power for X_1's Increment **(sr_1^2).** Consulting Soper (2025f) with $f^2 = 0.03$, $k_A = 2$, $k_B = 1$, $p = .05$, and $N = 85$, the power for X_1's increment is .3503. Thus, there is a 35.03% probability that the small effect found in the sample relative to X_1's increment to explaining the variance in weather test scores also exists in the population. This result led to a fail to reject H_0 decision.

Power for X_2's Increment **(sr_2^2).** Consulting Soper (2025f) with $f^2 = 0.07$, $k_A = 3$, $k_B = 1$, $p = .05$, and $N = 85$, the power for X_2's increment is .6695. Thus, there is a 66.95% probability that the small effect found in the sample relative to X_2's increment to explaining the variance in weather test scores also exists in the population, and there is a 66.95% probability that the decision to reject the corresponding null hypothesis is correct.

Power for X_5's Increment **(sr_5^2).** Consulting Soper (2025f) with $f^2 = 3.46$, $k_A = 4$, $k_B = 1$, $p = .05$, and $N = 85$, the power for X_5's increment is 1.0. Thus, there is a greater than 99% probability that the large effect found in the sample relative to X_5's increment to explaining the variance in weather test scores also exists in the population, and there is a greater than 99% probability that the decision to reject the corresponding null hypothesis is correct.

End-of-Section Commentary

Following are additional comments/insight relative to the Guided Example for the reader to consider.

1. The hierarchical regression results are based on the variable entry order X_3–X_4–X_1–X_2–X_5. If a different entry order is used, then with the exception of the final model, the results most likely will be different. The reader is encouraged to alter this entry order and compare the results to those presented in Tables 6.10 and 6.11.
2. Referencing Table 6.10, the reader will recognize that the results presented for Step 5 are the same as those that will result from a simultaneous regression analysis where all five IVs remain in the final data set.
3. Recall that as part of preliminary data screening, the added variable plots for X_1 and X_4 indicated that they did not have any relationship to the DV and hence should be excluded from the final data set. However, we chose to retain these IVs because our focus is on examining each variable's increment to R^2. This

decision was prudent for X_4 = Age. Although X_4 is not significant in the final model (see Step 5 in Table 6.10), its increment is significant (see Step 2 in Table 6.10 and Table 6.11). For X_1 = Total Flight Hours, though, neither its increment nor its presence in the final model is significant.

4. Accenting Comment 3, if we were conducting a simultaneous regression, we most likely would have excluded both X_1 and X_4 from the final data set. In doing so, the loss in the overall R^2 would have been minimal. For example, $R^2_{Y \cdot 12345}$ = .838 vs. $R^2_{Y \cdot 235}$ = .834, a difference of .004. In doing so, though, we would have lost information about each IV's increment to explaining *Y* variance when entered into the analysis via a specific entry order.
5. Table 6.12 compares the differences between the sr_i^2 values from simultaneous regression to those from hierarchical regression with respect to the Guided Example's data set. Observe from Table 6.12 that the contribution each IV makes to explaining *Y* variance appears to be weaker for simultaneous regression than for hierarchical regression. This is because in the former, X_i is being assessed when controlling for the effects of all the other IVs taken together, whereas in the latter X_i is being assessed when controlling only for those IVs that were present in the previous model. Also note that with simultaneous regression, the sum of the sr_i^2 values does not provide a full accountability of explained *Y* variance: R^2 = .583 vs. R^2 = .838. Once again, this is because of the redundancy among the IVs. For example: $r^2_{12} \approx .12$, $r^2_{23} \approx .23$, and $r^2_{35} \approx .10$.

TABLE 6.12 *A Comparison of the sr_i^2 Values between Simultaneous and Hierarchical Regression with Respect to the Data Set for Chapter 6's Guided Example*

	Simultaneous	Hierarchical
IV [a]	$sr_i^2 = R^2_{Y \cdot 123 \ldots i \ldots k} - R^2_{Y \cdot 123 \ldots (i) \ldots k}$ [b]	$sr_i^2 = R^2_{Y \cdot AB} - R^2_{Y \cdot A}$ [c]
X_3	$sr_3^2 = .005$	$sr_3^2 = .156$
X_3, X_4	$sr_4^2 = .002$	$sr_4^2 = .050$
X_3, X_4, X_1	$sr_1^2 = .002$	$sr_1^2 = .025$
X_3, X_4, X_1, X_2	$sr_2^2 = .014$	$sr_2^2 = .047$
X_3, X_4, X_1, X_2, X_5	$sr_5^2 = .560$	$sr_5^2 = .560$
Total	$R^2_{Y \cdot 34125} = .583$	$R^2_{Y \cdot 34125} = .838$

Note. *N* = 85. The DV is *Y* = Weather Test Scores.
[a] X_3 = Number of Weather Courses Taken, X_4 = Age, X_1 = Total Flight Hours, X_2 = Formal Education Background, X_5 = Aviation Safety Score. [b] The first R^2 term reflects *Y* regressed on all *k* IVs where *i* represents the IV under discussion. The second R^2 term reflects *Y* regressed on all *k* IVs except the one under discussion, which is represented by placing *i* in parentheses to denote that X_i is being omitted in the regression. For example, $sr_2^2 = R^2_{Y \cdot 12345} - R^2_{Y \cdot 1345}$, which denotes that X_2 is being partialled by X_1, X_3, X_4, and X_5. [c] Set A contains the IV(s) in the previous model and set B contains the IV entering the model at the current step. For example, for sr_2^2, set A contains X_3, X_4, and X_1, which were in the previous model, and set B contains X_2, which is entering the model at the current step. Thus, at this step of the analysis, X_2 is being partialled by X_3, X_4, and X_1: $sr_2^2 = R^2_{Y \cdot 3412} - R^2_{Y \cdot 341}$, or alternatively, $sr_2^2 = r^2_{Y2 \cdot 341}$, which represents the squared bivariate correlation between *Y* and X_2 after the effects of X_3, X_4, and X_1 have been partialled out.

TABLE 6.13 *Summary of Results of the Forward Addition Option for Statistical Regression Relative to Chapter 6's Guided Example Data Set*

Step-by-Step Analysis Results

Analysis	IVs in Model	IV Being Examined [a]	Cum. R^2 [b]	$I_i = sr_i^2$ [c]	df_I	F_I	p_i	AICc	BIC
Step 1		X_1	.00215	.00215	1, 83	0.18	.6736		
		X_2	.11961	.11961	1, 83	11.28	.0012		
		X_3	.15627	.15627	1, 83	15.37	.0002		
		X_4	.02149	.02149	1, 83	1.82	.1806		
		X_5	.80630	.80630	1, 83	345.50	< .0001	558.90	565.93
Step 2	X_5	X_1	.80715	.00085	1, 82	0.36	.5506		
		X_2	.83064	.02434	1, 82	11.78	.0009	549.69	558.96
		X_3	.82166	.01536	1, 82	7.06	.0095		
		X_4	.80663	.00033	1, 82	0.14	.7092		
Step 3	X_5, X_2	X_1	.83162	.00098	1, 81	0.47	.4949		
		X_3	.83432	.00368	1, 81	1.80	.1835	550.08	561.53
		X_4	.83154	.00090	1, 81	0.43	.5138		
Step 4	X_5, X_2, X_3	X_1	.83592	.00160	1, 80	0.78	.3797		
		X_4	.83638	.00206	1, 80	1.00	.3196		

Note. Final Model is $R^2_{Y \cdot 523}$ = .834, $F(3, 81)$ = 135.97, p < .0001. The second color reflects the "best" IV at each step relative to the other IVs currently present.
[a] X_1 = Total Flight Hours, X_2 = Formal Education Background, X_3 = Number of Weather Courses Taken, X_4 = Age, X_5 = Aviation Safety Score. [b] Cum. R^2 = the R^2 for the current model at each step of the analysis. [c] I_i = the increment in explained variance by the X_i entering the model at each step.

Coefficients (Parameter Estimates) of Final Model

Term	B_i	SE_i	$t(81)$	p_i	95% CI
Constant	3.46	5.32	0.65	.5175	[–7.13, 14.05]
X_5 = Aviation Safety Score	0.85	0.05	17.78	< .0001	[0.75, 0.94]
X_2 = Formal Education Background	0.76	0.30	2.49	.0149	[0.15, 1.36]
X_3 = Number of Weather Courses	0.28	0.21	1.34	.1836	[–0.13, 0.69]

Note. N = 85. Results based on α-to-enter = .20.

6. Table 6.13 contains a summary of the results from analyzing Chapter 6's Guided Example data set using a statistical regression strategy involving the forward addition option with α-to-enter = .20. Observe from Table 6.13 that the final model consists of three variables and the entry order is X_5–X_2–X_3, which is quite different than what we hypothesized. Also note that the analysis stops after Step 3 when X_3 enters because at Step 4 the p values of the remaining IVs are greater than α-to-enter = .20. However, because this parameter is researcher controlled, if α-to-enter = .35, then the final model is X_5–X_2–X_3–X_4, but if α-to-enter = .15 or smaller, then the final model is X_5–X_2 because at Step 3, X_3's p value would be too large. Moreover, if the selection criterion were AICc or BIC instead of α-to-enter, then the final model is X_5–X_2 because this is when AICc and BIC are at a minimum. (*Note:* The backward elimination and mixed options yield the same results as the forward addition option.)

TABLE 6.14 *Summary of Results of the Best Model Option for Statistical Regression Relative to Chapter 6's Guided Example Data Set*

Model [a]	Number of IVs in Model	Cum. R^2 [b]	*RMSE*	AICc	BIC
X_5	1	.8063	6.3192	558.902	565.933
X_3	1	1563	13.1886	683.982	691.014
X_2	1	.1196	13.4721	687.597	694.629
X_4	1	.0215	14.2030	696.579	703.610
X_1	1	.0021	14.3427	698.242	705.274
X_2, X_5	2	.8306	5.9447	549.689	558.960
X_3, X_5	2	.8217	6.1003	554.082	563.353
X_1, X_5	2	.8071	6.3437	560.734	570.005
X_4, X_5	2	.8066	6.3522	560.962	570.232
X_3, X_4	2	.2060	12.8721	681.025	690.296
X_2, X_3	2	.1878	13.0186	682.950	692.221
X_1, X_3	2	.1776	13.1003	684.013	693.284
X_1, X_2	2	.1506	13.3136	686.759	696.030
X_2, X_4	2	.1455	13.3528	687.259	696.529
X_1, X_4	2	.0238	14.2724	698.581	707.851
X_2, X_3, X_5	3	.8343	5.9159	550.082	561.535
X_1, X_2, X_5	3	.8316	5.9641	551.460	562.913
X_2, X_4, X_5	3	.8315	5.9655	551.500	562.954
X_3, X_4, X_5	3	.8241	6.0954	555.162	566.616
X_1, X_3, X_5	3	.8217	6.1372	556.324	567.778
X_1, X_4, X_5	3	.8074	6.3777	562.860	574.314
X_2, X_3, X_4	3	.2334	12.7256	680.296	691.750
X_1, X_3, X_4	3	.2314	12.7424	680.521	691.975
X_1, X_2, X_3	3	.2281	12.7697	680.884	692.338
X_1, X_2, X_4	3	.1781	13.1763	686.213	697.667
X_2, X_3, X_4, X_5	4	.8364	5.9158	551.340	564.919
X_1, X_2, X_3, X_5	4	.8359	5.9239	551.574	565.153
X_1, X_2, X_4, X_5	4	.8326	5.9833	553.269	566.848
X_1, X_3, X_4, X_5	4	.8242	6.1312	557.420	570.999
X_1, X_2, X_3, X_4	4	.2779	12.4280	677.535	691.114
X_1, X_2, X_3, X_4, X_5	5	.8383	5.9172	552.688	568.332

Note. $N = 85$. The DV is Y = Weather Test Scores. Best models relative to the number of IVs are given in a second color.
[a] X_1 = Total Flight Hours, X_2 = Formal Education Background, X_3 = Number of Weather Courses Taken, X_4 = Age, X_5 = Aviation Safety Score. [b] Cum. R^2 = the R^2 for the model being examined.

7. Table 6.14 contains a summary of the results from analyzing Chapter 6's Guided Example data set using a statistical regression strategy involving the best model option. Because there are five IVs, the best model option examines $2^5 - 1 = 31$ different models: five models with 1 IV, 10 models with 2 IVs, 10 models with 3 IVs, five models with 4 IVs, and one model with 5 IVs. As shown in Table 6.14, the "best "model, based on largest cumulative R^2, minimum *RMSE*, minimum AICc, and minimum BIC, are as follows: 1 IV = X_5; 2 IVs = X_2, X_5; 3 IVs = $X_2, X_3,$

X_5; 4 IVs = X_2, X_3, X_4, X_5; and 5 IVs = X_1, X_2, X_3, X_4, X_5. Note that the "best" 3-IV model is exactly what resulted from: (a) a simultaneous regression where X_1 and X_4 are excluded as part of preliminary data screening (see Comment 4 above), and (b) a statistical regression strategy involving the forward addition option with α-to-enter = .20 (see Comment 6 above).

8. The reader is reminded of the differences among the three MRC strategies:
   - The focus of simultaneous regression is on the final model and includes an examination of the squared semipartial correlation coefficients (sr_i^2) and regression coefficients (B_i) but does not provide a full accountability of the explained *Y* variance.
   - The focus of hierarchical regression is on the increment each IV makes in explaining *Y* variance relative to a predetermined variable entry order grounded in theory, past or preliminary studies, or professional experience, and provides a full accountability of explained *Y* variance.
   - The focus of statistical regression is on acquiring a model that best fits the given data set by relying solely on statistical parameters without concern for logic, causal priority of the IVs, or theory. This strategy is often viewed as a "fishing expedition" and hence could capitalize on chance relationships among the IVs or lead to different variable entry orders between different samples selected from the same population.

Chapter 6 Summary

6.2 Simultaneous Regression

- In simultaneous regression, all the IVs of a given data set are entered into the model at the same time without any regard for a variable entry order.
- A simultaneous regression strategy does not provide a full accountability of explained *Y* variance when the correlation among the IVs is not zero; that is, $R^2_{Y \cdot 123 \ldots k} \neq sr_1^2 + sr_2^2 + sr_3^2 + \ldots + sr_k^2$.
- From an inferential statistics perspective, the focus of simultaneous regression is examining the significance of the overall model ($R^2_{Y \cdot 123 \ldots k}$), the squared semipartial correlation (sr_i^2) coefficients, and the regression coefficients (B_i).

6.3 Hierarchical (Sequential) Regression

- In hierarchical regression, the IVs of a given data set are entered into the model one step at a time via a predetermined sequence or variable entry order. At each step of the analysis, the IV entering the model is examined for its increment in explaining *Y* variance relative to the IVs that were in the previous model.
- The variable entry order of a hierarchical regression analysis may be determined from theory, past studies or preliminary studies, or personal experience/expertise. The entry order also may be determined by considering the variables' presumed causal priority or their relevance with respect to the DV.
- A hierarchical regression strategy provides a full accountability of explained *Y* variance; that is, $R^2_{Y \cdot 123 \ldots k} = sr_1^2 + sr_2^2 + sr_3^2 + \ldots + sr_k^2$.

- The primary objective of a hierarchical regression strategy is to answer the question, "What is the incremental effect of each IV relative to the DV?" This question is answered by each IV's respective sr_i^2 at the time the IV enters the model. To determine the significance of the increment, we use the F test for the increment:

$$F_I = \frac{sr_i^2(N - k_A - k_B - 1)}{(1 - R_{Y \bullet AB}^2)k_B}$$

where sr_i^2 = increment for X_i, N = the overall sample size, k_A = the number of IVs in the previous model, k_B = the number of IVs entering the current model, and $R_{Y \bullet AB}^2$= the R^2 for the current model, which contains the IVs from the previous model and the IVs entering the model at the current step.
- The regression equation from the last step of a hierarchical regression analysis is equal to the overall model from a simultaneous regression analysis.
- The results of a hierarchical regression analysis may be presented using an APA (2020) compliant table as demonstrated in Table 6.2 and Table 6.10, or a researcher-derived table as demonstrated in Table 6.3 and Table 6.11.

6.4 Statistical (Stepwise) Regression

- In statistical regression, the IVs of a given data set are examined using any of four variable selection options a researcher believes is appropriate, and the IVs that are selected for the final model are determined by various selection criteria. The variable selection options are forward addition, backward elimination, mixed, and best model, and the variable selection criteria are α-to enter, α-to leave, AICc, and BIC.
- The forward addition option begins with no IVs in the model. Each IV then enters the model, one step at a time, and its increment to explained Y variance is examined in the presence of the IVs in the previous model. If the p value of the increment of the IV under consideration at each step is less than the preset α-to enter, then the IV enters the model. The analysis stops when the remaining IVs' respective increments have corresponding p values > α-to enter. Once an IV is added to the model it remains in the model. The final model also may be determined by minimum AICc or minimum BIC.
- The backward elimination option begins with all the IVs of a given data set in the model, and the IVs are examined for their respective contribution to explaining Y variance. The IV with the least contribution to R^2 and p > α-to-leave is removed. The model is then refitted with the IVs that remain in the model, and the process of removing one IV one step at a time is repeated until removing any additional IVs in the model does not result in a statistically better model fit. This occurs when the respective p values of the IVs in the model are less than α-to-leave, or equivalently, the remaining IVs outside the model have p values greater than α-to-leave. Once an IV leaves the model it remains excluded from the model. The final model also may be determined by minimum AICc or minimum BIC.
- The mixed option, which is called "stepwise" in SPSS, combines the forward addition and backward elimination options. The primary difference is an IV that

enters or is removed from the model at one step could respectively be removed from or reenter the model at a subsequent step after the model is refitted.

- The best model option examines all the different possible combinations involving the IVs of a given data set. For a given data set with k IVs, there are $2^k - 1$ combinations and they are examined groupwise by considering all possible models with 1 IV, 2 IVs, 3 IVs, and so forth. The "best" model for each group has the lowest *RMSE*, minimum AICc, and minimum BIC. The researcher then selects the final model.
- Statistical regression may be used as a variable selection strategy for preliminary research but is not encouraged for primary data analysis because it can exploit chance relationships or result in a different variable entry order for different samples selected from the same population. It also is suggested that the final model from a statistical regression analysis be subjected to a cross-validation (see Chapter 5).

Key Terms and Concepts

Akaike Information Criterion (AIC)
Akaike Information Criterion corrected (AICc)
α-to-enter
α-to-leave
Backward elimination option
Bayesian Information Criterion (BIC)
Best model option
Bidirectional elimination (in SPSS)
Consistency property of BIC
Delta R^2 (ΔR^2)
Forward addition option
F value of the increment (F_I)
Hierarchical regression
Incremental effect (I)
k_A
k_B
Mixed option
Over specified model
p value of the increment (p_I)
Presumed causal priority
Primary IVs
p value stopping rule
Relevance of the IVs
$R^2_{Y \bullet A}$
$R^2_{Y \bullet AB}$
Secondary IVs
Sequential regression
Simultaneous regression
sr_i^2
Statistical regression
Stepwise regression
Stepwise selection (in SPSS)
Underspecified model
Variable entry order

Review Exercises

A. Checking Conceptual Knowledge

In 1–10, choose the best answer among the choices provided.

1. Given Figure 6.6(a), if a hierarchical regression strategy is conducted with variable entry order X_1–X_2–X_3–X_4, what region(s) does sr_4^2 represent?
   a. *d*, *e*, and *f* b. *d* and *e* c. *f* d. *g*
2. Given Figure 6.6(a), if a simultaneous regression strategy is conducted, what region(s) does sr_3^2 represent?
   a. *d*, *e*, and *f* b. *d* c. *e* d. *f*

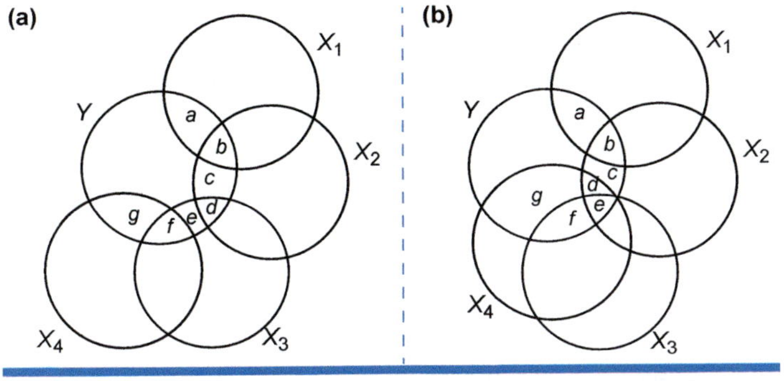

FIGURE 6.6 Venn diagram for Review Exercises Part A, Items 1–4.

3. Given Figure 6.6(b), if a simultaneous regression strategy is conducted, what region(s) does sr_3^2 represent?
   a. *e* and *f* b. *e* c. *f* d. None
4. Given Figure 6.6(b), if a hierarchical regression strategy is conducted with variable entry order X_1–X_2–X_3–X_4,what region(s) does sr_3^2 represent
   a. *e* and *f* b. *e* c. *f* d. None
5. Referencing Table 6.15, which statement is true?
   a. The forward addition option is used, and the best model is X_3–X_2.
   b. The backward elimination option is used, and the best model is X_3–X_2.
   c. The forward addition option is used, and the best model based on BIC is X_3.
   d. The best model option is used, and the best model is X_3–X_2.
6. Referencing Table 6.15, which statement is true?
   a. The forward addition option is used with α-to-enter = .15.
   b. The forward addition option is used with α-to-enter = .05.
   c. The backward elimination option is used with α-to-leave = .05.
   d. The backward elimination option is used with α-to-leave = .15.

TABLE 6.15 *Summary of MRC Analysis Results for Ch_6 Review Exercises, Part A Items 5–6*

Analysis	IV in Model	IV Being Examined [a]	Cum. R^2 [b]	$I_i = sr_i^2$ [c]	F_I	p_i	AICc	BIC
Step 1		X_1	.0136	.014	4.42	.0363		
		X_2	.0061	.006	1.96	.1619		
		X_3	.0319	.032	10.60	.0013	1766.73	1777.99
		X_4	.0091	.009	2.96	.0865		
Step 2	X_3	X_1	.0322	.0003	0.99	.3205		
		X_2	.0422	.0103	3.44	.0645	1765.34	1780.33
		X_4	.0321	.0002	0.66	.4172		

Note. N = 323.The results are from using a statistical regression strategy and the second color reflects the "best" IV at each step relative to the other IVs currently present.
[a] X_1 = Age, X_2 = Marital Status (Married vs. Not Married), X_3 = Years Held CFI Certificate, X_4 = Hours Dual Given. [b] Cum. R^2 = the R^2 for the current model at each step. [c] I_i = the increment in explained Y variance by the variable entering the model at each step.

7. In hierarchical regression, which statement is true?
   a. The most important IV is always entered first.
   b. The variable entry order is arbitrary and based on a computer algorithm.
   c. The variable entry order is determined by each IV's increment.
   d. The variable entry order is researcher determined.
8. If the respective zero-order squared correlation coefficients for a set of predictors is 0, then
   a. the final model of a simultaneous regression strategy will provide a full accountability of the explained *Y* variance.
   b. the final model of a simultaneous regression strategy will not provide a full accountability of the explained *Y* variance.
   c. the final model of a hierarchical regression strategy will not provide a full accountability of the explained *Y* variance.
   d. the final model of a simultaneous regression strategy will not be the same as the final model of a hierarchical simultaneous strategy.
9. Which of the following descriptions would be most likely associated with simultaneous regression?
   a. Systematic
   b. Scatter-shot
   c. Fishing expedition
   d. Wild goose chase
10. Among the various variable selection strategies presented in this chapter, which statement is NOT true?
    a. In statistical regression/forward addition with $k = 20$, early entry IVs may be removed if they are no longer significant in the presence of later entry IVs.
    b. Hierarchical regression's primary focus is testing the proportion of *Y* variance each IV makes after the amount of explained *Y* variance of other IVs currently in the model have been accounted for.
    c. Statistical regression is better suited for model-building than model-testing
    d. In statistical regression/best model, a researcher gets to determine the best model from a collection of different models based on all possible combinations involving the IVs.

B. Applying Procedural and Conceptual Knowledge

Use the following research description and the corresponding data set to conduct the activities given in Parts B1–B7. See Section 6.5 for guidance.

Consider a study that is modeled after Bell et al. (1995) who investigated the effect of pilot experience on aviation safety. For the current study, four factors are targeted:

- X_1 = Flight Hours, defined as pilots' total flight hours in all aircraft.
- X_2 = Ratings, defined as the number of ratings pilots hold with respect to all aircraft classifications.
- X_3 = Duration, defined as the number of years pilots have held a pilot certificate.
- X_4 = Malfunction History, defined as the number of in-flight malfunctions pilots experienced over their aviation careers.

The dependent measure is Y = Aviation Safety Index (ASI) scores, which refer to pilots' responses to a set of five safety-related statements that corresponded to behaviors associated with preflight preparation. Pilots are asked to self-assess the degree to which they performed each behavior relative to a 7-item bipolar adjective scale with "Never" at one end of the scale (1) and "Always" at the other end of the scale (7). Thus, the aggregate score could range from 5 to 35, with higher scores reflecting a more safety conscious pilot. The data set, which consists of hypothetical data, is given in the file "Ch_6 Review Exercises Data–Part B."

B1. Research Questions and Hypotheses. Write the research questions and corresponding hypotheses that pertain to the overall regression model and respective increments.

B2. Sample Size Planning. Conduct an a priori power analysis to see if the given data set is sufficiently large for the overall R^2 and each of the increments.

B3. Preliminary Data Screening. Perform regression diagnostics on the data set and then confirm that the final model is compliant with the underlying regression assumptions.

B4. Data Analysis. Conduct a hierarchical regression analysis of the final data set and summarize the results in both an APA-compliant table similar to Table 6.10 and a non-compliant APA table similar to Table 6.11. For this analysis, assume that: (a) the variable entry order is X_1–X_4–X_2–X_3; (b) the expected overall population R^2 = .30 and the expected increments are sr_1^2 = .10, sr_4^2 = .10, sr_2^2 = .05, sr_3^2 = .05; (c) the targeted IVs will collectively explain a significant proportion of the variance in ASI scores; and (d) there will be a significant predictive gain in the explained variance of ASI scores made by each IV at each step of the analysis. Also assume α = .05.

B5. Interpretation of Results. Interpret the results relative to the research questions and corresponding null hypotheses stated in Part B1.

B6. Post Hoc Considerations. Determine and comment on the effect sizes and power for the overall R^2 and respective increments.

B7. Commentary. Report and compare the results to those that would have been acquired from (a) a simultaneous regression, (b) a statistical regression involving the forward addition option with α-to-enter = .20, and (c) a statistical regression involving the best model option.

References

Aho, K., Derryberry, D., & Peterson, T. (2014). Model selection for ecologists: The worldviews of AIC and BIC. *Ecology*, *95*(3), 631–636. https://doi.org/10.1890/13-1452.1

American Psychological Association. (2020). *Publication manual of the American Psychological Association* (7th ed.). https://doi.org/10.1037/0000165-000

Bell, B. D., Robertson, C. L., & Wagner, G. S. (1995). Aviation safety as a function of pilot experience: Rationale or rationalization. *The Journal of Aviation/Aerospace Education & Research*, *5*(3), 27–38. https://commons.erau.edu/cgi/viewcontent.cgi?article=1154&context=jaaer

Bendel, R. B., & Afifi, A. A. (1977). Comparison of stopping rules in forward regression. *Journal of the American Statistical Association*, *72*, 46–53.

Burian, B. K. (2002). *The general aviation pilot weather knowledge and training study: Final report.* FAA Grant #00-G-020. https://www.tc.faa.gov/its/worldpac/grants/00g20a.pdf

Burnham, K. P., & Anderson, D. R. (2004). Multimodel inference: Understanding AIC and BIC in model selection. *Sociological Methods & Research*, *33*(2), 261–304. https://doi.org/10.1177/0049124104268644

Chowdhury, M. Z. I., & Turin, T. C. (2020). Variable selection strategies and its importance in clinical prediction modelling. *Family Medicine and Community Health*, *8*(1), 1–7. https://doi.org/10.1136/fmch-2019-000262

Cohen, J., Cohen, C., West, S. G., & Aiken., L. S. (2003). *Applied multiple regression/correlation analysis for the behavioral sciences* (3rd ed.). Lawrence Erlbaum Associates.

Harrell, F. E. (2001). *Regression modeling strategies with applications to linear models, logistic regression, and survival analysis*. Springer. https://link.springer.com/book/10.1007/978-1-4757-3462-1

Hosmer, D. W., Lemeshow, S., & Sturdivant, R. X. (2013). *Applied logistic regression* (3rd ed.). John Wiley & Sons, Inc.

Soper, D. (2025a). *Critical F-value calculator.* https://www.danielsoper.com/statcalc/calculator.aspx?id=4

Soper, D. (2025b). *F-value calculator for hierarchical multiple regression.* https://www.danielsoper.com/statcalc/calculator.aspx?id=14

Soper, D. (2025c). *p-value calculator for an F-test.* https://www.danielsoper.com/statcalc/calculator.aspx?id=7

Soper, D. (2025d). *A-priori sample size calculator for multiple regression.* https://www.danielsoper.com/statcalc/calculator.aspx?id=1

Soper, D. (2025e). *A-priori sample size calculator for hierarchical multiple regression.* https://www.danielsoper.com/statcalc/calculator.aspx?id=16

Soper, D. (2025f). *Post-hoc statistical power calculator for hierarchical multiple regression.* https://www.danielsoper.com/statcalc/calculator.aspx?id=17

Soper, D. (2025g). *Post-hoc statistical power calculator for multiple regression.* https://www.danielsoper.com/statcalc/calculator.aspx?id=9

Vrieze, S. I. (2012), Model selection and psychological theory: A discussion of the differences between the Akaike Information Criterion (AIC) and the Bayesian Information Criterion (BIC). *Psychological Methods*, *17*(2), 228–243. https://doi.org/10.1037/a0027127

Warner, R. M. (2008). *Applied statistics: From bivariate through multivariate techniques.* Sage.

Chapter 7
Analyzing Sets of Independent Variables

Student Learning Outcomes

After studying this chapter, you will be able to do the following:

1. Organize the IVs of a given data set into functional sets.
2. Perform and interpret the results from a simultaneous regression analysis using functional sets.
3. Perform and interpret the results from a hierarchical regression analysis using functional sets.
4. Recognize and address the various issues that might emerge when working with functional sets.
5. Conduct a hierarchical regression analysis for a research study that involves functional sets by engaging in the activities presented in Section 7.6.

7.1 Chapter Overview

In Chapter 6, we presented three variable selection strategies for multiple regression: simultaneous, hierarchical, and statistical regression, which included the forward addition, backward elimination, mixed, and best model options. Regardless of the strategy, though, the ultimate objective is to select those IVs from a given data set that yield the best fitting model relative to a specific variable selection criterion such as p values, AIC, and BIC. We now extend this discussion from selecting *individual* IVs to selecting *sets* of IVs. We begin by introducing the general concept of sets and the different types of sets that can be formed in statistics. As part of this discussion, we provide rationale and examples for considering and examining sets of IVs instead of individual IVs. We then focus our attention on analyzing sets of IVs using a simultaneous regression strategy followed by a hierarchical regression strategy. In this latter presentation, instead of imposing an a priori variable entry order as we did in Chapter 6, we impose an a priori *set entry order.* These presentations are then followed by a discussion of various issues that could emerge when analyzing sets of IVs. We conclude the chapter with an example of a research study that incorporates sets of IVs and analyzes them via hierarchical regression.

7.2 The Concept of Sets and Their Application to MRC

An Overview of Sets

In the most general sense, a set refers to a collection of well-defined objects or related items, which are formally called elements. Examples include a set of tires, a set of golf clubs, a set of knives, a set of books, and a set of songs. If a set has a specific number of elements, then it is called a finite set, whereas if it has an unlimited number of elements, then it is called an infinite set. For example, the set of vowels in the English language, namely, {*a*, *e*, *i*, *o*, *u*}, is a finite set but the set of natural

DOI: 10.4324/9781003506935-10

or counting numbers, {1, 2, 3, 4, ... }, is an infinite set. In this latter case, the established pattern among the list of elements continues without end as denoted by the use of the ellipsis (...). Note that a set can be described either verbally, for example, "the counting numbers from 1 to 10," or by listing its elements within braces such as {1, 2, 3, ..., 8, 9, 10}.

In addition to classifying sets by the number of elements they contain, sets are also classified by the role they play in a given situation. When examined from this context, there are two types of sets: *universal* and *subsets*. A universal set consists of all the elements that are to be considered for a particular discussion, and a subset consists of some, none, or all the elements contained in the universal set. For example, recall from Chapter 6 when we presented the best model option for statistical regression. In this context we defined the universal set as the set containing four IVs, $\{X_1, X_2, X_3, X_4\}$ and then presented all possible subsets where each subset consisted of 0 IVs (called the *empty set*), 1 IV, 2 IVs, 3 IVs, and the universal set itself.

Types of Sets in MRC

When applied to multiple regression, Cohen et al. (2003, pp. 162–164) classify sets as either *structural* or *functional*. A brief discussion of each type follows.

Structural Sets. A structural set is a set whose elements represent how a factor is formed or defined. To make this definition more meaningful, think of what is entailed with single-factor ANOVA. Recall that the "factor" in single-factor ANOVA is a construct such as attitudes, motivation, or self-efficacy, and the analysis involves examining multiple "levels" of the factor with respect to some dependent variable. As an example, consider a study that examines the effect of sleep deprivation on flight attendants' behavior. For this study, we will assume that a sample of flight attendants are randomly assigned to sleep deprivation periods of 0, 6, 12, and 18 hours, respectively, and then observed for the number of aggressive behaviors they exhibit when interacting with pilots and passengers in a controlled work environment. Examples of aggressive behaviors might include verbal "put downs," getting into an argument, or verbal interruptions. In this context, the factor is "sleep deprivation," and the levels of this factor are 0, 6, 12, and 18 hours. Thus, the set {0 hr., 6 hr., 12 hr., 18 hr.} is a structural set for sleep deprivation because the elements reflect how this factor is formed or defined.

Another example of a structural set involves a categorical variable such as participants' biological sex assigned at birth. Recall from Chapter 6 that we treated this IV as a comparison between two groups (Female vs. Male) and we coded the IV as 1 = Female and 0 = Male. Thus, similar to the ANOVA example presented above, the set {Female, Male}, or equivalently, {0, 1}, is a structural set because it defines the categorical factor "Biological Sex." Other examples of dichotomous categorical variables and their respective structural sets include "Type of Student" {Flight, Non-flight}, "CFI Training Source" {Civilian, Military}, and "Type of Airline" (Full-Service Carrier, Low-Cost Carrier}.

We also can extend this concept to categorical variables that consist of more than two groups such as "Airline Seat Classifications" {Economy, Premium Economy, Business, First Class}, "Types of Airports" {Large Hub, Medium Hub, Small Hub, Non-Hub}, and "Types of Pilot Licenses" (PPL, CPL, ATPL}. The difference is that unlike a dichotomous factor, which uses a single IV to compare two groups, a factor that involves a comparison between more than two groups requires more than one IV. We discuss various coding strategies for analyzing categorical IVs that consist of more than two groups in Chapter 8.

Functional Sets. A functional set is a set whose elements consist of variables that represent a specific attribute or characteristic. As an example, consider the term "demographics." In statistics, demographics refer to specific characteristics of human populations and generally include factors such as age, sex, educational background, race/ethnicity, religion, marital status, and household income. Note that each factor may be represented by an IV. For example, if we let X_1 = Age, X_2 = Sex, X_3 = Educational Background, X_4 = Race/Ethnicity, X_5 = Religion, X_6 = Marital Status, and X_7 = Household Income, and if we let set D = Demographics, then D = $\{X_1, X_2, X_3, X_4, X_5, X_6, X_7\}$. Thus, set D is a functional set because it consists of IVs that represent a common attribute or characteristic. Following are additional illustrations of research situations where IVs could be grouped into functional sets.

- In aviation human factors, researchers might be interested in pilots' *affective domain*, which consists of their attitudes, values, motivations, and emotions.
- In aviation safety, researchers might be interested in factors that are correlated with general aviation accidents or with runway incursions.
- In airport management, researchers might be interested in environmental challenges, airport security, and passenger screenings.

Example 7.1 *Organizing IVs into Functional Sets*

An aviation researcher is examining various factors related to Y = Number of Times Pilots Deviate from ATC Instructions When Taxiing. The targeted factors are listed below. Organize these factors into functional sets.

- X_1 = Hazardous Attitudes, measured by the Antiauthority and Impulsivity dimensions of Hunter's (2005) New-HAS, which is scored on a traditional 5-point Likert response scale where 1 = Strongly Disagree and 5 = Strongly Agree. Thus, higher scores reflect higher levels of each dimension.
- X_2 = General Self-Efficacy, measured by Chen et al.'s (2001) instrument, which is scored on a traditional 5-point Likert response scale. Higher scores reflect higher levels of self-efficacy.
- X_3 = Airport Experience, measured by a researcher-developed questionnaire and scored on a traditional 5-point Likert response scale. Higher scores reflect higher levels of experience.
- X_4 = Involved in Runway Incursion, where participants responded either "Yes" (coded 1) or "No" (coded 0) to being involved in a runway incursion.
- X_5 = Years as Pilot-in-Command (PIC).
- X_6 = Total Flight Hours.
- X_7 = Number of FAA Ratings.
- X_8 = Years' Experience with Modern Cockpit Tools such as electronic flight bags, moving map displays, and heads-up displays.
- X_9 = Age.
- X_{10} = Biological Sex Assigned at Birth (coded 1 = Female and 0 = Male).

Continued on next page...

Solution

The 10 IVs can be organized into four functional sets as follows:

- M = Pilots' Affective Domain, where M represents "mental state" IVs: $\{X_1, X_2\}$.
- E = Pilots' Airport Experience, where E represents "experience" IVs: $\{X_3, X_4\}$.
- F = Pilots' Flight Background, where F represents "flight" IVs: $\{X_5, X_6, X_7, X_8\}$.
- D = Pilot Demographics, where D represents "demographic" IVs: $\{X_9, X_{10}\}$.

Why Use Sets?

Organizing a data set into sets of IVs has several advantages over single variable selection strategies for determining the model of best fit or identifying the strongest predictors of a dependent variable. First, grouping similar IVs into sets facilitates understanding the analysis. Instead of having many single IVs to manage, the focus is now on a small number of sets consisting of IVs that are logically related and may be considered having equal importance with respect to their impact on the DV.

A second advantage is with respect to the number of hypothesis tests that have to be performed. For example, as summarized in Table 7.1, a simultaneous or hierarchical strategy involving the individual IVs given in Example 7.1 would require performing 10 separate regression analyses and testing 10 separate hypotheses to determine the respective sr_i^2 values. Using sets of IVs, however, requires four analyses and testing four hypotheses. Reducing the number of analyses/hypotheses also lowers the probability of committing a Type I error. Furthermore, because each set contains fewer IVs than the total number of individual IVs—in Example 7.1, k = 10 overall, but k = 2, 2, 4, 2, respectively, for sets M, E, F, D—power increases. So, echoing Cohen et al.'s (2003, pp. 185–186) edict about the benefit to grouping individual IVs into sets: "less is more."

As alluded to in the previous passage, when the IVs of a given data set have been organized into sets, the variable selection strategies presented in Chapter 6 for individual IVs may be generalized to sets of IVs. Thus, we may apply the strategies of simultaneous regression, hierarchical regression, or any of the options associated with statistical regression to analyzing sets of IVs. Because the results of statistical regression are governed by various selection criteria, which may be manipulated by the researcher, and can change when these criteria change, we restrict our discussion to simultaneous and hierarchical regression. Furthermore, we will focus on functional sets in this chapter, and then extend this discussion to structural sets in Chapter 8 when we introduce coding strategies for categorical variables involving more than two groups.

7.3 Using Simultaneous Regression to Analyze Sets of IVs

Generalizing from Single IVs to Sets of IVs: A Conceptual Development

Quick Review of Simultaneous Regression with Individual IVs. Recall from Chapters 3–6 that when working with individual IVs, a simultaneous regression strategy involves entering all k IVs into a regression model in no logical order. This results in the collective contribution the IVs make in explaining Y variance, namely, $R^2_{Y \cdot 123 \ldots k}$. Further recall that the collective contribution the k IVs make in explaining

TABLE 7.1 *Summary of Regression Analyses Needed for Determining the Squared Semipartial Correlation Coefficients for the IVs Given in Example 7.1*

	Simultaneous	**Hierarchical** [a]
Analysis	***Individual IVs***	
1	$sr_1^2 = R^2_{Y \cdot 123 \ldots 10} - R^2_{Y \cdot 234 \ldots 10}$	$sr_1^2 = R^2_{Y \cdot 1} = r^2_{Y1}$
2	$sr_2^2 = R^2_{Y \cdot 123 \ldots 10} - R^2_{Y \cdot 134 \ldots 10}$	$sr_2^2 = R^2_{Y(2 \cdot 1)} = R^2_{Y \cdot 12} - R^2_{Y \cdot 1}$
3	$sr_3^2 = R^2_{Y \cdot 123 \ldots 10} - R^2_{Y \cdot 1245 \ldots 10}$	$sr_3^2 = R^2_{Y(3 \cdot 12)} = R^2_{Y \cdot 123} - R^2_{Y \cdot 12}$
4	$sr_4^2 = R^2_{Y \cdot 123 \ldots 10} - R^2_{Y \cdot 1235 \ldots 10}$	$sr_4^2 = R^2_{Y(4 \cdot 123)} = R^2_{Y \cdot 1234} - R^2_{Y \cdot 123}$
5	$sr_5^2 = R^2_{Y \cdot 123 \ldots 10} - R^2_{Y \cdot 12346 \ldots 10}$	$sr_5^2 = R^2_{Y(5 \cdot 1234)} = R^2_{Y \cdot 12345} - R^2_{Y \cdot 1234}$
6	$sr_6^2 = R^2_{Y \cdot 123 \ldots 10} - R^2_{Y \cdot 123457 \ldots 10}$	$sr_6^2 = R^2_{Y(6 \cdot 12345)} = R^2_{Y \cdot 123456} - R^2_{Y \cdot 12345}$
7	$sr_7^2 = R^2_{Y \cdot 123 \ldots 10} - R^2_{Y \cdot 123 \ldots 689,10}$	$sr_7^2 = R^2_{Y(7 \cdot 123456)} = R^2_{Y \cdot 1234567} - R^2_{Y \cdot 123456}$
8	$sr_8^2 = R^2_{Y \cdot 123 \ldots 10} - R^2_{Y \cdot 123 \ldots 79,10}$	$sr_8^2 = R^2_{Y(8 \cdot 1234567)} = R^2_{Y \cdot 12345678} - R^2_{Y \cdot 1234567}$
9	$sr_9^2 = R^2_{Y \cdot 123 \ldots 10} - R^2_{Y \cdot 123 \ldots 8,10}$	$sr_9^2 = R^2_{Y(9 \cdot 12345678)} = R^2_{Y \cdot 123456789} - R^2_{Y \cdot 12345678}$
10	$sr_{10}^2 = R^2_{Y \cdot 123 \ldots 10} - R^2_{Y \cdot 123 \ldots 9}$	$sr_{10}^2 = R^2_{Y(10 \cdot 123456789)} = R^2_{Y \cdot 123 \ldots 10} - R^2_{Y \cdot 123456789}$
	Sets of IVs [b]	
1	$sR_M^2 = R^2_{Y \cdot MEFD} - R^2_{Y \cdot EFD}$	$sR_M^2 = R^2_{Y \cdot M} = R^2_{YM}$
2	$sR_E^2 = R^2_{Y \cdot MEFD} - R^2_{Y \cdot MFD}$	$sR_E^2 = R^2_{Y(E \cdot M)} = R^2_{Y \cdot ME} - R^2_{Y \cdot M}$
3	$sR_F^2 = R^2_{Y \cdot MEFD} - R^2_{Y \cdot MED}$	$sR_F^2 = R^2_{Y(F \cdot ME)} = R^2_{Y \cdot MEF} - R^2_{Y \cdot ME}$
4	$sR_D^2 = R^2_{Y \cdot MEFD} - R^2_{Y \cdot MEF}$	$sR_D^2 = R^2_{Y(D \cdot MEF)} = R^2_{Y \cdot MEFD} - R^2_{Y \cdot MEF}$

Note. See Chapters 4 and 6 for a review of the "dot" notation subscript.
[a] The variable entry order for single IVs is X_1–X_2–X_3–X_4–X_5–X_6–X_7–X_8–X_9–X_{10} and the set entry order for sets of IVs is M–E–F–D. [b] The multiple R is used for the squared semipartial correlation coefficients (i.e., sR^2) when there are multiple IVs involved.

Y variance also includes some redundancy among the IVs. As illustrated in the left pane of Figure 7.1(a), which is a replication of Figure 3.5 for the two IVs case, this redundancy is captured by region *c*.

For example, in the left pane of Figure 7.1(a), we can see that X_1's effect on *Y* encompasses regions *a* and *c*, and X_2's effect on *Y* encompasses regions *b* and *c*. So, part of X_1's contribution to explaining *Y* variance is being captured by X_2's contribution to explaining *Y* variance, and vice versa. Focusing on X_2, this means that to determine X_2's *unique* contribution to explaining *Y* variance—without X_1 having any influence on X_2—we must remove, or partial out, X_1's effect from X_2, but at the same time we do not remove X_1's effect from *Y*. As we observed in Chapter 3, another way of looking at this is to ask, "If we know the amount of information X_1 provides about *Y* variance without X_2 being present, then how much *additional* information about *Y* variance would we gain when X_2 enters the regression model in the presence of X_1?

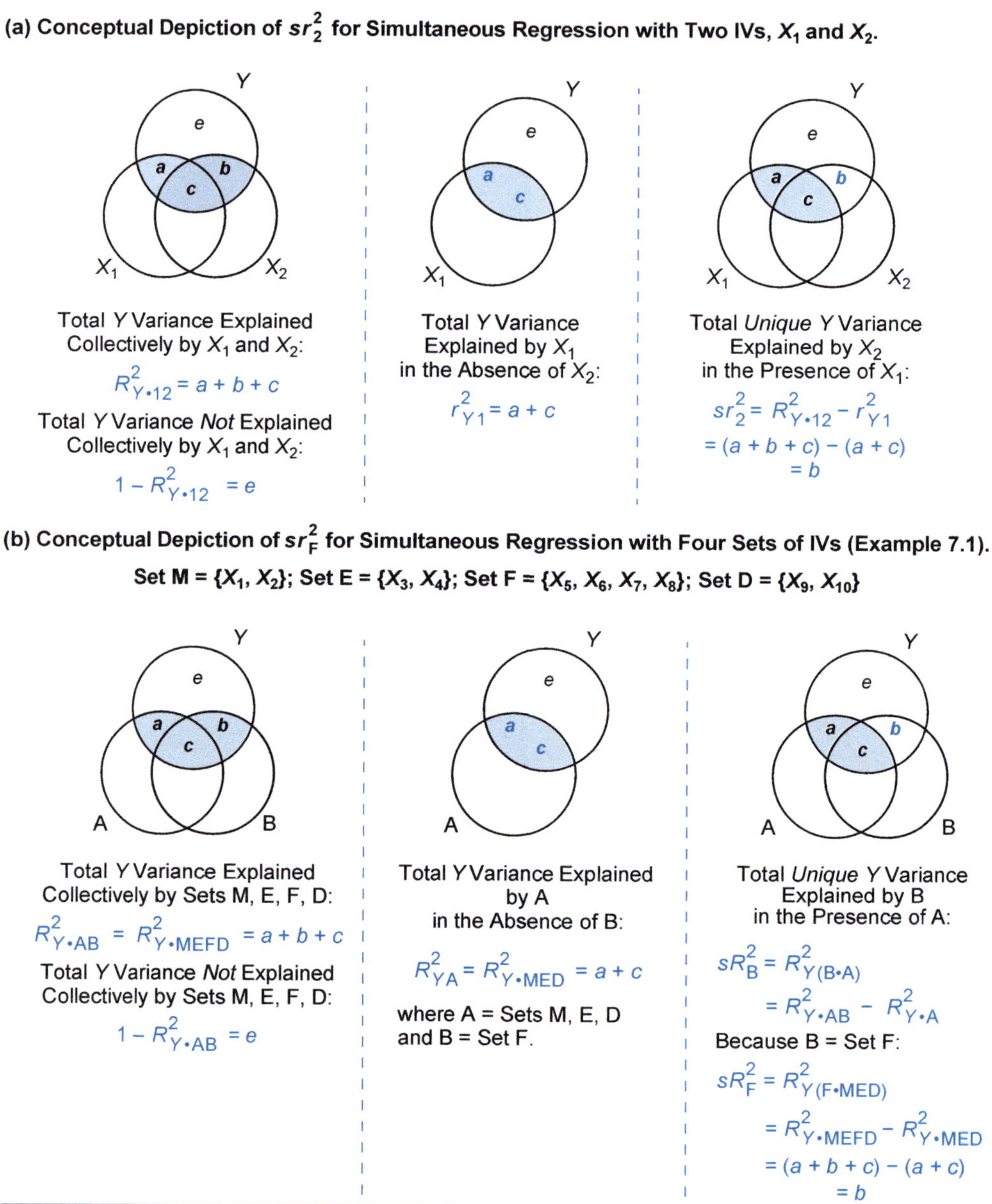

FIGURE 7.1 Venn diagram representation for the squared semipartial correlation coefficient with respect to simultaneous regression for (a) two IVs case and (b) sets of IVs

The answer to this question is illustrated conceptually in the middle and right panes of Figure 7.1(a). In the middle pane, we first imagine a model with X_1 present, which reflects the squared bivariate correlation between Y and X_1, namely, r^2_{Y1}. In the right pane, we then let X_2 enter the model in the presence of X_1, which leads to X_2's unique contribution to explaining Y variance, namely: $sr_2^2 = (a + b + c) - (a + c) = b$, or more specifically, $sr_2^2 = R^2_{Y \cdot 12} - r^2_{Y1}$. Generalizing this relationship to k IVs, the unique contribution each IV makes to explaining Y variance is equal to the difference between the overall R^2 and the R^2 that results from regressing Y on all the IVs except the IV under discussion; that is: $sr_i^2 = R^2_{Y \cdot 123 \ldots i \ldots k} - R^2_{Y \cdot 123 \ldots (i) \ldots k}$, where the i in parentheses in the second term denotes that X_i is being omitted from the regression. For example, in a simultaneous regression with four IVs, $sr_2^2 = R^2_{Y \cdot 1234} - R^2_{Y \cdot 134}$, which denotes that X_2 is being partialled by X_1, X_3, and X_4.

The reader also might recall that because of the shared variance among the IVs, a simultaneous regression does not provide a full accountability of the overall R^2. In other words, the sum of the respective sr_i^2 values from a simultaneous regression will not equal the overall R^2 (assuming nonzero correlation coefficients between all IVs). Once again, this is because of the "hole" that exists (region *c*) when performing a simultaneous regression: Because each IV is being partialled by all the other IVs, this region is never "counted" when examining the aggregate of the IVs' unique contribution to explaining Y variance.

Generalizing to Sets of IVs. With this review in mind, we now generalize from individual IVs to sets of IVs. To do so, we will consider the four sets from Example 7.1, M, E, F, D, where $k_M = 2$ IVs, $k_E = 2$ IVs, $k_F = 4$ IVs, and $k_D = 2$ IVs. To perform a simultaneous regression on these sets, we regress Y on all four sets at the same time, which results in $R^2_{Y \cdot MEFD}$. Although this result is the same as that from simultaneously regressing Y on the 10 individual IVs, resulting in $R^2_{Y \cdot 123 \ldots 10}$, the primary difference is with respect to the partials. For example, with 10 individual IVs, we must perform 10 separate regression analyses to determine each IV's unique contribution to explaining Y variance:

- $sr_1^2 = R^2_{Y \cdot 123 \ldots 10} - R^2_{Y \cdot 234 \ldots 10}$
- $sr_2^2 = R^2_{Y \cdot 123 \ldots 10} - R^2_{Y \cdot 134 \ldots 10}$
- $sr_3^2 = R^2_{Y \cdot 123 \ldots 10} - R^2_{Y \cdot 124 \ldots 10}$

...

- $sr_{10}^2 = R^2_{Y \cdot 123 \ldots 10} - R^2_{Y \cdot 123 \ldots 9}$

By organizing the 10 IVs into four functional sets, though, the unit of analysis for determining "unique contributions" changes from individual IVs to *sets* of IVs. To illustrate this, we use the same Venn diagram representation from Figure 7.1(a) except now we label the two circles that represented individual IVs as set "A" and set "B" as shown in Figure 7.1(b). Note that in this context, "A" and "B" are not specific sets but instead are generic labels for different sets of IVs. For example, we could let A = sets M, E, D, and B = set F as shown in the right pane of Figure 7.1(b). Alternatively, we could let A = sets M and E and B = sets F and D. The point is "A" and "B" are not actual sets but instead are labels used to represent the "real" sets.

Let's now examine the unique contribution set F makes to explaining Y variance, where set F = $\{X_5, X_6, X_7, X_8\}$ and represents variables associated with participants' flight background. In this context, the concept of "unique contribution" means we are interested in examining how much additional knowledge we gain in explaining Y variance when set F enters the model in the presence of all the other sets of IVs. When examining this situation conceptually relative to Figure 7.1(b), recognize that because set F is the set under discussion, "A" represents *all the other sets*, namely, sets M, E, D, and "B" represents set F. So, as shown in the middle pane of Figure 7.1(b), we have a regression model that contains sets M, E, and D, and in the right pane of Figure 7.1(b), set F enters this model in the presence of these other sets, which leads to set F's unique contribution to explaining *Y* variance:

$$sR_B^2 = R_{Y(B \cdot A)}^2 = R_{Y \cdot AB}^2 - R_{Y \cdot A}^2$$

$$sR_F^2 = R_{Y(F \cdot MED)}^2 = R_{Y \cdot MEFD}^2 - R_{Y \cdot MED}^2$$

Let's stop for a minute and consider what we just did here. We conceptually generalized the representation of sr^2 from two individual IVs to sets of IVs by imagining that the set under discussion (F) is a single IV and all the remaining sets (M, E, D) are encompassed by the other IV. Some readers also might recognize that we denoted the squared semipartial correlation coefficient using the multiple *R*; that is, sR^2 as opposed to sr^2. This is because we are now examining the partial variance for a set, which contains multiple IVs, instead of a single IV.

The reader also should not lose focus of region *c*. As is the case with individual IVs, when performing a simultaneous regression with sets of IVs, region *c* is still a "hole" and never gets counted in the aggregation of each set's respective sR_i^2 value. For example, when set F enters the model in the presence of sets M, E, and D as shown in the right pane of Figure 7.1(b), the effect of these latter three sets of variables is being removed from the four IVs of set F, but not from *Y*. We denote this generically as $R_{Y(B \cdot A)}^2$, which means we are examining the relationship between *Y* and B where the effect of A is being removed from B but not from *Y*. In the current case, where B = set F and A = sets M, E, D, we have $R_{Y(F \cdot MED)}^2$, which indicates that we are examining the relationship between set *Y* and F where the collective effect of sets M, E, and D are being removed (or partialled out) from set F but not from *Y*.

A Concrete Example

To demonstrate more explicitly how the squared semipartial correlation coefficients are determined in a simultaneous regression, the file "Ch_7 Sample Data for Sets of IVs" contains hypothetical data aligned to Example 7.1. Recall from Example 7.1 that set M = $\{X_1, X_2\}$, set E = $\{X_3, X_4\}$, set F = $\{X_5, X_6, X_7, X_8\}$, set D = $\{X_9, X_{10}\}$, and *Y* = the number of times pilots deviate from ATC instructions during taxiing.

To facilitate our discussion, we note the following with respect to the given data set: $R_{Y \cdot MEFD}^2 = .580$, $R_{Y \cdot EFD}^2 = .547$, $R_{Y \cdot MFD}^2 = .309$, $R_{Y \cdot MED}^2 = .509$, and $R_{Y \cdot MEF}^2 =$ .563. The reader is encouraged to confirm these results. The reader also is encouraged to reference Figure 7.1(b) where "A" represents the sets in the previous model and "B" represents the set entering the current model. A brief discussion of each set's sR^2 calculation follows, and a summary is provided in Table 7.2. The reader will note that this summary contains information about all the partial correlation coefficients (*sR*, sR^2, *pR*, and pR^2) as well as information about their significance, which is discussed following the current presentation. For the moment, it is important to recognize that the respective correlation coefficients are reported for each *set* and not for the individual IVs. Furthermore, although we include the *pR* and pR^2 values for completeness, we reserve our discussion of the full partial correlation coefficient and its square for Chapter 9.

TABLE 7.2 *Summary of Simultaneous Regression Results for the Data Set Given in the File "Ch_7 Sample Data for Sets of IVs"*

		Semipartial Correlations		Full Partial Correlations [b]		Significance		
Sets/IVs [a]	B_i	sR_B	sR^2_B	pR_B	pR^2_B	t_i [c]	Set-Wise F	p_i
Set M		.182	.033	.270	.073		3.93	.0227
X_1	0.043					2.11		.0375
X_2	−0.077					−1.21		.2292
Set E		.521	.271	.626	.392		32.26	< .0001
X_3	0.010					0.19		.8487
X_4	3.173					8.02		< .0001
Set F		.266	.071	.381	.145		4.23	.0033
X_5	−0.084					−3.38		.0010
X_6	0.0001					2.68		.0085
X_7	0.048					0.46		.6440
X_8	0.064					1.40		.1658
Set D		.130	.017	.197	.039		2.02	.1380
X_9	0.036					1.91		.0595
X_{10}	−0.175					−0.46		.6477

Note. $N = 111$. Y = Number of Times Pilots Deviate from ATC Instructions When Taxiing. $R^2_{Y\bullet MEFD} = .580$, $F(10, 100) = 13.79$, $p < .0001$.
[a] Set M = Mental State: {X_1 = Hazardous Attitudes scores, X_2 = Self-Efficacy scores}. Set E = Airport Experience: {X_3 = Airport Experience scores, X_4 = Involved in Runway Incursion (Yes/No)}. Set F = Flight Background: {X_5 = Years as PIC, X_6 = Total Flight Hours, X_7 = Number of FAA Ratings, X_8 = Years' Experience with Modern Cockpit Tools such as electronic flight bags, moving map displays, and heads-up displays}. Set D = Demographics: {X_9 = Age, X_{10} = Biological Sex Assigned at Birth}. [b] The subscript B refers to the set that is under discussion. For example, $sR^2_B = .033$ indicates that set M uniquely explains 3.3% of Y variance when examined in the presence of the other three sets of variables. [c] t_i refers to the regression coefficients, B_i.

Set M. To determine set M's unique contribution to explaining Y variance, we have a model where "A" = sets E, F, D, and "B" = set M as follows:

$$sR^2_M = R^2_{Y(M\bullet EFD)} = (R^2_{Y\bullet MEFD} - R^2_{Y\bullet EFD}) = (.580 - .547) = .033$$

Therefore, when examined in the presence of sets E, F, and D, the Affective Domain set, which contains two IVs, uniquely explains 3.3% of the variance in the number of times pilots deviate from ATC instructions during taxiing.

Set E. To determine set E's unique contribution to explaining Y variance, we have a model where "A" = sets M, F, D, and "B" = set E as follows:

$$sR^2_E = R^2_{Y(E\bullet MFD)} = (R^2_{Y\bullet MEFD} - R^2_{Y\bullet MFD}) = (.580 - .309) = .271$$

Therefore, when examined in the presence of sets M, F, and D, the Airport Experience set, which contains two IVs, uniquely explains 27.1% of the variance in the number of times pilots deviate from ATC instructions during taxiing.

Set F. To determine set F's, unique contribution to explaining Y variance we have a model where "A" = sets M, E, D, and "B" = set F as follows:

$$sR^2_F = R^2_{Y(F\bullet MED)} = (R^2_{Y\bullet MEFD} - R^2_{Y\bullet MED}) = (.580 - .509) = .071$$

Therefore, when examined in the presence of sets M, E, and D, the Flight Background set, which contains four IVs, uniquely explains 7.1% of the variance in the number of times pilots deviate from ATC instructions during taxiing.

Set D. To determine set D's unique contribution to explaining *Y* variance, we have a model where "A" = sets M, E, F, and "B" = set D as follows:

$$sR_D^2 = R_{Y(D \cdot MEF)}^2 = (R_{Y \cdot MEFD}^2 - R_{Y \cdot MEF}^2) = (.580 - .563) = .017$$

Therefore, when examined in the presence of sets M, E, and F, the Demographics set, which contains two IVs, uniquely explains 1.7% of the variance in the number of times pilots deviate from ATC instructions during taxiing.

The reader is reminded that like individual IVs, a simultaneous regression involving sets of IVs does not provide a full accountability of the explained *Y* variance. For example, although $R_{Y \cdot MEFD}^2 = .580$, the sum of the respective squared semipartial correlation coefficients is $sR_M^2 + sR_E^2 + sR_F^2 + sR_D^2 = .033 + .271 + .071 + .017 = .392$. Once again, this is because region *c* never gets counted in this summation.

Significance of the Partials

To determine if the partial variances (e.g., sR_i^2 values) are significant, we employ the same strategy we did in Chapter 6. We may either manually calculate *F* using an appropriate *F* formula and then consult the *F* table in Appendix B / Table B4,or we may use an online calculator such as Soper (2025a) and Soper (2025b) to determine the *F* value and its corresponding *p* value, respectively. As a point of information, the calculated *F* value in this context is commonly referred to as the set-wise *F* because it is for a specific set of IVs and not for the entire model or for an individual IV. To demonstrate each approach, our focus will be on determining the significance of the unique contribution to explaining *Y* variance set *F* = Flight Background makes, where $sR_F^2 = .071$.

***F* Formula Approach.** The *F* formula we use to manually calculate the set-wise *F* for a set of IVs is the same one we presented in Chapter 6 for individual IVs with some minor adjustments. Instead of sr_i^2, which represents X_i's unique contribution to explaining *Y* variance, we use sR_B^2 to denote the unique contribution of the set under discussion. (Remember: The "B" in this context does not refer to a specific set but instead is used as a generic label for the set under discussion.) Second, in the context of simultaneous regression involving sets of variables, k_A = the total number of IVs in *all* the sets *except* the set under discussion, and k_B = the total number of IVs within the targeted set. With these points in mind, the manual calculation of the set-wise *F* for the unique contribution to explaining *Y* variance for set F is:

$$F = \frac{sr_B^2(N - k_A - k_B - 1)}{(1 - R_{Y \cdot AB}^2)k_B}$$

$$F = \frac{.071(111 - 6 - 4 - 1)}{(1 - .580)(4)} = \frac{.071(100)}{(.420)(4)} = \frac{7.1}{1.68} = 4.22619$$

Consulting Appendix B / Table B4 for $F(4, 100)$ and $\alpha = .05$, $F_{Critical} = 2.46$. Therefore, the set-wise *F* for sR_F^2 is $F = 4.23$, which is significant for $p < .05$. (See also Table 7.2.)

Online Calculator Approach. The online calculator approach also is a two-step process in that we first determine the set-wise *F* using Soper (2025a) and then we determine its corresponding *p* value via Soper (2025b).

(1) Consulting Soper 2025(a):
- Number of predictors in set A = 6
- Number of predictors in set B = 4
- R^2 for set A = .509
- R^2 for sets A and B = .580
- Sample size = 111
- Click Calculate: $F = 4.22619$

(2) Consulting Soper 2025(b):
- Degrees of freedom 1 = 4
- Degrees of freedom 2 = 100
- *F*-value = 4.22619
- Click Calculate: $p = .0033$

The results from the two approaches are exactly the same with one exception: Soper (2025b) provides the exact *p* value vs. simply reporting $p < \alpha$ or $p > \alpha$, which is the case when using an *F* table. These results match what is reported in Table 7.2.

Recall from ANOVA that if the overall ANOVA model is significant, then we may conduct pairwise comparisons. On the other hand, if the model is not significant, then we do not examine any pairwise comparisons because of inflated Type I error rates. This concept, known as Fisher's protected *t* test, may be applied to a multiple regression analysis involving sets of IVs:

If a set is significant, then we may examine the variables within the set for significance; otherwise, we do not perform any post hoc examinations.

In our current example, set F consists of variables that represent pilots' flight background. Because this set is significant, we may examine each of the four variables for significance without incurring an inflated alpha level. Contrast this with set D = Demographics, which is not significant, $F(2, 100) = 2.02$, $p = .1380$, and therefore we do not examine the two IVs (X_9 and X_{10}) within this set for significance.

If a set is significant, then the focus of the within-set post hoc examination is with respect to the partial regression coefficients. These B_i values represent the effect the corresponding IVs have on the DV in the presence of all the other IVs in its home set as well as all the IVs in the other sets included in the model. For example, as reported in Table 7.2, of set F's four IVs, two are significant: $X_5 = -0.084$, $t(100) = -3.38$, $p = .0010$, and $X_6 = 0.0001$, $t(100) = 2.68$, $p = .0085$. Therefore, we may interpret the corresponding regression coefficients (as well as any of the other corresponding partials):

- $B_5 = -0.084$. Holding all other Flight Experience variables constant and holding all the IVs in the other three sets constant, for every 1-year increase as PIC, the number of times pilots deviate from ATC instructions when taxiing *decreases*, on average, 0.084 times. More concretely, for every 10-year increase as PIC, the number of times pilots deviate from ATC instructions when taxiing *decrease*, on average, approximately 1 time (–0.8).
- $B_6 = 0.0001$. Holding all other Flight Experience variables constant and holding all the IVs in the other three sets constant, for every 1-hour increase in total

flight hours, the number of times pilots deviate from ATC instructions when taxiing increase, on average, 0.0001 times. More concretely, for every 10,000-hour increase in flight hours, the number of times pilots deviate from ATC instructions when taxiing increase, on average, 1 time.

Similarly, because sets M and E are significant, we may examine the IVs in these sets as well. These include the regression coefficient for X_1 = Hazardous Attitudes from set M and the regression coefficient for X_4 = Involved in Runway Incursion (Yes/No) from set E.

- B_1 = 0.043. Holding all other Mental State variables constant and holding all the IVs in the other three sets constant, for every 1-unit increase in hazardous attitude scores, the number of times pilots deviate from ATC instructions when taxiing increase, on average, 0.043 times. More concretely, as pilots' antiauthority and impulsivity attitudes become stronger, the number of times they deviate from ATC instructions when taxiing increases.
- B_4 = 3.173. Pilots involved in a runway incursion (coded 1) have, on average, between 3 and 4 more deviations from not following ATC instructions when taxiing than pilots not involved in a runway incursion (coded 0).

In summary, and with respect to the variable selection process, we conclude that the results reported in Table 7.2 indicate that the final regression model contains X_1 from the Mental State set, X_4 from the Airport Experiences set, and X_5 and X_6 from the Flight Background set.

7.4 Using Hierarchical Regression to Analyze Sets of IVs

Generalizing from Individual IVs to Sets of IVs: A Conceptual Development

Recall from Chapter 6 that when working with individual IVs, hierarchical regression involves a specific variable entry order and, unlike simultaneous regression in which *Y* is regressed on all the IVs at the same time, it consists of a sequence of analyses relative to this entry order. Furthermore, when examining partial variances, unlike simultaneous regression where each IV gets partialled by all the other IVs, in hierarchical regression the IV entering the model at the current step gets partialled only by those IVs that were in the previous model. Generalizing this to sets of IVs, we employ the same procedure except now we impose a specific set entry order.

As an example, we once again consider the four sets from Example 7.1, but this time impose a set entry order of M–E–F–D. To perform a hierarchical regression, we sequentially regress *Y* on: set M; sets M and E; sets M, E, and F; and sets M, E, F, and D. These respective regressions result in cumulative R^2 values of $R^2_{Y \bullet M}$, $R^2_{Y \bullet ME}$, $R^2_{Y \bullet MEF}$, and $R^2_{Y \bullet MEFD}$ at each step of the analysis. Furthermore, the increment each set makes in explaining *Y* variance is the difference between each succeeding R^2 and the preceding R^2. For example, with set entry order M–E–F–D:

- $sR^2_M = R^2_{YM}$, where "A" is the empty set, and "B" is set M.
- $sR^2_E = R^2_{Y(E \bullet M)} = (R^2_{Y \bullet ME} - R^2_{Y \bullet M})$, where "A" is set M, and "B" is set E.
- $sR^2_F = R^2_{Y(F \bullet ME)} = (R^2_{Y \bullet MEF} - R^2_{Y \bullet ME})$, where "A" is sets M and E, and "B" is set F.
- $sR^2_D = R^2_{Y(D \bullet MEF)} = (R^2_{Y \bullet MEFD} - R^2_{Y \bullet MEF})$, where "A" is sets M, E, F, and "B" is set D.

A Concrete Example

To demonstrate more explicitly how the squared semipartial correlation coefficients are determined in a hierarchical regression, we will analyze the same hypothetical data we used previously for simultaneous regression, which is given in the file "Ch_7 Sample Data for Sets of IVs." Recall that this data set is aligned to Example 7.1 where set M = $\{X_1, X_2\}$, set E = $\{X_3, X_4\}$, set F = $\{X_5, X_6, X_7, X_8\}$, set D = $\{X_9, X_{10}\}$, and Y = the number of times pilots deviate from ATC instructions during taxiing. For the current application, we will use set entry order M–E–F–D. A brief discussion of each set's sR^2 calculation follows, and a summary is provided in Table 7.3.

Step 1. Set M enters the model in the absence of any other set and the cumulative R^2 is R^2_{YM} = .110, which is equal to sR^2_M. Thus, when examined in the presence of no other sets, the Affective Domain set, which contains two IVs, explains 11% of the variance in the number of times pilots deviate from ATC instructions during taxiing.

Step 2. Set E enters the model in the presence of set M and the cumulative R^2 is $R^2_{Y \cdot ME}$ = .503 and set E's increment is

$$sR^2_E = R^2_{Y(E \cdot M)} = (R^2_{Y \cdot ME} - R^2_{Y \cdot M}) = (.503 - .110) = .393$$

Therefore, the Airport Experience set, which contains two IVs, when examined in the presence of set M explains an additional 39.3% of the variance in the number of times pilots deviate from ATC instructions during taxiing.

Step 3. Set F enters the model in the presence of sets M and E and the cumulative R^2 is $R^2_{Y \cdot MEF}$ = .563 and set F's increment is

$$sR^2_F = R^2_{Y(F \cdot ME)} = (R^2_{Y \cdot MEF} - R^2_{Y \cdot ME}) = (.563 - .503) = .06$$

Therefore, the Flight Background set, which contains four IVs, when examined in the presence of sets M and E, explains an additional 6% of the variance in the number of times pilots deviate from ATC instructions during taxiing.

Step 4. Set D enters the model in the presence of sets M, E, and F and the cumulative R^2 is $R^2_{Y \cdot MEFD}$ = .580 and set D's increment is

$$sR^2_D = R^2_{Y(D \cdot MEF)} = (R^2_{Y \cdot MEFD} - R^2_{Y \cdot MEF}) = (.580 - .563) = .017$$

Therefore, the Demographics set, which contains two IVs, when examined in the presence of sets M, E, and F, explains an additional 1.7% of the variance in the number of times pilots deviate from ATC instructions during taxiing.

Significance of the Partials

To determine the significance of the partial variances, we once again either manually calculate the set-wise F and consult Appendix B / Table B4, or use Soper (2025a) and Soper (2025b) to determine the set-wise F and its corresponding p value, respectively. To manually calculate the set-wise F, we use the same F formula we used for simultaneous regression in the previous section. This is illustrated below for set F. In the current example, though, "A" = sets M and E and "B" = set F.

$$F = \frac{sr^2_B(N - k_A - k_B - 1)}{(1 - R^2_{Y \cdot AB})k_B} = \frac{.06(111 - 4 - 4 - 1)}{(1 - .563)(4)} = \frac{.06(102)}{(.437)(4)} = \frac{6.12}{1.748} = 3.50$$

TABLE 7.3 *Summary of Results of the Hierarchical Regression Analysis for the Data Set Given in the File "Ch_7 Sample Data for Sets of IVs" with Set Entry Order* M–E–F–D

Sets/IVs [a]	Set-Wise *F*				Significance of IVs		
	ΔR^2	*df*	*F*	*p*	B_i	t_i	p_i
Step 1	.110	2, 108	6.67	.0018	$B_1 = 0.09$	$t_1 = 3.52$	$p_1 = .0006$
"A" = { }					$B_2 = -0.03$	$t_2 = -0.39$	$p_2 = .7007$
"B" = set M = $\{X_1, X_2\}$							
Cum $R^2 = R^2_{Y \cdot M} = R^2_{YM} = .110$							
Step 2	.393	2, 106	41.91	< .0001	$B_1 = 0.05$	$t_1 = 2.66$	$p_1 = .0090$
"A" = set M = $\{X_1, X_2\}$					$B_2 = -0.07$	$t_2 = -1.18$	$p_2 = .2410$
"B" = set E = $\{X_3, X_4\}$					$B_3 = 0.06$	$t_3 = 1.31$	$p_3 = .1942$
Cum $R^2 = R^2_{Y \cdot ME} = .503$					$B_4 = 3.48$	$t_4 = 8.98$	$p_4 < .0001$
Step 3	.060	4, 102	3.50	.0101	$B_1 = 0.04$	$t_1 = 1.94$	$p_1 = .0555$
"A" = sets M, E = $\{X_1, X_2, X_3, X_4\}$					$B_2 = -0.11$	$t_2 = -1.76$	$p_2 = .0817$
"B" = set F = $\{X_5, X_6, X_7, X_8\}$					$B_3 = 0.01$	$t_3 = 0.28$	$p_3 = .7791$
Cum $R^2 = R^2_{Y \cdot MEF} = .563$					$B_4 = 3.06$	$t_4 = 7.74$	$p_4 < .0001$
					$B_5 = -0.05$	$t_5 = -2.70$	$p_5 = .0080$
					$B_6 = 0.0001$	$t_6 = 2.72$	$p_6 = .0077$
					$B_7 = 0.02$	$t_7 = 0.24$	$p_7 = .8077$
					$B_8 = 0.08$	$t_8 = 1.66$	$p_8 = .1005$
Step 4	.017	2, 100	2.02	.1380	$B_1 = 0.04$	$t_1 = 2.11$	$p_1 = .0375$
"A" = sets M, E, F = $\{X_1, X_2, X_3,$					$B_2 = -0.08$	$t_2 = -1.21$	$p_2 = .2292$
$X_4, X_5, X_6, X_7, X_8\}$					$B_3 = 0.01$	$t_3 = 0.19$	$p_3 = .8487$
"B" = set D = $\{X_9, X_{10}\}$					$B_4 = 3.17$	$t_4 = 8.02$	$p_4 < .0001$
Cum $R^2 = R^2_{Y \cdot MEFD} = .580$					$B_5 = -0.08$	$t_5 = -3.38$	$p_5 = .0010$
					$B_6 = 0.0001$	$t_6 = 2.68$	$p_6 = .0085$
					$B_7 = 0.05$	$t_7 = 0.46$	$p_7 = .6440$
					$B_8 = 0.06$	$t_8 = 1.40$	$p_8 = .1658$
					$B_9 = 0.04$	$t_9 = 1.91$	$p_9 = .0595$
					$B_{10} = -0.17$	$t_{10} = -0.46$	$p_{10} = 6477$

Note. $N = 111$. Y = Number of Times Pilots Deviate from ATC Instructions When Taxiing. The boxed results represent the IVs entering the analysis at each stage. If the corresponding IVs in the entering set are significant, then a second color is used for the B_i and t_i values in addition to the p_i values.

[a] "A" = set A from Figure 7.1 and represents the set in the previous model. "B" = set B from Figure 7.1 and represents the set entering the model at the current step. Set M = Mental State: {X_1 = Hazardous Attitudes, X_2 = Self-Efficacy}. Set E = Airport Experience: {X_3 = Years of Airport Experience, X_4 = Involved in Runway Incursion (Yes/No)}. Set F = Flight Background: {X_5 = Years as PIC, X_6 = Total Flight Hours, X_7 = Number of FAA Ratings, X_8 = Years' Experience with Modern Cockpit Tools (e.g., electronic flight bags, moving map displays, and heads-up displays)}. Set D = Demographics: {X_9 = Age, X_{10} = Biological Sex (Female or Male)}.

Consulting Appendix B / Table B4 for $F(4, 102)$ and $\alpha = .05$, the closest entries for critical F are $F(4, 100) = 2.46$ and $F(4, 150) = 2.43$. Therefore, the set-wise F for sR^2_F, namely, $F = 3.50$, is significant for $p < .05$. Note that we get a similar result when consulting Soper (2025a), but we also get a more specific p value from Soper (2025b), namely, $p = .0101$.

When performing a hierarchical regression with individual IVs, the reader might recall from Chapter 6 that it is not appropriate to interpret the B_i values at each step of a hierarchical regression to report the effect the respective IVs have on Y at each stage of the analysis. This is because the B_i values collectively reflect a "mixed bag" of regression coefficients that were acquired from separate analyses and therefore cannot be used to make any accurate predictions about the effect each IV has on the DV. However, when working with sets of IVs, if the set entering the analysis at the current step is significant, then it is appropriate to examine the IVs contained within this set for significance as we did in the previous section on simultaneous regression when we applied Fisher's protected *t*. Therefore, in Table 7.3, we boxed off the set of IVs entering the analysis at each step to separate them from the previous step, and placed in a second color those IVs within the set that are significant. For example, at Step 3 the four IVs of set F, $\{X_5, X_6, X_7, X_8\}$, are entering the analysis. Because set F is significant, we may examine these IVs without incurring an inflated Type I error, and as reported in Table 7.3, X_5 and X_6 are significant predictors when examined in the presence of sets M and E. More specifically, holding all other IVs within sets M, E, and F constant: (a) For every 10-year increase in Years as PIC, the number of times pilots deviate from ATC instructions when taxiing decrease, on average, one-half time, which is significant, $B_5 = -0.05$, $t(102) = -2.70$, $p = .0080$; and (b) For every 10000-hour increase in Total Flight Hours, the number of times pilots deviate from ATC instructions when taxiing increase, on average, one time, which is significant, $B_6 = 0.0001$, $t(102) = 2.72$, $p = .0077$.

In summary, based on the results reported in Table 7.3, it appears that with respect to the variable selection process, the final regression model contains X_1 from the Mental State set, X_4 from the Airport Experiences set, and X_5 and X_6 from the Flight Background set. We do not, however, retain any variables from the Demographics set. Note that this final model is the same as what was acquired from simultaneous regression, but the difference is we now have a full accountability of the explained Y variance relative to the set entry order, which was not the case for simultaneous regression:

$$\begin{aligned} R^2_{Y \cdot MEFD} &= sR^2_M + sR^2_E + sR^2_F + sR^2_D \\ &= .110 + .393 + .06 + .017 \\ &= .580 \end{aligned}$$

7.5 Additional Considerations When Working with Sets of IVs

The Potential for Multicollinearity

When organizing IVs into functional sets, it is reasonable to conclude that the IVs within a set are related to each other. For example, consider the four IVs in set F from our current discussion. This set is comprised of four flight experience variables: X_5 = Years as PIC, X_6 = Total Flight Hours, X_7 = Number of FAA Ratings, and X_8 = Years' Experience with Modern Cockpit Tools. Using the data from the "Ch_7 Sample Data for Sets of IVs" data set, the respective correlation coefficients and their squares are: $r_{56} = .4553$, $r^2_{56} = .2073$; $r_{57} = .4062$, $r^2_{57} = .1650$; $r_{58} = .3665$, $r^2_{58} = .1343$; $r_{67} = .4719$, $r^2_{67} = .2227$; $r_{68} = .3619$, $r^2_{68} = .1310$; and $r_{78} = .1708$, $r^2_{78} = .0292$. So, for example, with respect to X_6 and X_7, 22.27% of the variance in total flight hours is being explained by the number of FAA ratings and vice versa.

Although the *VIF* values for these two IVs are less than 1.5, the reader should be sensitive to the possibility of multicollinearity when working with sets of data. One common area where this might occur is between participants' age and a related variable. For example, consider X_5 = Years as PIC and X_9 = Age. The respective

*VIF*s are 2.9 and 2.5, and the correlation between these two IVs is $r_{59} = .7356$ with $r^2_{59} = .5411$. Thus, 54.11% of the variance in age is being explained by years as PIC, and vice versa. For our current example, though, this is not a concern because the two IVs are placed in different sets. Nevertheless, the reader should be sensitive to this possibility, and in the event of multicollinearity within a functional set, researchers must determine which IV is more important relative to the DV and research context and delete the other IV(s).

Issues Related to Statistical Significance

When examining sets of data, most of the time there will be consistency between the significance of a set and the IVs within the set, which is what resulted from the hypothetical data associated with Example 7.1. For example, sets M, E, and F are significant and at least one IV within each set also is significant, whereas neither set D nor any of the IVs in set D is significant. It is also possible, though, for an inconsistency to result between the significance of a set and the significance of its IVs. A brief discussion of the two ways in which this can occur follows.

Significant Set with No Significant IVs. One inconsistency between the significance of a set and the significance of its IVs is with respect to the situation where a set is significant but none of the IVs in the set is significant. As an example, let's perform another hierarchal regression on the hypothetical data from Example 7.1 but this time we use the set entry order E–F–M–D. Table 7.4 contains a summary of the results of this analysis.

Observe from Table 7.4 that at Step 3 of the analysis, set M enters the model in the presence of sets E and F. When the effects of these sets are partialled from M, set M is significant, $sR^2_{\text{M}} = .038$, $F(2, 102) = 4.43$, $p = .0143$, but the IVs in set M are not significant at the preset alpha level of $\alpha = .05$: $B_7 = 0.04$, $t(102) = 1.94$, $p = .0555$, and $B_8 = -0.11$, $t(102) = -1.76$, $p = .0817$. To reconcile a situation of this nature, we defer to Cohen et al. (2003, p. 189):

> A risky but not unreasonable resolution to this dilemma is to tentatively interpret as significant any IV whose *t* is almost large enough to meet the significance criterion: whatever is lost by the inflation of the Type I error is likely to be compensated by the reduction of Type II error and the resolution of the apparent inconsistency.

Cohen et al. also suggested that the *VIF*s of the IVs be reviewed for the possibility of high levels of correlation, which could result in multicollinearity. Following Cohen et al.'s guidance for the current situation, we make the following observations: (a) Of the two IVs in set M, we would tentatively interpret X_1 = Hazardous Attitudes as significant because its *t* value, $t(102) = 1.94$, is nearly large enough to meet the *t* critical value of $t_{\text{Critical}}(102) = 1.98$ for $\alpha = .05$. (b) The *VIF*s for X_1 and X_2 are each less than 1.25 so multicollinearity is not a concern.

Nonsignificant Set with At Least One Significant IV. The second inconsistency is the opposite of the first, namely, a set is not significant but at least one IV in the set is significant. Referring to Table 7.4, a concrete example of this situation is with respect to set D, which is not significant, $sR^2_{\text{D}} = .017$, $F(2, 100) = 2.02$, $p = .1380$. Focusing on X_9 = Age, where $t(100) = 1.91$, $p = .0595$, let's suppose that the preset alpha level is $\alpha = .06$ instead of $\alpha = .05$. If this were indeed the case, then

TABLE 7.4 *Summary of Results of the Hierarchical Regression Analysis for the Data Set Given in the File "Ch_7 Sample Data for Sets of IVs" with Set Entry Order* E–F–M–D

	Set-Wise *F*				Significance of IVs		
Sets/IVs [a]	ΔR^2	*df*	*F*	*p*	B_i [b]	t_i [b]	p_i [b]
Step 1	.456	2, 108	45.34	< .0001	$B_1 = 0.05$	$t_1 = 1.06$	$p_1 = .2898$
"A" = { }					$B_2 = 3.69$	$t_2 = 9.36$	$p_2 < .0001$
"B" = set E = $\{X_3, X_4\}$							
Cum $R^2 = R^2_{Y\cdot E} = R^2_{YE} = .456$							
Step 2	.069	4, 104	3.78	.0065	$B_1 = -0.01$	$t_1 = -014$	$p_1 = .8874$
"A" = set E = $\{X_3, X_4\}$					$B_2 = 3.22$	$t_2 = 7.96$	$p_2 < .0001$
"B" = set F = $\{X_5, X_6, X_7, X_8\}$					$B_3 = -0.05$	$t_3 = -2.66$	$p_3 = .0090$
Cum $R^2 = R^2_{Y\cdot EF} = .525$					$B_4 = 0.0002$	$t_4 = 3.17$	$p_4 = .0020$
					$B_5 = -0.0008$	$t_5 = -0.01$	$p_5 = .9942$
					$B_6 = 0.06$	$t_6 = 1.25$	$p_6 = .2139$
Step 3	.038	2, 102	4.43	.0143	$B_1 = 0.01$	$t_1 = 0.28$	$p_1 = .7791$
"A" = sets E, F					$B_2 = 3.06$	$t_2 = 7.74$	$p_2 < .0001$
= $\{X_3, X_4, X_5, X_6, X_7, X_8\}$					$B_3 = -0.05$	$t_3 = -2.70$	$p_3 = .0080$
"B" = set M = $\{X_1, X_2\}$					$B_4 = 0.0001$	$t_4 = 2.72$	$p_4 = .0077$
Cum $R^2 = R^2_{Y\cdot EFM} = .563$					$B_5 = 0.03$	$t_5 = 0.24$	$p_5 = .8077$
					$B_6 = 0.08$	$t_6 = 1.66$	$p_6 = .1005$
					$B_7 = 0.04$	$t_7 = 1.94$	$p_7 = .0555$
					$B_8 = -0.11$	$t_8 = -1.76$	$p_8 = .0817$
Step 4	.017	2, 100	2.02	.1380	$B_1 = 0.01$	$t_1 = 0.19$	$p_1 = .8487$
"A" = sets E, F, M					$B_2 = 3.17$	$t_2 = 8.02$	$p_2 < .0001$
= $\{X_3, X_4, X_5, X_6, X_7, X_8,$					$B_3 = -0.08$	$t_3 = -3.38$	$p_3 = .0010$
$X_1, X_2\}$					$B_4 = 0.0001$	$t_4 = 2.68$	$p_4 = .0085$
"B" = set D = $\{X_9, X_{10}\}$					$B_5 = 0.05$	$t_5 = 0.46$	$p_5 = .6440$
Cum $R^2 = R^2_{Y\cdot EFMD} = .580$					$B_6 = 0.06$	$t_6 = 1.40$	$p_6 = .1658$
					$B_7 = 0.04$	$t_7 = 2.11$	$p_7 = .0375$
					$B_8 = -0.08$	$t_8 = -1.21$	$p_8 = .2292$
					$B_9 = 0.04$	$t_9 = 1.91$	$p_9 = .0595$
					$B_{10} = -0.18$	$t_{10} = -0.46$	$p_{10} = 6477$

Note. $N = 111$. Y = Number of Times Pilots Deviate from ATC Instructions When Taxiing. The boxed results represent the IVs entering the analysis at each stage. If the corresponding IVs in the entering set are significant, then a second color is used for the B_i and t_i values in addition to the p_i values.
[a] "A" = set A from Figure 7.1 and represents the set in the previous model. "B" = set B from Figure 7.1 and represents the set entering the model at the current step. Set E = Airport Experience: $\{X_3$ = Years of Airport Experience, X_4 = Involved in Runway Incursion (Yes/No)$\}$. Set F = Flight Background: $\{X_5$ = Years as PIC, X_6 = Total Flight Hours, X_7 = Number of FAA Ratings, X_8 = Years' Experience with Modern Cockpit Tools (e.g., electronic flight bags, moving map displays, and heads-up displays)$\}$. Set M = Mental State: $\{X_1$ = Hazardous Attitudes, X_2 = Self-Efficacy$\}$. Set D = Demographics: $\{X_9$ = Age, X_{10} = Biological Sex (Female or Male)$\}$. [b] B_i, t_i, and p_i given in numerical order independent of X_i's subscript.

some readers might report that X_9 is a significant factor and interpret $B_9 = 0.04$. However, because set D is not significant, there is no protection from inflating the alpha level when examining the IVs within the set for significance. Additionally, some readers also might recall from an earlier statistics course that the inflated alpha level is calculated as $1 - (1 - \alpha)^k$ where k = the number of hypothesis tests that are conducted. Because set D has two IVs, two hypothesis tests would be conducted, and without being able to apply the Fisher's protected t to set D, the actual Type I error rate is $1 - (1 - \alpha)^k = 1 - (1 - .06)^2 = 1 - (.94)^2 = 1 - .8836 = .1164$.

This illustration warrants a word of caution: Although the computer-generated parameter estimates table (or equivalent) for a multiple regression analysis will provide the results associated with each IV, including corresponding *p* values, the statistics program neither knows what IVs belong to what set, nor does it invoke Fisher's protected *t*. So, as tempting as it might be to simply scan the *p* values to determine if an IV is significant, researcher's must first examine the significance of the sets.

An Alternative F Test

As demonstrated earlier, we used the same *F* test to determine the set-wise *F* for both simultaneous and hierarchical regression involving sets of IVs. Cohen et al. (2003, p. 171) refer to this *F* test as the Model 1 Error *F* and they contrast it with an alternative called the Model 2 Error *F* as shown below.

$$\text{Model 1 Error } F = \frac{sr_{\text{B}}^2(N - k_{\text{A}} - k_{\text{B}} - 1)}{(1 - R_{\text{Y}\cdot\text{AB}}^2)k_{\text{B}}}$$

$$\text{Model 2 Error } F = \frac{sr_{\text{B}}^2(N - k - 1)}{(1 - R_{\text{Y}\cdot\text{ABC}}^2)k_{\text{B}}}$$

The difference between the two formulas is with respect to the error term in the denominator and the corresponding degrees of freedom. For example, referencing Figure 7.1(b), note that region *e* represents the amount of *Y* variance not accounted for by sets A and B. Recall that in regression, this is considered the residual, or error, and with respect to the Venn diagram representation in Figure 7.1(b) is denoted as $1 - R_{\text{Y}\cdot\text{AB}}^2$. Because there are situations where region *c* does not provide any *unique* contribution to explaining *Y* variance, region *c* becomes part of the error. On the other hand, if region *c* does make a unique contribution to *Y* variance, then the error term becomes $1 - R_{\text{Y}\cdot\text{ABC}}^2$. To help decide which model to use, Cohen et al. (2003, p. 176) suggest researchers use Model 1 Error *F* for large *k* or when there is a variable- or set entry order, but use Model 2 Error *F* for large *N*, small *k*, and large sR^2s. For our purposes, we will use Model 1 Error *F*. Our choice notwithstanding, the reader is cautioned that some statistics programs, which provide modules or separate macros for calculating partial variances for simultaneous and hierarchical regression, use Model 2 Error *F* to report significance. For additional information, the reader is directed to Cohen et al. (pp. 174–176).

7.6 Analyzing Sets of IVs in Research: A Guided Example

We now present a guided example of how to conduct a hierarchical regression analysis for a research study that involves sets of data. This guided example is an extension of Chapter 6's Guided Example except we include additional IVs and organize them into functional sets. A description of the research follows.

> An aviation researcher conducted a study that examined the relationship between various factors and general aviation (GA) pilots' understanding of basic weather concepts. A random sample of N = 100 GA pilots was administered a general weather knowledge test (*Y*), which consisted of

100 dichotomously scored multiple-choice items. Thus, scores could range from 0 to 100, with higher scores reflecting greater understanding of basic weather concepts. As for the targeted factors, the pilots were asked to self-report the following information:

- X_1 = Average Flight Time per Month.
- X_2 = Total Flight Hours (since pilots began flying).
- X_3 = Years of Formal Education completed (12 = high school, 14 = 2 years of college/trade school, 16 = 4 years of college, 18 = master's degree, and more than 18 = post-master's).
- X_4 = Number of Formal Weather/Meteorological Courses completed.
- X_5 = Age.
- X_6 = Biological Sex (assigned at birth).
- X_7 = Annual Income (U.S. dollars).

The data are provided in the file "Ch_7 Guided Example Data." Let set F = Flight Experience: $\{X_1, X_2\}$, set E = Education Level: $\{X_3, X_4\}$, and set D = Demographics: $\{X_5, X_6, X_7\}$. Given these sets, a reasonable set entry order is F–E–D. The structure of our presentation follows Chapter 6's Guided Example and consists of six parts: Research Questions and Hypotheses, Sample Size Planning, Preliminary Data Screening, Data Analysis, Interpretation of Results, and Post Hoc Considerations.

Research Questions and Hypotheses

Recall there are two primary research questions and corresponding hypotheses for hierarchical regression. The first is related to the overall R^2 and the second is related to the increments in R^2. With respect to the increments, the reader is reminded that the focus is on *sets of IVs* and a corresponding *set entry order* and not on individual IVs and a variable entry order, which was the case in Chapter 6.

RQ 1. What is the total proportion of variance in weather knowledge test scores explained collectively by the three sets of IVs?

Hyp 1. The focus is on the collective contribution of the sets to *Y* variance.

- H_0: $\rho^2_{Y \cdot FED} = 0$ (The three sets of IVs collectively do not explain a significant proportion of *Y* variance in the population.)
- H_1: $\rho^2_{Y \cdot FED} > 0$ (The three sets of IVs collectively explain a significant proportion of *Y* variance in the population.)

RQ 2. When examined from a hierarchical perspective with set entry order F–E–D, what is the incremental knowledge gained in the explained variance of weather knowledge test scores at each step of the analysis? Because there are three sets, the corresponding hypothesis has three parts.

Hyp. 2a. The focus is on set F's increment to explaining *Y* variance in the population when it enters the model first and in the presence of no other sets.

- H_0: $\rho^2_{YF} = 0$ (Set F does not have a significant incremental gain in explained *Y* variance in the population when no other sets are in the model.)
- H_1: $\rho^2_{YF} > 0$ (Set F has a significant incremental gain in explained *Y* variance in the population when no other sets are in the model.)

Hyp. 2b. The focus is on set E's increment to explaining Y variance in the population when it enters the model second and in the presence of set F.

- H_0: $\rho^2_{Y \bullet FE} - \rho^2_{Y \bullet F} = 0$ (Set E does not have a significant incremental gain in explained Y variance in the population in the presence of set F.)
- H_1: $\rho^2_{Y \bullet FE} - \rho^2_{Y \bullet F} > 0$ (Set E has a significant incremental gain in explained Y variance in the population in the presence of set F.)

Hyp. 2c. The focus is on set D's increment to explaining Y variance in the population when it enters the model last and in the presence of sets F and E.

- H_0: $\rho^2_{Y \bullet FED} - \rho^2_{Y \bullet FE} = 0$ (Set D does not have a significant incremental gain in explained Y variance in the population in the presence of sets F and E.)
- H_1: $\rho^2_{Y \bullet FED} - \rho^2_{Y \bullet FE} > 0$ (Set D has a significant incremental gain in explained Y variance in the population in the presence of sets F and E.)

Sample Size Planning

When conducting an a priori power analysis for sample size planning, readers are reminded that estimates of the population parameters are guided by theory, prior research or preliminary studies, or professional experience. In the absence of this information, readers may recall that Cohen et al. (2003), with some hesitation, suggested respective effect sizes of "small," "medium," and "large" as follows: (a) .02, .13, and .26 for R^2 (p. 93); and (b) .02, .15, and .35 for the proportion of Y variance being explained by an IV or set of IVs that is not being explained by the other IVs in the model (p. 95). For our purposes, because the current Guided Example is an extension of Chapter 6's Guided Example, our estimates of the population parameters will be based in part on the results of Chapter 6's Guided Example.

Readers also might recall from Chapter 6 that when applied to hierarchical regression, sample size planning involves estimating the expected overall population R^2 and the expected increments in R^2 (i.e., sR_i^2) at each step of the analysis. Because we are now working with sets of variables, the unit of analysis is sets of IVs and the expected increments must include the number of IVs within each set. With this in mind, we now conduct an a priori power analysis for the overall R^2 and each of the three sets for the current study.

Overall R^2. Let's assume that we expect the seven targeted variables of the current example's data set, which we organized into three functional sets, will collectively explain 45% of the variance in Y = Weather Knowledge Test scores. As a result, $R^2_{Y \bullet FED} = .45$, which is equal to an effect size of $ES = f^2 = (.45) / (1 - .45) = (.45) / (.55) = 0.82$. Consulting Soper (2025c) with $f^2 = 0.82$, power = .8, number of IVs = 7, and $\alpha = .05$, the minimum sample size needed is $n^* = 26$. Because the given data set has $N = 100$, it is of sufficient size for having an 80% chance of finding a large effect ($f^2 = 0.82$) at a Type I error rate of 5%.

The Increments (sR_B^2). To determine the minimum sample size needed for the increments at each step of the analysis, we: (a) quantify the amount of explained Y variance we expect each set to make; (b) calculate the respective effect sizes; and (c) consult Soper (2025c) for Step 1 of the analysis, and Soper (2025d) for the remaining steps of the analysis. The results of this process are summarized in Table 7.5 and explained below.

TABLE 7.5 *Summary of the A Priori Power Analysis Results Involving the Increments for the Data Set Given in the File "Chapter 7's Guided Example Data"*

Sets [a]	k_A [b]	k_B [c]	$R^2_{Y\cdot AB}$ [d]	sR^2_B [e]	f^2 [f]	n^* [g]	Power for $N = 100$ [h]
Set F	0	2	.20	.20	0.25	42	.9957
Set E	2	2	.35	.15	0.23	47	.9915
Set D	4	3	.45	.10	0.18	69	.9433

Note. $N = 100$. The dependent variable is Y = Weather Test Scores, which could range from 0 to 100, with higher scores reflecting higher levels of understanding of basic weather concepts. Analyses based on $\alpha = .05$ and power = .8.
[a] Set F = Flight Experience: {X_1 = Average Flight Time per Month, X_2 = Total Flight Hours}. Set E = Education Level: {X_3 = Years of Formal Education, X_4 = Number of Formal Weather/Meteorological Courses}. Set D = Demographics: {X_5 = Age, X_6 = Biological Sex (Female or Male), X_7 = Annual Income (U.S. dollars)}. [b] k_A = Number of IVs in the previous model. [c] k_B = Number of IVs entering the current model. [d] $R^2_{Y\cdot AB}$ = Expected cumulative R^2 at each step of the analysis. [e] sR^2_B = Expected increment in R^2 for the set entering the model. [f] $f^2 = (sR^2_B\ /\ (1 - R^2_{Y\cdot AB})$ is the effect size for the increment: [g] Minimum sample size needed for the increment to be statistically significant. [h] Post hoc power of the increment based on the current study's sample size, $N = 100$, and the increment's effect size.

Step 1. Set F enters the model in the absence of any other set: (a) We expect set F to explain 20% of Y variance. (b) Given $sR^2_F = R^2_{YF} = .20$, the corresponding effect size is $f^2 = 0.25$. (c) Consulting Soper (2025c), the minimum sample size needed is $n^* = 42$.

Step 2. Set E enters the model in the presence of set F: (a) We expect set E's increment to explaining Y variance will be 15%. (b) Given $sR^2_E = .15$ and $R^2_{Y\cdot FE} = .35$, the effect size is $f^2 = 0.23$. (c) Consulting Soper (2025d), the minimum sample size needed is $n^* = 47$.

Step 3. Set D enters the model in the presence of sets F and E. (a) We expect set D's increment to explaining Y variance will be 10%. (b) Given $sR^2_D = .10$ and $R^2_{Y\cdot FED} = .45$, the effect size is $f^2 = 0.18$. (c) Consulting Soper (2025d), the minimum sample size needed is $n^* = 69$.

As a result, the current data set with $N = 100$ meets the minimum sample requirements for the three sets. This can be confirmed by examining the respective post hoc powers for each set based on a sample size of $N = 100$. We do this using Soper (2025f) for set F and Soper (2025e) for sets E and D. For example, as reported in the last column of Table 7.5, given a sample size of $N = 100$ and a preset alpha of $\alpha = .05$: (a) The probability of finding a significant effect at Step 1, where the increment for set F is $sR^2_F = .20$ and $f^2 = 0.25$, is 99.57%; (b) The probability of finding a significant effect at Step 2, where the increment for set E is $sR^2_E = .15$ and $f^2 = 0.23$ is 99.15%; and (c) The probability of finding a significant effect at Step 3, where the increment for set D is $sR^2_D = .10$ and $f^2 = 0.18$, is 94.33%. Because all post hoc powers are greater than .80, we conclude that the current study's sample size is sufficient.

Preliminary Data Screening

Our preliminary data screening will follow the structure and contents of what we did in Chapter 6's Guided Example. We also do *not* provide copies of any detailed summary tables, leverage plots, or residual plots but instead simply report the results from our statistics program.

Outlier Analysis. Examining the initial data set for outliers via Jackknife distances, zero outliers were flagged and therefore this is not a concern. (*Note:* Other outlier procedures might yield different results.)

Multicollinearity. The *VIFs* for each IV are less than 1.81 and therefore multicollinearity is not a concern.

Missing Data. The data set has no missing data.

Measurement Error. No IVs are measuring any type of psychological construct but instead reflect factual data.

Relevant IVs. To determine if the IVs are correctly specified, we examine each IV's added variable plot (see Chapter 5). The results of these plots suggest five of the seven IVs should be deleted from the final model, leaving only X_3 = Years of Formal Education and X_7 = Annual Income as relevant IVs. However, because we are organizing the IVs into functional sets and are performing a hierarchical regression with a specified set entry order, these results are not necessarily applicable. Therefore, we will keep all the IVs in the final model and report this decision as a study delimitation.

We now check to see if the model is compliant with the assumptions of regression.

Linearity. A scatter plot of the residuals vs. the predicted values, in conjunction with the corresponding kernel smoother line, confirms that the scores are randomly scattered around the 0-line and do not show any pronounced or systematic pattern. As a result, the final model is compliant with the linearity assumption.

Equal Variances (Homoscedasticity of the Residuals). The scatter plot of the residuals vs. the predicted values used to check the linearity assumption also confirms the equal variances assumption because the residuals appear to be approximately equally scattered above and below the 0-line without any pronounced pattern. Therefore, the final model is compliant with the equal variances assumption.

Independence of the Residuals. A scatter plot of the residuals vs. the case numbers, in conjunction with the corresponding kernel smoother line, confirms there is no systematic pattern and therefore the model is compliant with the independence assumption.

Normality of the Residuals. The *q-q* plot of the residuals shows that the data closely "hug" the normal line and that all the points are within the 95% confidence band. The corresponding Shapiro-Wilk Goodness of Fit test also is not significant, $W = .988$, $p = .4930$. So, the model is compliant with the normality assumption.

Because preliminary data screening did not result in any changes, the initial data set is the final data set that will be used for the primary analysis.

Data Analysis

We now conduct a hierarchical regression analysis for the final data set with set entry order F–E–D. The results are summarized in Table 7.6 (APA compliant) and Table 7.7 (non-APA compliant).

TABLE 7.6 *Summary of Results of the Hierarchical Regression Analysis for the Data Set Given in the File "Ch_7 Guided Example Data" with Set Entry Order F–E–D (APA Compliant)*

Sets/IVs [a]	B_i	95% CI for B_i [b]		$SE\ B_i$	β_i	$R^2_{Y\cdot AB}$ [c]	ΔR^2 [d]
		LL	*UL*				
Step 1 (Set F)						.318***	.318***
Constant	50.67	42.97	58.37	3.88			
X_1	0.48***	0.33	0.62	0.07	0.55***		
X_2	0.001	−0.001	0.003	0.001	0.07		
Step 2 (Sets F, E)						.408***	.090**
Constant	31.09**	12.88	49.30	9.17			
X_1	0.39***	0.24	0.54	0.08	0.45***		
X_2	0.003*	0.0005	0.005	0.001	0.21*		
X_3	1.37*	0.26	2.49	0.56	0.24*		
X_4	1.07	−0.43	2.57	0.76	0.14		
Step 2 (Sets F, E, D)						.832***	.424***
Constant	3.76	−7.17	14.69	5.50			
X_1	0.04	−0.06	0.14	0.05	0.05		
X_2	0.001	−0.001	0.002	0.001	0.04		
X_3	0.90**	0.29	1.51	0.31	0.16**		
X_4	0.48	−0.37	1.32	0.43	0.06		
X_5	−0.01	−0.15	0.12	0.07	−0.01		
X_6	−0.21	−2.87	2.44	1.34	−0.01		
X_7	0.001***	0.0007	0.0009	0.00005	0.82***		

Note. $N = 100$. The DV is Y = Weather Test Scores.
[a] Set F = Flight Experience: {X_1 = Average Flight Time per Month, X_2 = Total Flight Hours}. Set E = Education Level: {X_3 = Years of Formal Education, X_4 = Number of Formal Weather/ Meteorological Courses}. Set D = Demographics: {X_5 = Age, X_6 = Biological Sex (Female or Male), X_7 = Annual Income in U.S. dollars}. [b] CI = confidence interval; *LL* = lower limit; *UL* = upper limit. [c] $R^2_{Y\cdot AB}$ = the cumulative R^2 at each step of the analysis. [d] $\Delta R^2 = sR^2_B$ = the increment in explained Y variance for each set at each step of the analysis.
*$p < .05$. **$p < .01$. ***$p < .001$.

Interpretation of Results

We next interpret the results from Tables 7.6 and 7.7 relative to the RQs and corresponding hypotheses.

RQ 1 and Hyp 1. This research question relates to the overall R^2. The total proportion of variance in weather test scores explained collectively by the three sets is $R^2_{Y\cdot FED} = .832$, with an adjusted $R^2 = .819$. Thus, the three sets collectively explain 83.2% of the variance in weather test scores and when adjusted for bias, these sets collectively explained 81.9% of the variance in weather test scores. This result is significant, $F(7, 92) = 65.02$, $p < .0001$ and therefore we reject the corresponding null hypothesis and conclude that the three sets collectively explain a significant proportion of Y variance.

RQ 2. This research question relates to the increment in explained Y variance each set makes when it enters the model at each step of the analysis with set entry order F–E–D. Because there are three sets, there are three statistical hypotheses.

TABLE 7.7 *Summary of Results of the Hierarchical Regression Analysis for the Data Set Given in the File "Ch_7 Guided Example Data" with Set Entry Order F–E–D (Non-APA Compliant)*

Sets/IVs [a]	Set-Wise F				Significance of IVs		
	ΔR^2	*df*	*F*	*p*	B_i	t_i	p_i
Step 1	.318	2, 97	22.63	< .0001	$B_1 = 0.48$	$t_1 = 6.56$	$p_1 < .0001$
"A" = { }					$B_2 = 0.001$	$t_2 = 0.77$	$p_2 = .4420$
"B" = set F = $\{X_1, X_2\}$							
Cum. $R^2 = R^2_{Y \cdot F} = R^2_{YF} = .318$							
Step 2	.090	2, 95	7.22	.0012	$B_1 = 0.39$	$t_1 = 5.17$	$p_1 < .0001$
"A" = set F = $\{X_1, X_2\}$					$B_2 = 0.003$	$t_2 = 2.39$	$p_2 = .0186$
"B" = set E = $\{X_3, X_4\}$					$B_3 = 1.37$	$t_3 = 2.44$	$p_3 = .0164$
Cum. $R^2 = R^2_{Y \cdot FE} = .408$					$B_4 = 1.07$	$t_4 = 1.41$	$p_4 = .1612$
Step 3	.424	3, 92	77.40	< .0001	$B_1 = 0.04$	$t_1 = 0.82$	$p_1 = .4170$
"A" = sets F, E = $\{X_1, X_2, X_3, X_4\}$					$B_2 = 0.001$	$t_2 = 0.89$	$p_2 = .3752$
"B" = set D = $\{X_5, X_6, X_7\}$					$B_3 = 0.90$	$t_3 = 2.94$	$p_3 = .0042$
Cum. $R^2 = R^2_{Y \cdot FED} = .832$					$B_4 = 0.48$	$t_4 = 1.12$	$p_4 = .2665$
					$B_5 = -0.01$	$t_5 = -0.20$	$p_5 = .8401$
					$B_6 = -0.21$	$t_6 = -0.16$	$p_6 = .8738$
					$B_7 = 0.001$	$t_7 = 15.19$	$p_7 < .0001$

Note. $N = 100$. Y = Weather Test Scores. The boxed results represent the IVs entering the analysis at each stage. If the corresponding IVs in the entering set are significant, then a second color is used for the B_i and t_i values in addition to the p_i values.
[a] Set F = Flight Experience: $\{X_1$ = Average Flight Time per Month, X_2 = Total Flight Hours$\}$. Set E = Education Level: $\{X_3$ = Years of Formal Education, X_4 = Number of Formal Weather/Meteorological Courses$\}$. Set D = Demographics: $\{X_5$ = Age, X_6 = Biological Sex (Female or Male), X_7 = Annual Income in U.S. dollars$\}$. [b] $\Delta R^2 = sR^2_B$ = the increment in explained Y variance for each set at each step of the analysis; *df*, *F*, and *p* represent the increment.

Hyp 2a. As reported in Tables 7.6 and 7.7, at Step 1, when the two IVs of set F enter the model without any other sets present, set F explains 31.8% of the variance in weather test scores, which is significant, $sR^2_F = R^2_{YF} = .318$, $F(2, 97) = 22.63$, $p < .0001$. Because this result is significant, we reject the corresponding null hypothesis and conclude that set F makes a significant incremental gain (from 0) in explained Y variance in the absence of any other sets.

Given a significant set, we may examine set F's IVs without incurring an inflated alpha level. From Table 7.7, X_1 = Average Flight Time per Month is significant, $B_1 = 0.48$, $t(97) = 6.56$, $p < .0001$. Thus, holding all other IVs in the model constant, for every 10-hour increase in average flight time per month, weather test scores increase, on average, by approximately 5 points.

Hyp 2b. As reported in Tables 7.6 and 7.7, at Step 2, when the two IVs of set E enter the model in the presence of set F, the overall model is significant, $R^2_{Y \cdot AB} = R^2_{Y \cdot FE} = .408$, $F(4, 95) = 16.36$, $p < .0001$. Furthermore, set E's increment in explaining Y variance is 9%, which is significant, $sR^2_E = R^2_{Y(E \cdot F)} = (R^2_{Y \cdot FE} - R^2_{Y \cdot F}) = (.408 - .318) = .09$, $F(2, 95) = 7.22$, $p = .0012$. Because this increment is statistically significant, we reject the corresponding null hypothesis and conclude that set E makes a significant incremental gain in explained Y variance in the presence of set F.

Given a significant set, we may examine set E's IVs for significance without incurring an inflated alpha level. From Table 7.7, X_3 = Years of Formal Education is significant, $B_3 = 1.37$, $t(95) = 2.44$, $p = .0164$. Thus, holding all other IVs in the model constant, for every 1 additional year of formal education, weather test scores increase, on average, by approximately 1.4 points.

Hyp 2c. As reported in Tables 7.6 and 7.7, at Step 3, when the three IVs of set D enter the model in the presence of sets F and E, the overall model is significant, $R^2_{Y \cdot AB} = R^2_{Y \cdot FED} = .832$, $F(7, 92) = 65.03$, $p < .0001$. Furthermore, set D's increment in explaining Y variance is 42.4%, which is significant, $sR^2_D = R^2_{Y(D \cdot FE)} = (R^2_{Y \cdot FED} - R^2_{Y \cdot FE}) = (.832 - .408) = .424$, $F(3, 92) = 77.40$, $p < .0001$. Because this increment is statistically significant, we reject the corresponding null hypothesis and conclude that set D makes a significant incremental gain in explained Y variance in the presence of sets F and E.

Given a significant set, we may examine set D's IVs for significance without incurring an inflated alpha level. From Table 7.7, X_7 = Annual Income in U.S. dollars is significant, $B_7 = 0.001$, $t(92) = 15.19$, $p < .0001$. Thus, holding all other IVs in the model constant, for every $1,000-increase in annual salary, weather test scores increase, on average, by approximately 1 point.

The final regression model is

$$\hat{Y} = 0.04X_1 + 0.001X_2 + 0.90X_3 + 0.48X_4 - 0.01X_5 - 0.21X_6 + 0.001X_7 + 3.76$$

and the model is statistically significant, $R^2 = .832$, $F(7, 92) = 65.03$, $p < .0001$, Adj. $R^2 = .819$, $RMSE = 5.9$. Thus, the seven targeted IVs, which were organized into three functional sets, explains 83.2% of the variance in weather test scores. Although in the final model only two IVs are significant, when considered from a hierarchical perspective via sets, X_1 = Average Flight Time per Month is significant relative to participants' flight experience, X_3 = Years of Formal Education is significant relative to participants' education level, and X_7 = Annual Income in U.S. dollars is significant relative to participants' demographics.

Post Hoc Considerations

We now present the effect sizes and post hoc power values for the overall model and respective increments for the reader's benefit and interpretation. (*Note:* The labels of small, medium, and large effect sizes are relative to Cohen et al. (2003, pp. 93–95).

Effect Size. Following are the respective post hoc effect sizes for the overall model and set increments.

ES for $\boldsymbol{R^2_{Y \cdot FED}}$. $$f^2 = \frac{R^2}{1 - R^2} = \frac{.832}{1 - .832} = \frac{.832}{.168} = 4.95 \text{ (large effect)}$$

ES for $\boldsymbol{sR^2_F}$. $$f^2 = \frac{sR^2_F}{1 - R^2_{Y \cdot F}} = \frac{.318}{1 - .318} = \frac{.318}{.682} = 0.47 \text{ (large effect)}$$

ES for $\boldsymbol{sR^2_E}$. $$f^2 = \frac{sR^2_E}{1 - R^2_{Y \cdot FE}} = \frac{.09}{1 - .408} = \frac{.09}{.592} = 0.15 \text{ (medium effect)}$$

ES for $\boldsymbol{sR^2_D}$. $$f^2 = \frac{sR^2_D}{1 - R^2_{Y \cdot FED}} = \frac{.424}{1 - .832} = \frac{.424}{.168} = 2.52 \text{ (large effect)}$$

Power. Following are the respective post hoc powers for the overall model and set increments.

***Power of the Overall Model* ($R^2_{Y \cdot FED}$).** Consulting Soper (2025f) with $k = 7$, $R^2 = .832$, $p = .05$, and $N = 100$, the power for the overall model is 1.0, which we report as "greater than .99," because we do not make any claim of 100% certainty. Thus, there is a greater than 99% probability that the large effect found in the sample relative to the seven IVs, which were organized into three functional sets, also exists in the population. There also is a greater than 99% probability that the decision to reject the corresponding null hypothesis is correct.

***Power for Set F's Increment* (sR^2_F).** Because Step 1 of a hierarchical regression does not include any other IVs in the model other than those under discussion, we consult Soper (2025f) with $k = 2$, $R^2 = .318$, $p = .05$, and $N = 100$. This returns a power of .99999. As a result, there is a 99.999% probability that the large effect found in the sample for set F with no other IVs in the model also exists in the population. There also is a 99.999% probability that the decision to reject the corresponding null hypothesis is correct.

***Power for Set E's Increment* (sR^2_E).** Consulting Soper (2025e) with $f^2 = 0.15$, $k_A = 2$, $k_B = 2$, $p = .05$, and $N = 100$, the power for set E's increment is .9345. Thus, there is a 93.45% probability that the medium effect found in the sample for Set E in the presence of set F also exists in the population. There also is a 93.45% probability that the decision to reject the corresponding null hypothesis is correct.

***Power for Set D's Increment* (sR^2_D).** Consulting Soper (2025e) with $f^2 = 2.52$, $k_A = 4$, $k_B = 3$, $p = .05$, and $N = 100$, the power for set D's increment is 1.0. Thus, there is a greater than 99% probability that the large effect found in the sample for Set D in the presence of sets F and E also exists in the population. There also is a greater than 99% probability that the decision to reject the corresponding null hypothesis is correct.

Chapter 7 Summary

7.2 The Concept of Sets and Their Application to MRC

- A set consists of a collection of well-defined objects or related items, which are referred to as members or elements of the set, and is represented by enclosing the elements within braces.
- The empty set, { }, contains no elements, a finite set contains a countable number of elements, and an infinite set has an unlimited number of elements.
- A universal set consists of all the elements relative to a particular situation, and a subset consists of some, none, or all the elements contained in the universal set.
- When applied to MRC, a structural set consists of elements that represent how a factor is formed or defined, whereas a functional set consists of variables that represent a specific attribute or characteristic. An example of the former is numerical coding used to define the levels of a categorical variable, and an example of the latter is a set of IVs that represents a common characteristic such as demographics.
- When organizing data into functional sets, the unit of analysis changes from individual IVs to sets of IVs. This action (a) reduces the Type I error rate

because there are fewer hypotheses to be tested, and (b) it increases power because each set has fewer IVs than the total number of IVs overall, which promotes a "less is more" dictum: with fewer IVs, statistical power increases.

7.3 Analyzing Sets of IVs via Simultaneous Regression

- Analyzing sets of IVs using a simultaneous regression strategy is an extension of examining individual IVs in that the DV is regressed on all the sets at the same time.
- For the overall R^2 of a regression model, the subscript reflects the individual sets instead of the IVs. For example, given set R = $\{X_1, X_2, \ldots, X_6\}$, set S = $\{X_7, X_8, \ldots, X_{11}\}$, and set T = $\{X_{12}, X_{13}, \ldots, X_{22}\}$, then the overall R^2 is denoted $R^2_{Y\bullet RST}$.
- With respect to the partial variances, instead of analyzing a single IV in the presence of all the other IVs, the DV is regressed on a set of IVs in the presence of all the other sets. As an example, given sets R, S, and T, if the focus is the unique contribution set S makes in the presence of sets R and T, then A = sets R and T, B = set S, and this is denoted as $sR^2_S = (R^2_{Y\bullet RST} - R^2_{Y\bullet RT})$.
- The set-wise *F* for the partials may be calculated manually using the *F* formula given in Section 7.3 or by consulting Soper (2025a) and Soper (2025b). Regardless of the approach, k_A = the number of variables in set A, k_B = the number of variables in set B, and $R^2_{Y\bullet AB}$ is the R^2 for the overall model.
- When working with sets of IVs, we apply Fisher's protected *t*: If the unique contribution a set makes to explaining *Y* variance is significant, then we may examine the variables within the set for significance; otherwise, we do not perform any post hoc examinations.
- The focus of a within set post hoc examination is on the regression coefficients to determine if they have a significant effect on the DV.

7.4 Analyzing Sets of IVs via Hierarchical Regression

- Analyzing sets of IVs using a hierarchical regression strategy is an extension of examining individual IVs hierarchically. Instead of imposing a variable entry order, though, we impose a set entry order and perform a series of analyses one step at a time relative to the entry order.
- At each step of the analysis, we let set A = the set(s) of IVs that were in the model in previous step and set B = the set of IVs that is entering the model at the current step. For example, given sets R, S, and T, if the set entry order is T–S–R, then at Step 2, the focus is on set S's increment in explaining *Y* variance when it enters the model in the presence of set T; that is $sR^2_S = R^2_{Y(B\bullet A)} = R^2_{Y(S\bullet T)} = (R^2_{Y\bullet TS} - R^2_{Y\bullet T})$.
- The set-wise *F* for the partials may be calculated manually using the *F* formula given in Section 7.3 or by consulting Soper (2025a) and Soper (2025b). Regardless of the approach, k_A = the number of variables in set A, k_B = the number of variables in set B, and $R^2_{Y\bullet AB}$ is the cumulative R^2 for the model at the current stage.

- When working with sets of IVs, we apply Fisher's protected *t*: If a set's increment in explaining *Y* variance is significant, then we may examine the variables within the set for significance; otherwise, we do not perform any post hoc examinations.
- The focus of a within set post hoc examination is on the regression coefficients to determine if they have a significant effect on the DV at that step of the analysis.

7.5 Additional Considerations When Working with Sets of IVs

- When working with functional sets, there is the potential for multicollinearity within a set because the IVs that comprise a set generally are related to each other in some way and therefore researchers must be sensitive to this issue.
- If a set is significant but none of the IVs within the set is significant, then it is possible that some of the variables are highly correlated. One approach to resolving this situation is to follow Cohen et al.'s (2003, p. 189) suggestion.
- If a set is not significant but at least one of the IVs in the set is significant, any claim of significance for the IV(s) is subject to an inflated alpha level. Therefore, researchers are advised not to scan the output for *p* values that suggest a significant IV but instead focus on the significance of the set.
- Cohen et al. (2003) provide two different set-wise *F* formulas. The Model 1 Error *F*, which we used throughout this chapter, assumes that region *c* of Figure 7.1(b) does not make a unique contribution to explaining *Y* variance and therefore this region is included as part of the residual. The Model 2 Error *F* assumes that region *c* does make a unique contribution to *Y* variance and therefore is not part of the residual. Cohen et al. (p. 176) suggest using Model 1 Error *F* for large *k* or when there is a variable- or set entry order, but use Model 2 Error *F* for large *N*, small *k*, and large sR^2s.

Key Terms and Concepts

Elements of a set	Set
Finite set	Set entry order
Fisher's protected *t* test	Set-wise *F*
Functional set	Structural set
Infinite set	Subset
Model 1 Error *F*	Universal set
Model 2 Error *F*	

Review Exercises

A. Checking Conceptual Knowledge

In 1–10, choose the best answer among the choices provided.

1. A researcher is investigating the effect 12 IVs (e.g., number of passengers screened, education level and age of TSA agents, and airport location) have on the number of illegal items identified during passenger airport screenings. The

researcher's primary focus is to develop a prediction equation that consists of statistically significant predictors that could be used to help inform government policy on improving passenger screening and safety. What regression strategy would you recommend?

a. Simultaneous regression
b. Hierarchical regression
c Forward addition or backwards elimination options of statistical regression
d. Bivariate regression where each IV is examined independently.

2. Which of the following statements is NOT true?
   a. Organizing data into functional sets reduces the number of IVs that have to be analyzed.
   b. Organizing data into functional sets makes managing the IVs easier and more informative.
   c. Analyzing data that have been organized into functional sets increases statistical power.
   d. Analyzing data that have been organized into functional sets can lead to multicollinearity.

Use the following research description to answer Items 3–5.

Consider a study that examined the effect various factors had on the performance of $N = 100$ randomly selected TSA airport screeners (Y). Let's assume that after data collection, the researchers organize the data into the following sets: D = Demographic Factors, which consists of $n = 8$ IVs; P = Level of Professionalism, which consists of $n = 2$ IVs; E = Employment History, which consists of $n = 6$ IVs; and H = Health History, which consists of $n = 5$ IVs.

3. Suppose a simultaneous regression analysis is conducted. Which expression denotes the unique contribution of set E?
   a. $R^2_{Y \cdot HPE} - R^2_{Y \cdot HP}$ b. $R^2_{Y \cdot HPED} - R^2_{Y \cdot HPD}$
   c. $sr^2_E = R^2_{Y(E \cdot HP)}$ d. $sr^2_E = R^2_{YE \cdot HPD}$
4. Suppose a hierarchical regression analysis is conducted with set entry order H–P–E–D. Which expression denotes the increment of set E?
   a. $R^2_{Y \cdot HPED} - R^2_{Y \cdot HPD}$ b. $R^2_{Y \cdot HPE} - R^2_{Y(E \cdot HP)}$
   c. $sr^2_E = R^2_{Y(E \cdot HP)}$ d. $sr^2_E = R^2_{YE \cdot HP}$
5. Referencing the Venn diagram given in Figure 7.1(b), let's assume that A represents sets E and H, and B represents set P. If $R^2_{Y \cdot HPE} = .48$, $R^2_{YP_2} = .30$, $R^2_{Y \cdot EH} = .19$, and a hierarchical regression analysis is conducted, then $R^2_{Y(B \cdot A)}$ = ______.
   a. .30 b. .29 c. .18 d. .19
6. An airport manager at a local medium hub airport is interested in identifying key predictors of job performance among airport personnel, which includes all individuals responsible for the operation, maintenance, and servicing of the airport. The researcher who was hired to conduct the study focused on six IVs per the manager's request: X_1 = Age, X_2 = Education Level, X_3 = Sex, X_4 = Job Experience, X_5 = Training Hours, and X_6 = Performance Evaluations. The researcher

Continued on next page ...

organized these IVs into two functional sets—Demographics, where D = $\{X_1, X_2, X_3\}$; and Employment History, where E = $\{X_4, X_5, X_6\}$—and conducted a hierarchical multiple regression with set entry order D–E. Let's assume that the results show a significant increase in R^2 when set E enters the model. Which statement best describes the purpose for the given set entry order?

a. To control for the effects of set D while assessing the additional variance explained by set E.
b. To test the hypothesis that set D will completely account for the variance in job performance, leaving no room for set E.
c. To determine if there is an interaction between set D and set E that affects job performance.
d. To assess whether set D and set E contribute equally to the prediction of job performance.

7. Consider a study with 11 IVs organized into three functional sets where $k_M = 4$, $k_T = 6$, and $k_G = 1$. Let's assume $N = 111$ and a hierarchical multiple regression analysis is performed with the following results: $R^2_{Y \bullet M} = .04$, $R^2_{Y \bullet MT} = .16$, and $R^2_{Y \bullet MTG} = .20$. Which of the following most closely approximates the Model 1 Error set-wise F at Step 2?

a. $F(11, 99) = 2.25$
b. $F(6, 99) = 2.50$
c. $F(6, 100) = 3.33$
d. $F(6, 100) = 2.40$

8. In a study examining factors influencing aviation safety, researchers used functional sets and a hierarchical multiple regression strategy to analyze the impact of various predictors on the frequency of safety incidents. The model included three sets of predictors: set D = Demographic factors such as pilot age and experience level, set O = Operational factors such as weather conditions and aircraft maintenance, and set S = Safety Protocol Adherence factors such as compliance with safety checklists and adherence to procedural guidelines. The analysis was performed in three steps with set entry order D–O–S. If the results show a significant increase in explained variance at Step 3 but not at the other steps, then which statement accurately describes the results of this analysis?

a. Because set D entered first, it accounted for all the variability in safety incidents and set O did not provide any additional explanatory power.
b. Set O was the only significant set with set D showing no significant impact.
c. Set D was significant at Step 1, but the addition of set O at Step 2 significantly increased the explained variance in safety incidents, indicating that the operational factors added additional predictive knowledge beyond the demographic factors.
d. The results suggest that set S should be considered first before including set D and Set O in the model.

9. In a study examining factors influencing pilot performance, researchers use hierarchical regression to understand how different sets of variables contribute to the prediction of overall performance scores. Suppose the study includes three sets of predictor variables: set D = Demographic such as age and years of experience, set T = Technical Skills/Training factors such as hours of simulator training and type ratings, and set P = Psychological factors such as stress levels and cognitive abilities. If the researchers are interested in determining how each set of vari-

ables contributes to the variance in pilot performance scores when entered into the regression model sequentially, then which statement best describes the purpose of hierarchical multiple regression in this study?

a. To identify which individual variables from each set have the strongest relationship with pilot performance scores.
b. To understand how the three sets contribute incrementally to explaining the variance in pilot performance scores.
c. To evaluate the combined effect of all predictor variables on pilot performance scores without considering their hierarchical entry.
d. To compare the overall predictive power of pilot performance scores against a benchmark score from previous studies.

10. Consider a study with 20 IVs organized into three functional sets, where $k_U = 5$, $k_J = 2$, and $k_M = 13$. Let's assume $N = 113$ and a hierarchical multiple regression analysis is performed with the following results:
    - $R^2_{YU} = .041$, $F(5, 107) = 0.91$, $p = .4775$.
    - $R^2_{Y \bullet UJ} = .074$, $F(7, 105) = 1.20$, $p = .3093$, set-wise $F(2, 105) = 1.87$, $p = .1592$.
    - $R^2_{Y \bullet UJM} = .786$, $F(20, 92) = 16.92$, $p < .0001$, set-wise $F(13, 92) = 23.55$, $p < .0001$.

    Which statement is NOT correct relative to the final model?

    a. If a simultaneous analysis were performed instead of a hierarchical analysis, then the final results would be different.
    b. The incremental effect of set M in explaining *Y* variance is significant.
    c. One recommendation for future research would be to not consider the factors in set J in the presence of set U.
    d. Based on the set-entry order, it appears that the researcher underestimated the effect set M has on explaining *Y* variance.

B. Applying Procedural and Conceptual Knowledge

Use the following research description and the corresponding data set to conduct the activities given in Parts B1–B7. See Section 7.6 for guidance.

Consider a study modeled after Uhuegho (2017) who examined the safety climate, psychological strain, and demographics of aviation maintenance technicians (AMTs) at U.S. based aviation maintenance, repair, and overhaul (MRO) organizations. For the current study, the data are organized into three functional sets:

- Set S = Safety Climate consists of $n = 5$ IVs: X_1 = Recognition is participants' perceptions of being rewarded or recognized for doing safe work; X_2 = Safety Concern is participants' perceptions of their organization's concern for safety; X_3 = Supervision is participants' perceptions of their supervisors' expertise and the type of assistance they provide; X_4 = Feedback is participants' perceptions of the amount and quality of feedback they receive; and X_5 = Training is participants' perceptions of the training they received for their job and if it is adequate.
- Set P = Psychological Strain consists of $n = 2$ IVs: X_6 = Stress is participants' perceptions of the amount of stress they feel when doing their job; and X_7 = Psychological Distress is participants' perceptions of their general psychological

health with respect to depression, anxiety, social dysfunction, and psychosomatic symptoms.

- Set D = Demographics consists of n = 8 IVs: X_8 = Age; X_9 = Biological Sex (Female or Male); X_{10} = Marital Status (Married, Not married); X_{11} = Years of MRO Experience; X_{12} = Years Employed at Current MRO; X_{13} = English is Native Language (Yes or No); X_{14} = Years of Formal Education; and X_{15} = Race (White or Nonwhite).

The dependent variable is Y = Maintenance Errors, which measures the number of errors participants recall making, either self-detected or those flagged by their supervisors. This is assessed using an 11-item instrument measured on a traditional 5-point Likert response scale. Thus, scores could range from 11 to 55 with higher scores indicating higher occurrence of maintenance errors. The data set, which consists of hypothetical data, is given in the file "Ch_7 Review Exercises Data–Part B."

B1. Research Questions and Hypotheses. Write the research questions and corresponding hypotheses for the overall regression model and the respective increments for each set.

B2. Sample Size Planning. Conduct a power analysis to see if the given data set is sufficiently large for the overall R^2 and for the increments of each set.

B3. Preliminary Data Screening. Perform regression diagnostics on the data set and then confirm that the final model is compliant with the underlying regression assumptions.

B4. Data Analysis. Conduct a hierarchical regression analysis of the final data set and summarize the results in both an APA-compliant table similar to Table 7.6 and a non-compliant APA table similar to Table 7.7. Using Uhuegho (2017) as guidance: (a) the set entry order is S–P–D; (b) the expected amount of explained variance in the population is $R^2_{Y \bullet SPD}$ = .38, and the expected increments are sR^2_{YS} = .15, $sR^2_{Y(P \bullet S)}$ = .20, and $sR^2_{Y(D \bullet SP)}$ = .03; (c) the sets collectively explain a significant proportion of the variance in maintenance errors; and (d) there is a significant predictive gain in the explained variance of maintenance errors made by each set at each step of the analysis. Also assume α = .05.

B5. Interpretation of Results. Interpret the results relative to the research questions and corresponding null hypotheses stated in Part B1.

B6. Post Hoc Considerations. Determine and comment on the effect sizes and power for the overall R^2 and the respective increments for each set.

B7. Commentary. Provide any additional commentary that is meaningful with respect to the results.

References

Chen, G., Gully, S. M., & Eden, D. (2001). Validation of a new general self-efficacy scale. *Organizational Research Methods, 4*(1), 62–83. https://doi.org/10.1177/109442810141004

Cohen, J., Cohen, C., West, S. G., & Aiken., L. S. (2003). *Applied multiple regression/correlation analysis for the behavioral sciences* (3rd ed.). Lawrence Erlbaum Associates.

Hunter, D. R. (2005). Measurement of hazardous attitudes among pilots. *The International Journal of Aviation Psychology*, *15*(1), 23–43. https://doi.org/10.1207/s15327108ijap1501_2

Soper, D. (2025a). *F-value calculator for hierarchical multiple regression*. https://www.danielsoper.com/statcalc/calculator.aspx?id=14

Soper, D. (2025b). *p-value calculator for an F-test*. https://www.danielsoper.com/statcalc/calculator.aspx?id=7

Soper, D. (2025c). *A-priori sample size calculator for multiple regression*. https://www.danielsoper.com/statcalc/calculator.aspx?id=1

Soper, D. (2025d). *A-priori sample size calculator for hierarchical multiple regression* https://www.danielsoper.com/statcalc/calculator.aspx?id=16

Soper, D. (2025e). *Post-hoc statistical power calculator for hierarchical multiple regression* https://www.danielsoper.com/statcalc/calculator.aspx?id=17

Soper, D. (2025f). *Post-hoc statistical power calculator for multiple regression*. https://www.danielsoper.com/statcalc/calculator.aspx?id=9

Uhuegho, K. O. (2017). *Examining the safety climate of U.S. based aviation maintenance, repair, and overhaul (MRO) organizations* [Doctoral dissertation, Florida Institute of Technology]. Scholarship Repository. https://repository.fit.edu/cgi/viewcontent.cgi?article=1056&context=etd

Chapter 8
Coding Strategies for Nominal IVs

Student Learning Outcomes

After studying this chapter, you will be able to do the following:

1. Conduct and interpret the results of an MRC analysis involving a single, nominal IV that represents three or more dummy coded groups.
2. Conduct and interpret the results of an MRC analysis involving a single, nominal IV representing three or more effects coded groups (unweighted and weighted).
3. Conduct and interpret the results of an MRC analysis involving a single, nominal IV representing three or more contrast coded groups.
4. Conduct and interpret the results of an MRC analysis for a research study with a single, coded nominal IV by engaging in the activities presented in Section 8.6.

8.1 Chapter Overview

This chapter introduces three coding strategies for analyzing variables measured on a nominal scale. The reader might recall that such variables are commonly referred to in research as *group membership variables* because they involve studying two or more distinct groups representing people, objects, or events. We begin with a review of the concept of group membership. We then introduce the first of three coding strategies, *dummy coding*, which is used when it is plausible to make pairwise group comparisons to a designated *reference group*. We follow this discussion with the second coding strategy, *effects coding*, which involves identifying a *base group* so that all pairwise group comparisons are made with respect to the overall sample mean involving all the groups. As part of this discussion, we distinguish between *unweighted and weighted effects coding*. The last coding strategy we introduce is *contrast coding*, which enables researchers to construct planned pairwise group comparisons independent of using a reference or base group. We conclude the chapter with a guided example of a research study that consists exclusively of a nominal IV and compares the results from the three coding strategies. (*Note:* In Chapter 9, we examine data sets that contain a mix of nominal and continuous IVs.)

8.2 Group Membership, Single-Factor ANOVA, and Coded IVs

The Concept of Group Membership

The reader will recall that in aviation research, variables are classified generally as continuous or nominal (see Appendix A). A continuous variable consists of numerical data and is measured on either an interval or ratio scale, which makes some or all four of the basic operations of arithmetic (addition, subtraction, multiplication, division) meaningful. Examples include test scores, IQ, and temperature (interval scale), as well as age, annual income, and flight time (ratio scale). On the other hand, a nominal variable consists of non-numerical data and references distinct

DOI: 10.4324/9781003506935-11

characteristics or traits that are placed into categories. Examples include biological sex (e.g., female, male), employment status (e.g., unemployed, part-time, full-time, retired), pilot licenses (e.g., SPL, PPL, CPL, ATPL), runway incursions (e.g., OD, OE, PD, VPD), and flight-based weather conditions (e.g., VFR, MVFR, IFR, LIFR).

A nominal variable is often referred to in research as a group membership variable because it involves comparing two or more groups with respect to a dependent measure. For example, in a research study that examines the effect flight-based weather conditions have on the percentage of canceled flights, the DV = Percentage of Canceled Flights and the IV = Flight-Based Weather Conditions. If we define the IV as visual flight rules (VFR), marginal visual flight rules (MVFR), instrument flight rules (IFR), and low instrument flight rules (LIFR), then the IV is a group membership variable, and each flight rule represents a distinct category as defined by the Federal Aviation Administration (FAA, 2024). When examined from an ANOVA context, the reader might recognize that this group membership variable is denoted a *single-factor with four levels*, and the data may be analyzed via a single-factor ANOVA. Furthermore, if the overall ANOVA model is significant, then we may conduct pairwise comparisons between groups.

A Review of Single-Factor ANOVA

As a concrete example, consider the data set given in the file, "Ch_8 Coding Example Data." This data set consists of hypothetical data for a research study modeled after Turgut (2019), who examined, in part, differences in levels of professionalism among four aviation groups: airport managers (AMs); aviation maintenance technicians (AMTs); air traffic controllers (ATCs); and Airline Transport Pilots (ATPs). Professionalism was measured using Snizek's (1972) Hall's Professionalism Inventory (HPI) 25-item instrument that is scored on a traditional Likert response scale with responses ranging from 1 = Strongly Disagree to 5 = Strongly Agree. Thus, overall scores could range from 25 to 125, with higher scores denoting a higher level of professionalism. As a result, Y = Level of Professionalism and X = Aviation Groups, which, is a single factor with four levels. A summary of the results from analyzing this data set via a single-factor ANOVA is provided in Table 8.1.

As reported in Table 8.1, the overall ANOVA model is significant, $R^2 = .075$, $F(3, 153) = 4.1599$, $p = .0073$, and hence we may perform pairwise comparisons between the groups without incurring inflated Type I error rates. The reader might recall that given k groups, there are a total of $k(k-1)/2$ pairwise comparisons (and corresponding hypotheses). Thus, for the given data set, there are $4(3)/2 = 6$ pairwise comparisons. These are summarized in the "Tukey's HSD Post Hoc Comparisons" subtable of Table 8.1. Note that of the six pairwise comparisons, two are significant:

- AM vs. ATC. Airport managers averaged approximately 4 points higher on the HPI than air traffic controllers and this 4-point mean difference is significant, $M_{\text{Diff}} = 3.93$, $t(153) = 2.829$, $p = .0269$. Therefore, airport managers have a significantly higher level of professionalism than air traffic controllers.
- ATC vs. ATP. Air traffic controllers averaged 4.4 points *lower* on the HPI than airline transport pilots and this 4.4-point mean difference is significant, $M_{\text{Diff}} = -4.36$, $t(153) = -3.225$, $p = .0083$. Therefore, air traffic controllers have a significantly lower level of professionalism than airline transport pilots.

TABLE 8.1 *Summary of Single-Factor ANOVA and Follow-up Results for the Data Set Given in the File "Ch_8 Coding Example Data"*

ANOVA

Source	*SS*	*df*	*MS*	*F*	*p*
Between	468.9906	3	156.330	4.15990	.0073
Within (Error)	5749.8374	153	37.581		
Total	6218.8280	156			

Note. $N = 157$. $R^2 = \eta^2 = \frac{SS_{\text{Between}}}{SS_{\text{Total}}} = \frac{468.9906}{6218.8280} = .0754$.

Group Means and Standard Deviations

Group	*N*	*M*	*SD*
Airport Managers	37	83.81	6.57
Aviation Maintenance Technicians	38	82.68	6.64
Air Traffic Controllers	41	79.88	5.07
Airline Transport Pilots	41	84.24	6.19
Overall	157	82.65 [a]	6.31

Note. [a] Unweighted mean.

Tukey's HSD Post Hoc Comparisons

Group Comparisons	M_{Diff}	*t*(153)	95% CI	*p*
AM – AMT	1.13	0.796	[–2.55, 4.80]	.8563
AM – ATC	3.93	2.829	[0.32, 7.54]	.0269
AM – ATP	–0.43	–0.312	[–4.04, 3.18]	.9895
AMT – ATC	2.80	2.033	[–0.78, 6.39]	.1805
AMT – ATP	–1.56	–1.130	[–5.14, 2.03]	.6717
ATC – ATP	–4.36	–3.225	[–7.88, –0.85]	.0083

Note. AM = Airport Managers, AMT = Aviation Maintenance Technicians, ATC = Air Traffic Controllers, ATP = Airline Transport Pilots.

The Concept of Coded IVs

Given that our focus is on multiple regression/correlation, not single-factor ANOVA, the obvious question is, "How do we analyze nominal IVs in MRC?" To partly answer this question, let's review the strategy we used in previous chapters when there was a nominal IV that consisted of two groups.

- In Chapter 4, Example 4.1, X_3 = Flight Training Program compared two flight training programs relative to aviation safety index scores: Part 141 (coded 1) vs. Part 61 (coded 0).
- In Chapter 4, Example 4.2, X_2 = Corrected Vision compared pilots with corrected vision (coded 1 = Yes) vs. pilots who did not have corrected vision (coded 0 = No) relative to pilots' visual reaction times.
- In Chapter 4's Guided Example, X_4 = Airline Business Model compared low-cost carriers (coded 1) vs. full-service carriers (coded 0) relative to average air fares.

- In Chapter 7, Example 7.1, X_4 = Involved in Runway Incursion compared pilots who were involved in a runway incursion (coded 1 = Yes) vs. pilots who were not involved in a runway incursion (coded 0 = No) relative to the number of times pilots deviated from ATC instructions when taxiing.
- Chapters 5 and 6 included examples where an IV represented participants' biological sex assigned at birth, which we coded 1 = Female and 0 = Male.

From these examples, the reader should make three key observations: (a) the IVs are nominal and reflect two distinct groups; (b) we changed the data type of the IVs from nominal to continuous by using numerical codes (0 and 1) to represent the groups; and (c) although each nominal IV consists of two groups, only one quantitative IV is used to represent both groups.

Generalizing from these observations, it appears that to analyze nominal IVs in MRC: (a) the data type of the group membership variable is changed from nominal to continuous; (b) the continuous data used to represent the groups consist of numerical codes like 0 and 1; and (c) the number of IVs needed for this representation is equal to one fewer than the number of groups that comprise the group membership variable, and therefore, a nominal IV comprised of k groups requires $k - 1$ quantitative IVs to represent the k groups.

Applying these observations to our running example on professionalism, an MRC analysis of X = Aviation Group would require three quantitative IVs to represent the four groups. What's more, these $k - 1$ IVs represent a structural set (see Chapter 7), which is examined simultaneously. Independent of these observations there also is the presumption that the groups are mutually exclusive. So, for our running example on professionalism, a participant is assumed to be in one and only one of the four groups, which means, for example, that an airport manager cannot also be an airline transport pilot, or an airline transport pilot cannot also be an aviation maintenance technician.

Given this presentation, a logical follow-up question to the first one we posed is, "What numerical codes do we use to represent the groups?" The answer to this question is that it depends on the coding strategy—dummy, effects, and contrast—because each one uses a different set of codes, and these will be presented in the remaining sections of this chapter. The reader is advised, though, that regardless of the coding strategy used, the final regression model will yield the same results because the same data set is being used for all three coding strategies. The difference in the results among these strategies will be with respect to the interpretation of the regression coefficients in the final model. Each coefficient will answer a different question based on the coding system used.

Before we conclude this section, the reader is reminded that we will use the data set for our running example on professionalism, which is given in the file "Ch_8 Coding Example Data," for all three coding strategies presented in Sections 8.3, 8.4, and 8.5. We also will periodically refer to the results reported in Table 8.1 to contrast the differences between the results from a single-factor ANOVA vs. those from an MRC analysis. Lastly, the reader is informed that our focus in Chapter 8 is exclusively on nominal IVs and therefore we do *not* mix nominal IVs and continuous IVs. This latter discussion is reserved for Chapter 9.

8.3 Dummy Coding

The Concept of Dummy Coding

Dummy coding uses a dichotomous coding scheme of 0s and 1s, and all group comparisons are made to a reference group, which is coded with 0s and denoted as G_g, or more simply, the *gth group*. The reader will recognize that we previously applied the concept of dummy coding when we presented a two-group nominal IV in our earlier work as summarized in the bullet list of examples given in Section 8.2. For example, in Chapter 4 / Example 4.1, we coded the nominal IV, X_3 = Flight Training Program, 1 = Part 141 and 0 = Part 61. This is a dummy coding scheme where Part 61 is designated the reference group, and the focus is on comparing Part 141 to Part 61.

We now extend this concept to the running example on professionalism, where the nominal IV is X = Aviation Groups. Because X has four groups, we need three coded variables to examine it via MRC, and these variables are coded using 0s and 1s. We also must specify which group is the reference group. Table 8.2 provides several different coding charts that illustrate how X may be dummy coded using different reference groups.

Observe from Table 8.2 that we labeled the coded variables D_1, D_2, and D_3 to indicate that they are dummy codes for the IV. Further notice that each coding chart has a different reference group, which is coded with all 0s. This raises the question, "What group should be designated the reference group for dummy coding?" Although it does not matter which group serves as the reference group, Hardy (1993), as cited in Cohen et al. (2023, pp. 303–304), provided the following considerations about the reference group:

TABLE 8.2 *Dummy Coding Charts for the Running Example on Professionalism*

(a) ATP is reference group

Group	Coded Variables D_1	D_2	D_3
AM	1	0	0
AMT	0	1	0
ATC	0	0	1
ATP	0	0	0

(b) AM is reference group

Group	Coded Variables D_1	D_2	D_3
AMT	1	0	0
ATC	0	1	0
ATP	0	0	1
AM	0	0	0

(c) AMT is reference group

Group	Coded Variables D_1	D_2	D_3
AM	1	0	0
ATC	0	1	0
ATP	0	0	1
AMT	0	0	0

(d) ATC is reference group

Group	Coded Variables D_1	D_2	D_3
AM	1	0	0
ATP	0	1	0
AMT	0	0	1
ATC	0	0	0

Note. AM = Airport Managers (n = 37), AMT = Aviation Maintenance Technicians (n = 38), ATC = Air Traffic Controllers (n = 41), and ATP = Airline Transport Pilots (n = 41).

- It should serve as a useful and meaningful comparison.
- It should be well-defined and not a "wastebasket" category such as "Other."
- It should not have a small sample size relative to the other groups.

When applying these considerations to our running example, it would be reasonable to make ATP the reference group, which would result in an MRC analysis that compares the levels of professionalism of the other three groups to ATPs. Note that this selection is compliant with all three of Hardy's (1993) considerations: (a) Unlike the other three aviation groups, ATPs are required to maintain their qualifications by undergoing recurrent training and periodic check rides, which makes them a meaningful comparison group; (b) The ATP group is well-defined and does not reflect a "wastebasket" or "Other" category; and (c) The sample sizes of all four groups are approximately the same. As a result, we will use the coding chart in Table 8.2(a) to demonstrate an MRC analysis using dummy coding.

The Hypotheses Associated with Dummy Coding

From an inferential statistics perspective, the reader will recognize that the coding chart given in Table 8.2(a) implies the following set of hypotheses:

D_1 = Airport Managers vs. Airline Transport Pilots

- H_0: $\mu_{AM} - \mu_{ATP} = 0$ (There is no significant difference in mean HPI scores between airport managers and airline transport pilots.)
- H_1: $\mu_{AM} - \mu_{ATP} \neq 0$ (There is a significant difference in mean HPI scores between airport managers and airline transport pilots.)

D_2 = Aviation Maintenance Technicians vs. Airline Transport Pilots

- H_0: $\mu_{AMT} - \mu_{ATP} = 0$ (There is no significant difference in mean HPI scores between aviation maintenance technicians and airline transport pilots.)
- H_1: $\mu_{AMT} - \mu_{ATP} \neq 0$ (There is a significant difference in mean HPI scores between airline maintenance technicians and airline transport pilots.)

D_3 = Air Traffic Controllers vs. Airline Transport Pilots

- H_0: $\mu_{ATC} - \mu_{ATP} = 0$ (There is no significant difference in mean HPI scores between air traffic controllers and airline transport pilots.)
- H_1: $\mu_{ATC} - \mu_{ATP} \neq 0$ (There is a significant difference in mean HPI scores between air traffic controllers and airline transport pilots.)

Conducting and Interpreting the Results of a Dummy-Coded MRC Analysis

To conduct the analysis using MRC, we must first modify the given data set to include the dummy coded variables relative to the coding chart of Table 8.2(a). Table 8.3 contains a partial listing of this revised data set. Observe from Table 8.3 that D_1 = 1 denotes airport managers, D_2 = 1 denotes aviation maintenance technicians, and D_3 = 1 denotes air traffic controllers. Note further that the reference group (ATPs) is coded all 0s and is not represented by a dummy coded variable, which means there will not be a corresponding partial regression coefficient for this group in the regression model. Based on this coding scheme, D_1 represents a comparison between airport managers and ATPs, D_2 represents a comparison between AMTs and ATPs, and D_3 represents a comparison between ATCs and ATPs. These

TABLE 8.3 *Partial Listing of the Data Set Given in the File "Ch_8 Coding Example Data" Revised for Dummy Coded Variables*

Row	Y = HPI	X = Aviation	Coded Variables		
			D_1	D_2	D_3
1	81	AM	1	0	0
2	79	AM	1	0	0
3	93	AM	1	0	0
. . .	. . .	. . .	. . .	. . .	. . .
36	87	AM	1	0	0
37	87	AM	1	0	0
38	87	AMT	0	1	0
39	78	AMT	0	1	0
40	94	AMT	0	1	0
. . .	. . .	. . .	. . .	. . .	. . .
74	82	AMT	0	1	0
75	85	AMT	0	1	0
76	84	ATC	0	0	1
77	89	ATC	0	0	1
78	83	ATC	0	0	1
. . .	. . .	. . .	. . .	. . .	. . .
115	81	ATC	0	0	1
116	72	ATC	0	0	1
117	87	ATP	0	0	0
118	84	ATP	0	0	0
119	88	ATP	0	0	0
. . .	. . .	. . .	. . .	. . .	. . .
156	89	ATP	0	0	0
157	84	ATP	0	0	0

Note. AM = Airport Managers, AMT = Aviation Maintenance Technicians, ATC = Air Traffic Controllers, and ATP = Airline Transport Pilots.

respective comparisons are reflected in the regression model via corresponding B_i values. Lastly, to represent the overall effect of the four aviation groups on participants' level of professionalism (as measured by the HPI—higher scores reflect higher levels of professionalism), all the dummy coded variables must be in the regression equation. Therefore, when working with a coded set of variables that represents a nominal IV, we perform a simultaneous regression. Table 8.4 contains a summary of the results of an MRC analysis involving the dummy coded variables.

Interpreting the Overall Model. As reported in Table 8.4, the results of the overall model are $R^2 = .0754$, $F(3, 153) = 4.1599$, $p = .0073$, and the regression equation is, $\hat{Y} = -0.43D_1 - 1.56D_2 - 4.36D_3 + 84.24$. Thus, 7.54% of the variance in HPI scores is being explained collectively by the four aviation groups (airport managers, aviation maintenance technicians, air traffic controllers, and airline transport pilots), and this amount of explained variance is significant. Note that these results (sans the regression equation) are equivalent to those from the single-factor ANOVA as reported in the ANOVA subtable of Table 8.1.

TABLE 8.4 *Summary of MRC Results for the Running Example on Professionalism Using Dummy Coding with ATP as the Reference Group*

ANOVA

Source	*SS*	*df*	*MS*	*F*	*p*
Between	468.9906	3	156.330	4.1599	.0073
Within (Error)	5749.8374	153	37.581		
Total	6218.8280	156			

Note. $N = 157$. $R^2 = .0754$.

Summary of Partials

			Semipartial Correlations		Full Partial Correlations		Significance	
Factor [a]	B_i	SE_{Bi}	sr_i	sr_i^2	pr_i	pr_i^2	$t_i(153)$	p_i
Constant	84.24	0.96					87.99	< .0001
D_1 = AM	−0.43	1.39	−.0245	.0006	−.0245	.0006	−0.31	.7558
D_2 = AMT	−1.56	1.38	−.0877	.0077	−.0911	.0083	−1.13	.2603
D_3 = ATC	−4.36	1.35	−.2506	.0628	−.2522	.0636	−3.22	.0015

Note. $N = 157$. Y = Level of Professionalism, measured by Snizek's (1972) Hall's Professionalism Inventory (HPI).
[a] AM = Airport Managers ($n = 37$), AMT = Aviation Maintenance Technicians ($n = 38$), ATC = Air Traffic Controllers ($n = 41$).

Although our statistics program provides information about the significance of the overall model, it is informative from an instructional perspective to present the corresponding F formula used to determine the significance of R^2. This F formula is the same one we presented in previous chapters except we modify it to reflect that we are working with groups:

$$F = \frac{R^2(N-k-1)}{(1-R^2)k} \rightarrow F = \frac{R^2(N-g)}{(1-R^2)(g-1)}$$

Applying this F formula to our running example:

$$F = \frac{R^2(N-g)}{(1-R^2)(g-1)} = \frac{.0754(157-4)}{(1-.0754)(4-1)} = \frac{.0754(153)}{(.9246)(3)} = \frac{11.5362}{2.7738} = 4.1599$$

Determining Group Means. When we substitute into the regression equation the numerical codes from the coding chart, we get each group's mean HPI score. For example, based on the coding chart in Table 8.2(a), the mean HPI score for airport managers is found by entering 1, 0, 0 into the regression equation for D_1, D_2, and D_3, respectively, and then performing the arithmetic. This procedure is demonstrated below for the four groups.

- $M_{AM} = -0.43(1) - 1.56(0) - 4.36(0) + 84.24 = -0.43 + 84.24 = 83.81$
- $M_{AMT} = -0.43(0) - 1.56(1) - 4.36(0) + 84.24 = -1.56 + 84.24 = 82.68$
- $M_{ATC} = -0.43(0) - 1.56(0) - 4.36(1) + 84.24 = -4.36 + 84.24 = 79.88$
- $M_{ATP} = -0.43(0) - 1.56(0) - 4.36(0) + 84.24 = 84.24$

Note that the respective group means are equivalent to those reported in Table 8.1 from the single-factor ANOVA. Also recall that because the reference group is coded all 0s, it is not associated with an IV and therefore does not have a corresponding regression coefficient (B_i). However, as demonstrated above, the reference group is represented in the regression equation as the regression constant (B_0). Thus, for dummy coding, the regression constant of the overall regression equation is equal to the mean of the reference group.

Interpreting the Regression Coefficients (B_i). In each of the arithmetic operations we performed to get the respective group means for the coded IVs, we added each variable's coefficient to the regression constant. This can be seen in the last operation prior to getting the mean. For example, the last operation for airport managers is "–0.43 + 84.24," and for AMTs it is "–1.56 + 84.24." This pattern leads to the following general relationship:

> *The mean of the group represented by the coded IV (X_i) is equal to the sum of X_i's coefficient (B_i) and the regression constant (B_0).*

Using symbols to denote this relationship:

$$M_i = B_i + B_0$$

Because the regression constant in a dummy coded regression equation is the mean of the reference group, denoted as the *g*th group, we substitute M_g for B_0:

$$M_i = B_i + M_g$$

Lastly, we subtract M_g from both sides of the equation:

$$M_i - M_g = B_i$$

As a result, the unstandardized regression coefficients, B_i, represent the difference in group means between a targeted group and the designated reference group with respect to the DV. Applying this relationship to our running example:

- $B_1 = -0.43$. The difference in mean professionalism scores between Airport Managers and Airline Transport Pilots is 83.81 – 84.24 = –0.43. Thus, holding all other variables constant, airport managers have a lower level of professionalism, on average, than ATPs. As reported in the Summary of Partials subtable of Table 8.4, this difference in group means is *not* significant, $t(153) = -0.31$, $p = .7558$. (*Note:* It is not necessary to preface the interpretation of B_i with "holding all other IVs constant" because no other IVs whose effects must be partialled are in the model. We do so, though, to get students in the habit of using this phrase.)
- $B_2 = -1.56$. The difference in mean professionalism scores between Aviation Maintenance Technicians and Airline Transport Pilots is 82.68 – 84.24 = –1.56. Thus, holding all other variables constant, AMTs have a lower level of professionalism, on average, than ATPs. As reported in the Summary of Partials subtable of Table 8.4, this difference in group means is *not* significant, $t(153) = -1.13$, $p = .2603$.
- $B_3 = -4.36$. The difference in mean professionalism scores between Air Traffic Controllers and Airline Transport Pilots is 79.88 – 84.24 = –4.36. Thus, holding all other variables constant, ATCs have a lower level of professionalism, on average, than ATPs. As reported in the Summary of Partials subtable of Table 8.4, this difference in group means is significant, $t(153) = -3.22$, $p = .0015$.

These results are equivalent to what is reported in Table 8.1 for the singe-factor ANOVA. For example, focusing exclusively on Tukey's post hoc pairwise comparisons with respect to ATPs: (a) AM – ATP = –0.433, (b) AMT – ATP = –1.560, and (c) ATC – ATP = –4.366. Lastly, and as a point of information for completeness, although the *t* test results for B_i are provided as part of our statistics program's output, they also can be manually calculated by the formula, $t_i = B_i / SE_i$. For example, from Table 8.4, $B_1 = -0.43$ and $SE_1 = 1.39$. Therefore, $t_1 = -0.43/1.39 = -0.31$.

Interpreting the Squared Semipartial Correlation Coefficients (sr_i^2). Recall from Chapters 4 and 6 that the squared semipartial correlation coefficient for simultaneous regression requires two separate regression analyses. The first involves regressing *Y* on all the IVs, which represents the overall R^2, and the second involves regressing *Y* on all the IVs *except* the one being targeted. The difference between these two R^2 values is sr_i^2. More generally,

$$sr_i^2 = R^2_{Y \bullet 123\ldots i \ldots k} - R^2_{Y \bullet 123\ldots (i) \ldots k}$$

where the *i* in the subscript of the second term is placed in parentheses to signify that the targeted IV (i.e., X_i) is being omitted in the regression. The resulting sr_i^2 value represents the amount of unique *Y* variance explained by X_i after all the other IVs have been partialled from X_i but not from *Y*. When applying this concept to an MRC analysis involving dummy coded variables, we follow the same procedure. Thus, with respect to our running example:

- $sr_1^2 = R^2_{Y \bullet 123} - R^2_{Y \bullet 23} = .0754 - .0748 = .0006$
- $sr_2^2 = R^2_{Y \bullet 123} - R^2_{Y \bullet 13} = .0754 - .0677 = .0077$
- $sr_3^2 = R^2_{Y \bullet 123} - R^2_{Y \bullet 12} = .0754 - .0126 = .0628$

As is the case for simultaneous regression, the sum of the sr_i^2 values does not equal the overall R^2: .0006 + 0077 + .0628 = .0711, whereas $R^2_{Y \bullet 123} = .0754$.

To interpret these sr_i^2 values, we must remember that each dummy coded IV represents a comparison between two groups as reflected in the corresponding coding chart. For our running example, D_1 = AM vs. ATP, D_2 = AMT vs. ATP, and D_3 = ATC vs. ATP. As a result, when conducting the second regression analysis that does not include the targeted IV, we are omitting a group comparison, not an IV. Therefore, the interpretation of sr_i^2 from a dummy coded MRC analysis must reflect this exclusion of a group comparison. Figure 8.1 provides an illustration of how to make this interpretation with respect to sr_3^2 for our running example.

Observe from Figure 8.1 that when we calculate sr_3^2, we effectively remove D_3 from the model as depicted by crossing out the respective row and column. When we do this, we lose the comparison between air traffic controllers and airline transport pilots. This leaves only two pairwise comparisons: AM vs. ATP and AMT vs. ATP. Because sr_3^2 corresponds to D_3, which is the pairwise comparison between ATC and ATP, by excluding D_3 from the analysis, the corresponding sr_3^2 value of .0628 represents the proportion of explained *Y* variance that has been "lost." More specifically, $sr_3^2 = .0628$ is the proportion of explained *Y* variance that $R^2_{Y \bullet 123}$ is reduced by when we exclude the ATC vs. ATP comparison from the analysis.

Group	**Coded Variables** D_1	D_2	D_3
Airport Manager (AM)	1	0	0
Aviation Maintenance Technician (AMT)	0	1	0
Air Traffic Controller (ATC)	0	0	1
Airline Transport Pilot (ATP)	0	0	0

D_1 = AM vs. ATP
D_2 = AMT vs. ATP
D_3 = ATC vs. ATP

$$sr_3^2 = R^2_{Y\bullet 123} - R^2_{Y\bullet 12}$$

sr_3^2 = [Proportion of Y variance collectively explained by D_1, D_2, D_3] minus [Proportion of Y variance collectively explained by D_1, D_2]

= .0754 − .0126

= **.0628** Proportion of explained Y variance that is "lost" by not including D_3.

FIGURE 8.1 Graphical illustration of how to interpret sr_3^2 from the running example on professionalism.

Cohen et al. (2003) refer to these sr_i^2 values as "distinctions" because one group is being distinguished from another group with respect to a dependent variable. Thus, when we apply this term to $sr_3^2 = .0628$, we would say that the *ATC-ATP distinction accounts for 6.28% of the variance in HPI scores.* Similarly, $sr_1^2 = .0006$ is interpreted as the AM-ATP distinction accounts for 0.06% of the variance in HPI scores, and $sr_2^2 = .0077$ is interpreted as the AMT-ATP distinction accounts for 0.77% of the variance in HPI scores. Remember, though, that the AM-ATP distinction of .0006 is the amount of Y variance that $R^2_{Y\bullet 123}$ is reduced by when the AM vs. ATP comparison is omitted, and the ATM-ATP distinction of .0077 is the amount of Y variance that $R^2_{Y\bullet 123}$ is reduced by when the AMT vs. ATP comparison is omitted. Thus, the term "distinction," when used in this context, refers to a reduction in total Y variance attributed to excluding a specific group comparison.

As for the significance of these sr_i^2 values, we rely on what is reported from our statistics program. For example, as reported in Table 8.4, $sr_1^2 = .0006$ is not significant, $t(153) = -0.31$, $p = .7558$, $sr_2^2 = .0077$ is not significant, $t(153) = -1.13$, $p = .2603$, and $sr_3^2 = .0628$ is significant, $t(153) = -3.22$, $p = .0015$. For completeness, though, these t values may be calculated manually using the t formula we presented in Chapter 4.

$$t_i = sr_i\sqrt{\frac{N-k-1}{1-R^2}}$$

The reader will recall that this t test may be applied to determine the significance of all corresponding partial coefficients, B_i, sr_i, sr_i^2, pr_i, and pr_i^2. As an example, we will use this formula to get the t value for sr_3^2. To do so, though, we first need to get sr_3. Note that $sr_3 = \sqrt{sr_3^2} = \sqrt{.0628} = \pm\ .2506$. Because the corresponding regression coefficient (B_3) from the regression equation is negative, $sr_3 = -.2506$. We now solve for t_3:

$$t_3 = sr_3\sqrt{\frac{157-3-1}{1-.0754}} = -.2506\sqrt{\frac{153}{.9246}}$$

$$= -.2506\sqrt{165.47} = -.2506(12.864) = -3.2236$$

The result matches what is reported in Table 8.4. Furthermore, from Appendix B / Table B1 for $df = 153$ and $\alpha = .05$, $t_{\text{Critical}} \approx \pm 1.976$. Therefore, $t(153) = -3.2236$ is statistically significant, $p < .05$. For more accurate t_{Critical} and p values, the reader is directed to Soper (2025a) and Soper (2025b). For example, consulting these respective sources, $t_{\text{Critical}} = \pm 1.97559$ and $p = .0016$.

Additional Observations

Dummy Coding via MRC vs. Single-Factor ANOVA. Some readers might question the use of dummy coding via MRC when a single-factor ANOVA appears to be simpler and easier to conduct. After all, as reported in Tables 8.1 and 8.4, respectively, the overall ANOVA model is the same for both analyses, and Tukey's post hoc pairwise comparisons include the same results obtained from dummy coding. Although there is some merit to this question, concomitantly there also should be some appreciation for the utility of MRC, namely, whatever ANOVA brings to the table MRC also brings to the table *plus more*. For example, in addition to directly acquiring differences in group means from the regression coefficients (B_i), we also can directly acquire each group's mean by substituting the dummy codes into the regression equation and performing the resulting arithmetic. An MRC analysis also enables us to examine partial variances, which provide information about the proportion of Y variance the respective paired-group distinctions explain, and it enables us to include other variables, both nominal and quantitative, as part of the analysis. We discuss this latter situation in Chapter 9.

Pairwise Comparisons Independent of the Reference Group. As we indicated earlier, dummy coding is an appropriate coding strategy for research settings that have a well-defined reference group, and it makes sense to compare all the other groups to this designated reference group. Given this situation and independent of the initial set of RQs, researchers also might be interested in comparing differences in group means between two groups, neither of which is the reference group. For example, let's assume in our running example that independent of the a priori RQs and hypotheses—which are focused on making comparisons to ATPs—we also are interested in comparing the levels of professionalism between airport managers and air traffic controllers. Because neither group is the designated reference group, there are two options: (a) recode the IVs with a different reference group, or (b) manually calculate to corresponding t test to see if the difference is significant. A brief discussion of each option follows.

Recode the IVs. The first option is to recode the data set by making either AM or ATC the reference group. As shown in the first subtable of Table 8.5, the reference group in the modified coding chart (b) is ATC. The modified coding chart precipitates a change in the data set, namely, we swap the codes for ATC and ATP such that ATC gets coded 0 0 0 and ATP gets coded 0 0 1. A summary of the results for the analysis using this modified coding chart is included in Table 8.5. As reported in the ANOVA subtable of Table 8.5, the overall model is significant, $R^2 = .0754$, F(3,

TABLE 8.5 *Summary of Results from Modifying the Dummy Coding Chart for the Running Example on Professionalism to Compare AM vs. ATC*

(a) Original coding chart relevant to the study's initial RQs and hypotheses

Group	D_1	D_2	D_3
	Coded Variables		
AM	1	0	0
AMT	0	1	0
ATC	0	0	1
ATP	0	0	0

(b) Modified coding chart with ATC as the reference group to compare AMs vs. ATCs

Group	D_1	D_2	D_3
	Coded Variables		
AM	1	0	0
AMT	0	1	0
ATP	0	0	1
ATC	0	0	0

Note. AM = Airport Manager, AMT = Aviation Maintenance Technician, ATC = Air Traffic Controller, and ATP = Airline Transport Pilot.

ANOVA

Source	*SS*	*df*	*MS*	*F*	*p*
Between	468.9906	3	156.330	4.1599	.0073
Within (Error)	5749.8374	153	37.581		
Total	6218.8280	156			

Note. $N = 157$. $R^2 = .0754$.

Summary of Partials

Factor [a]	B_i	SE_{Bi}	sr_i	sr_i^2	pr_i	pr_i^2	$t_i(153)$	p_i
			Semipartial Correlation		**Full Partial Correlation**		**Significance**	
Constant	79.88	0.96					83.43	< .0001
D_1 = AM	3.93	1.39	.2200	.0484	.2229	.0497	2.83	.0053
D_2 = AMT	2.81	1.38	.1581	.0250	.1622	.0263	2.03	.0438
D_3 = ATP	4.36	1.35	.2506	.0628	.2522	.0636	3.22	.0015

Note. $N = 157$. Y = Level of Professionalism, measured by Snizek's (1972) Hall's Professionalism Inventory (HPI).
[a] AM = Airport Managers ($n = 37$), AMT = Aviation Maintenance Technicians ($n = 38$), ATC = Air Traffic Controllers ($n = 41$).

153) = 4.1599, $p = .0073$. This is equivalent to was reported earlier because we are analyzing the same data set. From the Summary of Partials subtable, the resulting regression equation is, $\hat{Y} = 3.93D_1 + 2.81D_2 + 4.36D_3 + 79.88$. Focusing on $B_1 = 3.93$, which reflects the AM vs. ATC comparison, the mean HPI score for airport managers is, on average, 3.93 points higher than the mean HPI score of air traffic controllers, and this difference is significant, $t(153) = 2.83$, $p = .0053$. Furthermore, the AM-ATC distinction is .0484, which means that the total amount of explained variance in HPI scores is reduced by 4.84% if we exclude comparing airport managers vs. air traffic controllers. The reader will note that although the overall model is the same for the two coding charts, the regression equations are different. This is what we meant when we wrote earlier that it does not matter what group is the reference group; the differences in the final model will be with respect to the regression coefficients.

Manually Calculate t. An alternative approach to testing the significance of the difference in group means where neither group is the reference group, is to use the following t test formula:

$$t = \frac{B_i - B_j}{\sqrt{(SE_{Y-\hat{Y}})^2\left(\frac{1}{n_i} + \frac{1}{n_j}\right)}}$$

where B_i = the regression coefficient of the first group; B_j = the regression coefficient of the second group; $SE_{Y-\hat{Y}}$ = the standard error of the estimate, which is equal to the root mean square error ($RMSE$); n_i = the sample size of the first group; and n_j = the sample size of the second group. To apply this formula to compare airport managers to air traffic controllers, we use the results from the initial MRC analysis based on the initial coding chart (see Table 8.4) and let the first group be AM and the second group be ATM:

$$t = \frac{B_i - B_j}{\sqrt{(SE_{Y-\hat{Y}})^2\left(\frac{1}{n_i} + \frac{1}{n_j}\right)}} = \frac{-0.43-(-4.36)}{\sqrt{(6.13)^2\left(\frac{1}{37} + \frac{1}{41}\right)}}$$

$$= \frac{-0.43 + 4.36}{\sqrt{(37.5769)(0.027 + 0.0244)}}$$

$$= \frac{3.93}{\sqrt{(37.5769)(0.0514)}}$$

$$= \frac{3.93}{\sqrt{1.9315}} = \frac{3.93}{1.3898} = 2.83$$

Note that this t value is equivalent to what is reported in Table 8.5 for D_1, which reflects the results of the analysis after we recoded the IVs.

Preselected Coding Systems. When conducting an MRC analysis with nominal IVs, it is tempting to not code the nominal IV but instead simply run the analysis with the uncoded single group membership variable. As an illustration, consider our running example where X is a single nominal IV with four groups: airport managers (AMs), aviation maintenance technicians (AMTs), air traffic controllers (ATCs), and airline transport pilots (ATPs). Instead of representing X using three dummy coded IVs as we did, if we use our statistics program to regress Y on X *directly*, the overall results are the same as before, R^2 = .0754, $F(3, 153)$ = 4.1599, p = .0073, but the regression equation is quite different:

$$\hat{Y} = 1.16\text{AM} + 0.03\text{AMT} - 2.78\text{ATC} + 82.65$$

Note that the B_i and B_0 values are completely different from what we obtained using dummy coded IVs. This is because our statistics program's default coding scheme for nominal IVs is *unweighted effects coding*, which we discuss in the next section. Therefore, readers are advised to learn what their statistics program's default coding scheme is for nominal IVs. This also goes for online survey tools such as *SurveyMonkey* or *Qualtrics*. Readers are advised to understand how data for nominal IVs are being recorded by these tools because many use counting numbers (1, 2, 3, ...) to distinguish between groups. For example, it is common for a dichotomy such as sex to be coded 1 = Female and 2 = Male (or vice versa), and for our running example, the four groups might be coded 1 = AM, 2 = AMT, 3 = ATC, and 4 = ATP.

To illustrate the problems with alternative coding schemes for a dichotomy, consider the hypothetical data given in the file "Ch_8 Alternative Coding Data for Sex." This data set contains three different coding schemes for the nominal IV, "Biological Sex Assigned at Birth" as follows: (a) X_1 is coded 1 = Female and 0 = Male; (b) X_2 is coded 1 = Female and 2 = Male; and (c) X_3 is coded 2 = Female and 0 = Male. The DV is Y = Instructor Rating, and the research objective is to determine the difference between female and male flight students' perceptions of their CFI's instructional capabilities; higher ratings equate to higher perceptions. Let's now consider the differences in the results among these coding schemes. Before doing so, remember that the overall results will be identical regardless of the coding scheme because we are analyzing the same data. For the current example, the overall results are R^2 = .064, $F(1, 71) = 4.858$, $p = .0308$.

Regress Y on X Where 1 = Female and 0 = Male. This is our standard dummy coding scheme, and the regression equation is $\hat{Y} = 9.94X + 56.06$. As a result, B_1 = 9.94 means that female flight students' instructor ratings are, on average, 9.94 points higher than those of male flight students'. Furthermore, B_0 = 56.06 represents the mean of Males, and when we substitute the numerical code for Females (1) into the regression equation, the mean for Females is 9.94 + 56.06 = 66. The reader will note that these results are correct.

Regress Y on X Where 1 = Female and 2 = Male. The regression equation for this alternative coding scheme is $\hat{Y} = -9.94X + 75.94$. Because the numerical code for Males (2) is greater than that for Females (1), B_1 is interpreted from the perspective of Males vs. Females. More specifically, the instructor ratings for male flight students are, on average, 9.94 points *lower* than those of female flight students, which is correct. However, look at the regression constant (B_0). Recall that B_0 represents the y coordinate of the y intercept, which is the point where the regression line crosses the Y axis. This occurs at the point (0, y), which means that $x = 0$. For the current coding scheme, though, at no point will $x = 0$ because there is no case for "Sex" = 0 and therefore the regression constant is bogus and not interpretable.

Regress Y on X Where 2 = Female and 0 = Male. The regression equation for this alternative coding scheme is $\hat{Y} = 4.97X + 56.06$. Notice that the regression constant, B_0 = 56.06, correctly represents the mean score for Males, which is the reference group. However, also notice that the regression coefficient, B_1 = 4.97, cannot be correctly interpreted as the difference in group means between female and male flight students. Recall that the regression coefficient (B) represents the slope of the regression equation and is interpreted generically as "for every a 1-unit change in X, there is a B-unit change in Y." For a dichotomy, this 1-unit change in X occurs when we go from a code of 0 to a code of 1, and the corresponding change in Y is the difference between the means of the two groups. By coding one group 0 and another group 2, there is no longer a 1-unit change in X, but instead there is a *one-half-unit* change. So, instead of B_1 = 9.94, it is reflected in the regression equation as B_1 = 4.97. As a result, this coding scheme yields the correct B_0 but B_i is one-half of what it should be.

8.4 Effects Coding

The Concept of Effects Coding

Another coding strategy for representing nominal variables quantitatively in an MRC analysis is effects coding. Like dummy coding, effects coding uses $g - 1$ IVs to represent g groups, but unlike dummy coding, effects coding uses a *trichotomous* coding scheme that consists of 0, 1, and –1. Effects coding also does not have a reference group but instead uses what Cohen et al. (2003) call a base group, which we refer to as the sacrificial group. This is because unlike dummy coding where all g groups are represented in the regression equation—coded IVs are represented by regression coefficients (B_i), and the reference group is represented by the regression constant (B_0)—the base group has no presence in the regression equation. A consequence of this absence is that the final model will not include a corresponding significance test involving the base group. As a result, the guiding criterion for most researchers when selecting a base group is to select the group they are *least interested in*; that is, the one they are willing to "sacrifice." The reader also is informed that although the base group has no presence in the final model, the mean of the base group still may be determined by substituting the codes for the IVs into the regression equation and performing the resulting arithmetic as we did for dummy coding.

An effects coding strategy also does not compare individual group means to each other. Instead, all group comparisons are made with respect to the overall mean of the sample, known as the *mean of the group means*, or grand mean. Thus, research situations that warrant an effects coding scheme are those where researchers are interested in how each group's mean differs from the overall sample's mean and not relative to each other or to a reference group. For example, consider a study that examines how pilot performance—defined as the measure of deviation, in degrees, while following the glide slope on an instrument landing system—differs across three simulator environments: basic flight simulator, high-fidelity simulator, and virtual reality simulator. If the objective of the study is to compare pilot performance across each simulator type (e.g., basic flight vs. high fidelity, basic flight vs. virtual reality, and high-fidelity vs. virtual reality), then dummy coding would be appropriate and a reasonable reference group would be the basic flight simulator. However, if the objective of the study is to assess the effectiveness of each simulator type, then effects coding would be appropriate because the results would help the researcher understand the extent to which each simulator type deviates from the overall mean performance across *all* simulator types. This could be useful in pinpointing which simulator environment is associated with better or worse performance relative to the overall average.

An effects coding strategy can be accomplished using either an *unweighted* or *weighted* approach. The criterion that guides which approach to use is the sample sizes of the groups. Regardless of their respective sizes in the sample, if the targeted groups of a study are presumed to be of equal size in the parent population, then the former is appropriate, but if the presumption is that group sizes in the population are disproportional, then the latter is appropriate. In practice, the difference between the two approaches is in their representation and interpretation of the coded variables, which reflect the absence or presence of "weights." This will be demonstrated as part of our discussion of each strategy.

Unweighted Effects Coding

As noted above, unweighted effects coding is used when the targeted groups of a study are presumed to be of equal size in the parent population, regardless of their respective sizes in the sample. Unweighted effects coding uses 1 and 0 to distinguish the coded IVs that are represented in the regression equation, but the base group, which will not have any presence in the regression equation, is coded –1.

To illustrate this coding scheme, let's assume that for our running example we are interested in the differences in the levels of professionalism among the three *non-pilot* groups: airport managers, aviation maintenance technicians, and air traffic controllers. More specifically, we endeaavor to examine the extent to which each non-pilot group deviates from the overall sample's unweighted mean HPI score across all four groups. As a result, the base group is ATP. This is illustrated in Table 8.6 where subtable 8.6(a) contains the corresponding coding chart and subtable 8.6(b) contains a partial listing of the data set that shows how the groups representing the nominal IV are coded. From an inferential statistics perspective, the reader will recognize that the coding chart given in subtable 8.6(a) implies the following set of hypotheses:

E_1 = Airport Managers vs. Overall Sample

- H_0: $\mu_{AM} - \mu_{UWG} = 0$ (There is no significant difference between airport managers' mean HPI score and the unweighted grand mean.)
- H_1: $\mu_{AM} - \mu_{UWG} \neq 0$ (There is a significant difference between airport managers' mean HPI score and the unweighted grand mean.)

E_1 = Aviation Maintenance Technicians vs. Overall Sample

- H_0: $\mu_{AMT} - \mu_{UWG} = 0$ (There is no significant difference between airline maintenance technicians' mean HPI score and the unweighted grand mean.)
- H_1: $\mu_{AMT} - \mu_{UWG} \neq 0$ (There is a significant difference between airline maintenance technicians' mean HPI score and the unweighted grand mean.)

E_3 = Air Traffic Controllers vs. Overall Sample

- H_0: $\mu_{ATC} - \mu_{UWG} \neq 0$ (There is no significant difference between air traffic controllers' mean HPI score and the unweighted grand mean.)
- H_1: $\mu_{ATC} - \mu_{UWG} \neq 0$ (There is a significant difference between air traffic controllers' mean HPI score and the unweighted grand mean.)

TABLE 8.6 *Example of Unweighted Effects Coding for the Running Example on Professionalism*

(a) Unweighted Effects Coding Chart

	Coded Variables		
Group	E_1	E_2	E_3
Airport Manager (AM)	1	0	0
Aviation Maintenance Technician (AMT)	0	1	0
Air Traffic Controller (ATC)	0	0	1
Airline Transport Pilot (ATP)	–1	–1	–1

Table 8.6 Continued on next page ...

TABLE 8.6 *Example of Unweighted Effects Coding for the Running Example on Professionalism (Continued)*

(b) Partial Listing of the Data Set That Reflects Unweighted Effects Coding

			Coded Variables		
Row	***Y* = HPI**	***X* = Aviation**	E_1	E_2	E_3
1	81	AM	1	0	0
2	79	AM	1	0	0
3	93	AM	1	0	0
...	...	...	...	...	...
36	87	AM	1	0	0
37	87	AM	1	0	0
38	87	AMT	0	1	0
39	78	AMT	0	1	0
40	94	AMT	0	1	0
...	...	...	...	...	...
74	82	AMT	0	1	0
75	85	AMT	0	1	0
76	84	ATC	0	0	1
77	89	ATC	0	0	1
78	83	ATC	0	0	1
...	...	...	...	...	...
115	81	ATC	0	0	1
116	72	ATC	0	0	1
117	87	ATP	–1	–1	–1
118	84	ATP	–1	–1	–1
119	88	ATP	–1	–1	–1
...	...	...	...	...	...
156	89	ATP	–1	–1	–1
157	84	ATP	–1	–1	–1

Note. AM = Airport Managers, AMT = Aviation Maintenance Technicians, ATC = Air Traffic Controllers, and ATP = Airline Transport Pilots.

Interpreting the Results of an MRC Analysis with Unweighted Effects Coding

Table 8.7 contains a summary of the results of a simultaneous regression analysis involving the unweighted effects coded variables for our running example. As reported in Table 8.7, the overall results of the MRC analysis with the unweighted effects coded IVs are equivalent to those obtained from both the single-factor ANOVA (Table 8.1) and from the dummy coded MRC analysis (Table 8.4): R^2 = .0754, $F(3, 153) = 4.1599$, $p = .0073$. Thus, 7.54% of the variance in HPI scores is being explained collectively by the four aviation groups, and the proportion of explained variance is significant.

Further observe from Table 8.7 that the regression equation for the overall model is, $\hat{Y} = 1.16E_1 + 0.03E_2 - 2.78E_3 + 82.65$. When we substitute into this regression equation the numerical codes from the corresponding coding chart given in subtable 8.6(a), we get each group's mean HPI score as shown below.

- $M_{AM} = 1.16(1) + 0.03(0) - 2.78(0) + 82.65 = 1.16 + 82.65 = 83.81$
- $M_{AMT} = 1.16(0) + 0.03(1) - 2.78(0) + 82.65 = 0.03 + 82.65 = 82.68$
- $M_{ATC} = 1.16(0) + 0.03(0) - 2.78(1) + 82.65 = -2.78 + 82.65 = 79.87$
- $M_{ATP} = 1.16(-1) + 0.03(-1) - 2.78(-1) + 82.65 = -1.16 - 0.03 + 2.78 + 82.65 = 84.24$

TABLE 8.7 *Summary of MRC Results for the Running Example on Professionalism Using Unweighted Effects Coding with ATP as the Base Group*

ANOVA

Source	***SS***	***df***	***MS***	***F***	***p***
Between	468.9906	3	156.330	4.1599	.0073
Within (Error)	5749.8374	153	37.581		
Total	6218.8280	156			

Note. $N = 157$. $R^2 = .0754$.

Summary of Partials

			Semipartial Correlations [b]		**Full Partial Correlations** [b]		**Significance**	
Factor [a]	B_i	SE_B	sr_i	sr_i^2	pr_i	pr_i^2	$t_i(153)$	p_i
Constant	82.65	0.49					168.76	< .0001
E_1 = AM	1.16	0.86	.1039	.0108	.1077	.0116	1.34	.1830
E_2 = AMT	0.03	0.86	.0027	7.4e–6	.0028	8.0e-6	0.03	.9721
E_3 = ATC	–2.78	0.84	–.2583	.0667	–.2594	.0673	–3.32	.0011

Note. $N = 157$. Y = Level of Professionalism, measured by Snizek's (1972) Hall's Professionalism Inventory (HPI).
[a] AM = Airport Managers ($n = 37$), AMT = Aviation Maintenance Technicians ($n = 38$), ATC = Air Traffic Controllers ($n = 41$). [b] "e–6" is abbreviated scientific notation for 10^{-6}. Thus, 7.4e–6 = .0000074.

Once again—except for M_{ATC}= 79.87 as opposed to 79.88, which is the result of rounding—these group means are equivalent to those obtained from both the single-factor ANOVA and the dummy coded MRC analysis. However, when compared to dummy coding, the results for the partials and their corresponding interpretations for unweighted effects coding are different. A brief discussion follows.

Interpreting the Regression Coefficients (B_i). As noted previously, unlike dummy coding, which compares individual group means to the mean of a reference group, effects coding makes comparisons between a group's mean and the grand mean, which is equal to the "mean of the means." When applied to an unweighted effects coding strategy, the grand mean is the overall sample's unweighted mean. With respect to our running example, the respective means for the four groups are: M_{AM} = 83.81, M_{AMT} = 82.68, M_{ATC} = 79.88, and M_{ATP} = 84.24. Therefore, the unweighted grand mean is

$$M_{\text{Unweighted}} = \frac{M_{AM} + M_{AMT} + M_{ATC} + M_{ATP}}{g}$$

$$= \frac{83.81 + 82.68 + 79.88 + 84.24}{4} = \frac{330.61}{4} = 82.6525$$

Observe from the unweighted mean formula that although the group sizes are different ($n_{AM} = 37$, $n_{AMT} = 38$, $n_{ATC} = 41$, $n_{ATP} = 41$), each group has equal weight in the calculation of the grand mean. In other words, each group has equal importance, and the grand mean is simply the average of all the groups: Add each group's mean and divide the sum by the total number of groups, g. This is similar to a grading system where all assessments contribute equally to the final grade. For example: class participation = 25%, homework assignments = 25%, midterm exam = 25%, and final exam = 25%.

Let's now focus on the regression equation. Note that the regression constant, B_0 = 82.65, is equal to the unweighted grand mean of the sample. Further observe that when we examine the difference between each group's mean and the grand mean, the difference is represented by the unstandardized regression coefficients, B_i, in the regression equation. The deviation between a group mean and the grand mean is called its effect (Pedhazur, 1997, p. 363) and this is why the coding scheme is called *effects coding*. So, from Table 8.7, given E_1 represents Airport Managers, E_2 represents Aviation Maintenance Technicians, and E_3 represents Air Traffic Controllers, we make the following interpretations of each B_i's effect.

- B_1 = 83.81 – 82.65 = 1.16. Holding all other variables constant, Airport Managers scored, on average, 1.16 points higher on the HPI when compared to the overall performance of the four groups taken collectively. This difference is not significant, $t(153) = 1.34$, $p = .1830$.
- B_2 = 82.68 – 82.65 = 0.03. Holding all other variables constant, AMTs scored, on average, 0.03 points higher on the HPI when compared to the overall performance of the four groups taken collectively. This difference is not significant, $t(153) = 0.03$, $p = .9721$.
- B_3 = 79.87 – 82.65 = –2.78. Holding all other variables constant, ATCs scored, on average, 2.78 points *lower* on the HPI when compared to the overall performance of the four groups taken collectively. This difference is significant, $t(153) = -3.32$, $p = .0011$.

Therefore, among the three non-pilot groups and relative to the overall sample's unweighted grand mean, the results suggest that airport managers have the highest level of professionalism and air traffic controllers have the lowest level of professionalism, as measured by Snizek's (1972) Hall's Professionalism Inventory (HPI).

Comparisons Not Represented by the Regression Coefficients. Independent of interpreting comparisons directly from the regression equation, other comparisons also may made from the results of this analysis.

Comparing the Base Group to the Grand Mean. To determine the difference between the mean of the base group (Airline Transport Pilots) and the overall sample's unweighted grand mean, we simply subtract the two means: 84.24 – 82.65 = 1.59. Thus, ATPs scored on average 1.59 points higher on the HPI when compared to the overall performance of all four groups. Some readers might claim that this difference is higher than that of airport managers, but these readers are reminded that this comparison is being ignored because the focus of the analysis is exclusively on the non-pilot groups. Furthermore, because the base group has no presence in the regression equation, to determine if this difference is significant, we recode the data set with a different base group. Doing so with airport managers as the base group, the result for ATPs is $t(153) = 1.90$, $p = .0590$. Thus, the 1.59-point difference is not significant. (*Note:* For the interested reader, an alternative to recoding is to manually calculate the t value using the t formula presented by Cohen et al., 2003, p. 326).

Comparing A Coded Group to the Base Group. Another comparison that can be made is to compare one of the groups represented by a coded IV to the base group. For example, focusing on air traffic controllers and airline transport pilots, we know the difference in their level of professionalism is 79.87 – 84.24 = –4.37. Thus, ATCs scored on average 4.37 points lower on the HPI than ATPs. To determine if this 4.37-point difference—or more generally, the difference between any coded group and the base group—is significant, we direct the reader to the *t* formula presented in Cohen et al. (2003, p. 327).

Comparing Any Two Groups Neither of Which is the Base Group. Similar to dummy coding, it also is possible to compare differences between any two groups neither of which is the base group. For example, comparing Airport Managers to ATCs, we know the difference in their respective levels of professionalism is 83.81 – 79.87 = 3.94. Thus, Airport Managers scored, on average, nearly 4 points higher on the HPI than ATCs. To determine if this difference—or more generally, the difference between any two groups, neither of which is the base group—is significant, we use the same *t* test we presented for dummy coding.

Interpreting the Squared Semipartial Correlation Coefficients (sr_i^2). As demonstrated earlier, the unweighted grand mean of a sample in an effects coded MRC analysis is calculated by adding the respective group means and then dividing this sum by the total number of groups, *g*. Therefore, any differences between a targeted group (e.g., Airport Managers) and the unweighted grand mean of the sample includes the targeted group. This suggests that when we examine the unique contribution a targeted group makes to explaining *Y* variance, this is effectively the same as distinguishing between the targeted group and the collective contribution of the remaining groups. As a result, sr_i^2 is the unique contribution this distinction makes to explaining *Y* variance. Applying this interpretation to the sr_i^2 values of our running example:

- $sr_1^2 = R^2_{Y \bullet 123} - R^2_{Y \bullet 23} = .0754 - .0646 = .0108$. Therefore, 1.08% of the variance in HPI scores is being explained by the airport mangers' "effect," which represents the distinction between Airport Managers vs. the other three groups (AMTs, ATCs, ATPs).
- $sr_2^2 = R^2_{Y \bullet 123} - R^2_{Y \bullet 13} = .0754 - .0754 = 0$. Therefore, 0% of the variance in HPI scores is being explained by the aviation maintenance technicians' "effect," which represents the distinction between AMTs vs. the other three groups (AMs, ATCs, ATPs).
- $sr_3^2 = R^2_{Y \bullet 123} - R^2_{Y \bullet 12} = .0754 - .0087 = .0667$. Therefore, 6.67% of the variance in HPI scores is being explained by the air traffic controllers' "effect," which represents the distinction between ATCs vs. the other three groups (AMs, AMTs, ATPs).

Weighted Effects Coding

As noted earlier, weighted effects coding is used when it is presumed that the group sizes in the parent population are disproportionate. Although similar in application, weighted effects coding is distinguished from unweighted effects coding in two ways. First, a weighted effects coding scheme uses a weighted mean, which incorporates each group's sample size into the calculation. This weighted mean is calculated by: (a) multiplying each group's mean by its sample size, (b) adding the products, and (c) dividing this sum by the sum of each group's sample size. For example, given the four groups from our running example, the weighted grand mean is calculated as follows:

$$M_{\text{Weighted}} = \frac{n_1M_1 + n_2M_2 + n_3M_3 + n_4M_4}{n_1 + n_2 + n_3 + n_4}$$

$$= \frac{37(83.81) + 38(82.68) + 41(79.88) + 41(84.24)}{37 + 38 + 41 + 41}$$

$$= \frac{3100.97 + 3141.84 + 3275.08 + 3453.84}{157}$$

$$= \frac{12971.73}{157} = 82.62248$$

Observe from the formula how the sample sizes are applied to account for the differences in group sizes. These weights adjust the effects coding scheme to reflect the proportions or importance of each level more accurately. This is analogous to a grading system that uses a weighted approach to reflect the importance of various skills and knowledge needed for safe flight operations. For example, a pilot training program might assign a weight of 50% to flight skills, 30% to theoretical knowledge (e.g., aerodynamics, navigation, etc.), and 20% to safety procedures and emergency handling. If a student scores 85% on flight skills, 75% on theoretical knowledge, and 90% on safety procedures, then the final grade is calculated as follows: (.85)(.50) + (.75)(.30) + (.90)(.20) = .425 + .225 + .18 = .83, or 83%, which generally equates to a letter grade of B.

Because each group, relative to its sample size, is making a different contribution to the calculation of the weighted grand mean, the result should be different than the unweighted grand mean. Although the two grand means for our running example are indeed not equal (e.g., $M_{\text{Weighted}} = 82.62248$ vs. $M_{\text{Unweighted}} = 82.6525$), the difference is slight (.03). This is because the sample sizes of the groups are nearly the same and hence their respective contributions to the calculation of the weighted grand mean are nearly identical.

The second difference between weighted and unweighted effects coding is with respect to the coding itself. Instead of using –1 to represent the base group across the coded IVs, weighted effects coding uses a negative fraction that expresses the ratio between each group's sample size (n_i) and the sample size of the base group (n_{Base}). Applying this to our running example where the base group is ATP ($n = 41$), the respective codes for the three coded IVs are $E_1 = -37/41$, $E_2 = -38/41$, and $E_3 = -41/41$, This is illustrated in Table 8.8. Observe that the coding chart in subtable (a) displays the fraction, but the data table in subtable (b) contains the decimal equivalent to six decimal places. This is because our statistics program cannot accept data expressed in fractional form. Also note from the coding chart that if a group's sample size is equal to that of the base group, then the base group's code is –1.

From an inferential statistics perspective, the hypotheses associated with a weighted effects coding strategy are similar to those we presented for unweighted effects coding. The primary difference is that comparisons are made between the respective means of the groups assigned a coded IV vs. the weighted sample mean.

TABLE 8.8 *Example of Weighted Effects Coding for the Running Example on Professionalism*

(a) Weighted Effects Coding Chart

Group	Coded Variables		
	E_1	E_2	E_3
Airport Manager (AM)	1	0	0
Aviation Maintenance Technician (AMT)	0	1	0
Air Traffic Controller (ATC)	0	0	1
Airline Transport Pilot (ATP)	–37/41	–38/41	–41/41

(b) Partial Listing of the Data Set Given in the File "Ch_8 Coding Example Data" That Reflects Weighted Effects Coding

Row	Y = HPI	X = Aviation	Coded Variables		
			E_1	E_2	E_3
1	81	AM	1	0	0
2	79	AM	1	0	0
3	93	AM	1	0	0
...	...	...	...	...	...
36	87	AM	1	0	0
37	87	AM	1	0	0
38	87	AMT	0	1	0
39	78	AMT	0	1	0
40	94	AMT	0	1	0
...	...	...	...	...	...
74	82	AMT	0	1	0
75	85	AMT	0	1	0
76	84	ATC	0	0	1
77	89	ATC	0	0	1
78	83	ATC	0	0	1
...	...	...	...	...	...
115	81	ATC	0	0	1
116	72	ATC	0	0	1
117	87	ATP	–0.902439	–0.926829	–1
118	84	ATP	–0.902439	–0.926829	–1
119	88	ATP	–0.902439	–0.926829	–1
...	...	...	...	...	...
156	89	ATP	–0.902439	–0.926829	–1
157	84	ATP	–0.902439	–0.926829	–1

Note. AM = Airport Managers, AMT = Aviation Maintenance Technicians, ATC = Air Traffic Controllers, and ATP = Airline Transport Pilots.

Table 8.9 contains a summary of the results of a simultaneous regression analysis involving the weighted effects coded variables for our running example with ATP as the base group. Note the following with respect to these results.

- The overall model is exactly the same as the respective models from single-factor ANOVA, dummy coding, and unweighted effects coding.

TABLE 8.9 *Summary of MRC Results for the Running Example on Professionalism Using Weighted Effects Coding with ATP as the Base Group*

ANOVA

Source	*SS*	*df*	*MS*	*F*	*p*
Between	468.9906	3	156.330	4.1599	.0073
Within (Error)	5749.8374	153	37.581		
Total	6218.8280	156			

Note. N = 157. R^2 = .0754.

Summary of Partials

			Semipartial Correlations		Full Partial Correlations [b]		Significance	
Factor [a]	B_i	SE_B	sr_i	sr_i^2	pr_i	pr_i^2	$t_i(153)$	p_i
Constant	82.62	0.49					168.88	< .0001
E_1 = AM	1.19	0.88	.1047	.01096	.1083	.01172	1.35	.1801
E_2 = AMT	0.06	0.86	.0055	.00003	.0056	3.14e-5	0.07	.9448
E_3 = ATC	–2.75	0.82	–.2594	.06729	–.2605	.06784	–3.34	.0011

Note. N = 157. Y = Level of Professionalism, measured by Snizek's (1972) Hall's Professionalism Inventory (HPI).
[a] AM = Airport Managers (n = 37), AMT = Aviation Maintenance Technicians (n = 38), ATC = Air Traffic Controllers (n = 41). [b] "e–5" is abbreviated scientific notation for 10^{-5}. Thus, 3.14e–5 = .0000314.

- The regression equation for the overall model is $\hat{Y} = 1.19E_1 + 0.06E_2 - 2.75E_3 + 82.62$. Note that there is not a pronounced difference between this equation and that from unweighted effects coding. This is because the group sizes between the two approaches are nearly the same. Notice also that B_0 = 82.62, which is the weighted grand mean.
- When we substitute into this regression equation the numerical codes from the corresponding coding chart (Table 8.8a), we get each group's mean HPI score, and these group means are equivalent to those from the previous analyses. (*Note.* M_{ATC} = 79.88 if the regression coefficients and constant are carried out to three decimal places. We also rounded the numerical codes for ATP to two decimal places.)
  - » $M_{AM} = 1.19(1) + 0.06(0) - 2.75(0) + 82.62 = 1.19 + 82.62 = 83.81$
  - » $M_{AMT} = 1.19(0) + 0.06(1) - 2.75(0) + 82.62 = 0.06 + 82.62 = 82.68$
  - » $M_{ATC} = 1.19(0) + 0.06(0) - 2.75(1) + 82.62 = -2.75 + 82.62 = 79.87$
  - » $M_{ATP} = 1.19(-0.90) + 0.06(-0.93) - 2.75(-1) + 82.62 = -1.07 - 0.06 + 2.75 + 82.62 = 84.24$
- The interpretations of the partials as well as the significance of group comparisons that are not represented by the regression coefficients follow those from unweighted effects coding, except the results are relative to the *weighted* grand mean.

8.5 Contrast Coding

The Concept of Contrast Coding

One of the inherent "features" of both dummy coding and effects coding is the concept of *fixed-group comparison*. For example, given a set of coded IVs that represents the respective groups of a nominal IV: (a) dummy coding compares the respective *Y* means of each coded IV to the *Y* mean of a reference group, and (b) effects coding compares the respective *Y* means (unweighted or weighted) of each coded IV to the overall sample *Y* mean (i.e., the grand mean). Although the coding strategy follows the RQs and hypotheses, neither coding strategy provides any flexibility to these default comparisons. Contrast coding, on the other hand, provides researchers with the ability to create planned comparisons between groups that are not fixed. As an illustration, Table 8.10 contains three different contrast coding charts with respect to our running example.

Observe from Table 8.10 that each coding chart represents a set of contrasts relative to a specific theme that reflect differences in the DV where *Y* = Levels of Professionalism as measured by the HPI. For example, the set of contrasts in Model 1 are related to job focus, the set of contrasts in Model 2 are related to job location, and the set of contrasts in Model 3 are related to safety. Further notice that the codes in each chart contain positive and negative fractions and 0. For each coded IV (C_i), the comparisons are between positively coded groups vs. negatively coded groups, and groups coded 0 are not included in the comparison.

For example, in Model 1: C_1 compares the two groups whose job is primarily airport related (AM and ATC) vs. the two groups whose job is primarily aircraft related (AMT and ATP), C_2 compares the two groups that focus primarily on the airport (AM vs. ATC), and C_3 compares the two groups that focus primarily on the aircraft (AMT

TABLE 8.10 *Contrast Coding Charts for the Running Example on Professionalism*

(a) Model 1: Job Focus

	Coded Variables		
Group	C_1	C_2	C_3
AM	1/2	1/2	0
AMT	–1/2	0	1/2
ATC	1/2	–1/2	0
ATP	–1/2	0	–1/2

(b) Model 2: Job Location

	Coded Variables		
Group	C_1	C_2	C_3
AM	1/3	1/4	1/2
AMT	–2/3	1/4	0
ATC	1/3	1/4	–1/2
ATP	0	–3/4	0

(c) Model 3: Safety Related

	Coded Variables		
Group	C_1	C_2	C_3
AM	1/4	0	0
AMT	–3/4	0	1/2
ATC	1/4	1/2	0
ATP	1/4	–1/2	–1/2

Note. AM = Airport Managers (n = 37), AMT = Aviation Maintenance Technicians (n = 38), ATC = Air Traffic Controllers (n = 41), and ATP = Airline Transport Pilots (n = 41).

vs. ATP). Similar comparisons are reflected in the Model 2 and Model 3 coding charts. In Model 2, C_1 compares two groups whose job location is at the airport proper (AM and ATC) vs. one group whose job location is at an off-airport property or outside the airport's authority. In Model 3, C_1 compares the three groups with a concern for the safe operation of the aircraft (AM, ATC, ATP) vs. the one group that is concerned with maintaining an airworthy aircraft (AMT).

These examples highlight the utility of contrast coding. In each model, specific research questions and corresponding hypotheses based on substantive theory can be stated in terms of expected mean differences between groups or between combinations of groups without making default comparisons to a reference group or the overall sample. In other words, contrast coding enables us to test specific—and planned—hypotheses of interest rather than hypotheses that are a consequence of using a default coding scheme such as dummy or effects coding.

Preparing A Contrast Coding Chart

Understanding the Rules. The inherent flexibility of using contrast coding to analyze nominal variables via MRC comes with a cost, and that cost is the challenge in constructing the codes. As is the case for dummy and effects coding, given g groups, $g - 1$ coded variables are needed to fully represent the g groups. The challenge is with respect to the total number of possible contrasts. According to Cohen et al. (2003, p. 335), there are

$$1 + \frac{3^g - 1}{2} - 2^g$$

possible contrasts, where g = the number of groups. So, for our running example where $g = 4$, we need to prepare a coding chart with three coded IVs but there are

$$[1 + \frac{3^4 - 1}{2} - 2^4] = [1 + \frac{81 - 1}{2} - 16] = [1 + \frac{80}{2} - 16] = [1 + 40 - 16] = 25$$

possible contrasts of group means possible. For example, given four groups, A, B, C, D, possible contrasts include A vs. B, AB vs. C, ABC vs. D, BCD vs. A, and so forth. To underscore this, consider a study with $g = 5$ groups. The total number of possible contrasts is 171, which is nearly 7 times more than it is with four groups. Although many of these contrasts will be pointless, they nevertheless are excessive, and this makes it more challenging to prepare a coding chart that is compliant with two essential rules.

Rule 1. The codes assigned to a contrast, C_i, must add to zero. This is confirmed by simply adding the codes in each column of the coding chart. As an illustration of this rule, consider Model 1 from Table 8.10(a):

- $C_1 = \left(\frac{1}{2}\right) + \left(\frac{-1}{2}\right) + \left(\frac{1}{2}\right) + \left(\frac{-1}{2}\right) = 0$
- $C_2 = \left(\frac{1}{2}\right) + 0 + \left(\frac{-1}{2}\right) + 0 = 0$
- $C_3 = 0 + \left(\frac{1}{2}\right) + 0 + \left(\frac{-1}{2}\right) = 0$

The reader will note that Models 2 and 3 in Table 8.10 are also compliant with Rule 1.

Rule 2. The contrasts must be orthogonal, which means that they are unrelated to or independent of each other. This is confirmed mathematically by adding the cross products, $C_i \times C_j$, of the contrasts across all g groups. If the sum is zero, then the two contrasts that comprise the cross product are orthogonal. As an illustration of this rule, once again consider Model 1 from Table 8.10(a):

- C_1 and C_2: $\left(\frac{1}{2}\right)\left(\frac{1}{2}\right) + \left(\frac{-1}{2}\right)(0) + \left(\frac{1}{2}\right)\left(\frac{-1}{2}\right) + \left(\frac{-1}{2}\right)(0) = \left(\frac{1}{4}\right) + 0 + \left(\frac{-1}{4}\right) + 0 = 0$
- C_1 and C_3: $\left(\frac{1}{2}\right)(0) + \left(\frac{-1}{2}\right)\left(\frac{1}{2}\right) + \left(\frac{1}{2}\right)(0) + \left(\frac{-1}{2}\right)\left(\frac{-1}{2}\right) = 0 + \left(\frac{-1}{4}\right) + 0 + \left(\frac{1}{4}\right) = 0$
- C_2 and C_3: $\left(\frac{1}{2}\right)(0) + (0)\left(\frac{1}{2}\right) + \left(\frac{-1}{2}\right)(0) + (0)\left(\frac{-1}{2}\right) = 0 + 0 + 0 + 0 = 0$

Because the respective sums of all cross products are zero, the contrast coding chart of Model 1 is compliant with Rule 2. The reader will note that the contrast coding chart of Model 2 in Table 8.10(b) is also compliant with Rule 2. This is not the case for Model 3, though, because Rule 2 fails when examining C_1 and C_3 and when examining C_2 and C_3. As a result, Models 1 and 2 represent valid contrasts because they both satisfy Rules 1 and 2, but Model 3 does not, and therefore the initial set of hypotheses that correspond to Model 3's coding chart must be reconsidered.

Constructing the Codes. To illustrate how to construct a contrast coding chart, we will focus on the contrast coding chart of Model 2 in Table 8.10(b) and work backwards by examining the codes for each coded IV. We will then generalize the patterns that emerge from our observations.

***C_1*.** We first observe that the contrast involves three groups and therefore the denominator of the codes is 3. We next recognize that the first part of the contrast consists of *two* groups (AM, ATC) and the second part of the contrast consists of *one* group (AMT). As a result, the two groups that comprise the first part of the contrast are each coded $\frac{1}{3}$, the single group that comprises the second part of the contrast is coded $\frac{-2}{3}$, and the group that is not involved in the contrast (ATP) is coded 0.

***C_2*.** We first observe that C_2 has no groups coded 0, which means that all four groups are involved in the contrast and therefore the denominator of the codes is 4. We next recognize that the first part of the contrast consists of *three* groups (AM, AMT, ATC) and the second part of the contrast consists of *one* group (ATP). As a result, the three groups that comprise the first part of the contrast are each coded $\frac{1}{4}$, and the single group that comprises the second part of the contrast is coded $\frac{-3}{4}$.

***C_3*.** We first observe that the contrast involves two groups and therefore the denominator of the codes is 2. We next recognize that the first part of the contrast consists of *one* group (AM) and the second part of the contrast consists of *one* group (ATC). As a result, the single group that comprises the first part of the contrast is coded $\frac{1}{2}$, the single group that comprises the second part of the contrast is coded $\frac{-1}{2}$, and the two groups not involved in the contrast (AMT, ATP) are each coded 0.

Generalizing from these observations, if we let

- z = the total number of groups involved in the contrast,
- x = the number of groups that comprise the first part of the contrast,
- y = the number of groups that comprise the second part of the contrast, and
- w = the number of groups not involved in the contrast,

then the codes for a contrast are $\frac{y}{z}$ and $\frac{-x}{z}$, or alternatively, $\frac{-y}{z}$ and $\frac{x}{z}$, and $w = 0$. This is graphically illustrated in Figure 8.2.

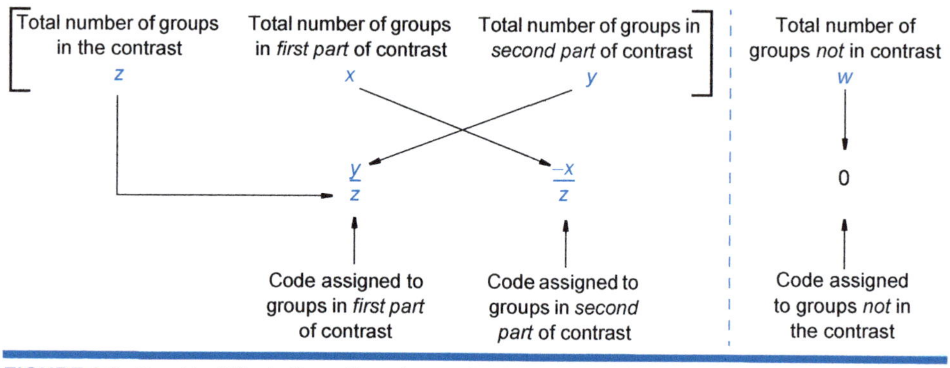

FIGURE 8.2 Graphical Illustration of how to construct contrast codes.

It does not matter which contrast is assigned the positive or negative sign because the same comparison is being made. Although the corresponding regression coefficient, B_i, which represents the difference between the groups of the contrast will be opposite in sign depending on which groups are assigned positive and negative values, the significance of B_i will be the same, and its interpretation will be made accordingly. Let's apply these guidelines to the coding chart for Model 1 in Table 8.10(a).

***C*₁.** All four groups are involved in the contrast, so $z = 4$; two groups (AM, ATC) comprise the first part of the contrast, so $x = 2$; and two groups (AMT, ATP) comprise the second part of the contrast, so $y = 2$. As a result, AM and ATC are each coded $\frac{y}{z} = \frac{2}{4} = \frac{1}{2}$, and AMT and ATP are each coded $\frac{-x}{z} = \frac{-2}{4} = \frac{-1}{2}$.

C_2. The contrast is between two groups, so $z = 2$; the first part of the contrast consists of one group (AM), so $x = 1$; the second part of the contrast consists of one group (ATC), so $y = 1$; and two groups (AMT, ATP) are not part of the contrast. Therefore, AM is coded $\frac{y}{z} = \frac{1}{2}$, ATC is coded $\frac{-x}{z} = \frac{-1}{2}$, and AMT and ATP are each coded $w = 0$.

C_3. The codes for C_3 are equivalent to those for C_2.

The Hypotheses Associated with Contrast Coding

From an inferential statistics perspective, the hypotheses for an MRC analysis involving contrast coding reflect the contrast given by the corresponding codes. If only two groups are involved in the contrast, then the hypotheses are equivalent to a *t* test of independent sample means. For example, in Table 8.10 / Model 2, C_3 is comparing airport managers vs. air traffic controllers and hence the null hypothesis is simply, H_0: $\mu_{AM} = \mu_{ATC}$, or equivalently, H_0: $\mu_{AM} - \mu_{ATC} = 0$, which states there is no significant difference in HPI scores between the two groups. If, however, one part of the contrast consists of two or more groups, then the unweighted average of the groups' means is used. For example, in Table 8.10 / Model 2, C_1 is comparing two groups (AM and ATC) to one group (AMT). Therefore, the null hypothesis is

$$H_0: \frac{\mu_{AM} + \mu_{ATC}}{2} = \mu_{AMT}, \text{ or equivalently, } H_0: \frac{\mu_{AM} + \mu_{ATC}}{2} - \mu_{AMT} = 0,$$

which states there is no significant difference in mean HPI scores between the combined group of airport managers and air traffic controllers vs. aviation maintenance technicians.

Although the construction of these null hypotheses is implied by the coding chart, it is instructional to demonstrate how they are derived mathematically, which involves creating a *weighted linear combination of population means*. The procedure is as follows: (a) multiply each group's population mean by the group's code, (b) add the products, (c) set the sum to zero, and (d) simplify the equation. Applying this procedure to C_1 from Table 8.10 / Model 2:

$$\frac{1}{3}\mu_{AM} - \frac{2}{3}\mu_{AMT} + \frac{1}{3}\mu_{ATC} + (0)\mu_{ATP} = 0$$

$$\frac{\mu_{AM}}{3} - \frac{2\mu_{AMT}}{3} + \frac{\mu_{ATC}}{3} + 0 = 0$$

$$\mu_{AM} - 2\mu_{AMT} + \mu_{ATC} = 0$$

$$\mu_{AM} + \mu_{ATC} = 2\mu_{AMT}$$

$$\frac{\mu_{AM} + \mu_{ATC}}{2} = \mu_{AMT}$$

$$\text{or } \frac{\mu_{AM} + \mu_{ATC}}{2} - \mu_{AMT} = 0$$

Here are the statistical hypotheses for the Model 1 coding chart given in Table 8.10.

C_1 = (Airport Managers and Air Traffic Controllers) vs. (Aviation Maintenance Technicians and Airline Transport Pilots)

- H_0: $(\mu_{AM} + \mu_{ATC}) - (\mu_{AMT} + \mu_{ATP}) = 0$ (There is no significant difference in the mean HPI scores between the combined group of airport managers and air traffic controllers vs. the combined group of aviation maintenance technicians and airline transport pilots.)
- H_1: $(\mu_{AM} + \mu_{ATC}) - (\mu_{AMT} + \mu_{ATP}) \neq 0$ (There is a significant difference in the mean HPI scores between the combined group of airport managers and air traffic controllers vs. the combined group of aviation maintenance technicians and airline transport pilots.)

C_2 = Airport Managers vs. Air Traffic Controllers

- H_0: $\mu_{AM} - \mu_{ATC} = 0$ (There is no significant difference in the mean HPI scores between airport managers and air traffic controllers.)
- H_1: $\mu_{AM} - \mu_{ATC} \neq 0$ (There is a significant difference in the mean HPI scores between airport managers and air traffic controllers.)

C_3 = Aviation Maintenance Technicians vs. Airline Transport Pilots

- H_0: $\mu_{AMT} - \mu_{ATP} = 0$ (There is no significant difference in the mean HPI scores between aviation maintenance technicians and airline transport pilots.)
- H_1: $\mu_{AMT} - \mu_{ATP} \neq 0$ (There is a significant difference in the mean HPI scores between aviation maintenance technicians and airline transport pilots.)

Interpreting the Results of An MRC Analysis Involving Contrast Coding

Table 8.11 contains a summary of the results of a simultaneous regression analysis involving the contrast coded variables for our running example relative to Table 8.10 / Model 1.

TABLE 8.11 *Summary of MRC Results for the Running Example on Professionalism Using Contrast Coding with Respect to the Model 1 Coding Chart in Table 8.10*

ANOVA

Source	*SS*	*df*	*MS*	*F*	*p*
Between	468.9906	3	156.330	4.1599	.0073
Within (Error)	5749.8374	153	37.581		
Total	6218.8280	156			

Note. $N = 157$. $R^2 = .0754$.

Summary of Partials

			Semipartial Correlations		Full Partial Correlations		Significance	
Factor [a]	B_i	SE_B	sr_i	sr_i^2	pr_i	pr_i^2	$t_i(153)$	p_i
Constant	82.65	0.49					168.76	< .0001
C_1	−1.62	0.98	−.1284	.0165	−.1327	.0176	−1.65	.1003
C_2	3.93	1.39	.2200	.0484	.2229	.0497	2.83	.0053
C_3	−1.56	1.38	−.0877	.0077	−.0911	.0083	−1.13	.2603

Note. $N = 157$. Y = Level of Professionalism, measured by Snizek's (1972) Hall's Professionalism Inventory (HPI).
[a] See Model 1 coding chart in Table 8.10(a) for the specific contrasts.

Interpreting the Overall Model. As reported in Table 8.11, the overall results of the MRC analysis involving the contrast coded IVs are equivalent to those obtained from all previous MRC analyses for the running example on professionalism: $R^2 = .0754$, $F(3, 153) = 4.1599$, $p = .0073$. Thus, 7.54% of the variance in HPI scores is being explained collectively by the four aviation groups, and the proportion of explained variance is significant. Further observe from Table 8.11 that the regression equation for the overall model is, $\hat{Y} = -1.62C_1 + 3.93C_2 - 1.56C_3 + 82.65$. When we substitute into this regression equation the numerical codes from the corresponding coding chart (Table 8.10 / Model 1), we get each group's mean HPI score:

- $M_{AM} = -1.62(0.5) + 3.93(0.5) - 1.56(0) + 82.65 = -0.81 + 1.965 + 82.65 = 83.805$
- $M_{AMT} = -1.62(-0.5) + 3.93(0) - 1.56(0.5) + 82.65 = 0.81 - 0.78 + 82.65 = 82.68$
- $M_{ATC} = -1.62(0.5) + 3.93(-0.5) - 1.56(0) + 82.65 = -0.81 - 1.965 + 82.65 = 79.875$
- $M_{ATP} = -1.62(-0.5) + 3.93(0) - 1.56(-0.5) + 82.65 = 0.81 + 0.78 + 82.65 = 84.24$

Once again, these group means are equivalent to those obtained from all previous MRC analyses for the running example. Also note that the regression constant, $B_0 = 82.65$, is the unweighted grand mean.

Interpreting the Regression Coefficients (B_i). The unstandardized regression coefficients for contrast coding represent comparisons between two group means with respect to the DV, and these comparisons are reflected in the corresponding chart.

- $B_1 = -1.62$ is the difference in group means between airport managers and air traffic controllers taken collectively as one group vs. aviation maintenance technicians and airline transport pilots taken collectively as the second group. Furthermore, as presented earlier in our discussion of constructing hypotheses, when two or more groups comprise one part of the contrast, the unweighted average of the groups' means is used. For example:

$$B_1 = \frac{M_{AM} + M_{ATC}}{2} - \frac{M_{AMT} + M_{ATP}}{2}$$

$$= \frac{83.81 + 79.88}{2} - \frac{82.68 + 84.24}{2}$$

$$= \frac{163.89}{2} - \frac{166.92}{2} = 81.845 - 83.46 = -1.615$$

Therefore, the merged group of airport managers and air traffic controllers averaged 1.62 points *lower* on the HPI than the merged group of aviation maintenance technicians and airline transport pilots. This implies that, collectively, the two groups whose job is focused primarily on the airport (AM and ATC) have a lower level of professionalism, as measured by the HPI, than the two collective groups whose job is focused primarily on the aircraft (AMT and ATP). This result is not significant, $t(153) = -1.65$, $p = .1003$.

- $B_2 = 3.93$ is the difference in group means between airport managers and air traffic controllers: Airport managers averaged 3.93 points higher on the HPI than air traffic controllers. This implies that between the two groups whose job focuses primarily on the airport, airport managers have a higher level of professionalism than air traffic controllers. This result is significant, $t(153) = 2.83$, $p = .0053$.
- $B_3 = -1.56$ is the difference in group means between aviation maintenance technicians and airline transport pilots: Aviation maintenance technicians averaged 1.56 points *lower* on the HPI than airline transport pilots. This implies that between the two groups whose job focuses primarily on the aircraft, AMTs have a *lower* level of professionalism than ATPs. This result is not significant, $t(153) = -1.13$, $p = .2603$.

Interpreting the Squared Semipartial Correlation Coefficients (sr_i^2). The squared semipartial correlation coefficients is the proportion of *Y* variance explained by contrast C_i. These coefficients are reported in Table 8.11.

- $sr_1^2 = .0165$. This coefficient indicates that 1.65% of the variance in HPI scores is explained by C_1, which represents differences in the level of professionalism between groups with jobs primarily focused on the airport (AM, ATC) vs. groups with jobs primarily focused on the aircraft (AMT, ATP). This is not significant, $t(153) = -1.65$, $p = .1003$.
- $sr_2^2 = .0484$. This coefficient indicates that 4.84% of the variance in HPI scores is explained by C_2, which represents differences in the level of professionalism between the two groups with jobs primarily focused on the airport (AM vs. ATC). This is significant, $t(153) = 2.83$, $p = .0053$.
- $sr_3^2 = .0077$. This coefficient indicates that 0.77% of the variance in HPI scores is explained by C_3, which represents differences in the level of professionalism between the two groups with jobs primarily focused on the aircraft (AMT vs. ATP). This is not significant, $t(153) = -1.13$, $p = .2603$.

An Alternative Approach to Constructing Contrast Codes

Although the procedure we presented and graphically illustrated in Figure 8.2 for constructing contrast codes is preferable, we present two alternative approaches for completeness.

Excluding *z*. One alternative approach is to exclude *z*, which is equal to the total number of groups involved in the contrast and serves as the denominator for the codes. The procedure is as follows: Let x = the number of groups that comprise the first part of the contrast, and y = the number of groups that comprise the second part of the contrast. The code assigned to the first set of groups is $\frac{1}{x}$, and the code assigned to the second set of groups is $\frac{-1}{y}$.

As an example, let's assume we have five groups, G_1, G_2, G_3, G_4, G_5, and we want to contrast (G_1, G_2, G_3) vs. (G_4, G_5).

- Using our preferred approach to constructing contrast codes (Figure 8.2), $z = 5$, $x = 3$, and $y = 2$. Therefore, the three groups in the first part of the contrast are each assigned $\frac{2}{5}$, and the two groups in the second part of the contrast are each assigned $\frac{-3}{5}$. This is summarized in Table 8.12 in Column C_{1a}.
- Using the excluding *z* approach, the code assigned to the first set of three groups is $\frac{1}{3}$, and the code assigned to the second set of two groups is $\frac{-1}{2}$. This is summarized in Table 8.12 for Column C_{1b}.

As reported in Table 8.12, both sets of codes represent the same contrast, but the "excluding *z*" approach does not yield codes with the same denominator.

Coding with Integers Instead of Fractions. A second alternative approach is to represent the codes using integers instead of fractions by multiplying the codes by their least common denominator (LCD). For example, focusing on the excluding *z* contrast from Table 8.12, this contrast is represented mathematically as

$$\left[\frac{1}{3}(M_1) + \frac{1}{3}(M_2) + \frac{1}{3}(M_3)\right] - \left[\frac{1}{2}(M_4) + \frac{1}{2}(M_5)\right],$$

which is equivalent to

$$\left[\frac{M_1 + M_2 + M_3}{3}\right] - \left[\frac{M_4 + M_5}{2}\right]$$

and the LCD = 6. If we multiply either expression by 6, then C_{1b}'s mathematical representation is now $(2M_1 + 2M_2 + 2M_3) - (3M_4 + 3M_5)$, which transforms the frac-

TABLE 8.12 *Different but Equivalent Codes for the Same Contrast*

	Coded Variables		
Group	C_{1a} (via Figure 8.2)	C_{1b} (Excluding *z*)	C_{1c} (Transformed to Integers) [a]
G_1	2/5	1/3	2
G_2	2/5	1/3	2
G_3	2/5	1/3	2
G_4	−3/5	−1/2	−3
G_5	−3/5	−1/2	−3

Note. [a] Transformation to integers was performed by multiplying the codes in C_{1a} and C_{1b} by their respective LCD (5 for C_{1a} and 6 for C_{1b}).

tional codes to integers as reflected for contrast C_{1c} in Table 8.12. Note that we get the same result if we transform the codes given for C_{1a} in Table 8.12, which is our preferred method, by multiplying each code by its LCD, which is 5. Although transforming the fractional codes for contrasts C_{1a} and C_{1b} to integers is mathematically correct, this transformation has an effect on the regression coefficients, B_i.

To illustrate the magnitude of this effect, we return to the coding chart given in Table 8.10/Model 2 and do the following: (a) transform the fractional codes to integers by multiplying the codes for C_1, C_2, and C_3 by their respective LCDs of 3, 4, and 2, and (b) conduct a simultaneous regression using the transformed codes. Table 8.13 contains the original coding chart for Model 2, the transformed chart with integer codes, and a summary of the results of an MRC analysis based on the original codes and the transformed codes. The reader is reminded there is no difference in the results between the overall models because the same data set is being analyzed. Thus, the ANOVA table given in Table 8.13 is the same for both analyses.

TABLE 8.13 *Summary of MRC Results for the Running Example on Professionalism Using Contrast Coding with Respect to the Model 2 Coding Chart from Table 8.10 Transformed to Integer Codes*

Model 2: Coding Chart (see Table 8.10)

	Coded Variables (Original)			*Coded Variables (Transformed to Integers)*		
Group	C_1	C_2	C_3	C_1	C_2	C_3
AM	1/3	1/4	1/2	1	1	1
AMT	–2/3	1/4	0	–2	1	0
ATC	1/3	1/4	–1/2	1	1	–1
ATP	0	–3/4	0	0	–3	0

Note. AM = Airport Managers (n = 37), AMT = Aviation Maintenance Technicians (n = 38), ATC = Air Traffic Controllers (n = 41), and ATP = Airline Transport Pilots (n = 41).

ANOVA

Source	*SS*	*df*	*MS*	*F*	*p*
Between	468.9906	3	156.330	4.1599	.0073
Within (Error)	5749.8374	153	37.581		
Total	6218.8280	156			

Note. N = 157. R^2 = .0754.

Summary of Parameter Estimates

	Original Codes				**Integer Codes**			
			Significance				**Significance**	
Factor [a]	B_i	SE_{Bi}	$t_i(153)$	p_i	B_i [b]	SE_{Bi}	$t_i(153)$	p_i
Constant	82.65	0.49	168.76	< .0001	82.65	0.49	168.76	< .0001
C_1	–0.84	1.21	–0.69	.4899	–0.28	0.40	–0.69	.4899
C_2	–2.12	1.11	–1.90	.0590	–0.53	0.28	–1.90	.0590
C_3	3.93	1.39	2.83	.0053	1.97	0.69	2.83	.0053

Note. N = 157. Y = Level of Professionalism.
[a] See coding chart for specific contrasts. [b] B_1 is one-third of the original coefficient, B_2 is one-fourth of the original coefficient, and B_3 is one-half of the original coefficient.

Focusing on the Summary of Parameter Estimates subtable in Table 8.13, look at the differences in the regression coefficients between the results from using the original fractional codes vs. the results from using the integer codes. The regression coefficients for the analysis involving the integer codes are reduced by the *reciprocal of the factor* that was used to transform the original codes. For example, the original codes for C_1 were multiplied by the LCD of 3 to transform them to integers. The effect this transformation has on B_1 is that the coefficient is now one-third of the original coefficient; that is, $\frac{1}{3}(-0.84) = -0.28$. A similar relationship occurs for C_2 (the regression coefficient $B_2 = -0.53$ is one-fourth of the original) and for C_3 (the regression coefficient $B_3 = 1.97$ is one-half of the original). Also notice that the respective standard errors are impacted in the same way. What is not impacted, though, is the significance of the regression coefficients (and the related partials). As a result, the reader is cautioned that although transforming the original fractional contrast codes to integers maintains significance, the magnitude of the regression coefficients changes, which will impact corresponding interpretations.

8.6 Applying Coding Strategies in Research: A Guided Example

We now present a guided example of how to conduct a MRC analysis for a research study that involves a nominal IV with four groups. For completeness, we will code the IV using dummy coding, unweighted effects coding, and contrast coding and apply these respective coding strategies to the same data set. A description of the research setting follows.

> An airline consultant examined the flying public's general opinions of the four largest airlines in the U.S. as defined by seat capacity. One of the factors the consultant considered as possibly influencing a person's opinion of the airlines was seat classification. For example, it is conceivable that a person who generally flies first class might have a more positive opinion of the airlines than a person who generally flies economy. Opinions were assessed on a 22-item questionnaire scored on a traditional 5-point Likert response scale. Thus, scores could range from 22 to 110, with higher scores reflecting more positive opinions. The classes of services were Economy (E), Premium Economy (PE), Business Class (BC), and First Class (FC), and participants were asked to specify the type of seat they generally purchase when flying domestically.

The corresponding data set, which consists of hypothetical data is provided in the file "Ch_8 Guided Example Data." The structure of our presentation follows that of previous chapters' Guided Examples and consists of six parts: Research Questions and Hypotheses, Sample Size Planning, Preliminary Data Screening, Data Analysis, Interpretation of Results, and Post Hoc Considerations.

Research Questions and Hypotheses

The RQs and corresponding hypotheses differ among the three coding strategies. Therefore, we present separate parts for each coding strategy. In doing so, though, we provide abridged versions of the RQs and hypotheses because of word and page constraints. As a result, the onus is on the reader to "fill in the blanks."

Dummy Coding. We designate First Class as the reference group.

RQ 1 and Hyp 1. What is the difference in mean opinion scores between participants who generally purchase E seats and participants who generally purchase FC seats?

- H_0: $\mu_E - \mu_{FC} = 0$ (No significant difference in mean opinion scores.)
- H_1: $\mu_E - \mu_{FC} \neq 0$ (There is a significant difference in mean opinion scores.)

RQ 2 and Hyp 2. What is the difference in mean opinion scores between participants who generally purchase PE seats and participants who generally purchase FC seats?

- H_0: $\mu_{PE} - \mu_{FC} = 0$ (No significant difference in mean opinion scores.)
- H_1: $\mu_{PE} - \mu_{FC} \neq 0$ (There is a significant difference in mean opinion scores.)

RQ 3 and Hyp 3. What is the difference in mean opinion scores between participants who generally purchase BC seats and participants who generally purchase FC seats?

- H_0: $\mu_{BC} - \mu_{FC} = 0$ (No significant difference in mean opinion scores.)
- H_1: $\mu_{BC} - \mu_{FC} \neq 0$ (There is a significant difference in mean opinion scores.)

Unweighted Effects Coding. We designate First Class as the base group.

RQ 1 and Hyp 1. What is the difference between the mean opinion score of participants who generally purchase E seats vs. the unweighted grand mean?

- H_0: $\mu_E - \mu_{UWG} = 0$ (No significant difference in mean opinion scores.)
- H_1: $\mu_E - \mu_{UWG} \neq 0$ (There is a significant difference in mean opinion scores.)

RQ 2 and Hyp 2. What is the difference between the mean opinion score of participants who generally purchase PE seats vs. the unweighted grand mean?

- H_0: $\mu_{PE} - \mu_{UWG} = 0$ (No significant difference in mean opinion scores.)
- H_1: $\mu_{PE} - \mu_{UWG} \neq 0$ (There is a significant difference in mean opinion scores.)

RQ 3 and Hyp 3. What is the difference between the mean opinion score of participants who generally purchase BC seats vs. the unweighted grand mean?

- H_0: $\mu_{BC} - \mu_{UWG} = 0$ (No significant difference in mean opinion scores.)
- H_1: $\mu_{BC} - \mu_{UWG} \neq 0$ (There is a significant difference in mean opinion scores.)

Contrast Coding. The corresponding coding chart is derived from the following hypotheses and is given later in Table 8.14, which contains a summary of the results.

RQ 1 and Hyp 1. What is the difference in mean opinion scores between participants who generally purchase E, PE or FC seats vs. participants who generally purchase BC seats?

- H_0: $\frac{\mu_E + \mu_{PE} + \mu_{FC}}{4} - \frac{3\mu_{BC}}{4} = 0$ (No significant difference in mean opinion scores.)
- H_1: $\frac{\mu_E + \mu_{PE} + \mu_{FC}}{4} - \frac{3\mu_{BC}}{4} \neq 0$ (There is a significant difference in mean opinion scores.)

RQ 2 and Hyp 2. What is the difference in mean opinion scores between participants who generally purchase E or FC seats vs. participants who generally purchase PE seats?

- H_0: $\frac{\mu_E + \mu_{FC}}{2} - \mu_{PE} = 0$ (No significant difference in mean opinion scores.)
- H_1: $\frac{\mu_E + \mu_{FC}}{2} - \mu_{PE} \neq 0$ (There is a significant difference in mean opinion scores.)

RQ 3 and Hyp 3. What is the difference in mean opinion scores between participants who generally purchase E seats vs. participants who generally purchase FC seats?

- H_0: $\mu_E - \mu_{FC} = 0$ (No significant difference in mean opinion scores.)
- H_1: $\mu_E - \mu_{FC} \neq 0$ (There is a significant difference in mean opinion scores.)

Sample Size Planning

For coded nominal IVs, sample size planning focuses on the overall R^2 and the respective sr_i^2 values. Because we conduct a simultaneous regression on the coded IVs, the sample size needed for each sr_i^2 is the same. Following what we presented in Chapter 4's Guided Example, we use Soper (2025c) for the overall R^2 and Soper (2025d) for the sr_i^2 values. We also use Cohen et al.'s (2003) definitions for a medium effect for R^2 (p. 93) and a medium effect for sr_i^2 (p. 95).

- Overall R^2. Given $ES = f^2 = 0.13$, power = .8, IVs = 3, and $\alpha = .05$: $n^* = 88$.
- sr_i^2. Given $ES = f^2 = 0.15$, power = .8, $k_{\text{Set A}} = 2$, $k_{\text{Set B}} = 1$, and $\alpha = .05$: $n^* = 56$.

Therefore, we will need at least $n^* = 88$ cases to have an 80% probability of finding a significant medium effect for the overall R^2 and each sr_i^2. The given data set has $N = 171$, so we have met this minimum sample size.

Preliminary Data Screening

Our preliminary data screening follows the structure and contents of what we did in Chapter 7's Guided Example. We also do *not* provide copies of any detailed summary tables, leverage plots, or residual plots but instead simply report the results from our statistics program.

Outlier Analysis. Examining the initial data set for outliers via Jackknife distances, zero outliers are flagged and therefore this is not a concern. (*Note:* Other outlier procedures might yield different results.) The reader will recognize that this result applies to all three coding strategies because they are based on the same data set.

Multicollinearity. For dummy coding, the *VIF*s are no greater than 1.45; for unweighted effects coding, the *VIF*s are no greater than 1.62; and for contrast coding, the *VIF*s are no greater than 1.01. Therefore, multicollinearity is not a concern.

Missing Data. The data set has no missing data.

Measurement Error. The IVs are not measuring any psychological construct, but instead they are numerical codes for the nominal IV "seat classification."

Relevant IVs. To determine if the IVs are correctly specified, we examine each IV's added variable plot (see Chapter 5). The results of these plots indicate that all coded IVs for each coding strategy are relevant and therefore will remain in the model. (*Note*: Recall that if this were not the case and a coded IV is deleted, then the coding chart must be revised to reflect the absence of this group comparison.)

We now check to see if the model is compliant with the regression assumptions.

Linearity. A scatter plot of the residuals vs. the predicted values, in conjunction with the corresponding kernel smoother line, confirms that the scores are randomly scattered around the 0-line and do not show any pronounced or systematic pattern. As a result, the respective final models are compliant with the linearity assumption.

Equal Variances (Homoscedasticity of the Residuals). The scatter plot of the residuals vs. the predicted values used to check the linearity assumption also confirms the equal variances assumption because the residuals appear to be approximately equally scattered above and below the 0-line without any pronounced pattern. Therefore, the respective final models are compliant with the equal variances assumption.

Independence of the Residuals. A scatter plot of the residuals vs. the case numbers, in conjunction with the corresponding kernel smoother line, confirms there is no systematic pattern and therefore the respective models are compliant with the independence assumption.

Normality of the Residuals. The respective *q-q* plots of the residuals for the three coding strategies show that the data closely "hug" the normal line and that all the points are within the 95% confidence band. The corresponding Shapiro-Wilk Goodness of Fit tests also are not significant. Therefore, the respective models are compliant with the normality assumption.

Because preliminary data screening did not result in any changes, the initial data set is the final data set that will be used for the primary analysis.

Data Analysis

We now conduct a simultaneous regression analysis for each of the coding strategies. The results for each coding strategy are summarized in Table 8.14.

Interpretation of Results

We next interpret the results from Table 8.14 relative to the RQs and corresponding hypotheses for each coding strategy. The reader will first recognize that for the overall model, the four seat classifications collectively explain 31.08% of the variance in opinion scores, which is significant, $R^2 = .3108$. $F(3, 167) = 25.1085$, $p < .0001$.

Dummy Coding. As reported in Table 8.14, all group comparisons made to First Class are significant. More specifically:

D_1 = E. Holding all other IVs constant, participants who generally purchase Economy seats scored, on average, approximately 12 points *lower* ($B_1 = -11.88$) on the opinion questionnaire vs. those who generally purchase First Class seats. This is significant, $t(167) = -6.74$, $p < .0001$. Therefore, we reject the corresponding null hypothesis and conclude Economy participants have a significantly lower opinion of the four largest airlines than First Class participants. Furthermore, the Economy–First Class distinction ($sr_1^2 = .1876$) accounts for 18.76% of explained variance in

TABLE 8.14 *Summary of MRC Results for Chapter 8's Guided Example*

Coding Charts

Group	Dummy Coding D_1	D_2	D_3	Unweighted Effects Coding E_1	E_2	E_3	Contrast Coding C_1	C_2	C_3
Economy	1	0	0	1	0	0	1/4	1/3	1/2
Premium Economy	0	1	0	0	1	0	1/4	−2/3	0
Business Class	0	0	1	0	0	1	−3/4	0	0
First Class	0	0	0	−1	−1	−1	1/4	1/3	−1/2

ANOVA

Source	*SS*	*df*	*MS*	*F*	*p*
Between	5377.910	3	1792.64	25.1085	< .0001
Within (Error)	11923.049	167	71.40		
Total	17300.959	170			

Note. $N = 171$. $R^2 = .3108$.

Group Means and Standard Deviations

Group	*N*	*M*	*SD*
Economy	45	71.24	8.70
Premium Economy	42	69.21	7.29
Business Class	37	77.65	9.24
First Class	47	83.13	8.53
Overall	171	75.31 [a]	10.09

Note. [a] Unweighted mean

Summary of Partials

Factor [a]	B_i	SE_{Bi}	Semipartial Correlations sr_i	sr_i^2	Full Partial Correlations [b] pr_i	pr_i^2	Significance $t_i(167)$	p_i
Dummy Coding (First Class is Reference Group)								
Constant	83.13	1.23					67.45	< .0001
D_1 = E	−11.88	1.76	.4331	.1876	.4626	.2140	−6.74	< .0001
D_2 = PE	−13.91	1.79	.4982	.2482	.5146	.2648	−7.75	< .0001
D_3 = BC	−5.48	1.86	.1895	.0359	.2225	.0495	−2.95	.0036
Unweighted Effects Coding (First Class is Base Group)								
Constant	75.31	0.65					116.07	< .0001
E_1 = E	−4.06	1.10	.2368	.0561	.2744	.0753	−3.69	.0003
E_2 = PE	−6.09	1.13	.3473	.1206	.3859	.1489	−5.41	< .0001
E_3 = BC	2.34	1.18	.1277	.0163	.1520	.0231	1.99	.0485
Contrast Coding (See Coding Chart Above)								
Constant	75.31	0.65					116.07	< .0001
C_1	−3.12	1.57	.1277	.0163	.1520	.0231	−1.99	.0485
C_2	7.97	1.57	.3254	.1059	.3650	.1332	5.07	< .0001
C_3	−11.88	1.76	.4331	.1876	.4626	.2140	−6.74	< .0001

Note. $N = 171$.
[a] E = Economy, PE = Premium Economy, BC = Business Class.

the opinion scores, which means that if we were to omit this comparison, then the total amount of explained *Y* variance is reduced by 18.76%.

D_2 = PE. Holding all other IVs constant, participants who generally purchase Premium Economy seats scored, on average, nearly 14 points *lower* ($B_2 = -13.91$) on the opinion questionnaire vs. those who generally purchase First Class seats. This is significant, $t(167) = -7.75$, $p < .0001$. Therefore, we reject the corresponding null hypothesis and conclude that Premium Economy participants have a significantly lower opinion of the four largest airlines than First Class participants. Furthermore, the Premium Economy–First Class distinction ($sr_2^2 = .2482$) accounts for 24.82% of explained variance in the opinion scores, which means that if we were to omit this comparison then the total amount of explained *Y* variance is reduced by 24.82%

D_3 = BC. Holding all other IVs constant, participants who generally purchase Business Class seats scored, on average, about 5.5 points *lower* ($B_3 = -5.48$) on the opinion questionnaire vs. those who generally purchase First Class seats. This is significant, $t(167) = -2.95$, $p = .0036$. Therefore, we reject the corresponding null hypothesis and conclude that Business Class participants have a significantly lower opinion of the four largest airlines than First Class participants. Furthermore, the Business Class–First Class distinction ($sr_3^2 = .0359$) accounts for 3.59% of explained variance in the opinion scores, which means that if we were to omit this comparison, then the total amount of explained *Y* variance is reduced by 3.59%.

Unweighted Effects Coding. As reported in Table 8.14, all group comparisons made to the overall unweighted grand mean are significant. More specifically:

E_1 = E. Holding all other IVs constant, participants who generally purchase Economy seats scored, on average, approximately 4 points *lower* ($B_1 = -4.06$) on the opinion questionnaire when compared to the unweighted mean of the overall sample. This is significant, $t(167) = -3.69$, $p = .0003$. Therefore, we reject the corresponding null hypothesis and conclude that Economy participants have a significantly lower opinion of the four largest airlines when compared to all four groups in the sample. Furthermore, $sr_1^2 = .0561$ indicates that 5.61% of the variance in opinion scores is being explained by the Economy "effect," which represents the distinction between Economy vs. the other three groups (Premium Economy, Business Class, and First Class).

E_2 = PE. Holding all other IVs constant, participants who generally purchase Premium Economy seats scored, on average, about 6 points *lower* ($B_2 = -6.09$) on the opinion questionnaire when compared to the unweighted mean of the overall sample. This is significant, $t(167) = -5.41$, $p < .0001$. Therefore, we reject the corresponding null hypothesis and conclude that Premium Economy participants have a significantly lower opinion of the four largest airlines when compared to all four groups in the sample. Furthermore, $sr_2^2 = .1206$ indicates that 12.06% of the variance in opinion scores is being explained by the Premium Economy "effect," which represents the distinction between Premium Economy vs. the other three groups (Economy, Business Class, and First Class).

E_3 = BC. Holding all other IVs constant, participants who generally purchase Business Class seats scored, on average, $B_3 = 2.34$ points *higher* on the opinion questionnaire when compared to the unweighted mean of the overall sample. This is significant, $t(167) = 1.99$, $p = .0485$. Therefore, we reject the corresponding null hypothesis and conclude that Business Class participants have a significantly higher opinion of the four largest airlines when compared to all four groups in the sample. Furthermore, sr_3^2 = .0163 indicates that 1.63% of the variance in opinion scores is being explained by the Business Class "effect," which represents the distinction between Business Class vs. the other three groups (Economy, Premium Economy, and First Class).

Contrast Coding. As reported in Table 8.14, all designated contrasts are significant. More specifically:

***C_1*.** The merged group of participants who generally purchase Economy, Premium Economy, or First Class seats scored, on average, approximately 3 points *lower* (B_1 = –3.12) on the opinion questionnaire vs. those who generally purchase Business Class seats. This is significant, $t(167) = -1.99$, $p = .0485$. Therefore, we reject the corresponding null hypothesis and conclude that participants of the designated combined groups have a significantly lower opinion of the four largest airlines than Business Class participants. Furthermore, sr_1^2 = .0163 indicates that 1.63% of the variance in opinion scores is explained by C_1, which represents differences in opinions of the airlines between non-business passengers (i.e., those who generally do not purchase business class seats) and business passengers.

***C_2*.** The merged group of participants who generally purchase Economy or First Class seats scored, on average, about 8 points higher (B_2 = 7.97) on the opinion questionnaire vs. those who generally purchase Premium Economy seats. This is significant, $t(167) = 5.07$, $p < .0001$. Therefore, we reject the corresponding null hypothesis and conclude that participants of the designated combined groups have a significantly higher opinion of the four largest airlines than Premium Economy participants. Furthermore, sr_2^2 = .10.59 indicates that 10.59% of the variance in opinion scores is explained by C_2, which represents differences in opinions of the airlines between the group of passengers who generally purchase the least- and most expense seats vs. non-business passengers who generally purchase seats that are a little more expensive than Economy but not as expensive as First Class.

***C_3*.** This result is the same as D_1 from dummy coding.

Post Hoc Considerations

Discussions related to post hoc considerations were presented in earlier chapters and consisted of effect sizes for the overall model and sr_i^2 values (Chapters 3–7), 95% confidence intervals for B_i values (Chapters 2, 4, 5), and post hoc power for the overall model and sr_i^2 values (Chapters 4–7). The application of these considerations for the current Guided Example follows directly from what we presented earlier, and readers are directed to these past chapters for guidance.

Chapter 8 Summary

8.2 Group Membership, Single-Factor ANOVA, and Coded IVs

- A continuous variable consists of numerical data, whereas a nominal variable consists of non-numerical data.
- A nominal variable is often referred to as a categorical or group membership variable because it represents different categories of objects or groups.

- A single-factor ANOVA is a statistical procedure for comparing group means and involves a single nominal IV (called a "factor") that represents three or more groups (called "levels" of the factor). If the overall ANOVA model is significant, then pairwise group comparisons are examined.
- Nominal IVs are analyzed in MRC by changing the data type from nominal to continuous and using a set of numerically coded IVs to represent the categories. A nominal IV with k groups is fully represented by $k - 1$ coded IVs, and there are three primary coding strategies that can be used: dummy, effects, and contrast.

8.3 Dummy Coding

- A dummy coding strategy assigns coded IVs 0 or 1: If the IV represents one of the groups, then it is assigned 1; otherwise, it is assigned 0. A coding chart is used to reflect the respective codes for the coded IVs.
- A dummy coded scheme is used when it is plausible to make pairwise group comparisons with respect to a reference group, which gets coded with all 0s in the coding chart. Because such comparisons involve exactly two groups (a group coded 1 vs. the reference group), corresponding hypotheses are presented similar to those of an independent-samples *t* test.
- An MRC analysis involving a dummy coded nominal IV is conducted via simultaneous regression.
- In the final regression equation: (a) B_i represents the difference in group means between a group represented by one of the coded IVs and the reference group; (b) the regression constant, B_0, is equal to the mean of the reference group; and (c) substituting the numerical codes into the regression equation and performing the subsequent arithmetic yield each group's mean.
- The sr_i^2 values are referred to as the "distinction" between groups represented by a coded IV and the reference group and are interpreted as the amount of explained variance that is omitted from the overall R^2 if a particular distinction is not included in the analysis.
- To determine the significance of pairwise comparisons independent of the reference group, the data set either is recoded or a *t* test can be conducted by manually calculating *t* using the formula given in Section 8.3.
- Readers are cautioned about the use of preselected coding systems that may be used for dummy coding.

8.4 Effects Coding

- An effects coding strategy uses a base group, which has no presence in the regression equation and therefore any analysis involving the base group requires recoding the data set with a different base group. Furthermore, all group comparisons are made with respect to the overall sample mean, called the grand mean, which is the mean of the means, and the deviation between a group mean and the grand mean is referred to as its "effect." The grand mean may be calculated using either an unweighted or weighted approach.
- In an unweighted effects coding strategy, coded IVs are assigned 0, 1 or –1. If the IV represents one of the groups, it is assigned 1; if it does not represent

one of the groups, it is assigned 0; and if it is the designated base group, it is assigned –1.

- In a weighted effects coding strategy, the base group is assigned the numerical code $-a/b$ where a = the sample size of a coded group and b = the sample size of the base group. If the sample sizes of all groups are equal, then the fraction defaults to –1.
- An MRC analysis involving an effects coded nominal IV is conducted via simultaneous regression.
- In the final regression equation: (a) B_i represents the difference in group means between a group represented by one of the coded IVs and the grand mean; (b) the regression constant, B_0, is equal to the grand mean (unweighted or weighted); and (c) substituting the numerical codes into the regression equation and performing the subsequent arithmetic yields each group's mean.
- Several alternative group comparisons independent of the regression equation were also presented.
- The sr_i^2 values represent the amount of explained Y variance attributed to each coded IV's "effect," which is the distinction between the coded IV and all the other groups in the sample.

8.5 Contrast Coding

- A contrast coding strategy enables researchers to construct planned comparisons, called contrasts, between any groups to reflect specific hypotheses grounded in theory, prior research, or preliminary studies.
- A set of contrasts must be compliant with two rules: (a) The codes assigned to a contrast must add to 0, and (b) the contrasts must be orthogonal.
- The codes for a contrast are assigned $\frac{y}{z}$ and $\frac{-x}{z}$, where z = the total number of groups involved in the contrast, x = the number of groups that comprise the first part of the contrast, and y = the number of groups that comprise the second part of the contrast. Any group not assigned to a contrast is coded 0.
- An alternative coding scheme that excludes z assigns $\frac{1}{x}$ to the code for the first part of the contrast and $\frac{-1}{y}$ to the code for the second part of the contrast.
- Contrast codes may be transformed to integers, but this will affect the magnitude of the B_i values.
- If any part of a contrast consists of two or more groups, then that part of the contrast represents the unweighted mean of the groups' means.
- In the final regression equation: (a) B_i represents the difference in group means as reflected in the corresponding coding chart; (b) the regression constant, B_0, is equal to the unweighted grand mean; and (c) substituting the numerical codes into the regression equation and performing the subsequent arithmetic yields each group's mean.
- The sr_i^2 values represent the proportion of Y variance explained by the respective contrasts.

Key Terms and Concepts

Base group
Coding chart
Continuous variable
Contrast coding
Dummy coding
Effects coding
G_g
Grand mean
Group distinction (dummy coding)
Group effect (effects coding)
Group membership variable
gth group
Nominal variable
Orthogonal contrasts
Planned comparisons
Reference group
Rule 1 for contrast coding
Rule 2 for contrast coding
Sacrificial group
Single-factor ANOVA
Unweighted effects coding
Unweighted mean
Weighted effects coding
Weighted mean

Review Exercises

A. Checking Conceptual Knowledge

In 1–10, choose the best answer among the choices provided.

1. Let's assume you are analyzing the effect of different aircraft types on fuel efficiency using a regression model with dummy coding, and the aircraft types are coded as Small (0 0), Medium (1 0), and Large (0 1). If the regression equation is $\hat{Y} = -10X_1 + 5X_2 + 50$, then what is the estimated mean fuel efficiency for Large aircraft?
   a. 50 b. 55 c. 45 d. 60
2. A dummy-coded MRC analysis involving a nominal IV with four groups is conducted and the results are $R^2_{Y \bullet 123} = .039$, $R^2_{Y \bullet 23} = .02$, $R^2_{Y \bullet 13} = .011$, $R^2_{Y \bullet 12} = .001$. Which statement accurately represents sr_1^2?
   a. When comparing Group 1 to the reference group, Group 1 explains 2% of Y variance.
   b. Group 1 uniquely explains 2% of Y variance.
   c. If the Group 1-reference group distinction is not considered, then the overall amount of explained Y variance is reduced by 2%.
   d. There is not enough information to determine sr_1^2.
3. An unweighted effects coded MRC analysis involving a nominal IV with four groups is conducted and the results are $R^2_{Y \bullet 123} = .0475$, $sr_1^2 = .0034$, $sr_2^2 = .0263$, $sr_3^2 = .0181$. Which statement is accurate?
   a. Group 1 explains 0.34% of Y variance.
   b. The Group 2 distinction explains 2.63% of Y variance.
   c. The Group 3 effect explains 1.81% of Y variance.
   d. The results are spurious because the sum of $sr_i^2 > R^2_{Y \bullet 123}$.
4. Consider the RQ, "To what extent do the four phases of flight impact passengers' perceptions of their overall flight experience?" Which statement accurately reflects the contrast coding chart (next page) relative to this RQ?

Continued on next page ...

Phases of Flight	C_1	C_2	C_3
Takeoff (T)	1/2	1/2	1/4
Climb (CL)	−1/2	0	1/4
Cruise (CR)	0	0	−3/4
Landing (L)	0	−1/2	1/4

a. The null hypothesis for C_3 is H_0: $\frac{\mu_T + \mu_{CL} + \mu_L}{4} - \mu_{CR} = 0$
b. The analysis does not contrast the first three phases to Landing.
c. $B_3 = \frac{M_T + M_{CL} + M_L}{4} - 3M_{CR}$
d. The RQ cannot be answered completely because at least one pair of contrasts is not orthogonal.

Use the following research description to answer Items 5–7.

Consider an MRC analysis that examines the relationship between aircraft type and passenger comfort among business executives. Aircraft type is a nominal IV with four dummy-coded groups: Gulfstream G350 (1 0 0), Bombardier Challenger 605 (0 1 0), and Airbus ACJ319neo (0 0 1). The reference group is Cessna Citation Longitude. Passenger comfort refers to cabin size, digital connectivity, maximum range, and dining facilities and is measured using a 20-item questionnaire scored on a traditional 5-point Likert response scale. The regression equation is $\hat{Y} = 3D_1 - 5D_2 - 10D_3 + 92$.

5. Which of the following statements is correct?
   a. The aircraft that received the highest mean comfort level is Cessna.
   b. All the aircraft types have a mean comfort score that is less than 92.
   c. The mean comfort score for Airbus is 82.
   d. D_2 indicates that the difference in mean scores between Gulfstream and Bombardier is 5 points.
6. Which is the correct interpretation of D_3?
   a. For every 1-point increase in favor of Airbus, comfort scores decrease on average by 10 points.
   b. The mean comfort score for Airbus is on average 10 points lower than the mean comfort score for Cessna.
   c. The mean comfort score for Airbus is on average 10 points lower than the overall sample mean.
   d. The negative coefficient indicates that D_3 is not statistically significant.
7. Let's assume that the study uses unweighted effects coding with Bombardier as the base group. If we were to apply the given regression equation to this assumption, then which of the following statements is NOT correct?
   a. The unweighted sample mean is 92.
   b. The mean comfort score for Cessna is on average 5 points lower than the unweighted sample mean.
   c. When compared to all four aircraft types, participants perceive Airbus as having the least passenger comfort.
   d. If a weighted effects coding scheme is used, the resulting equation would be exactly the same.

Use the following research description to answer Items 8–10.

Consider the RQ, "What is the effect of different types of flight routes on passenger satisfaction?" Type of flight route (and duration) is a nominal IV with four groups: G1 = Domestic Short-Haul (< 3 hours), G2 = Domestic Long-Haul (≥ 3 hours), G3 = International Short-Haul (< 3 hours), and G4 = International Long-Haul (≥ 3 hours). Passenger satisfaction is measured on a dichotomously scored 10-item questionnaire with higher scores reflecting higher satisfaction. Let's assume that $n_1 = 84$, $n_2 = 42$, $n_3 = 37$, $n_4 = 8$, and an MRC analysis is conducted for the nominal IV, which is coded: G_1 (1 0 0), G_2 (0 1 0), G_3 (0 0 1), G_4 (−1 −1 −1). The results are $R^2 = .048$, $F(3, 167) = 2.774$, $p = .0429$, $\hat{Y} = 0.10E_1 - 0.32E_2 + 0.28E_3 + 7.37$.

8. Which of the following statements is NOT correct?
   a. B_1 indicates that Group 1's mean satisfaction score is higher than the grand mean.
   b. B_2 indicates that the mean satisfaction score for Group 2 is on average 0.32 points lower than the grand mean.
   c. The unweighted mean satisfaction score for the overall sample is 7.37.
   d. Because B_3's coefficient is the largest among the coded groups, Group 3 has the highest mean score among the coded groups.
9. Which statement reflects a potential flaw to the way the study is conducted?
   a. The coding strategy is not appropriate because of disparate sample sizes.
   b. The group means, which could be calculated prior to conducting the MRC analysis, are too close to each other to result in any statistical differences.
   c. The durations given for the international flight routes are incorrect.
   d. There are no flaws.
10. Which of the following is an accurate estimate of the weighted group mean?
    a. 7.37 b. 7.40 c. 7.31 d. 7.47

B. Applying Procedural and Conceptual Knowledge

Use the following research description and the corresponding data set to conduct the activities given in Parts B1–B7. See Section 8.6 for guidance.

Consider a study that examines the effect of four major airlines—American, Delta, Southwest, and United—on air carrier arrival delays, defined as the number of minutes between scheduled and actual arrival times. A flight is considered delayed if it arrives at the gate 15 or more minutes after the scheduled arrival time (Bureau of Transportation Statistics, 2025). The delays are air carrier related and independent of other conditions that can be attributed to arrival delays, including extreme weather, the national aviation system (NAS), late-arriving aircraft, and security. The data set is given in the file "Ch_8 Review Exercises Data–Part B." This data set was acquired from the Bureau of Transportation Statistics (2025) website and manipulated separately for each of the targeted airlines with respect to "all airports" from "August 2024 to July 2025." The data were then modified for instructional purposes and therefore are considered hypothetical.

B1. Research Questions and Hypotheses. Write the respective sets of research questions and corresponding hypotheses for the three coding strategies. For contrast coding, assume: C_1 = American–Delta–United vs. Southwest, C_2 = American vs. Delta, and C_3 = American–Delta vs. United.

B2. Sample Size Planning. Conduct a power analysis to see if the given data set is sufficiently large for the overall R^2 and the sr_i^2 values.

B3. Preliminary Data Screening. Perform regression diagnostics on the data set and then confirm that the respective final models for each coding strategy are compliant with the underlying regression assumptions.

B4. Data Analysis. Conduct a simultaneous regression analysis for each of the coding strategies and summarize the results in a table similar to Table 8.14.

B5. Interpretation of Results. Interpret the results relative to the RQs and corresponding null hypotheses stated in Part B1 separately for each coding strategy.

B6. Post Hoc Considerations. Determine and comment on the effect sizes, 95% CIs for the B_i values, and power for the overall R^2 and the respective increments for each coding strategy.

B7. Commentary. Provide any additional commentary that is meaningful with respect to the results.

References

Bureau of Transportation Statistics (2025). *Airline on-time statistics and delay causes.* https://www.transtats.bts.gov/OT_Delay/OT_DelayCause1.asp

Cohen, J., Cohen, C., West, S. G., & Aiken., L. S. (2003). *Applied multiple regression/correlation analysis for the behavioral sciences* (3rd ed.). Lawrence Erlbaum Associates.

Federal Aviation Administration (2024, March 21). *Airman's information manual,* Chapter 7: Safety of Flight, Section 7-1-7. Categorical Outlooks. https://www.faa.gov/air_traffic/publications/atpubs/aim_html/chap7_section_1.html

Hardy, M. A. (1993). *Regression with dummy variables.* Sage Publications, Inc.

Pedhazur, E. J. (1997). *Multiple regression in behavioral research: Explanation and prediction* (3rd ed.). Harcourt Brace.

Snizek, W. (1972). Hall's professionalism scale: An empirical assessment. *American Sociological Review, 37*, 109–114. https://doi.org/10.2307/2093498

Soper, D. (2025a). Student *t-value calculator.* https://www.danielsoper.com/statcalc/calculator.aspx?id=10

Soper, D. (2025b). *p-value calculator for a student t-test.* https://www.danielsoper.com/statcalc/calculator.aspx?id=8

Soper, D. (2025c). *A-priori sample size calculator for multiple regression.* https://www.danielsoper.com/statcalc/calculator.aspx?id=1

Soper, D. (2025d). *A-priori sample size calculator for hierarchical multiple regression* https://www.danielsoper.com/statcalc/calculator.aspx?id=16

Turgut, R. T. (2019). *Aviation professionalism: Examining the concept of professionalism within and between major subgroups of the aviation industry* [Doctoral dissertation, Florida Institute of Technology]. Scholarship Repository. https://repository.fit.edu/etd/39/

Part D
Additional Applications of MRC

Chapter 9
Combining Nominal and Continuous IVs, ANCOVA, and Interactions

Student Learning Outcomes

After studying this chapter, you will be able to do the following:

1. Conduct and report the results of a multiple regression analysis involving a combination of nominal and continuous IVs.
2. Conduct an analysis of covariance (ANCOVA) using multiple regression.
3. Conduct a 2 × 2 factorial ANOVA using multiple regression.
4. Conduct a two-way interaction analysis either indirectly from a misspecified ANCOVA model or directly as articulated by the research question.
5. Conduct an MRC analysis for a research study that has a combination of nominal and continuous IVs by engaging in the activities presented in Section 9.5.

9.1 Chapter Overview

This chapter extends the discussion on analyzing nominal variables from Chapter 8. We begin by presenting an example of a research setting that consists of more than one nominal IV and a combination of nominal and continuous IVs. As part of this presentation, we introduce the concepts of *adjusted group means* and *centering IVs*. We then introduce the concept of *analysis of covariance* (*ANCOVA*), which is a specific statistical strategy for controlling the effect a continuous IV, called the *covariate*, has on a nominal IV (group membership) with respect to the dependent variable. As part of this presentation, we demonstrate how to use MRC to conduct an ANCOVA, including checking for compliance to the *homogeneity of regression* assumption, which distinguishes a *valid ANCOVA model* from a *misspecified ANCOVA model*. We follow this discussion with an introduction to the concept of interactions by presenting and interpreting *two-way interaction* plots of *nominal-by-nominal interactions* and *nominal-by-continuous interactions*. Included in this discussion is the concept of *attribute-treatment interaction* (*ATI*). We conclude the chapter with a guided example that demonstrates how to use MRC to analyze data from a research study that consists of a mix of nominal and continuous IVs.

9.2 Combining Nominal and Continuous IVs

In Chapter 8, we introduced different strategies for conducting an MRC analysis involving a nominal IV—dummy, effects (unweighted and weighted), and contrast coding—and to facilitate understanding, we restricted this discussion to a single nominal IV. In most research settings, though, data sets that undergo an MRC analysis generally are not restricted to a single nominal IV but instead include several nominal IVs or a mix of nominal and quantitative variables.

As an example, consider a study that examines differences in knowledge of basic aviation weather concepts among four distinct groups: air traffic controllers

DOI: 10.4324/9781003506935-13

(ATCs), airline transport pilots (ATPs), certified flight instructors (CFIs), and aviation meteorology professors (AMPs). In addition to this group membership variable, participants also reported their biological sex assigned at birth (Female or Male) and age (in years). As a result, we now have a data set that consists of two nominal IVs (aviation group and biological sex) and one continuous IV (age). The dependent variable is Y = Weather Knowledge Scores, which represent the total number of items answered correctly from a preestablished set of 100 dichotomously scored multiple-choice items. Thus, scores could range from 0 to 100, with higher scores reflecting greater understanding of basic weather concepts.

Given the composition of the four targeted groups, our interest is in examining the classical "theory" vs. "practice" comparisons and therefore we will dummy code the aviation groups with AMPs as the reference group (coded 0). We also will dummy code biological sex with Male as the reference group. The respective coding charts for the two nominal IVs are given in Table 9.1, and the corresponding data set, which contains hypothetical data, is given in the file, "Ch_9 Combining Nominal and Continuous IVs Example Data." We will assume that the data were collected randomly from a sample of N = 100 full-time employees across the four aviation groups and that each group was of equal size ($n_1 = n_2 = n_3 = n_4 = 25$).

Treating the Nominal IVs Separately

Before we demonstrate how to analyze the given data set via MRC, let's first examine the nominal IVs separately. The results of these analyses are summarized in Table 9.2. As reported in Table 9.2(a), when Y is regressed on the dummy coded IVs representing the four aviation groups, the overall model is significant, $R^2 = .18$, $F(3, 96) = 7.0159$, $p = .0003$, and the regression equation is $\hat{Y} = 3.56D_1 + 22.96D_2 + 16.56D_3 + 61.76$. Because this analysis is not in the presence of any other IVs, when we substitute the numerical codes from the coding chart given in Table 9.1 into the model, we get the "true" group means and B_i reflects "true" group differences.

- Air Traffic Controller (coded 1 0 0). $M_{ATC} = 65.32$, and $B_1 = 3.56$ indicates that when holding all other IVs constant, air traffic controllers' mean weather knowledge score is, on average, 3.56 points higher than that of aviation meteorology professors and this difference is not significant, $t(96) = 0.62$, $p = .5385$.
- Airline Transport Pilot (coded 0 1 0). $M_{ATP} = 84.72$, and $B_2 = 22.96$ indicates that when holding all other IVs constant, airline transport pilots' mean weather knowledge score is, on average, approximately 23 points higher than that of aviation meteorology professors and this difference is significant, $t(96) = 3.98$, $p = .0001$.

TABLE 9.1 *Coding Charts for the Two Nominal IVs from the Data Set "Ch_9 Combining Nominal and Continuous IVs Example Data"*

Coding Chart for Aviation Group

Group	D_1	D_2	D_3
Air Traffic Controller (ATC)	1	0	0
Airline Transport Pilot (ATP)	0	1	0
Certified Flight Instructor (CFI)	0	0	1
Aviation Meteorology Professor (AMP)	0	0	0

Coding Chart for Biological Sex

Group	X_1
Female	1
Male	0

TABLE 9.2 *Summary of Results for Separate MRC Analyses Involving Dummy Coded IVs for (a) Aviation Groups and (b) Biological Sex with Respect to the Data Set Given in the File "Ch_9 Combining Nominal and Continuous IVs Example Data"*

(a) Regressing *Y* on the Dummy Coded IVs for Aviation Group

ANOVA

Source	*SS*	*df*	*MS*	*F*	*p*
Between	8752.43	3	2917.48	7.0159	.0003
Within (Error)	39920.48	96	415.84		
Total	48672.91	99			

Note. $N = 100$. $R^2 = .18$.

Parameter Estimates

Factor [a]	B_i	SE_i	95% CI	$t_i(96)$ [b]	p_i
Constant	61.76	4.08	[53.66, 69.86]	15.14	< .0001
D_1 = ATC	3.56	5.77	[−7.89, 15.01]	0.62	.5385
D_2 = ATP	22.96	5.77	[11.51, 34.41]	3.98	.0001
D_3 = CFI	16.56	5.77	[5.11, 28.01]	2.87	.0050

Note. $N = 100$; $n = 25$ per group. Y = Weather knowledge scores.
[a] ATC = Air Traffic Controller, ATP = Airline Transport Pilot, CFI = Certified Flight Instructor, where Aviation Meteorology Professor (AMP) is the reference group. [b] Two-tailed t critical with $df = 96$ and $\alpha = .05$ is ±1.985.

(b) Regressing *Y* on the Dummy Coded IV for Biological Sex

ANOVA

Source	*SS*	*df*	*MS*	*F*	*p*
Between	216.09	1	216.09	0.4370	.5101
Within (Error)	48456.82	98	494.457		
Total	48672.91	99			

Note. $N = 100$. $R^2 = .004$.

Parameter Estimates

Factor [a]	B_i	SE_i	95% CI	$t_i(98)$ [b]	p_i
Constant	71.06	3.14	[64.82, 77.30]	22.60	< .0001
X_1 = Female	2.94	4.45	[−5.89, 11.77]	0.66	.5101

Note. $N = 100$; $n = 50$ per group. Y = Weather knowledge scores.
[a] Male is the reference group. [b] Two-tailed t critical with $df = 98$ and $\alpha = .05$ is ±1.9845.

- Certified Flight Instructor (coded 0 0 1). $M_{CFI} = 78.32$, and $B_3 = 16.56$ indicates that when holding all other IVs constant, flight instructors' mean weather knowledge score is, on average, 16.56 points higher than that of aviation meteorology professors and this difference is significant, $t(96) = 2.87$, $p = .0050$.
- Aviation Meteorology Professor (coded 0 0 0). This is the reference group and $M_{AMP} = B_0 = 61.76$, which is the mean weather knowledge score for aviation meteorology professors.

Similarly, as reported in Table 9.2(b), when Y is regressed on the dummy coded IV for the two biological sex groups, the overall model is not significant, $R^2 = .004$, $F(1, 98) = 0.4370$, $p = .5101$, and the regression equation is $\hat{Y} = 2.94X_1 + 71.06$. Because this analysis is not in the presence of any other IVs, when we substitute the numerical codes from the coding chart given in Table 9.1 into the model, we get the "true" group means and B_i reflects "true" group differences.

- Female (coded 1). $M_{Female} = 74.00$, and $B_1 = 2.94$ indicates that female participants' mean weather knowledge score is, on average, approximately 3 points higher than that of male participants and this difference is not significant, $t(98) = 0.66$, $p = .5101$.
- Male (coded 0). This is the reference group and $M_{Male} = B_0 = 71.06$, which is the mean weather knowledge score for males.

As noted above, the results of these analyses yield "true" group means when we substitute the respective numerical codes into the regression equation and complete the arithmetic, and the regression coefficients reflect the "true" mean differences between groups that reflect the coding scheme used. For our example, because we used dummy coding, B_i represents the mean difference between a coded group and the reference group. What we mean by the word "true" is that because each nominal IV is being analyzed solo as we did in Chapter 8, there are no other IVs in the model whose effects have to be statistically controlled, and hence we do not have to partial out the effect of other IVs from the nominal IV under discussion. Further recall from Chapter 8 that although we opt to qualify the interpretation of B_i with "holding all other IVs constant," this also is not necessary (see the note in Chapter 8 when we first interpret B_i.) As a result, we may interpret the respective B_i and B_0 values *unconditionally*. Let's now see what happens when we include other IVs into the analysis.

Combining Nominal and Continuous IVs

The reader will recall that we previously conducted MRC analyses that involved both nominal and continuous IVs. For example, in Chapter 4 / Example 4.1, we examined the effect three IVs, X_1 = Total Flight Hours, X_2 = Number of FAA Ratings, and X_3 = Flight Training Program, have on Y = Aviation Safety Scores (ASI). For this example, X_3 was a nominal IV that compared two flight training programs, Part 141 (coded 1) vs. Part 61 (coded 0). The regression equation was $\hat{Y} = -0.0004X_1 + 1.23X_2 + 2.49X_3 + 9.77$, and we interpreted B_3 as a comparison of mean ASI scores between the Part 141 group and the Part 61 group: "When holding X_1 = Total Flight Hours and X_2 = Number of FAA Ratings constant, pilots who received flight training under a Part 141 program averaged approximately 2.5 points higher than pilots who received flight training under a Part 61 program." Other examples included Chapter 4 / Example 4.2, Chapter 4's Guided Example, Chapter 7 / Example 7.1, and various examples in Chapters 5 and 6 that consisted of an IV representing biological sex.

The difference between these earlier examples and the current discussion is that the nominal IVs were dichotomous, whereas for the current discussion one of the nominal IVs consists of more than two groups. To demonstrate this difference, let's conduct a simultaneous regression with all the variables contained in the current data set (i.e., D_1, D_2, D_3, X_1, X_2). A summary of the results of this analysis is given in Table 9.3.

TABLE 9.3 *Summary of Results of MRC Analysis Involving Nominal and Continuous IVs for the Data Set "Ch_9 Combining Nominal and Continuous IVs Example Data"*

ANOVA

Source	*SS*	*df*	*MS*	*F*	*p*
Between	11283.969	5	2256.79	5.6738	.0001
Within (Error)	37388.941	94	397.75		
Total	48672.910	99			

Note. $N = 100$. $R^2 = .232$.

Parameter Estimates

Factor [a]	B_i	SE_i	95% CI	$t_i(94)$ [b]	p_i
Constant	42.04	8.78	[24.61, 59.46]	4.79	< .0001
D_1 = ATC	6.44	5.87	[–5.22, 18.09]	1.10	.2755
D_2 = ATP	23.10	5.68	[11.82, 34.38]	4.07	< .0001
D_3 = CFI	17.76	5.70	[6.43, 29.08]	3.11	.0025
X_1 = Bio. Sex	4.30	4.13	[–3.91, 12.50]	1.04	.3013
X_2 = Age	0.40	0.17	[0.06, 0.74]	2.36	.0204

Note. $N = 100$. Y = Weather knowledge scores.
[a] ATC = Air Traffic Controller ($n = 25$), ATP = Airline Transport Pilot ($n = 25$), and CFI = Certified Flight Instructor ($n = 25$), where Aviation Meteorology Professor (AMP) is the reference group ($n = 25$). Bio. Sex = Biological sex assigned at birth ($n = 50$ per sex group). Age is given in years.
[b] Two-tailed t critical with $df = 94$ and $\alpha = .05$ is ±1.9855.

As reported in Table 9.3, the overall regression model is significant, $R^2 = .232$, $F(5, 94) = 5.6738$, $p = .0001$, and the regression equation is $\hat{Y} = 6.44D_1 + 23.10D_2 + 17.76D_3 + 4.30X_1 + 0.40X_2 + 42.04$. At first glance, some readers might presume that something is amiss with the regression equation. For example:

- The respective regression coefficients for the three coded aviation group IVs representing differences in group means between each coded group and the reference group are different from what is reported in Table 9.2: $B_1 = 6.44$ vs. 3.56, $B_2 = 23.10$ vs. 22.96, and $B_3 = 17.76$ vs. 16.56.
- The regression coefficient for the coded biological sex IV representing the difference in group means between female and male participants is different from what is reported in Table 9.2: $B_4 = 4.30$ vs. 2.94.
- The regression constant is not equal to the mean of either the reference group for the Aviation Group or for Biological Sex as reported in Table 9.2: 42.04 vs. 61.76 for aviation meteorology professors, and 71.06 for male participants.

The reason for the differences in the B_i and B_0 values between what is reported in Table 9.2 vs. Table 9.3 is because we are now combining a nominal IV with three groups, a nominal IV with two groups, and a continuous IV. As a result, all the IVs are being examined in the presence of each other and therefore we cannot interpret the regression coefficients and regression constant unconditionally as we did earlier. They must be interpreted conditionally to reflect that each effect is a *partialled effect*, which means that we have statistically removed the effects of the other IVs from the nominal IV. Once again, this is not any different than what we did in our earlier examples that involved dichotomous IVs. With this in mind, the interpretations for the partials reported in Table 9.3 follows.

- B_1 = 6.44. When holding all other IVs in the model constant, air traffic controllers' mean weather knowledge score is, on average, 6.44 points higher than that of aviation meteorology professors, and this difference is not significant, $t(94) = 1.10$, $p = .2755$.
- B_2 = 23.10. When holding all other IVs in the model constant, airline transport pilots' mean weather knowledge score is, on average, 23.10 points higher than that of aviation meteorology professors, and this difference is significant, $t(94) = 4.07$, $p < .0001$.
- B_3 = 17.76. When holding all other IVs in the model constant, certified flight instructors' mean weather knowledge score is, on average, 17.76 points higher than that of aviation meteorology professors, and this difference is significant, $t(94) = 3.11$, $p = .0025$.
- B_4 = 4.30. When holding all other IVs in the model constant, female participants' mean weather knowledge score is, on average, 4.30 points higher than that of male participants, and this difference is not significant, $t(94) = 1.04$, $p = .3013$.
- B_5 = 0.40. When holding all other IVs in the model constant, for every 1-year increase in age, weather knowledge scores increase on average by 0.40 points. More concretely, for every 10-year increase in age, weather knowledge scores increase on average by 4 points. Thus, older participants have a greater understanding of basic weather concepts than younger participants, and this difference is significant, $t(94) = 2.36$, $p = .0204$.
- B_0 = 42.04. Recall that the regression constant is the *y*-value of the intercept, which occurs when $x = 0$. Thus, the mean weather knowledge score for male, aviation meteorology professors who are 0 years old is 42.04 points. The reader will note that this interpretation is meaningless for the given research context.

As illustrated by this example, when conducting an MRC analysis that involves a mix of nominal and continuous IVs, the regression coefficients that represent the coded IVs of a nominal variable will not, in general, equal "true" mean group differences. This is because analyses that yield "true" group mean differences do not control for the effect of the other IVs' presence in the model. So, emphasizing what we wrote above, by including additional nominal or continuous IVs in a data set, we cannot make unconditional comparisons of group means as we did when we had a single categorical variable. Instead, we must make conditional comparisons based on the other IVs in the model. In other words, we must partial out the effect of the additional IVs by holding them statistically constant when we make group comparisons and interpret the results.

Given this change in perspective, one might ask, "How do we test a nominal IV for statistical significance when its coded IVs are combined with other nominal or continuous IVs?" The answer to this question is grounded in hierarchical regression involving sets of IVs. To demonstrate this, we apply what we presented in Chapter 7 to the current example with respect to determining the significance of the nominal IV "Aviation Group" follows.

1. Let set A = the set of variables that are not under discussion. For the current demonstration this consists of biological sex and age. Thus, set A = $\{X_4, X_5\}$.
2. Let set B = the set of variables that are under discussion. For the current demonstration, this consists of the coded IVs for the aviation group. Thus, set B = $\{D_1, D_2, D_3\}$.
3. Regress Y on set A. The overall model is significant, $R^2_{Y\bullet A} = .063$, $F(2, 97) = 3.2596$, $p = .0426$, and the regression equation is $\hat{Y} = 3.28X_1 + 0.45X_2 + 52.36$. (*Note:* Cohen et al., 2003, p. 343, denote this the reduced model.)
4. Regress Y on sets A and B. The overall model is significant, $R^2_{Y\bullet AB} = .232$, $F(5, 94) = 5.6738$, $p = .0001$, and the regression equation is $\hat{Y} = 6.44D_1 + 23.10D_2 + 17.76D_3 + 4.30X_1 + 0.40X_2 + 42.04$. (*Note:* Cohen et al., 2003, p. 342, denote this the full model.)
5. Examine the significance of the increment. $sR^2_B = R^2_{Y\bullet AB} - R^2_{Y\bullet A} = .232 - .063 = .169$. To determine if this gain in prediction is significant, we either calculate the corresponding F using the formula provided in Chapter 8 / Section 8.3, or consult Soper (2025a, 2025b, 2025c). We choose the latter and get the following results: (a) $F(3, 94) = 6.89$, (b) $F_{\text{Critical}}(3, 94) = 2.70$, and (c) $p = .0003$. Therefore, when the coded IVs for the aviation group enter the model in the presence of biological sex and age, the aviation group explains an additional 16.9% of the variance in weather knowledge scores and this gain in prediction is significant.
6. If sR^2_B is significant, then examine the nominal IV separately. Given a significant increment, we may now regress Y on the coded IVs for the aviation group separately from the other IVs without incurring inflated alpha levels. The results of this follow-up analysis are what we reported in Table 9.2(a) and now the interpretations of the regression coefficients may be made unconditionally.

The Concept of Adjusted Group Means

In addition to reporting the conventional results of an MRC analysis consisting of more than one nominal IV or a mix of nominal and continuous IVs, it also is useful to report adjusted group means. The concept of adjusted group means is similar to the concept of a partialled effect. As noted in the foregoing paragraphs—as well as in our earlier presentations in previous chapters—we always preface the interpretation of the regression coefficients from a regression model that consists of more than one predictor with the phrase "when holding all other IVs in the model constant." This phrase signals to the reader that each B_i represents a partialled effect where the effects of the other IVs in the model have been removed from the IV under discussion. In other words, the B_i that corresponds to the IV under discussion has been statistically "adjusted" to account for the effect of the other IVs in the model.

In a similar vein, an adjusted group mean reflects a group's mean score on the dependent variable that has been modified to control for the presence of other IVs in the model. As an example, consider Chapter 8's Guided Example, which examined airline passengers' general opinions of the four largest airlines in the U.S. The predictor was seat classification, a nominal IV with four groups—Economy, Premium Economy, Business Class, and First Class—and participants were asked to specify the type of seat they generally purchase when flying domestically. Thus, the corresponding RQ is, "What effect does seat classification have on passengers' opinions of the airlines?" An alternative RQ might be, "To what extent do the types of seats passengers generally purchase when flying domestically predict passengers' opinions of the airlines?"

Given its presentation in Chapter 8, the focus of the study was on a single nominal IV. However, if we were to expand this study, we might also consider participants' (a) annual income expressed as a continuous IV or via income categories, (b) job status such as full-time vs. part-time, and (c) marital status such as married vs. not married. When considered from this perspective, the respective *raw* or *unadjusted* mean opinion scores associated with each seat classification group are now being influenced by these other IVs in the model. As a result, it would be beneficial to adjust the raw group means to account for the influence of these other IVs. Furthermore, by reporting the adjusted group means, the reader is able to better assess the impact these other IVs are having on the targeted nominal IV with respect to the DV.

The approach used to determine adjusted group means depends on the data types of the IVs involved, and there are two different scenarios: One is an analysis that involves strictly nominal IVs and the second is an analysis that involves a mix of nominal and continuous IVs. A brief discussion of each scenario follows.

Adjusted Means Involving an Analysis Strictly with Nominal IVs. To demonstrate the concept of adjusted group means with respect to an MRC analysis that involves only nominal IVs, we will focus on the two nominal IVs of our running example. More specifically, our interest is understanding the effect biological sex has on the four aviation groups' respective weather knowledge scores. More concretely, as reported earlier, we know that these respective group means are $M_{\text{ATC}} = 65.32$, $M_{\text{ATP}} = 84.72$, $M_{\text{CFI}} = 78.32$, and $M_{\text{AMP}} = 61.76$. The question now is to what extent will these means change when we examine them in the presence of biological sex?

Because we are dealing strictly with nominal IVs, the first step in determining the adjusted means is to combine the IVs' respective coding charts into a single chart. In other words, we merge the four aviation groups (ATC, ATP, CFI, AMP) with the two biological sex groups (Female, Male), which produces a single coding chart that contains 4 × 2 = 8 unique combinations of the two nominal IVs as shown in the shaded portion of Table 9.4.

We next regress Y on the three coded IVs of the aviation group and the single coded IV of biological sex. This yields the regression equation $\hat{Y} = 4.89D_1 + 23.55D_2 + 17.30D_3 + 3.70X_1 + 59.25$. Because each B_i represents a partialled effect, to determine the adjusted group means we simply substitute the codes for each combined group into the regression equation and complete the arithmetic. For example, when controlling for the effect of biological sex, the respective adjusted mean weather knowledge scores for male and female air traffic controllers are:

$$\begin{aligned}\hat{Y}_{\text{ATC-Male}} &= 4.89D_1 + 23.55D_2 + 17.30D_3 + 3.70X_1 + 59.25\\ &= 4.89(1) + 23.55(0) + 17.30(0) + 3.70(0) + 59.25\\ &= 4.89 + 59.25\\ &= 64.14\end{aligned}$$

$$\begin{aligned}\hat{Y}_{\text{ATC-Female}} &= 4.89D_1 + 23.55D_2 + 17.30D_3 + 3.70X_1 + 59.25\\ &= 4.89(1) + 23.55(0) + 17.30(0) + 3.70(1) + 59.25\\ &= 4.89 + 3.70 + 59.25\\ &= 67.84\end{aligned}$$

TABLE 9.4 *Adjusted Means for the Nominal IVs from the Data Set Given in the File "Ch_9 Combining Nominal and Continuous IVs Example Data"*

Group [a]	D_1	D_2	D_3	X_1 [b]	N	Adj. Mean [c]
ATC–Male	1	0	0	0	17	64.14
ATC–Female	1	0	0	1	8	67.84
ATP–Male	0	1	0	0	12	82.80
ATP–Female	0	1	0	1	13	86.50
CFI–Male	0	0	1	0	13	76.55
CFI–Female	0	0	1	1	12	80.25
AMP–Male	0	0	0	0	8	59.25
AMP–Female	0	0	0	1	17	62.95

Note. $N = 100$. Y = Weather knowledge scores.
[a] ATC = Air Traffic Controller, ATP = Airline Transport Pilot, CFI = Certified Flight Instructor. AMP = Aviation Meteorology Professor, which is the reference group. [b] X_1 = Biological sex. [c] Adjusted means are *predicted* means acquired by substituting the numerical codes from the combined coding chart into the regression equation.

Note that these adjusted means do not equal the original group mean (M_{ATC} = 65.32) when examined in the absence of biological sex. Thus, when controlling for the effect of biological sex, male ATCs' adjusted mean score was less than 65.32, whereas female ATCs' adjusted mean score was higher than 65.32. A summary of the complete adjusted mean calculations for this example are included in Table 9.4.

Adjusted Means Involving an Analysis with Nominal and Continuous IVs. To demonstrate the concept of adjusted group means for an MRC analysis that involves a mix of nominal and continuous IVs, we will focus on an analysis that involves the nominal IV for aviation groups and the continuous variable X_2 = Age from our running example. Note that because we now have a continuous IV, we cannot merge coding charts and then calculate the adjusted group means directly by substituting the respective numerical codes into the regression equation as we did in the previous scenario because we do not code continuous IVs. There is a work-around, though, which entails the concept of centering continuous IVs. To center an IV, we create a new IV (e.g., X–Cntr) and then subtract each observed X_i score by the mean of the IV; that is: $X_i - M_X$.

Applying this procedure to X_2 = Age from our running example, first observe that the mean age is $M_2 = 41.18$. As a result, the corresponding centered IV for the first three rows of the given data set are: (a) Row 1, X_2–Cntr = 31 – 41.18 = –10.18; (b) Row 2, X_2–Cntr = 29 – 41.18 = –12.18; and (c) Row 3, X_2–Cntr = 45 – 41.18 = 3.82. To understand the effect centering a continuous IV has on the results of an MRC analysis, let's conduct two separate analyses for our running example as follows:

Regress Y on D_1, D_2, D_3, X_2

$R^2 = .223$, $F(4, 95) = 6.8165$, $p < .0001$

$\hat{Y} = 4.86D_1 + 22.43D_2 + 16.89D_3 + 0.39X_2 + 45.42$

Regress Y on D_1, D_2, D_3, X_2–Cntr

$R^2 = .223$, $F(4, 95) = 6.8165$, $p < .0001$

$\hat{Y} = 4.86D_1 + 22.43D_2 + 16.89D_3 + 0.39X_{2\text{–Cntr}} + 61.49$

The reader will note that the respective overall models, including the unstandardized regression coefficients, B_i, are exactly the same. However, the respective regression constants, B_0, are different as highlighted in a second color, and it is this difference in B_0 values that provides a work-around for acquiring adjusted group means.

To understand why this is the case, let's interpret the regression coefficient for D_1 from the first model where we used an uncentered X_2. Because the group membership variable is dummy coded, $B_1 = 4.86$ indicates that when holding all other IVs in the model constant, the mean difference in weather knowledge scores between air traffic controllers and aviation meteorology professors (the reference group) is 4.86 points. In other words, ATCs' mean weather knowledge score is, on average, nearly 5 points higher than AMPs' mean weather knowledge score. Recall that when using the phrase "when holding all other variables in the model constant," we effectively are zeroing out the other IVs. This means that when applied to X_2, we are zeroing out Age and claiming that the mean difference between ATC and AMP scores is occurring at Age = 0. The problem, though, is that the range of ages in the data set is from 23 to 54.

When centering a continuous IV, however, note that each centered score is a *deviation score*—it represents how far the original score deviates from the mean. As a result, the sum of these deviation scores is 0 and hence its mean also is 0. So, focusing on the second regression model where $M_{\text{Age–Cntr}} = 0$, the interpretations of the respective B_i values are appropriate relative to the given research setting because, unlike the first regression model, "holding X_2–Cntr constant" is now meaningful. With a dummy coding strategy, this implies that the regression constant, $B_0 = 61.49$, is the adjusted mean of the reference group. This also implies that we may determine adjusted group means for an MRC analysis involving a mix of nominal and continuous IVs directly by substituting the numerical codes for the coded IVs into the regression equation where the continuous IVs are centered. So, for our current example, the adjusted group means are:

- ATC (coded 1 0 0). $M_{\text{ATC-Adj.}} = 4.86(1) + 22.43(0) + 16.89(0) + 61.49 = 66.35$
- ATP (coded 0 1 0). $M_{\text{ATP-Adj.}} = 4.86(0) + 22.43(1) + 16.89(0) + 61.49 = 83.92$
- CFI (coded 0 0 1). $M_{\text{CFI-Adj.}} = 4.86(0) + 22.43(0) + 16.89(1) + 61.49 = 78.38$
- AMP (coded 0 0 0). $M_{\text{AMP-Adj.}} = 4.86(0) + 22.43(0) + 16.89(0) + 61.49 = 61.49$

A summary of these adjusted means along with corresponding raw means is given in Table 9.5.

TABLE 9.5 Adjusted Means for the Aviation Groups When Controlling for Age

Group	Raw Mean [a]	Adj. Mean [b]
Air Traffic Control (ATC)	65.32	66.35
Airline Transport Pilot (ATP)	84.72	83.92
Certified Flight Instructor (CFI)	78.32	78.38
Aviation Meteorology Professor (AMP)	61.76	61.49

Note. $N = 100$; $n = 25$ per group. Y = Weather knowledge scores.
[a] Raw means = each group's mean weather knowledge score independent of participants' age. [b] Adjusted means are each group's mean weather knowledge score adjusted for differences in participants' age. Adjusted means acquired by substituting the numerical codes from the coding chart into the regression equation where X_2 = Age is centered.

The reader will recognize that when calculating adjusted means, we ignore the continuous IV because it is being held constant. The reader also will note that the adjusted group means differ from the respective raw group means because the former have been adjusted for group differences in participants' age. More specifically, the raw means reflect mean weather knowledge scores without controlling for the effect of participants' age, whereas the adjusted means are the predicted mean weather knowledge scores when controlling for the effect of age.

9.3 Analysis of Covariance (ANCOVA)

The Concept of ANCOVA

Throughout Chapter 8 and thus far in the current chapter, our focus has been on analyzing nominal IVs via MRC. In Chapter 8, we restricted the discussion to a single IV that represented three or more groups, and in the previous section we removed this restriction and included additional IVs, both nominal and continuous. Regardless of the data types or the coding strategy used to code the nominal IVs, the ultimate goal has been to compare group means on the DV to determine if their differences are statistically significant.

In this section, we continue this discussion of comparing group means by introducing a statistical strategy known as analysis of covariance, or ANCOVA for short. ANCOVA is similar to single-factor ANOVA in that both strategies examine differences on a DV across different groups. With ANCOVA, though, a second factor is included in the model. This second factor, called a covariate, is a continuous variable that is not of primary interest, but its presence could influence the difference in group means with respect to the DV and therefore we want to control for its effect.

For example, consider a study that examines differences in pilots' risk assessment. In this example, Y = Level of Risk, which we will measure using Hunter's (2006) Risk Perception–Other, and the group membership variable is Pilots, which we classify into four groups: Test Pilots (TPs), Military Pilots (MPs), Airline Transport Pilots (ATPs), and Helicopter Pilots (HPs). In the current context, the RQ is, "What are the differences in risk assessment scores among the four pilot groups?" and the appropriate data analysis strategy is ANOVA.

Let's now suppose that we believe older pilots might be more risk averse whereas younger pilots might tend to be more risk prone. This implies that pilots' age could impact risk assessment. As a result, we would like to control for the effect age has on risk assessment scores and decide to include pilots' age (in years) in the analysis. As a result, age is considered a covariate, and the RQ is now, "*When controlling for the effect of pilots' age*, what are the differences in risk assessment scores among the four pilot groups?" and the appropriate data analysis strategy is now ANCOVA. From this example, note that the covariate is *not* the primary focus of the study, but it could influence the results of the study because of its possible effect on the DV. Therefore, the role of ANCOVA is to statistically control for the covariate's effect.

To understand conceptually what we mean by "statistically control for the covariate's effect," we reintroduce the reader to the *squared full partial correlation coefficient* (pr^2), which we first presented in Chapter 3 / Section 3.4. The reader might recall that pr^2 is similar to the squared semipartial correlation coefficient (sr^2) in that

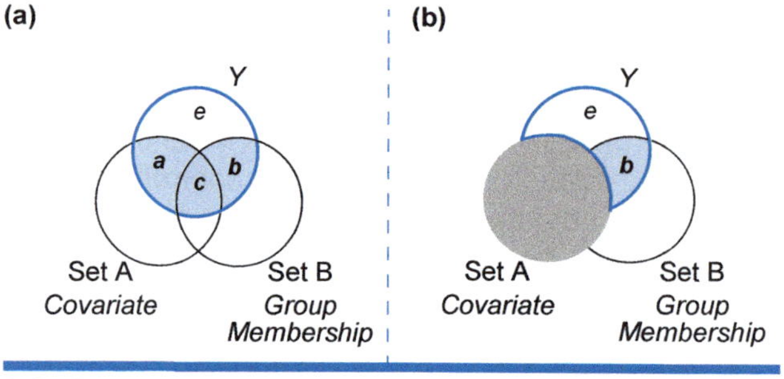

FIGURE 9.1 Graphical illustration of the concept "controlling for the effect of a covariate."

both represent the amount of *Y* variance uniquely explained by an IV. The difference between the two is that with pr^2, the effects of all the other IVs in the regression model are removed *completely* from *the IV under discussion and the DV*, whereas with sr^2, the effects of all the other IVs are removed only from the IV under discussion. (This is why pr^2 is referred to as the squared *full* partial correlation coefficient.) The reason for this reintroduction is because the concept of ANCOVA may be graphically illustrated via the concept of pr^2, which we presented in Figure 3.7. To reflect an ANCOVA perspective, Figure 9.1 contains a modified version of Figure 3.7, where *Y*, denoted by the blue circle, is the dependent variable, set A represents the covariate, and set B represents the coded group membership variable.

Focusing on Figure 9.1(a), regions *a*, *b*, *c*, and *e* represent 100% of the variance in *Y*: regions *a*, *b*, and *c* reflect the amount of variance in *Y* being explained collectively by the covariate and group membership sets, and region *e* is the amount of *unexplained variance* in *Y* (i.e., *e* is the residual, or error.) Also observe in Figure 9.1(a) that the covariate set affects the DV via regions *a* and *c*, and group membership via region *c*. As a result, to remove the effect of the covariate requires removing set A completely from the model.

Now focus on Figure 9.1(b) where we conceptually remove the covariate set by blackening out set A. (The same effect can be obtained by placing your finger tip over the entire circle for set A.) When doing this, the variance in the DV that was being explained by the covariate is now completely removed. Also observe in Figure 9.1(b) that the circle representing the variability in *Y* is no longer a full circle but instead is crescent shape and only two regions remain: region *b*, which is the amount of *Y* variance being explained by group membership, and region *e*, which is the amount of unexplained *Y* variance.

Because the influence of the covariate has been removed *completely* from both the DV and group membership, region *b* provides a truer accountability of the group membership's contribution to explaining *Y* variance. The reader might recall from Chapter 3 that this reflects the squared (full) partial correlation coefficient:

$$pr^2 = \frac{b}{b + e}$$

More specifically, when applied to Figure 9.1(b), when *controlling for the effect of the covariate*, the amount of *Y* variance explained by group membership is

$$pr_{\text{B}}^2 = \frac{R_{\text{Y•AB}}^2 - R_{\text{YA}}^2}{1 - R_{\text{YA}}^2} = \frac{sR_{\text{B}}^2}{1 - R_{\text{YA}}^2}$$

Thus, ANCOVA completely removes (or *partials out*) the effect the covariate has on both the DV and group membership. Many statistics programs provide an ANCOVA routine as part of their ANOVA package, and this routine performs the analysis automatically based on user inputs. Rather than demonstrate how to use an ANCOVA routine, we instead present an example of how to conduct and interpret the results of an ANCOVA via MRC.

Conducting an ANCOVA via MRC: The Research Setting

To introduce how to conduct an ANCOVA via an MRC analysis, we present a modified version of the weather knowledge study we presented earlier except now we are interested in examining the effectiveness of four instructional strategies for teaching aviation meteorology: classroom lectures, case study, independent study, and simulation. These instructional strategies are defined as follows: (a) *classroom lecture* involves traditional didactic instruction and is the control group, (b) *case study* involves students reading and discussing actual case studies of weather-related events pilots experienced as published in the Aviation Safety Reporting System (ASRS) database, (c) *independent study* involves students learning about weather by engaging in self-study under the direction of their teacher, and (d) *simulation* involves placing students in a flight simulator and exposing them to various flight scenarios comprising many different weather events.

The DV is Y = scores on the Test of Basic Weather Knowledge for Aviators, which consists of 40 multiple-choice items prepared by an aviation meteorologist and designed to assess operational weather knowledge that should be mastered, at a minimum, by VFR-only pilots. All items are dichotomously scored and therefore scores could range from 0 to 40, with higher scores reflecting a higher understanding of weather knowledge. The study will be conducted during a 5-week unit on aviation weather, and the goal is to compare the respective mean scores on the DV for the case study, independent study, and simulation groups to that of the classroom group.

To mitigate any potential scheduling conflicts, students are permitted to self-select the section they want to enroll in, which makes the research design *quasiexperimental*. However, because of this self-selection process, it is possible that students' prior weather knowledge might differ significantly across the groups, which is a *group-equivalency* issue. To control for this, we will administer the Test of Basic Weather Knowledge for Aviators as a *pre-assessment* prior to the start of the 5-week unit. Although this will address the group-equivalency issue, the administration of a pre-assessment represents a *testing threat to internal validity*. If we do not control for this threat, then it could have an impact on the final results because it would be unclear if the results of the post-assessment truly reflect differences in the instructional approaches or if they are due to students' pre-exposure to the assessment items.

So, although we are not interested in how the groups fare on the pre-assessment except from a group-equivalency perspective, we need to control for the effect of the pre-assessment. This is where ANCOVA is an appropriate strategy. By treating the pre-assessment as a covariate, we will be able to statistically control for the effect the pre-assessment has on the post-assessment scores, which is the dependent variable. As a result, the overall RQ is, "*When controlling for the effect of the pre-assessment*, to what extent do the respective post-assessed group mean scores for case study, independent study, and simulation differ from the post-assessed mean score for classroom lecture?"

A copy of the corresponding data set, which consists of hypothetical data, is given in the file, "Ch_9 ANCOVA Example Data." Based on the given research description, the DV is Y = Weather Knowledge Scores; the covariate is X_1 = Pre-Assessment, which are participants' pre-assessed weather knowledge scores measured prior to implementing the study; and X_2 = Group Membership, which is coded using a dummy coding strategy with D_1 = Case Study, D_2 = Independent Study, D_3 = Simulation, and "Classroom" as the reference group. Furthermore, and with respect to Figure 9.1, set A is equal to $\{X_1\}$ and set B = $\{D_1, D_2, D_3\}$.

To confirm group equivalency, we run a single-factor ANOVA for the pre-assessment scores. As reported in Table 9.6, the results of this analysis yield a nonsignificant model, R^2 = .04, $F(3, 56)$ = 0.8638, p = .4653. Thus, the four groups are statistically equivalent with respect to their knowledge of aviation weather prior to implementing the treatment, which is teaching the 5-week unit on aviation meteorology via four different instructional strategies. Group equivalency is further confirmed from Tukey's HSD Post Hoc Comparisons subtable in Table 9.6.

Conducting an ANCOVA via MRC: Checking Assumptions

Before conducting the analysis, we first check to see if the assumptions are met. When conducting an ANCOVA via MRC, the assumptions include those for any MRC analysis. Recall from Chapter 5 that these assumptions are *linearity*, *equal variances* (homoscedasticity of the residuals), *independence* of the residuals, and *normality* of the residuals. In addition, we also engage in preliminary data screening by examining the data set for *outliers*, *multicollinearity*, *missing data*, *measurement error*, and *relevant IVs* and make any adjustments as needed. For the given data set, we note the following when we simultaneously regress Y on X_1, D_1, D_2, D_3:

- Preliminary data screening yields (a) no outliers, (b) no multicollinearity (all *VIF*s < 1.6), (c) no missing data, (d) a reliability factor of α = .929 based on the Kudor-Richardson 21 formula for dichotomously scored data, and (e) all IVs are relevant based on their respective added variable plots. (*Note:* A *KR*-21 calculator is available online at https://www.easycalculation.com/statistics/kuder-richardson-formula21-calculator.php.)
- All regression assumptions are satisfied: (a) linearity and equal variances are confirmed by a scatter plot of the residuals by predicted values; (b) independence is confirmed by a scatter plot of residuals by case numbers; and (c) normality is confirmed by a *q-q* plot of the residuals coupled with the results of the Shapiro-Wilk Goodness of Fit test, W = .988, p = .8298.

Independent of the MRC assumptions, there are two assumptions specific to ANCOVA. The first is with respect to the data types: the DV must be a single, continuous variable; the covariate(s) must be continuous; and group membership, which

TABLE 9.6 *Summary of Single-Factor ANOVA and Follow-up Results to Confirm Group Equivalency for the Data Set Given in the File "Ch_9 ANCOVA Example Data"*

ANOVA

Source	*SS*	*df*	*MS*	*F*	*p*
Between	44.18333	3	14.7278	0.8638	.4653
Within (Error)	954.80000	56	17.0500		
Total	998.98333	59			

Note. $N = 60$. $\eta^2 = \frac{SS_{\text{Between}}}{SS_{\text{Total}}} = \frac{44.18333}{998.98333} = .0442$.

Group Means and Standard Deviations

Group		*N*	*M*	*SD*
Case Study		15	19.93	3.94
Independent Study		15	20.27	4.92
Simulation		15	19.00	3.89
Classroom		15	18.07	3.65
	Overall	60	19.32 [a]	4.11

Note. [a] Unweighted mean.

Tukey's HSD Post Hoc Comparisons

Group Comparisons	M_{Diff}	$t(56)$	95% CI	*p*
Case Study – Ind. Study	–0.333	–0.221	[–4.33, 3.66]	.9961
Case Study – Simulation	0.933	0.619	[–3.06, 4.93]	.9256
Case Study – Classroom	1.867	1.238	[–2.13, 5.86]	.6056
Ind. Study – Simulation	1.267	0.840	[–2.73, 5.26]	.8351
Ind. Study – Classroom	2.200	1.459	[–1.79, 6.19]	.4687
Simulation – Classroom	0.933	0.619	[–3.06, 4.93]	.9256

is the research factor, must be a coded nominal scale variable. The reader will note that the given data set is compliant with this assumption.

ANCOVA also has an underlying presumption known as homogeneity of regression, which is a component of MRC's linearity assumption. The homogeneity of regression assumption requires that the relationship between the covariate and the DV be the same for all values of the group membership variable. From an analysis perspective, this implies that the *interaction* between the covariate and the group membership variable relative to the dependent variable is not significant. Although we discuss interactions in Section 9.4, the reader might recall from factorial ANOVA that interactions are carried by product terms: Each numerical value of one IV is multiplied by each numerical value of a second IV (for each individual case) to produce a third variable, which is the product of the two IVs. So, with respect to our current example, the interaction between the covariate (X_1) and the group membership variable, which is represented by the dummy coded IVs, D_1, D_2, and D_3, is the product of these variables, namely, X_1D_1, X_1D_2, and X_1D_3. The homogeneity of regression assumption may be confirmed statistically, graphically, or algebraically. A discussion of each approach follows.

TABLE 9.7 *Summary of Results of the Hierarchical Regression Analysis for the Data Set Given in the File "Ch_9 ANCOVA Example Data"*

Sets/IVs [a]	ΔR^2	Set-Wise F *df*	*F*	*p*	Coefficient and Constants B_i	t_i	p_i
Step 1: Covariate	.294	1, 58	24.13	< .0001	$B_0 = 15.98$	$t_0 = 4.96$	$p_0 < .0001$
					$B_1 = 0.80$	$t_1 = 4.91$	$p_1 < .0001$
"A" = { }							
"B" = set A = $\{X_1\}$							
$R^2_{Y \cdot A} = .294$							
$F(1, 58) = 24.13, p < .0001$							
***Step 2: Covariate* + Group**	.416	3, 55	26.30	< .0001	$B_0 = 14.52$	$t_0 = 6.70$	$p_0 < .0001$
					$B_1 = 0.94$	$t_1 = 8.54$	$p_1 < .0001$
"A" = set A = $\{X_1\}$					$B_2 = -2.15$	$t_2 = -1.71$	$p_2 = .0924$
"B" = set B = $\{D_1, D_2, D_3\}$					$B_3 = -6.73$	$t_3 = -5.33$	$p_3 < .0001$
$R^2_{Y \cdot AB} = .710$					$B_4 = 4.19$	$t_4 = 3.37$	$p_4 = .0014$
$F(4, 55) = 33.75, p < .0001$							
Step 3: Covariate + Group + Interaction	.019	3, 52	1.22	.3117	$B_0 = 12.08$	$t_0 = 2.66$	$p_0 = .0104$
					$B_1 = 1.07$	$t_1 = 4.35$	$p_1 < .0001$
"A" = sets A, B = $\{X_1, D_1, D_2, D_3\}$					$B_2 = -3.90$	$t_2 = -0.60$	$p_2 = .5507$
"B" = set C = $\{X_1D_1, X_1D_2, X_1D_3\}$					$B_3 = -4.65$	$t_3 = -0.78$	$p_3 = .4370$
$R^2_{Y \cdot ABC} = .729$					$B_4 = 13.51$	$t_4 = 2.12$	$p_4 = .0392$
$F(7, 52) = 20.00, p < .0001$					$B_5 = 0.08$	$t_5 = 0.22$	$p_5 = .8238$
					$B_6 = -0.12$	$t_6 = -0.38$	$p_6 = .7041$
					$B_7 = -0.50$	$t_7 = -1.47$	$p_7 = .1479$

Note. $N = 60$. Y = Post-Assessed Weather Test Scores. The boxed results represent the IVs entering the analysis at each step. If the corresponding IVs in the entering set are significant, then a second color is used for the B_i and t_i values in addition to the p_i values. F and p values are from Soper (2025a and 2025c).

[a] Set A = represents the set in the previous model. Set B represents the set entering the model at the current step of the analysis. X_1 = Pre-assessed Weather Knowledge Scores. Set B = the dummy coded group membership variable where D_1 = Case Study, D_2 = Independent Study, D_3 = Simulation, and the reference group is Classroom Instruction.

Confirming the Homogeneity of Regression Assumption Statistically. To confirm the homogeneity of regression assumption statistically, we conduct a hierarchical regression using sets of variables as we did in Chapter 7. More specifically, we let set A equal the covariate(s), set B equal the group membership variable, and set C equal the interaction between the covariate and the coded group membership variables. If the increment of the interaction set, sR^2_C, is not significant, then the homogeneity of regression assumption is satisfied. Table 9.7 contains a summary of this analysis as applied to our current example.

Focusing on Step 3 in Table 9.7, the reader will note that the increment the interaction set makes is $sR^2_C = R^2_{Y \cdot ABC} - R^2_{Y \cdot AB} = .729 - .710 = .019$. Thus, when examined in the presence of the covariate and group membership (i.e., when holding these IVs constant), the interaction between the covariate and the coded group membership variables explains an additional 1.9% of the variance in the post-assessed weather knowledge scores. Consulting Soper (2025a, 2025b, 2025c), this predictive gain is *not* significant, $F(3, 52) = 1.22$, $F_{\text{Critical}}(3, 52) = 2.78$, $p = .3117$, and therefore the homogeneity of regression assumption is met. (*Note:* We will return to Table 9.7 to discuss Steps 1 and 2 after we complete our presentation on the homogeneity of regression assumption.)

Confirming the Homogeneity of Regression Assumption Graphically. The homogeneity of regression assumption may be confirmed graphically by using the graphing routine of a statistics program to examine the scatter plot where the DV is on the vertical axis, the covariate is on the horizontal axis, and group membership is the overlay. This plot is given in Figure 9.2 with respect to our current example. In Figure 9.2(a), the scatter plot contains the observed scores and corresponding regression line for each group, whereas in Figure 9.2(b), we removed the observed scores so we can focus solely on the respective regression lines.

As illustrated in Figure 9.2(b), none of the regression lines "cross" within the scope of the sample. Although some readers might argue (correctly) that the regression line for Simulation eventually will cross with the regression lines for Classroom and Case Study, this will occur outside the scope of the observed scores in the sam-

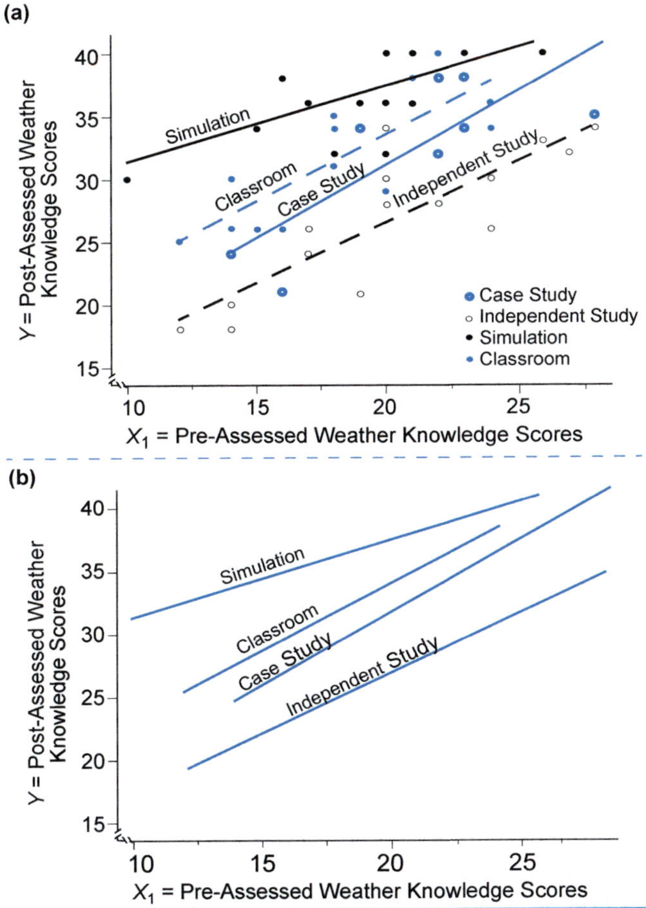

FIGURE 9.2 Graphical illustration that shows compliance with the homogeneity of regression assumption for ANCOVA with respect to the running example on instructional strategies.

ple, which for X_1 ranges from 10 to 28. Because the lines do not cross, this means that with respect to the DV *the levels of one factor (group membership) are the same across the levels of the second factor (pre-assessment scores)*. This implies that differences in group means are consistent across pre-assessment scores. For example, regardless of participants' pre-assessment scores: (a) the Simulation group's post-assessed scores, on average, are higher than those of the other groups; (b) the Classroom group's post-assessed scores, on average, are higher than those of the Case Study and Independent Study groups; and (c) the Case Study group's post-assessed scores are higher than those of the Independent Study group.

Applying this to the homogeneity of regression assumption, Figure 9.2 confirms graphically that *the relationship between the covariate and the DV is the same for all values of the nominally-coded research factor*. Given this graphical perspective, it is easy to see why the homogeneity of regression assumption is often referred to as parallelism of slopes: Because none of the lines cross within the range of the covariate, the lines are considered "parallel," even if their slopes might be slightly different. Thus, if the graphs of the regression lines do not intersect within the range of the covariate, then the homogeneity of regression assumption is met.

Confirming the Homogeneity of Regression Assumption Algebraically. To confirm the homogeneity of regression assumption algebraically, we derive the equations of the respective regression lines and then examine their slopes. If the slopes are equal or similar, then we presume the lines do not cross. However, it is prudent to graph the equations of the lines to confirm that they do not cross.

To derive the equations of the regression lines, we begin with the regression equation of the full model, which contains the covariate, group membership, and interaction sets. As reported in Table 9.7 / Step 3, this equation is $\hat{Y} = 1.07X_1 - 3.90D_1 - 4.65D_2 + 13.51D_3 + 0.08X_1D_1 - 0.12X_1D_2 - 0.50X_1D_3 + 12.08$. We then take advantage of the data type restriction of ANCOVA. Because ANCOVA requires a group membership variable to be the research factor, we are able to derive the regression equations by substituting into the full model the numerical codes for the coded IVs representing the nominal variable and completing the algebra. This is demonstrated below for each instructional group.

Case Study ($D_1 = 1, D_2 = 0, D_3 = 0$)

$$
\begin{aligned}
\hat{Y} &= 1.07X_1 - 3.90D_1 - 4.65D_2 + 13.51D_3 + 0.08X_1D_1 - 0.12X_1D_2 - 0.50X_1D_3 + 12.08 \\
&= 1.07X_1 - 3.90(1) - 4.65(0) + 13.51(0) + 0.08X_1(1) - 0.12X_1(0) - 0.50X_1(0) + 12.08 \\
&= 1.07X_1 - 3.90 + 0.08X_1 + 12.08 \\
&= 1.07X_1 + 0.08X_1 - 3.90 + 12.08 \\
&= 1.15X_1 + 8.18
\end{aligned}
$$

Therefore, the regression equation for Case Study is $\hat{Y}_{\text{Case Study}} = 1.15X_1 + 8.18$.

Independent Study ($D_1 = 0, D_2 = 1, D_3 = 0$)

$$
\begin{aligned}
\hat{Y} &= 1.07X_1 - 3.90D_1 - 4.65D_2 + 13.51D_3 + 0.08X_1D_1 - 0.12X_1D_2 - 0.50X_1D_3 + 12.08 \\
&= 1.07X_1 - 3.90(0) - 4.65(1) + 13.51(0) + 0.08X_1(0) - 0.12X_1(1) - 0.50X_1(0) + 12.08 \\
&= 1.07X_1 - 4.65 - 0.12X_1 + 12.08 \\
&= 1.07X_1 - 0.12X_1 - 4.65 + 12.08 \\
&= 0.95X_1 + 7.43
\end{aligned}
$$

Therefore, the regression equation for Ind. Study is $\hat{Y}_{\text{Ind. Study}} = 0.95X_1 + 7.43$.

Simulation (D_1 = 0, D_2 = 0, D_3 = 1)

$$\begin{aligned}\hat{Y} &= 1.07X_1 - 3.90D_1 - 4.65D_2 + 13.51D_3 + 0.08X_1D_1 - 0.12X_1D_2 - 0.50X_1D_3 + 12.08\\ &= 1.07X_1 - 3.90(0) - 4.65(0) + 13.51(1) + 0.08X_1(0) - 0.12X_1(0) - 0.50X_1(1) + 12.08\\ &= 1.07X_1 + 13.51 - 0.50X_1 + 12.08\\ &= 1.07X_1 - 0.50X_1 + 13.51 + 12.08\\ &= 0.57X_1 + 25.59\end{aligned}$$

Therefore, the regression equation for Simulation is $\hat{Y}_{\text{Simulator}} = 0.57X_1 + 25.59$.

Classroom (D_1 = 0, D_2 = 0, D_3 = 0)

$$\begin{aligned}\hat{Y} &= 1.07X_1 - 3.90D_1 - 4.65D_2 + 13.51D_3 + 0.08X_1D_1 - 0.12X_1D_2 - 0.50X_1D_3 + 12.08\\ &= 1.07X_1 - 3.90(0) - 4.65(0) + 13.51(0) + 0.08X_1(0) - 0.12X_1(0) - 0.50X_1(0) + 12.08\\ &= 1.07X_1 + 12.08\end{aligned}$$

Therefore, the regression equation for Classroom is $\hat{Y}_{\text{Classroom}} = 1.07X_1 + 12.08$.

Observe from these derivations that each equation is in *slope-intercept* form, $y = mx + b$, where m is the slope and b is the y coordinate of the y intercept $(0, y)$. Further observe that the respective slopes are not equal, but they are "close" to each other:

Group	Slope
Case Study	1.15
Ind. Study	0.95
Simulation	0.57
Classroom	1.07

With the exception of Simulation, the differences among the other slopes are less than 0.2 units, which suggests that the regression lines for these three groups are nearly parallel. For Simulation, though, its slope differs from the slopes of the other three groups by 0.58 units, 0.38 units, and 0.50 units, respectively, which suggests that the regression line for Simulation might cross one of the other lines.

To manually graph these lines, we (a) substitute values for X_1 into each equation and solve for $\hat{Y}$, (b) plot the resulting $(x_1, \hat{y})$ points on the x–y axis, and (c) draw a line through the points. Although mathematically, exactly one and only one straight line can be drawn through two given points, we plot a third point as a check to confirm that all three points lie on the same line. Figure 9.3 shows the graphs of these equations based on the three points for X_1 = 0, 10, and 20. Note from Figure 9.3 that these graphs are consistent with those generated by our statistics program's graphing routine (as they should be). Further note that although the Simulation line intersects the Case Study line and nearly intersects the Classroom line, these occur *outside* the range of the sample data, which for X_1 is between 10 and 28. Therefore, the homogeneity of regression assumption is satisfied.

Conducting an ANCOVA via MRC: The Valid ANCOVA Model

Assuming all other assumptions are satisfied, if the homogeneity of regression assumption also is met, then we discard the nonsignificant interaction set (set C) and focus exclusively on the model that consists of the covariate and group membership variables. This model is now considered the valid ANCOVA model and becomes the primary focus of the analysis. For our current example, the model from Step 2 of the hierarchical regression from Table 9.7 is the valid ANCOVA model and

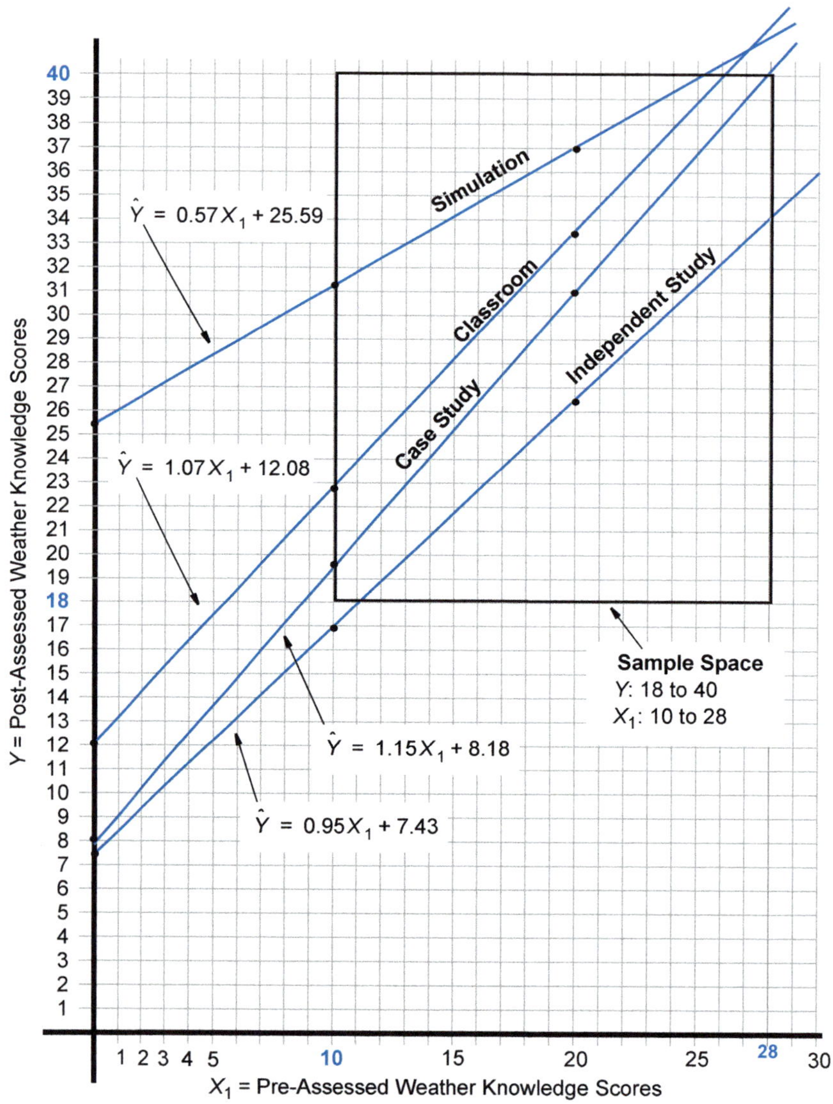

FIGURE 9.3 Graphs of the regression equations derived algebraically by substituting into the full model the numerical codes of the coded IVs that represent the group membership variable for the current example.

the regression equation is, $\hat{Y} = 0.94X_1 - 2.15D_1 - 6.73D_2 + 4.19D_3 + 14.52$. Furthermore, as reported in Table 9.7 / Step 2, $R^2_{Y\cdot AB} = .710$, $F(4, 55) = 33.75$, $p < .0001$, and $sR^2_B = .416$, $F(3, 55) = 26.30$, $p < .0001$, which indicates that when holding the covariate constant, group membership provides a significant predictive gain in explaining the variance in the post-assessed weather knowledge scores. In other words, the valid ANCOVA model reflects a significant *main effect* with respect to the means of the four instructional strategies.

Given a significant main effect, we may now report the differences in group means relative to the coding scheme. Because the overall purpose of ANCOVA is to *control for the effect of the covariate*, this entails reporting the *adjusted group means* as well as the corresponding pr^2. However, we cannot simply substitute the numerical codes for the coded IVs into this regression equation because it contains the covariate, which is a continuous IV. The reader will recall, though, that the work-around to this issue is to center the continuous IV. For our current example, when we center the covariate, X_1, the regression equation is, $\hat{Y} = 0.94X_1\text{–Cntr} - 2.15D_1 - 6.73D_2 + 4.19D_3 + 32.64$, and we may now calculate the adjusted group means by substituting the codes into the equation and performing the arithmetic. Table 9.8 contains a summary of the adjusted group means.

As reported in the Group Means subtable of Table 9.8, when we remove the effect of exposure to the pre-assessment and compare the *adjusted means* to the *unadjusted means*: Case Study decreases by 0.58 points, Independent Study decreases by 0.89 points, Simulation increases by 0.3 points, and Classroom increases by 1.17 points. Based on these differences, it appears that exposure to the pre-assessment had the greatest influence on the Classroom group. Now look

TABLE 9.8 *Summary of Results for Group Comparisons Relative to the Valid ANCOVA Model*

Group Means and Standard Deviations

Group	$M_{\text{Pre-Assess}}$ [a]	$M_{\text{Unadjusted}}$ [b]	M_{Adjusted} [c]	M_{Diff} [d]
Case Study	19.93	31.07	30.49	– 0.58
Independent Study	20.27	26.80	25.91	–0.89
Simulation	19.00	36.53	36.83	+0.30
Classroom	18.07	31.47	32.64	+1.17

Note. $N = 60$; $n = 15$ per group. Y = Scores on the 40-item dichotomously scored Test of Basic Weather Knowledge for Aviators. $\hat{Y} = 0.94X_1\text{-Cntr} - 2.15D_1 - 6.73D_2 + 4.19D_3 + 32.64$.

[a] $M_{\text{Pre-Assess}}$ = mean pre-assessment scores (X_1) used to confirm group equivalency a priori. [b] $M_{\text{Unadjusted}}$ = mean Y scores unadjusted for the influence of the pre-assessment. [c] M_{Adjusted} = mean Y scores when freed of the pre-assessment effect. [d] $M_{\text{Diff}} = M_{\text{Adjusted}} - M_{\text{Unadjusted}}$.

Coded vs. Reference Group Comparisons Relative to Unadjusted Means

Group Comparisons	M_{Diff}	SE	$t(56)$	95% CI	p
Case Study – Classroom	–0.40	1.87	–0.21	[–5.36, 4.56]	.997
Ind. Study – Classroom	–4.67	1.87	–2.49	[–9.63, 0.29]	.072
Simulation – Classroom	5.07	1.87	2.71	[0.11, 10.03]	.043

Note. $N = 60$; $n = 15$ per group. Comparisons based on Tukey's HSD.

Coded vs. Reference Group Comparisons Relative to Adjusted Means

Group Comparisons	M_{Diff}	SE	$t(56)$	95% CI	Cohen's d	p
Case Study –	–2.15	1.26	–1.71	[–5.48, 1.18]	–0.63	.327
Ind. Study – Classroom	–6.73	1.26	–5.33	[–10.07, –3.39]	–1.98	< .001
Simulation – Classroom	4.19	1.24	3.37	[0.90, 7.49]	1.23	.007

Note. $N = 60$; $n = 15$ per group. Comparisons and p values based on Tukey's HSD, which resulted from our statistics program's ANCOVA routine.

at the last two subtables in Table 9.8 where comparisons are made between the groups represented by coded IVs and the reference group, first for unadjusted means, and then for adjusted means (see also Table 9.7).

- Case Study's unadjusted mean score is 0.40 points *lower* than Classroom's unadjusted mean score and this difference is not significant, $t(56) = -0.21$, $p = .997$. When freed of the covariate's influence, though, Case Study's adjusted mean score is now 2.15 points *lower* than Classroom's adjusted mean score, and this difference also is not significant, $t(56) = -1.71$, $p = .327$.
- Independent Study's unadjusted mean score is 4.67 points *lower* than Classroom's unadjusted mean score and this difference is not significant, $t(56) = -2.49$, $p = .072$. When freed of the covariate's influence, though, Independent Study's adjusted mean score is now 6.73 points *lower* than Classroom's adjusted mean score and this difference *is* significant, $t(56) = -5.33$, $p < .001$.
- Simulation's unadjusted mean score is 5.07 points *higher* than Classroom's unadjusted mean score and this difference is significant, $t(56) = 2.71$, $p = .043$. When freed of the covariate's influence, though, Simulation's adjusted mean score is now 4.19 points higher than Classroom's adjusted mean score and this difference also is significant, $t(56) = 3.37$, $p = .007$.

The reader should also note that the respective 95% CIs are more precise (narrower width) for the adjusted mean differences than they are for the unadjusted mean differences and hence the former have higher accuracy in parameter estimates (AIPEs). Based on these results, the answer to the study's RQ with respect to the effectiveness of each "alternative" instructional strategy for teaching aviation meteorology compared to traditional classroom instruction: (a) Case Study is not significantly different as an instructional strategy than that of Classroom, (b) Independent Study is significantly less effective as an instructional strategy than Classroom, and (c) Simulation is significantly more effective as an instructional strategy than Classroom.

Lastly, we also report the effect size of the valid ANCOVA model. As presented earlier, this is equal to the squared (full) partial correlation coefficient:

$$pr_B^2 = \frac{R^2_{Y \cdot AB} - R^2_{Y \cdot A}}{1 - R^2_{Y \cdot A}} = \frac{sR_B^2}{1 - R^2_{Y \cdot A}} = \frac{.416}{1 - .294} = \frac{.416}{.706} = .589$$

Thus, 58.9% of the variance in the post-assessed weather knowledge scores, when adjusted for the effect of the covariate, is being explained by the "treatment condition," which is comprised of the four instructional strategies.

Conducting an ANCOVA via MRC: A Misspecified ANCOVA Model

As presented earlier, the purpose for conducting an ANCOVA is to control for the effect of a covariate. In doing so, though, there is an underlying presumption that the covariate will not have a significant interaction with group membership. Unfortunately, though, we will not know if this presumption holds until after we check for the homogeneity of regression assumption. This begs the question, "What do we do if this assumption is *not* satisfied?"

To answer this question, we present our running example involving instructional strategies for teaching aviation meteorology, but we modified the data set so that the homogeneity of regression assumption is not satisfied. A copy of the data set is given in the file, "Invalid ANCOVA Example Data." Focusing solely on the increment of the interaction set, we note the following:

- $R^2_{Y \bullet AB} = .619$, $F(4, 55) = 22.3792$, $p < .0001$.
- $R^2_{Y \bullet ABC} = .675$, $F(7, 52) = 15.3969$, $p < .0001$.
- $sR^2_C = .675 - .619 = .056$. Consulting Soper (2025a, 2025b, 2025c), this predictive gain is significant, $F(3, 52) = 2.92$, $F_{Critical}(3, 52) = 2.78$, $p = .0426$.

As a result, the homogeneity of regression assumption is not satisfied.

When confronted with such a situation where all other assumptions of ANCOVA are met except the homogeneity of regression assumption, the ANCOVA model is considered *invalid*, or *misspecified*, and ANCOVA is no longer the appropriate statistical strategy for analyzing the data set. Instead, we must change the statistical strategy so that the focus is on examining the interaction. This involves revising the RQs/hypotheses to be commensurate with an interaction study, and analyzing the full model, which includes the covariate, the coded IVs representing group membership, and the interaction terms. In short, what starts out as a covariance study ends up becoming an *interaction study*, which is what we discuss next.

9.4 Examining Interactions

The Concept of Interactions

As part of our ANCOVA presentation, we briefly, and informally, introduced the concept of interactions by presenting an example of both a valid and a misspecified ANCOVA model. In doing so, we learned that the key difference between them is that the former satisfies the homogeneity of regression assumption, but the latter does not. We also learned that compliance to this assumption requires a nonsignificant interaction between the covariate and group membership factor, which is the "treatment" in experimental research studies. When examined graphically, a valid ANCOVA model presumes that the slopes of the respective regression lines for group membership are either equal or similar, which results in either parallel lines or lines that do not cross within the range of the covariate.

The key to understanding the concept of interactions is grounded in the question: *Do the levels of one factor operate the same or differently across the levels of a second factor with respect to the DV?* If the answer is "the same," then there is no interaction, but if the answer is "differently," then an interaction is present, which means that one factor is influencing the relationship between the other factor and the DV. To facilitate an understanding of this concept, let's examine the *interaction plots* illustrated in Figure 9.4.

The Plots in Figures 9.4(a), 9.4(b), and 9.4(c). For these three plots, the reader is advised of the following: (a) the plots reflect hypothetical situations involving GA pilots; (b) the plots consist of *line segments*, and the endpoints of each segment, shown as closed circles, reflect the mean score for each group; and (c) the plots reflect the possible results from examining a 2 × 2 *nominal-by-nominal interaction*, which traditionally is examined via factorial ANOVA. Later in this section we will examine a 2 × 2 factorial ANOVA via an MRC analysis.

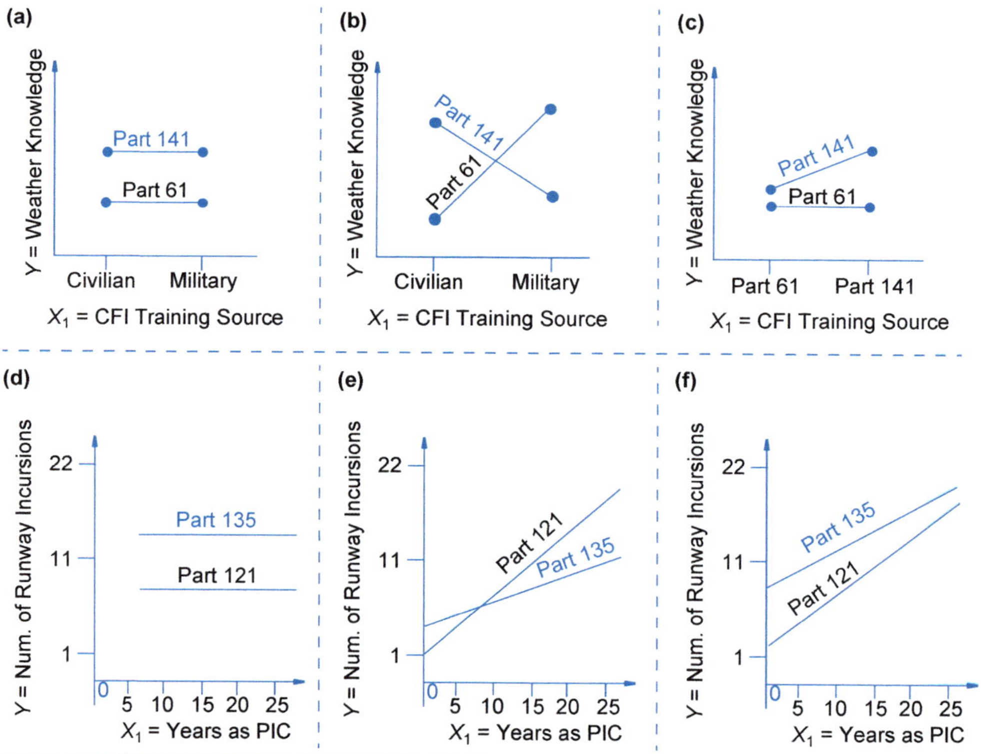

FIGURE 9.4 Graphical illustrations of different types of interactions. (*Note:* PIC = Pilot-in-Command.)

The research context involves two factors, X_1 = CFI Training Source and X_2 = Flight Training Program, and each factor has two levels. The two CFI training sources are Civilian and Military, and the two flight training programs are Part 61 and Part 141. The DV is Y = Weather Knowledge. For this discussion, our interest is in understanding the effect CFI Training Source has on the relationship between Flight Training Program and the DV. Thus, the underlying research question associated with this interest is, "Do the levels of Flight Training Program operate the same or differently with respect to weather knowledge scores under the levels of CFI Training Source?" A more succinct way in which to pose this question is, "What is the effect of X_1 on the X_2–Y relationship?" Following is a brief discussion of the possible scenarios for answering the RQ.

- In Figure 9.4(a), the line segments are parallel, which denotes zero interaction. Based on the way the line segments are structured, we can see that the relationship between Flight Training Program and the DV does not change across the two levels of X_1. More specifically, regardless of the source of CFI training (civilian or military), GA pilots who receive flight training under a Part 141 program *always* have higher weather knowledge than GA pilots who receive flight training under a Part 61 program. As a result, *the two levels of Flight Training Program are operating the same with respect to weather knowledge across the two levels CFI Training Source*, and therefore X_1 has no effect on the X_2–Y relationship.

- In Figure 9.4(b), the line segments "cross," which denotes a disordinal interaction. This means that the two levels of Flight Training Program are operating *differently* with respect to weather knowledge across the two levels of CFI Training Source. More specifically, this plot shows that GA pilots from a Part 141 flight training program have higher weather knowledge if their CFIs receive civilian-based training, whereas GA pilots from a Part 61 program have higher weather knowledge if their CFIs receive military-based training. As a result, X_1 influences, or moderates, the X_2–Y relationship and is therefore considered a moderating variable in this context because it has a substantial effect on the effectiveness of the flight training program with respect to weather knowledge.
- In Figure 9.4(c), the line segments are neither parallel nor "cross," which denotes an ordinal interaction. Because the line segments do not cross, the result is the same as that of a zero interaction: the levels of one factor are operating *the same* under the levels of the second factor with respect to the DV. Similar to the zero interaction model, X_1 has no effect on the X_2–Y relationship because regardless of their CFIs' training source, GA pilots from a Part 141 flight training program *always* have higher weather knowledge vs. GA pilots from a Part 61 program.

The Plots in Figures 9.4(d), 9.4(e), and 9.4(f). For these three plots, the reader is advised of the following: (a) the plots reflect hypothetical situations involving Part 121 air carrier pilots and Part 135 air charter pilots; (b) the plots are the result of an MRC analysis where Y is regressed on X_1, X_2, and their interaction (X_1X_2), and therefore the lines are regression lines derived from the regression equation; and (c) the plots reflect the possible results from examining a *nominal-by-continuous interaction*, which we will discuss shortly.

The research context consists of two factors, X_1 = Years as PIC, which is a continuous variable and ranges from 0 (less than 1 year) to 25, and X_2 = Type of Operations, a nominal factor with two levels, Part 121 and Part 135. The DV is Y = Number of Runway Incursions, which ranges from 0 to 22. For this discussion, we presume that as years as PIC increase, the number of runway incursions also will increase because pilots will have more opportunities to be involved in one. (*Note*: An argument also can be made for an inverse relationship.) We also believe that the relationship between X_1 and Y might be moderated by the type of operations (air carrier vs, air charter) and therefore the research objective is to examine the interaction between X_1 and X_2. More specifically, and using the language we developed previously, our interest is in understanding the extent to which X_2 = Type of Operations moderates the X_1–Y relationship: Does the relationship between Years as PIC and the number of runway incursions differ for Part 121 and Part 135 pilots?

- In Figure 9.4(d), the regression lines are parallel and hence there is *zero interaction*. This means that the relationship between X_1 = Years as PIC and X_2 = Type of Operations is consistent across the range of X_1. More specifically, throughout the range of X_1, Part 135 pilots *always* have more runway incursions, on average, than Part 121 pilots, and therefore X_2 is not a moderating variable because it has no effect on the X_1–Y relationship. Furthermore, because the plot consists of regression lines, this finding also implies that the regression coefficients (B_i), which are slopes, for the respective regression equations for Part 121 and Part 135 are equal.

- In Figure 9.4(e), the lines cross indicating a *disordinal interaction*, which implies that the relationship between X_1 and Y changes as a function of X_2. This change in the relationship between X_1 and Y occurs at approximately X_1 = 7 years where the number of runway incursions differ between Part 121 and Part 135 pilots. More specifically: Part 135 pilots with fewer than 7 years as PIC have more runway incursions, on average, than Part 121 pilots, but after 7 years as PIC, Part 121 pilots have more runway incursions, on average, than Part 135 pilots. As a result, X_2 = Type of Operation is a moderating variable because it alters the X_1–Y relationship. This also implies that the slopes (B_i) of the lines are not equal.
- In Figure 9.4(f), the lines are neither parallel nor cross within the range of X_1 and therefore there is an *ordinal interaction*, which means that the relationship between X_1 and Y is consistent across the two levels of X_2: Part 135 pilots *always* have a higher number of runway incursions, on average, than Part 121 pilots regardless of the number of years as PIC. As a result, X_2 is not a moderating variable. Notice, though, that if we were to extend Years as PIC beyond the range of X_1, then the interaction eventually will become disordinal. For nominal-by-continuous interactions, this emphasizes that an interaction is considered ordinal if the point of intersection is outside the range of the continuous IV. This plot also implies that the respective slopes of the lines are not equal.

The Concept of Moderation. As implied in our descriptions of Figure 9.4(b, e), a moderating variable influences (i.e., "moderates") the relationship between an IV and the DV such that the effect of the IV on the DV changes based on the level of the moderating variable. In interaction studies that involve group membership, the nominal IV is usually considered the moderating variable. For example, in Figure 9.4(e), the moderating variable is the X_2 = Type of Operation. If, however, an interaction study involves two nominal IVs, as is the case for the disordinal interaction in Figure 9.4(b), then either IV may be considered the moderator because both have the potential to moderate the relationship between the other IV and the DV. In this latter scenario, when deciding which IV to designate the moderator, the decision should be grounded in and commensurate with the RQ. For example, in Figure 9.4(b), we designated X_1 = CFI Training Source as the moderating IV because our research interest is in understanding the effect CFI Training Source has on the relationship between Flight Training Program and the DV.

The Concept of "One Size Fits All." An underlying question inherent in all interaction studies is: *Does one size fit all?* If an interaction effect is present, then the answer is "no." For example, in Figure 9.4(b), neither civilian-trained nor military-trained CFIs is the right "fit" for teaching weather to GA pilots from a Part 61 or Part 141 flight training program. Furthermore, when considered from a diagnostic perspective, the plot in Figure 9.4(b) suggests that GA pilots from a Part 141 program are better placed with civilian-trained CFIs, whereas GA pilots from a Part 61 program are better placed with military-trained CFIs. So, here, one size does not fit all.

The same also can be said for the nominal-by-continuous plot of Figure 9.4(e). Across the range of years as PIC, the number of runway incursions is not consistent between the two types of operations. The predicted results are different for the groups depending on which side of X_1 = 7 years as PIC they are on. Thus, one size does not fit all because the X_1–Y relationship changes as a function of X_2. This concept of "one size does not fit all" is sometimes reported in the literature as *different slopes for different folks*. For example, in both Figures 9.4(b) and 9.4(e), the two lines cross resulting in a disordinal interaction.

A Final Comment. When examining interactions, the reader is reminded that although a disordinal interaction implies an interaction effect, this does not necessarily imply that the effect is statistically significant. This is similar to examining differences between group means: A difference in two group means does not necessarily imply that the difference is significant. To determine if an interaction effect is significant, we need to conduct an interaction analysis, which we present next using an MRC strategy. We restrict our presentation to *two-way interactions*, which consist of two factors, and focus only on nominal-by-nominal and nominal-by-continuous interactions. Readers interested in either *three-way interactions* or *continuous-by-continuous interactions* are directed to Jaccard et al. (2003) and Cohen et al. (2003, pp. 255–301), respectively.

Examining Nominal-by-Nominal Interactions

To motivate this discussion, we present a scaled-down study modeled after Adamson et al. (2010) that examines the effect two factors have on GA pilots' flight simulator performance. The first factor is X_1 = FAA Rating, defined as whether pilots achieved VFR or IFR proficiency, and the second factor is X_2 = Flight Training Program, defined as Part 61 or Part 141. The DV, Y = Flight Simulator Performance, is defined as a single composite score acquired from four flight tasks: (a) accuracy of executing ATC-assigned headings, altitudes, radios; (b) scanning cockpit instruments for engine malfunction (reaction time to report malfunction); (c) avoiding conflicting air traffic (flying to the diagonally opposite quadrant of the traffic and maximizing distance safely); and (d) executing a visual approach to landing (deviations from ideal course). Higher composite scores reflect higher performance. Our interest is understanding the effect X_1 has on the X_2–Y relationship, and hence X_1 is designated the moderating variable. A copy of the data set is given in the file "Ch_9 Nominal-by-Nominal Interaction Example Data."

Analyzing the Data via Factorial ANOVA. Because X_1 and X_2 are nominal factors, and each factor consists of two levels, this study reflects a traditional *2 × 2 factorial ANOVA*. Some readers might recall that this analysis includes developing a groups-mean table as well as an interaction plot. Although we will analyze the given data set via an MRC analysis, we present a summary of the results of the corresponding 2 × 2 factorial ANOVA in Figure 9.5 for comparison purposes. The reader will note the following from Figure 9.5:

- *Row effect* = 2.75. This is the difference in mean composite scores between the IFR and VFR groups independent of the second factor. Thus, the IFR group's mean score is 2.75 points higher than that of the VFR group.
- *Column effect*. This is the difference in mean composite scores between the Part 61 and Part 141 groups independent of the first factor. Thus, the Part 61 group's mean score is 1.03 points higher than that of the Part 141 group.
- *Interaction effect*. This is the numerical difference in group means between the levels of the factors, which captures the extent to which the levels of Factor B are operating the same or differently with respect to the DV under the levels of

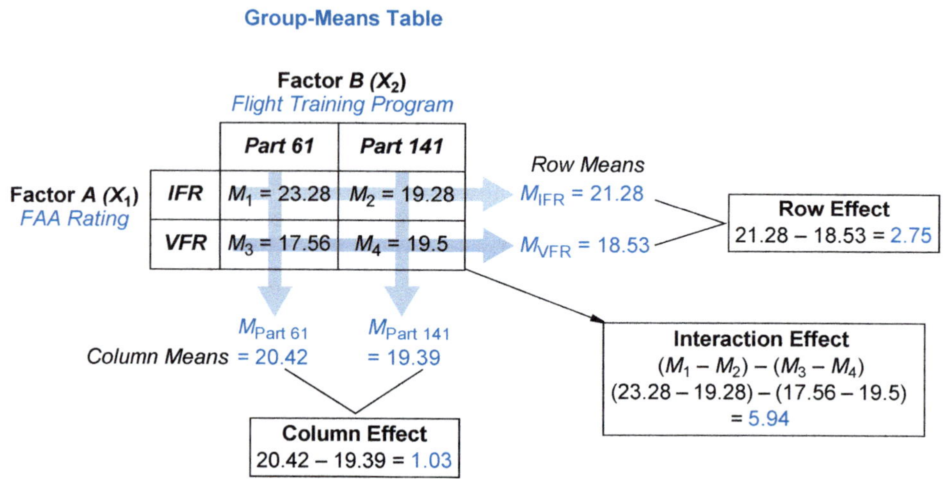

ANOVA Summary Table

Source	*SS*	*df*	*MS*	*F*	*p*
Between Groups	157.0764	3	52.3588	3.20	.0364
• *A* = FAA Rating	68.0625	1	68.0625	4.16	.0498
• *B* = Flight Training Program	9.5069	1	9.5069	0.58	.4516
• Interaction (*A* × *B*)	79.5069	1	79.5069	4.86	.0348
Within Groups (Error)	523.8333	32	16.3698		
Total	680.9097	35			

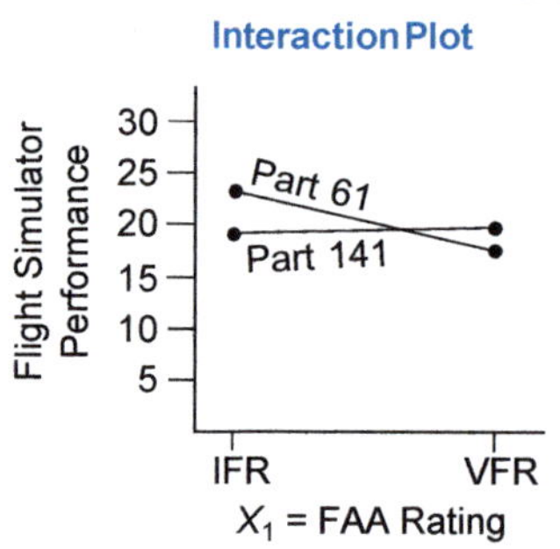

FIGURE 9.5 Results of a 2 × 2 factorial ANOVA for the data set given in the file "Ch_9 Nominal-by-Nominal Interaction Example Data."

Factor *A*. If the numerical representation of the interaction is 0, then the lines are parallel. Because it is not zero, this implies there is an interaction effect.

- *The ANOVA summary table.* The model is significant, $F(3, 32) = 3.20$, $p = .0364$.
- *The shaded part of the ANOVA summary table.* This shows the partitioning of the between-groups variance. Note that: (a) Factor *A*, the row effect, is significant, $F(1, 32) = 4.16$, $p = .0498$; (b) Factor *B*, the column effect, is not significant, $F(1, 32) = 0.58$, $p = .4516$; and (c) the interaction effect ($A \times B$) is significant, $F(1, 32) = 4.86$, $p = .0348$.

- *The interaction plot.* The interaction is disordinal, which indicates that X_1 has an effect on the relationship between X_2 = Flight Training Program and Y = Fight Simulator Performance: Pilots from a Part 61 program have higher performance if they are IFR rated, whereas pilots from a Part 141 program have higher performance if they are VFR rated.

Prior to presenting the corresponding MRC analysis, we remind the reader that the row and column effects are collectively referred to as "main effects," and that a significant main effect in the presence of a significant interaction effect could be spurious. This is because the level of one factor depends on the levels of a second factor, which implies that the effect on the DV changes from one level to the next, and therefore there is no consistent main effect.

Analyzing the Data via Multiple Regression. To analyze a 2 × 2 factorial ANOVA via MRC, we use contrast coding as presented in the coding chart given in Table 9.9. (*Note:* The reader might want to review contrast coding from Chapter 8.) The first contrast, C_1, represents the row effect because it is comparing the IFR vs. VFR groups regardless of the flight training program. The second contrast, C_2, represents the column effect because it is comparing the Part 61 vs. Part 141 groups regardless of FAA rating. The last contrast, C_3, is the product of C_1 and C_2, which represents the interaction between the two factors. A summary of the results from regressing Y on C_1, C_2, and C_3 simultaneously is included in Table 9.10.

As reported in the ANOVA subtable in Table 9.10, the overall regression model is significant, $F(3, 32) = 3.20$, $p = .0364$, and the regression equation is $\hat{Y} = 2.75C_1 + 1.03C_2 + 5.94C_3 + 19.90$. Recall from our previous work with nominally coded IVs, that if we substitute the numerical codes from the coding chart into this regression equation, we get the corresponding group means. For example:

- M_{G1} (IFR–Part 61) = 2.75(0.5) + 1.03(0.5) + 5.94(0.25) + 19.90 = 23.275
- M_{G2} (IFR–Part 141) = 2.75(0.5) + 1.03(–0.5) + 5.94(–0.25) + 19.90 = 19.275
- M_{G3} (VFR–Part 61) = 2.75(–0.5) + 1.03(0.5) + 5.94(–0.25) + 19.90 = 17.555
- M_{G4} (VFR–Part 141) = 2.75(–0.5) + 1.03(–0.5) + 5.94(0.25) + 19.90 = 19.495

TABLE 9.9 *Contrast Coding Chart for the MRC Analysis of a 2 × 2 Factorial ANOVA with Respect to the Data Set Given in the File "Ch_9 Nominal-by-Nominal Interaction Example Data"*

Group [a]	C_1	C_2	C_3
G1: IFR – Part 61	1/2	1/2	1/4
G2: IFR – Part 141	1/2	–1/2	–1/4
G3: VFR – Part 61	–1/2	1/2	–1/4
G4: VFR – Part 141	–1/2	–1/2	1/4

$N = 36$; $n = 9$ per group.
[a] IFR = Instrument Flight Rules; VFR = Visual Flight Rules. Each group reflects GA pilots with either an IFR or VFR rating who received their flight training under either a Part 61 program or a Part 141 program.

TABLE 9.10 *Summary of Results for MRC Analysis of a 2 × 2 Factorial ANOVA with Respect to the Data Set Given in the File "Ch_9 Nominal-by-Nominal Interaction Example Data"*

ANOVA

Source	*SS*	*df*	*MS*	*F*	*p*
Between	157.0764	3	52.3588	3.20	.0364
Within (Error)	523.8333	32	16.3698		
Total	680.9097	35			

Note. $N = 36$. $R^2 = .231$.

Summary of Partials

			Semipartial Correlations		Full Partial Correlations		Significance	
Factor [a]	B_i	SE_B	sr_i	sr_i^2	pr_i	pr_i^2	$t_i(32)$	p_i
Constant	19.90	0.67					29.52	< .0001
C_1	2.75	1.35	.3162	.1000	.3391	.1150	2.04	.0498
C_2	1.03	1.35	.1183	.0140	.1334	.0178	0.76	.4516
C_3	5.94	2.70	.3418	.1168	.3630	.1318	2.20	.0348

Note. $N = 36$. Y = Flight Simulator Performance; higher scores relate to higher performance.
[a] See the coding chart for the specific contrasts.

These group means are equivalent to those given in the Group-Means Table of Figure 9.5. Note that the regression constant and regression coefficients also provide the following specific information related to the main and interaction effects:

- The regression constant, $B_0 = 19.90$, is the unweighted grand mean:

$$\frac{23.275 + 19.275 + 17.555 + 19.495}{4} = \frac{79.6}{4} = 19.90$$

- $B_1 = 2.75$ represents the first contrast and it is equal to the row effect, which is the difference in group means between the IFR and VFR groups independent of the flight training program. This difference is significant, $t(32) = 2.04$, $p = .0498$. Thus, when holding all other variables in the model constant, there is a significant difference in mean flight simulator performance scores between GA pilots who are IFR vs. VFR rated. IFR rated pilots scored, on average, 2.75 points higher than VFR rated pilots. (*Note:* Recall that this main effect might be spurious given the significant interaction effect.)
- $B_2 = 1.03$ represents the second contrast and it is equal to the column effect, which is the difference in group means between the Part 61 and Part 141 groups independent of FAA rating. This difference is not significant, $t(32) = 0.76$, $p = .4516$. Thus, when holding all other variables in the model constant, there is no significant difference in mean flight simulator performance scores between GA pilots who receive flight instruction from a Part 61 vs. a Part 141 program.
- $B_3 = 5.94$ represents the third contrast and it is equal to the interaction effect, which is the mean difference between the two research factors. This is significant, $t(32) = 2.20$, $p = .0348$. Thus, when holding all other variables in the model constant, the two levels of flight training program with respect to flight simulator performance scores are operating significantly differently under the two levels of FAA rating.

Lastly, the results of the MRC analysis also enable us to quantify the amount of explained variance associated with each effect. For example, focusing on the interaction effect in the Summary of Partials subtable of Table 9.10, sr_3^2 = .1168, which means that 11.68% of the variance in flight simulator performance scores is being explained by the interaction. This amount of explained *Y* variance represents the *strength of the interaction effect*. As demonstrated by this example, conducting a 2 × 2 factorial ANOVA via multiple regression is a more efficient and more informative approach than the traditional factorial ANOVA strategy.

Examining Nominal-by-Continuous Interactions

For this discussion, we use the same DV, *Y* = Flight Simulator Performance, from Adamson et al. (2010), but now apply the theoretical model of Yerkes-Dodson (1908), which posits that a certain amount of stress is needed for optimal performance, whereas too little or too much stress is counterproductive. This relationship between stress and performance is commonly displayed as an upside-down parabola as shown in Figure 9.6. Our interest in the current example is to examine the interaction effect with respect to flight simulator performance between two variables, X_1 = Self-Efficacy and X_2 = Stress, defined as follows:.

- X_1 = Self-Efficacy is a predictor variable defined as scores on Chen et al.'s (2001) 8-item instrument scored on a traditional 5-point Likert response scale ranging from 1 = Strongly Disagree to 5 = Strongly Agree. Thus, scores can range from 8 to 40, with higher scores reflecting higher levels of self-efficacy.
- X_2 = Stress is a nominal variable and consists of three levels, Low, Medium, and High, which reflect manipulated conditions such as those related to environmental factors, time constraints, and unexpected failures. This factor is represented using two dummy coded variables, D_1 = High and D_2 = Low, with Medium as the reference group. The reason we select Medium as the reference group is because as shown in Figure 9.6, this is the stress level where, theoretically, optimal performance occurs, and our interest is in comparing performance at low and high stress levels to stress at optimal performance.

Participants are randomly assigned to one of the three stress conditions. As a result, this is considered an *experimental research design*. The different stress conditions are collectively regarded as *treatment*, and self-efficacy is considered a personological characteristic or *attribute*. Thus, this example reflects what is referred to in the literature as an Attribute-Treatment Interaction study, or ATI, and the research interest is understanding the extent to which the three stress conditions moderate

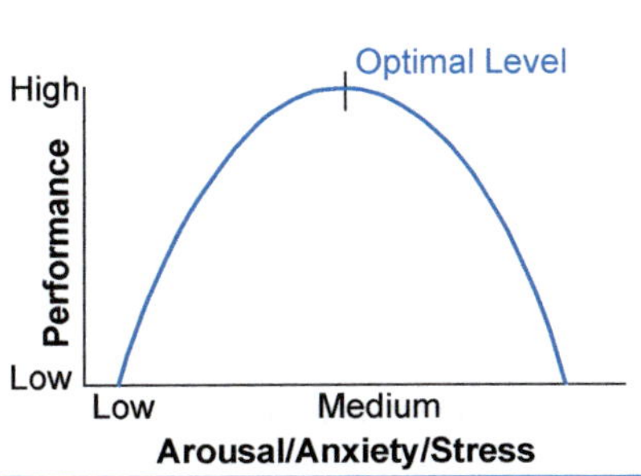

FIGURE 9.6 The Yerkes-Dodson (1908) curve.

the relationship between self-efficacy and performance. A copy of the data set is given in the file "Ch_9 Nominal-by-Continuous Interaction Example Data."

To examine the interaction effect, we conduct a hierarchical regression analysis that involves two sets: set A = $\{X_1, D_1, D_2\}$ and set B = $\{X_1D_1, X_1D_2\}$. In the first step, we regress Y on set A. In the context of an interaction analysis, this model is referred to as the main effects model, although as noted earlier, Cohen et al. (2003) labeled it as the *reduced model*. In the second step, we regress Y on sets A and B simultaneously (this is the *full model*) and then examine the increment the interaction set makes in explaining Y variance. In the context of an interaction analysis, this is sometimes referred to as product-term analysis because the interaction is being carried by multiplying the IVs associated with each factor. Table 9.11 contains a summary of this analysis. Observe the following from Table 9.11:

- The main effects model is significant, $R^2_{Y \bullet A} = .791$, $F(3, 113) = 142.30$, $p < .0001$, and the regression equation is $\hat{Y} = 0.21X_1 + 2.04D_1 + 19.04D_2 + 13.42$. Thus, the three IVs collectively explain 79.1% of the variance in flight simulator performance scores. Before we interpret the results of the individual IVs, though, we must first examine the interaction because, as noted earlier, a significant main effect in the presence of a significant interaction effect could be spurious.
- The full model is significant, $R^2_{Y \bullet AB} = .806$, $F(5, 111) = 91.99$, $p < .0001$, and the regression equation is $\hat{Y} = -0.01X_1 - 10.41D_1 + 13.22D_2 + 0.45X_1D_1 + 0.21X_1D_2 + 19.34$. Thus, the three IVs and two interaction terms collectively explain 80.6% of the variance in flight simulator performance scores.

TABLE 9.11 *Summary of Results of the Hierarchical Regression Analysis for the Data Set Given in the File "Ch_9 Nominal-by-Continuous Interaction Example Data"*

	Set-Wise F				Coefficients and Constants		
Sets/IVs [a]	ΔR^2	*df*	*F*	*p*	B_i	t_i	p_i
Step 1: (Main Effects)	.791	3, 113	142.3	< .0001	$B_0 = 13.42$	$t_0 = 7.23$	$p_0 < .0001$
"A" = { }					$B_1 = 0.21$	$t_1 = 3.21$	$p_1 = .0017$
"B" = set A = $\{X_1, D_1, D_2\}$					$B_2 = 2.04$	$t_2 = 2.21$	$p_2 = .0288$
$R^2_{Y \bullet A} = .791$					$B_3 = 19.04$	$t_3 = 19.28$	$p_3 < .0001$
$F(3, 113) = 142.30$, $p < .0001$							
Step 2: (Full Model)	.015	2, 111	4.29	.0160	$B_0 = 19.34$	$t_0 = 6.44$	$p_0 < .0001$
"A" = set A = $\{X_1, D_1, D_2\}$					$B_1 = -0.01$	$t_1 = -0.12$	$p_1 = .9076$
"B" = set B = $\{X_1D_1, X_1D_2\}$					$B_2 = -10.41$	$t_2 = -2.39$	$p_2 = .0187$
$R^2_{Y \bullet AB} = .806$					$B_3 = 13.22$	$t_3 = 3.24$	$p_3 = .0016$
$F(5, 111) = 91.99$, $p < .0001$					$B_4 = 0.45$	$t_4 = 2.92$	$p_4 = .0043$
					$B_5 = 0.21$	$t_5 = 1.42$	$p_5 = .1590$

Note. $N = 117$. Y = Flight Simulator Performance Scores. The boxed results represent the IVs entering the analysis at each step. If the corresponding IVs in the entering set are significant, then a second color is used for the B_i and t_i values in addition to the p_i values. *F* and *p* values are from Soper (2025a and 2025c).

[a] "A" represents the set in the previous model. "B" represents the set entering the model at the current step of the analysis. X_1 = Self-Efficacy scores. D_1 = High Stress, D_2 = Low Stress, and the reference group is Medium Stress.

- The interaction set explains an additional 1.5% of the variance in flight simulator performance scores, which is significant, $sR^2_B = R^2_{Y•AB} - R^2_{Y•A} = .806 - .791 = .015$, $F(2, 111) = 4.29$, $p = .0160$. This predictive gain in explained Y variance implies that a significant interaction is present.

To understand the nature of the interaction effect, we focus on the slopes of the regression lines that comprise the interactions. To do this, we substitute into the full model's regression equation the numerical codes assigned to the coded IVs D_1 and D_2, which represent "treatment," and then perform the necessary algebraic manipulations. The reader will recall that we demonstrated this procedure earlier as part of our presentation on the homogeneity of regression assumption for ANCOVA. Applying this procedure to the current example, the regression equations and the interpretation of their slopes follow (the reader is encouraged to confirm these results).

- High Stress: $\hat{Y}_{High} = 0.44X_1 + 8.93$. Holding all other IVs constant, for every 1-unit increase in self-efficacy scores, flight simulator performance scores are predicted to change, on average, 0.44 points under high stress conditions.
- Low Stress: $\hat{Y}_{Low} = 0.20X_1 + 32.56$. Holding all other IVs constant, for every 1-unit increase in self-efficacy scores, flight simulator performance scores are predicted to change, on average, 0.20 points under low stress conditions.
- Medium Stress: $\hat{Y}_{Medium} = -0.01X_1 + 19.34$. Holding all other IVs constant, for every 1-unit increase in self-efficacy scores, flight simulator performance scores are predicted to change, on average, –0.01 points under medium stress conditions. Note that this slope is practically zero, which means the graph of this line, for all practical purposes, is horizontal.

Focusing on the group comparisons governed by the coding scheme, the slopes of the regression equations for Low Stress vs. Medium Stress are dissimilar (0.20 vs. –0.01), but there is greater discrepancy between the slopes for High Stress vs. Medium Stress (.44 vs. –0.01). This latter observation suggests with respect to flight simulator performance, there might be a significant interaction between these two stress conditions across the levels of self-efficacy. To determine the significance of these interactions, we examine the respective regression coefficients for the product terms given in the full model (see Table 9.11).

- $B_4 = 0.45$ is the difference in slopes between the High and Medium stress conditions, $0.44 - (-0.01) = 0.44 + 0.01 = 0.45$. Furthermore, as reported in Table 9.11, when holding all other variables in the full model constant, this 0.45-unit slope difference is significant, $t(111) = 2.92$, $p = .0043$.
- $B_5 = 0.21$ is the difference in slopes between the Low and Medium stress conditions, $0.20 - (-0.01) = 0.20 + .01 = 0.21$. Furthermore, as reported in Table 9.11, when holding all other variables in the full model constant, this 0.21-unit difference is not significant, $t(111) = 1.42$, $p = .1590$.

From this analysis, we conclude there is indeed a significant interaction effect between self-efficacy scores and the three stress conditions ("treatment") with respect to flight simulator performance scores. More specifically, the results indicate there is a significant *disordinal interaction* between High and Medium stress conditions such that they are operating differently with respect to flight simulator performance across the range of self-efficacy scores. This is illustrated in Figure 9.7.

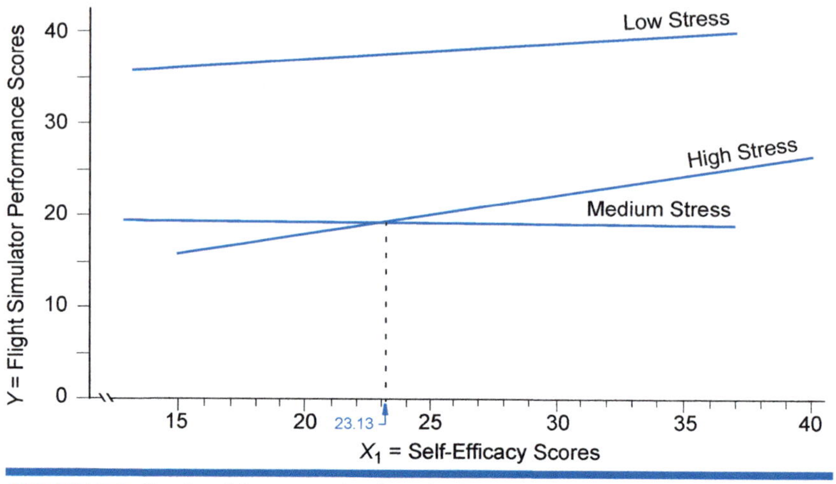

FIGURE 9.7 Graphs of the regression lines for the interaction set for the data set given in the file "Ch_9 Nominal-by-Continuous Interaction Example Data."

When examining a disordinal interaction involving a continuous variable, interpretations are made relative to the point of intersection. To determine this point, we set the equations of the two lines equal to each other and solve for the IV as follows:

$$\text{Given } \hat{Y}_{High} = 0.44X_1 + 8.93 \text{ and } \hat{Y}_{Medium} = -0.01X_1 + 19.34$$
$$0.44X_1 + 8.93 = -0.01X_1 + 19.34$$
$$0.44X_1 + 0.01X_1 = 19.34 - 8.93$$
$$0.45X_1 = 10.41$$
$$X_1 = \frac{10.41}{0.45} = 23.13$$

This X_1 value also may be calculated directly from the two regression equations by

$$\frac{a_1 - a_2}{b_2 - b_1}$$

where a_1 is the intercept of the first equation, a_2 is the intercept of the second equation, b_2 is the regression coefficient of the second equation, and b_1 is the regression coefficient of the first equation. Applying this formula to the current example:

$$X_1 = \frac{a_1 - a_2}{b_2 - b_1} = \frac{8.93 - 19.34}{-0.01 - 0.44} = \frac{-10.41}{-0.45} = 23.13$$

Observe from Figure 9.7 that for $X_1 < 23.13$, participants in the Medium stress condition have higher predicted flight simulator performance scores than those in the High stress condition. However, for $X_1 > 23.13$, this relationship reverses such that participants in the High stress condition have higher predicted flight simulator performance scores than those in the Medium stress condition. In other words, "different slopes for different folks." As a result, X_2 = Stress is a moderating variable because it moderates the X_1–Y relationship with respect to High and Medium stress conditions. X_2 does not, however, moderate the relationship between X_1 and Y with respect to low and medium stress conditions because flight performance is consistent across the entire range of self-efficacy scores with Low stress conditions *always* having higher flight performance than Medium stress conditions.

This finding also confirms the positive influence of self-efficacy reported in the literature: If participants *believe*—regardless of their actual ability—that they have the capability to do something, then when placed in a high stress environment, they will have higher performance than those placed in a medium stress environment. Notice also from Figure 9.7 that across the entire range of self-efficacy scores, flight performance is higher for the Low stress condition than it is for the other two conditions. This result is contrary to the Yerkes-Dodson curve, which posits that too little stress ("Low") or too much stress ("High") can lead to less-than-optimal performance when compared to "Medium" stress.

In addition to the significant disordinal interaction illustrated in Figure 9.7, also observe that self-efficacy scores closer to the point of intersection differ *less* in their flight simulator performance scores compared to self-efficacy scores that are farther from the point of intersection. This introduces the concept of regions of significance as reported by Johnson and Neyman (1936), which was extended by Potthoff (1964). Without going into any detail—the interested reader is encouraged to consult the cited references including Johnson and Fay (1950)—the Johnson-Neyman technique calculates a delta (Δ) such that when it is added to or subtracted from the point of intersection, the encompassed area is considered the region of nonsignificance. This region represents where differences in the DV between the groups are not significant, and outside this region on either side lies the region of significance, where the differences in the DV are significant. This is illustrated in Figure 9.8 for the current example. Applying the Johnson-Neyman technique to the current example, we expect differences in predicted flight simulation performance scores between High and Medium stress conditions to be significant for self-efficacy scores of ($< 23.13 - \Delta$) and ($> 23.13 + \Delta$), but not for self-efficacy scores within the interval from ($23.13 - \Delta$) to ($23.13 + \Delta$).

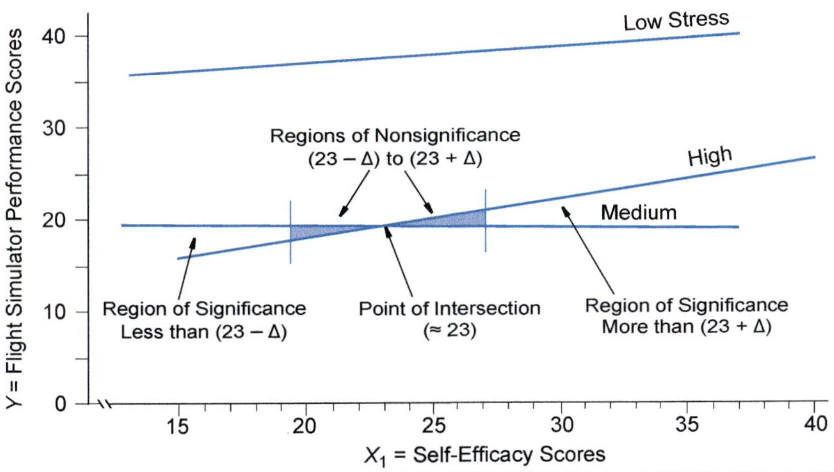

FIGURE 9.8 Demonstration of Johnson and Neyman's (1936) regions of significance applied to the disordinal interaction for the data set given in the file "Ch_9 Nominal-by-Continuous Interaction Example Data."

End-of-Section Commentary

- As demonstrated by the current example, unlike nominal-by-nominal interactions where the focus is on differences in group means, the focus on nominal-by-continuous interactions is on differences in slopes for the regression lines that represent the different treatment conditions.
- To examine the significance between the difference in slopes for the high and low stress conditions, neither of which is the reference group, we either recode the IVs or manually calculate *t* as presented in Chapter 8.
- When continuous IVs are included in an interaction model, they tend to increase multicollinearity. This can be resolved by centering the IVs. For the current example, the *VIF*s for D_1, D_2, X_1D_1, and X_1D_2 are between 23 and 32. When centering X_1 and then creating new product terms with X_1-Cntr, the *VIF*s for D_1, D_2, X_1-CntrD_1, and X_1-CntrD_2 are between 1.3 and 2.15. Centering has no change on the regression coefficients for X_1 or the product terms. (*Note*: Dummy coded variables generally are not centered.)
- For a significant interaction effect, the *strength* of the effect is equal to the increment the interaction set makes to explaining *Y* variance. Thus, for the current example, the strength of the interaction effect is sR^2_B =.015; it explains an additional 1.5% of the variance in flight simulator performance scores.
- There is a difference between an interaction study conducted *indirectly* as a result of a misspecified ANCOVA model vs. an interaction study that is conducted *directly*. In the former, the objective is controlling the effect of a covariate, but noncompliance with the homogeneity of regression assumption leads to an interaction study, whereas in the latter, the objective is examining the extent to which an IV moderates the relationship between another IV and the DV. For both approaches, though, an interaction is present if the increment in explained *Y* variance made by the interaction set is significant.

9.5 Combining Nominal and Continuous IVs in Research: A Guided Example

We now present a guided example of how to conduct a hierarchical regression analysis for a research study that involves both nominal and continuous IVs. This Guided Example is a hypothetical situation motivated by Hofer et al. (2006) who examined the effectiveness of the X-ray Object Recognition Test (ORT) as a pre-employment assessment tool for improving airport security screening. The X-ray ORT was developed to measure how well people can cope with X-ray images in which threat items such as "guns, sharp objects, hunt and blunt instruments, highly inflammable substances, explosives, chemicals, and other prohibited items" (p. 3) are superimposed by other objects in passenger bags. A description of the research setting follows.

> A sample of current TSA airport security screeners was selected from various commercial service / large hub airports across the USA. The sample was then partitioned into three groups based on participants' background:
>
> - Group 1 consists of participants who had no previous experience in X-ray screening prior to being hired as an airport security screener. These partici-

pants were administered the X-ray ORT as part of their pre-employment assessment and scored in the 90th percentile. At the time of the study, this group had 1 year of airport security screening experience. We denote this group as "X-ray."

- Group 2 consists of participants who had some previous X-ray screening experience prior to being hired as an airport security screener but were *not* administered the X-ray ORT as part of their pre-employment assessment. At the time of the study, this group had 1 year of airport security screening experience. We denote this group as "No X-ray."
- Group 3 consists of airport security screeners who currently have at least 5 years of experience, but at the time of hire had no previous X-ray screening experience and were not subject to any type of pre-employment assessment independent of TSA's training and assessment requirements. We denote this group as "Control."

Because screeners' age could influence the results—for example, vision acuity and mental stamina generally diminish with age—participants' age was treated as a covariate. To measure their ability to detect prohibited items from X-ray images, participants were administered the Prohibited Items Test (PIT), which requires screeners to decide within 10 seconds whether the bag that is displayed on the screen contains one or more prohibited items. For this Guided Example, we will assume a maximum of 98 prohibited items were hidden in passenger bags and hence scores could range from 0, where a screener detects 0 prohibited items, to 98, where a screener detects all 98 prohibited items. Thus, higher scores reflect higher screening ability.

The corresponding data set, which consists of hypothetical data, is given in the file, "Ch_9 Guided Example Data" with Y = PIT scores, X_1 = Age, and X_2 = the dummy coded group membership variable where Control is the reference group. The structure of our presentation follows that of previous chapters' Guided Examples and consists of six parts: Research Questions and Hypotheses, Sample Size Planning, Preliminary Data Screening, Data Analysis, Interpretation of Results, and Post Hoc Considerations.

Research Questions and Hypotheses

Because participants' age is being treated as a covariate, the statistical strategy is ANCOVA, and our interest is in comparing mean PIT scores between the designated groups after the group means have been adjusted for participants' age. The research questions and corresponding hypotheses that follow reflect this situation.

RQ 1 and Hyp 1. When freed of the influence of age, what is the mean difference in PIT scores between participants in the X-ray group vs. participants in the Control group?

- H_0: $\mu_{\text{X-ray-Adj}} - \mu_{\text{Control-Adj}} = 0$ (When adjusted for age, no significant difference in mean PIT scores between the X-ray and Control groups.)
- H_1: $\mu_{\text{X-ray-Adj}} - \mu_{\text{Control-Adj}} \neq 0$ (When adjusted for age, there is a significant mean difference in PIT scores between the X-ray and Control groups.)

RQ 2 and Hyp 2. When freed of the influence of age, what is the mean difference in PIT scores between participants in the No X-ray group vs. participants in the Control group?

- H_0: $\mu_{\text{No X-ray-Adj}} - \mu_{\text{Control-Adj}} = 0$ (When adjusted for age, no significant difference in mean PIT scores between the No X-ray and Control groups.
- H_1: $\mu_{\text{No X-ray-Adj}} - \mu_{\text{Control-Adj}} \neq 0$ (When adjusted for age, there is a significant mean difference in PIT scores between the No X-ray and Control groups.)

Sample Size Planning

Because we are conducting an ANCOVA via an MRC analysis, the focus of sample size planning is on the overall R^2 and the increment of the (presumed) valid ANCOVA model. We consult Soper (2025d) and Soper (2025e), respectively, and assume medium effects.

- Overall R^2. Given $ES = f^2 = 0.13$, power = .8, IVs = 3, and $\alpha = .05$: $n^* = 88$.
- sR_i^2. Given $ES = f^2 = 0.15$, power = .8, $k_{\text{Set A}} = 1$, $k_{\text{Set B}} = 2$, and $\alpha = .05$: $n^* = 68$.

Therefore, we need at least 88 cases to have an 80% probability of finding a significant medium effect for the overall R^2 and at least 68 cases to have an 80% probability of finding a significant medium effect for the increment. The given data set has $N = 52$ cases, which reduces the probability of finding significant medium effects.

Preliminary Data Screening

Our preliminary data screening follows the structure and contents of what we did in Chapter 8's Guided Example. We also do *not* provide copies of any detailed summary tables, leverage plots, or residual plots but instead simply report the results from our statistics program.

Outlier Analysis. Examining the initial data set consisting of the covariate (X_1) and the two dummy coded IVs (D_1 and D_2) for outliers via Jackknife distances, zero outliers are flagged and therefore this is not a concern. (*Note:* Other outlier procedures might yield different results.)

Multicollinearity. Regressing Y on the covariate (X_1) and the two dummy coded IVs (D_1 and D_2), the *VIF*s are no greater than 1.30, and therefore multicollinearity is not a concern.

Missing Data. The data set has no missing data.

Measurement Error. The IVs are not measuring any psychological construct.

Relevant IVs. To determine if the three targeted IVs are correctly specified, we examine each IV's added variable plot (AVP, see Chapter 5). The results of these plots indicate that all three variables are *not* relevant. However, the reader is reminded that we are conducting an ANCOVA, which means that it is possible the ANCOVA model might be misspecified. Therefore, we will not eliminate any IVs until we check for compliance with the homogeneity of regression assumption.

We now check to see if the model is compliant with the assumptions of regression, including the homogeneity of regression assumption.

Linearity. A scatter plot of the residuals vs. the predicted values, in conjunction with the corresponding kernel smoother line, confirms that the scores are randomly scattered around the 0-line and do not show any pronounced or systematic pattern. This is confirmed by a second scatter plot of the residuals vs. the observed Y values. As a result, the model is compliant with the linearity assumption.

Equal Variances (Homoscedasticity of the Residuals). The scatter plot of the residuals vs. the predicted values used to check the linearity assumption also confirms the equal variances assumption because the residuals appear to be approximately equally scattered above and below the 0-line without any pronounced pattern. Therefore, the model is compliant with the equal variances assumption.

Independence of the Residuals. A scatter plot of the residuals vs. the case numbers, in conjunction with the corresponding kernel smoother line, confirms there is no systematic pattern and therefore the model is compliant with the independence assumption.

Normality of the Residuals. The *q-q* plot of the residuals shows that the data closely "hug" the normal line and all the points are within the 95% confidence band. The corresponding Shapiro-Wilk Goodness of Fit test also is not significant, W = .968, p = .1766. Therefore, the model is compliant with the normality assumption.

Homogeneity of Regression. To check this assumption, we create an interaction set where the IVs are the products of the covariate and the coded IVs representing group membership and examine the increment the interaction set makes to explaining Y variance. The reader will note the following:

- Set A = $\{X_1, D_1, D_2\}$. $R^2_{Y \cdot A}$ = .03, $F(3, 48)$ = 0.4998, p = .6842.
- Set B = $\{X_1D_1, X_1D_2\}$. $R^2_{Y \cdot AB}$ = .236, $F(5, 46)$ = 2.85, p = .0253.
- sR^2_B = .236 – .03 = .206, $F(2, 46)$ = 6.20, p = .0041, $F_{Critical}$ = 3.1996.

The increment the interaction set makes to explaining Y variance is significant, which means that the ANCOVA model is *not* compliant with the homogeneity of regression assumption, resulting in a misspecified ANCOVA model. As a result, we must forego the ANCOVA and now focus on an interaction study. This changes our initial focus from examining differences in group means to examining differences in slopes. This also leads to a revised general RQ: "To what extent does X_2 moderate the relationship between X_1 and Y?" and the corresponding null hypothesis is that X_2 does not significantly moderate the X_1–Y relationship. As a result, our interest is now in determining if the relationship between Age and PIT scores is the same or different across the range of ages for the three groups.

Data Analysis

To conduct the interaction analysis, we examine the full model by regressing Y simultaneously on X_1, D_1, D_2, X_1D_1, X_1D_2. A summary of the results is provided in Table 9.12. As reported in Table 9.12, the interaction model is significant, $R^2_{Y \cdot AB}$ = .236, $F(5, 46)$ = 2.85, p = .0253, and the regression equation is $\hat{Y} = 1.03X_1 + 28.34D_1 + 92.51D_2 - 0.56X_1D_1 - 2.35X_1D_2 + 3.66$. Note, though, that the *VIF*s for this model range from 23 to 30 for the coded IVs and product terms, indicating high multicollinearity. Therefore, we center X_1, create new product terms using X_1-Cntr, and rerun the analysis. The results, which are summarized in the second subtable of Table 9.12, yield the following regression equation:

TABLE 9.12 *Summary of Results for the Interaction Model for Chapter 9's Guided Example*

Coefficients (Parameter Estimates): X_1 Not Centered

Term	*B*	*SE*	*t*(46)	*p*	*VIF*
Constant	3.66	19.97	0.18	.8554	
X_1 = Age	1.03	0.46	2.26	.0287	2.73
D_1 = X-ray	28.34	28.54	0.99	.3260	23.15
D_2 = No X-ray	92.51	28.94	3.20	.0025	30.05
X_1D_1 = Interaction 1	–0.56	0.66	–0.84	.4035	22.79
X_1D_2 = Interaction 2	–2.35	0.68	–3.43	.0013	28.98

Coefficients (Parameter Estimates): X_1 Centered

Term	*B*	*SE*	*t*(46)	*p*	*VIF*
Constant	46.20	4.30	10.73	< .0001	
X_1-Cntr	1.03	0.46	2.26	.0287	2.73
D_1 = X-ray	5.19	6.71	0.77	.4429	1.28
D_2 = No X-ray	–4.49	6.00	–0.75	.4579	1.29
X_1-Cntr × D_1	–0.56	0.66	–0.84	.4035	1.89
X_1-Cntr × D_2	–2.35	0.68	–3.43	.0013	1.82

Note. $N = 52$. Y = Prohibited Items Test (PIT) scores. Overall $R^2 = .236$, $F(5, 46) = 2.85$, $p = .0253$. Group membership was dummy-coded with Control as the reference group.

$$\hat{Y} = 1.03X_1\text{-Cntr} + 5.19D_1 - 4.49D_2 - 0.56X_1\text{-Cntr}D_1 - 2.35X_1\text{-Cntr}D_2 + 46.20$$

Focusing on the interaction terms' regression coefficients, observe the following:

- $B_4 = -0.56$ is the difference in slopes between the regression line for the X-ray group and the regression line for the Control group. As reported in Table 9.12, this difference is not significant, $t(46) = -0.84$, $p = .4035$, which implies that an interaction effect is not present between these groups.
- $B_5 = -2.35$ is the difference in slopes between the regression line for the No X-ray group and the regression line for the Control group. As reported in Table 9.12, this difference is significant, $t(46) = -3.43$, $p = .0013$, which implies that an interaction effect is present between these groups.

To visually examine these results, we prepare an interaction plot by deriving each group's regression equation and graphing it as we did for an earlier example. For these derivations, we use the regression equation from the full model *without X_1 centered* so that the horizontal axis is comprised of the observed X_1 scores. Following what we have done previously, the regression equations are given below, and the graphs of their lines are shown in Figure 9.9. (*Note*: Although we could use our statistics program's graphing utility to generate these graphs, we opt to do this manually for instructional purposes. The reader is encouraged to confirm these results.)

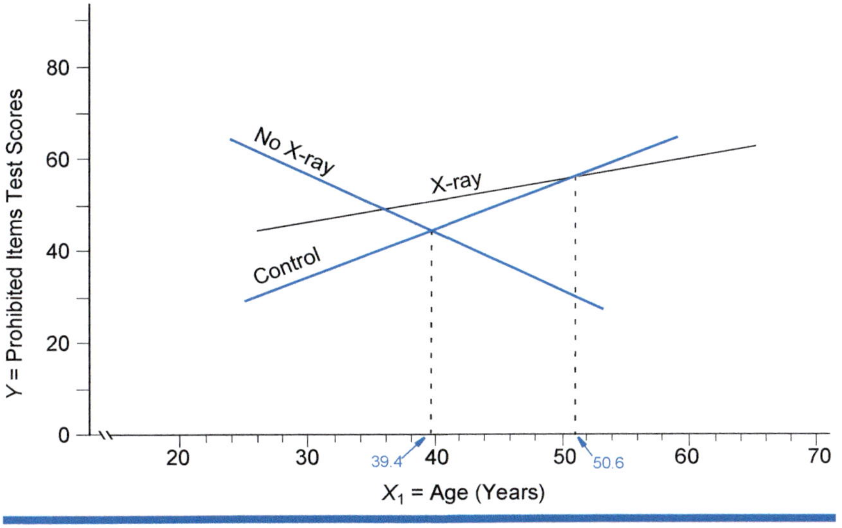

FIGURE 9.9 Interactions plots for Chapter 9's Guided Example.

- X-ray (coded 1 0). $\hat{Y}_{X\text{-ray}} = 0.47X_1 + 32$. Holding all other variables constant, for every 1-year increase in age, PIT scores are predicted to change (increase), on average, 0.47 points for participants with no previous X-ray screening experience and who were administered the X-ray ORT as part of their pre-employment assessment.
- No X-ray (coded 0 1). $\hat{Y}_{\text{No X-ray}} = -1.32X_1 + 96.17$. Holding all other variables constant, for every 1-year increase in age, PIT scores are predicted to change (decrease), on average, −1.32 points for participants with some previous x-ray screening experience prior to being hired as an airport security screener but were not administered the X-ray ORT as part of their pre-employment assessment.
- Control (coded 0 0). $\hat{Y}_{\text{Control}} = 1.03X_1 + 3.66$. Holding all other variables constant, for every 1-year increase in age, PIT scores are predicted to change (increase), on average, 1.03 points for participants with at least 5 years of experience but had no previous X-ray screening experience and were not subject to any type of pre-employment assessment independent of TSA's training and assessment requirements.

For confirmation purposes, note that the difference in slopes between the X-ray and Control groups is 0.47 − 1.03 = −0.56, which is B_4, and the difference in slopes between the No X-ray and Control groups is −1.32 − 1.03 = −2.35, which is B_5.

Interpretation of Results

We first observe from Figure 9.9 that the significant interaction between the No X-ray and Control groups is disordinal and the point of intersection is $X_1 = 39.4$ years old: Participants younger than 39.4 years old are predicted to detect significantly more prohibited items if they are in the No X-ray group vs. the Control group; however, participants older than 39.4 years old are predicted to detect significantly more prohibited items if they are in the Control group vs. the No X-ray group. As a result, these two levels of X_2 = Group Membership are not operating the same across the range of ages and therefore X_2 moderates the X_1–Y relationship with respect to these two groups. From a practical perspective, this interaction provides TSA with guidance for hiring airport security screeners: When presented with two candidates—one with at least 5 years of airport security screening experience vs. one with some previous X-ray screening experience who did not take the X-ray ORT—the age of 40 years old can be a critical decision point. To apply the Johnson-Neyman procedure, the significant interaction regions are younger than (40 – Δ) years and older than (40 + Δ years.)

We next observe from Figure 9.9 that the non-significant interaction between the X-ray and Control groups also is disordinal and the point of intersection is X_1 = 50.6 years old: Participants younger than 50.6 years old are predicted to detect more prohibited items if they are in the X-ray group vs. the Control group; however, participants older than 50.6 years old are predicted to detect more prohibited items if they are in the Control group vs. the X-ray group. As a result, these two levels of X_2 = Group Membership are not operating the same across the range of ages and therefore X_2 is a moderating variable because it moderates the X_1–Y relationship with respect to these two groups. Although this result is not significant, it nevertheless might have practical significance. For example, the interaction suggests that when compared to experienced airport security screeners (Control), the X-ray ORT could be an effective pre-employment assessment tool for identifying candidates younger than 51 years old as effective airport security screeners.

The last observation we make from Figure 9.9 is the interaction between the X-ray and No X-ray groups, neither of which is the reference group. This interaction is disordinal, but to determine if it is significant, we either recode the IVs with a different reference group or manually calculate t as presented in Chapter 8. This is left for the reader to do. (*Hint*: The point of intersection is approximately X_1 = 36 years old, and the corresponding regression coefficient is B = 1.79, $t(46)$ = 2.54, p = .0145.)

Post Hoc Considerations

We note the following post-hoc considerations.

- The effect size of the overall interaction model is

$$f^2 = \frac{R^2}{1 - R^2} = \frac{.236}{1 - .236} = \frac{.236}{.764} = 0.31$$

  Per Cohen et al. (2003, p. 93), this may be considered a large effect.

- The strength of the interaction effect is sR^2_B = .206. It explains an additional 20.6% of the variance in the number of detected prohibited items (PIT scores). The corresponding effect size is

$$f^2 = \frac{sR^2_B}{1 - R^2_{Y \cdot AB}} = \frac{.206}{.764} = 0.27$$

  Per Cohen et al. (2003, p. 95), this may be considered a medium effect.

- Consulting Soper (2025f), the post hoc power of the overall interaction model is .848. Thus, there is an 84.8% probability that the significant interaction effect found in the sample also exists in the population, and there is an 84.8% probability that the decision to reject the null hypothesis is correct.

- Consulting Soper (2025g), the post hoc power of the increment made by the interaction set is .897. Thus, there is an 89.7% probability that the significant interaction effect found in the sample also exists in the population, and there is an 89.7% probability that the decision to reject the null hypothesis is correct.

Final Thoughts/Commentary

From this Guided Example, the reader will recognize that what began initially as a covariance study became an interaction study. This example also illustrates one of the hallmarks (or glamours) of research: You never know what you are going to find. The misspecified ANCOVA model led to an interaction investigation, which identified a relationship that might have been overlooked or not detected. For example, as noted earlier, when considered from the perspective of hiring airport security screeners, we now know that the relationship between screeners' age and the number of prohibited items they can detect changes across the different groups, with age 40 being critical when comparing No X-ray and Control groups. This information, coupled with selective use of the X-ray ORT as a pre-employment assessment tool could help inform or guide hiring decisions.

Chapter 9 Summary

9.2 Combining Nominal and Continuous IVs

- When analyzing a data set with a mix of continuous IVs and nominal IVs with three or more groups, the regression coefficients for the coded IVs do not equal "true" differences in group means relative to the coding scheme because they now are in the presence other variables. Therefore, their interpretations are conditional and must be prefaced with the phrase, "when holding all other variables in the model constant."
- To test for the significance of nominal IVs representing three or more groups, we use hierarchical regression, where set A consists of all the IVs except the coded IVs being tested, and set B consists of the targeted coded IVs. $R^2_{Y \cdot A}$ is equal to the results of the reduced model, and $R^2_{Y \cdot AB}$ is equal to the results of the full model. If sR^2_B is significant. then we may regress Y on the coded IVs to determine their significance and interpret the regression coefficients unconditionally as differences in group means.
- When analyzing a data set with a mix of continuous IVs and nominal IVs with three or more groups, it is informative to also report adjusted group means, which reflect differences in group means that have been adjusted for the presence of other IVs in the model.
- For data sets strictly with nominal IVs, adjusted group means are obtained by combining the coding charts into a single chart, substituting the coded IVs into the regression equation, and completing the arithmetic.
- For data sets with continuous and nominal IVs, adjusted group means are obtained by first centering the continuous IVs, which makes their mean equal to zero. The coded IVs may then be substituted into the regression equation to calculate the adjusted group means.

9.3 Analysis of Covariance (ANCOVA)

- ANCOVA is a statistical strategy that controls for the effect of a covariate, which is an IV that is not of primary interest for the study but could have an impact on the results. ANCOVA fully partials the effect of the covariate from both the targeted IVs as well as from the DV and therefore it is informative to report the results of the corresponding pr^2.
- An ANCOVA requires the DV and covariate to be quantitative, and the research factor of interest to be measured on a nominal scale. The data set also must be compliant with the homogeneity of regression assumption, which requires a nonsignificant interaction between the covariate and research factor. This can be examined statistically, graphically, or algebraically.
- To conduct an ANCOVA via multiple regression, we perform a hierarchical analysis using sets of IVs: set A = the covariate(s), set B = the coded IVs representing group membership, and set C = the interactions between set A and set B. The interaction IVs are product terms that consist of each IV in set A being multiplied by each IV in set C. If set C's increment (i.e., sR^2_C) is not significant, then the homogeneity of regression assumption is satisfied and the ANCOVA model, which consists of sets A and B, is valid.
- If set C's increment is significant, then the ANCOVA model is misspecified and the analysis must transition to an interaction study.

9.4 Examining Interactions

- The underlying concept of an interaction is manifested in the following question: Given two factors, do the levels of one factor operate the same or differently across the levels of the second factor?
- The focus of an interaction is on slopes of lines. If the slopes are equal, then the lines are parallel and there is no interaction because the levels of one factor are operating the same (consistently) across the levels of the second factor. If the lines cross within the scope of the sample data, then there is a disordinal interaction and the levels of one factor are operating differently under the levels of the second factor. If the lines are not parallel but do not cross within the scope of the sample data, then there is an ordinal interaction and the levels of one factor are operating the same across the levels of the second factor.
- Two-way interactions involve two factors and when considered by their data types include nominal-by-nominal interactions, nominal-by-continuous interactions, and continuous-by-continuous interactions.
- Although commonly examined by factorial ANOVA, a 2 × 2 nominal-by-nominal interaction may be examined via an MRC analysis using contrast coding, where Contrast 1 represents the row effect, Contrast 2 is the column effect, and Contrast 3 is the interaction effect, which is the product of C_1 and C_2. When interpreting the resulting regression equation, B_1 is the difference in group means between (Groups 1 and 2) vs. (Groups 3 and 4), B_2 is the difference in group means between (Groups 1 and 3) vs. (Groups 2 and 4), and B_3 is the interaction effect.

- In a nominal-by-continuous interaction, the interaction model consists of the continuous IV(s), the coded IVs, and their products. From the regression equation, the regression coefficients for the product terms represent the differences in slopes relative to the coding scheme being used for the nominal factor. For example, if dummy coding is used, then the regression coefficients reflect the difference in slopes between each group represented by a coded IV and the reference group.
- If a nominal-by-continuous interaction analysis consists of a disordinal interaction, then it is informative to report the point of intersection and interpret the regions of significance relative to the Johnson-Neyman procedure.
- In a nominal-by-nominal interaction, either of the IVs may be the moderating variable depending on the focus of the research study as articulated by the RQ. In a nominal-by-continuous interaction, the nominal IV is generally the moderating variable, and the objective is to understand how the relationship between the continuous IV and the DV changes across the different levels of the nominal IV.

Key Terms and Concepts

Adjusted group means
Analysis of covariance (ANCOVA)
Attribute-Treatment Interaction (ATI)
Centering the continuous IV
Conditional comparisons
Covariate
Disordinal interaction
Full Model
Homogeneity of regression
Johnson-Neyman technique
Main effects model
Moderating variable
Moderation
Nominal-by-continuous interaction
Nominal-by-nominal interaction
Ordinal interaction
Parallelism of slopes
Product-term analysis
Reduced model
Regions of nonsignificance
Regions of significance
Three-way interaction
Two-way interaction
Unconditional comparisons
Valid ANCOVA model
Zero interaction

Review Exercises

A. Checking Conceptual Knowledge

In 1–10, choose the best answer among the choices provided.

1. Which of the following statements is accurate with respect to interactions?
   a. The graphs of the regression lines either will intersect or not intersect.
   b. The respective slopes of two regression lines of a disordinal interaction are equal.
   c. Given two nonparallel regression lines, there is always a point of intersection.
   d. A disordinal interaction is always statistically significant.

2. Consider a study that examines differences in a runway incursion training program with respect to the number of runway incursions committed at a particular airport. The training program, designed for pilots, consists of four groups: "No Training," "Basic Training," "Intermediate Training," and "Advanced Training." Because pilot experience, defined as years as PIC, could influence the results, it is treated as a covariate. Which statement is true?
   a. Adjusted group means represent the average number of runway incursions for each training program without considering pilot experience.
   b. Adjusted group means reflect the estimated number of runway incursions for each training program while statistically controlling for differences in pilot experience.
   c. Adjusted group means are used to determine the effect pilot experience has on the number of runway incursions for the respective training programs.
   d. Adjusted group means are used to determine if the respective training programs are equally effective regardless of pilot experience.
3. Consider the interaction plot in Figure 9.10. One interpretation of this graph is:
   a. There is no interaction effect.
   b. If years as PIC were to increase beyond 25 years, then there would be a significant disordinal interaction.
   c. The graph depicts a disordinal interaction because the lines are not parallel.
   d. Regardless of years as PIC, Part 121 operations have fewer runway incursions than Part 135 operations.

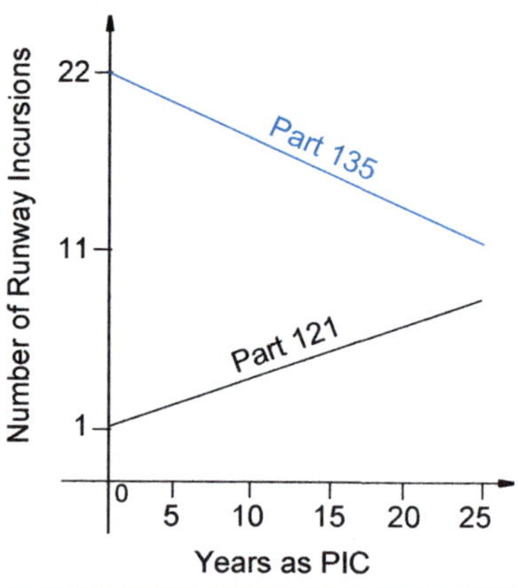

FIGURE 9.10 Interaction plot for Review Exercises Part A-3.

4. Consider a study that consists of a continuous DV, a continuous IV, and a group membership factor with four levels, which are coded using dummy coding. Let's assume that the results of a hierarchical regression analysis are as follows:
   - $R^2_{Y \bullet A} = .959437$, $F(4, 35) = 206.96$, $p < .0001$, set A = $\{X_1, D_1, D_2, D_3\}$.
   - $R^2_{Y \bullet AB} = .959843$, $F(7, 32) = 109.27$, $p < .0001$, set B = $\{X_1D_1, X_1D_2, X_1D_3\}$.
   - $sR^2_B = .000406$, $F(3, 32) = 0.1078$, $F_{Critical} = 2.90$.

   Which of the following statements is accurate?

Continued on next page ...

a. The results suggest an interaction effect.
b. The results suggest a significant interaction effect.
c. The results suggest a valid ANCOVA model.
d. The results suggest a misspecified ANCOVA model.

5. A researcher is examining the relationship between X_1 = Job Stress and Y = Systolic blood pressure, which is the "top" number of a blood pressure reading. It is her belief that "high stress occupations result in higher systolic numbers whereas low stress occupations result in lower systolic numbers." The researcher is considering an interaction study that involves randomly assigning participants to one of three groups: Counseling, Exercise, and Control (neither counseling nor exercise) and surmises that when compared to the Control group, the impact of job stress on blood pressure will be less for the Counseling and Exercise groups. Which of the following statements is accurate?
   a. Job stress is the moderating variable.
   b. The treatment is the moderating variable.
   c. Job stress is a covariate.
   d. The study reflects a nominal-by-nominal interaction.

6. Which interaction plot given in Figure 9.11 corresponds to the following group-means table for a 2 × 2 factorial ANOVA?

		Type of Delay	
		Air Carrier	***NAS***
Airline	***American***	$M_1 = 68$	$M_2 = 64$
	Delta	$M_3 = 26$	$M_4 = 46$

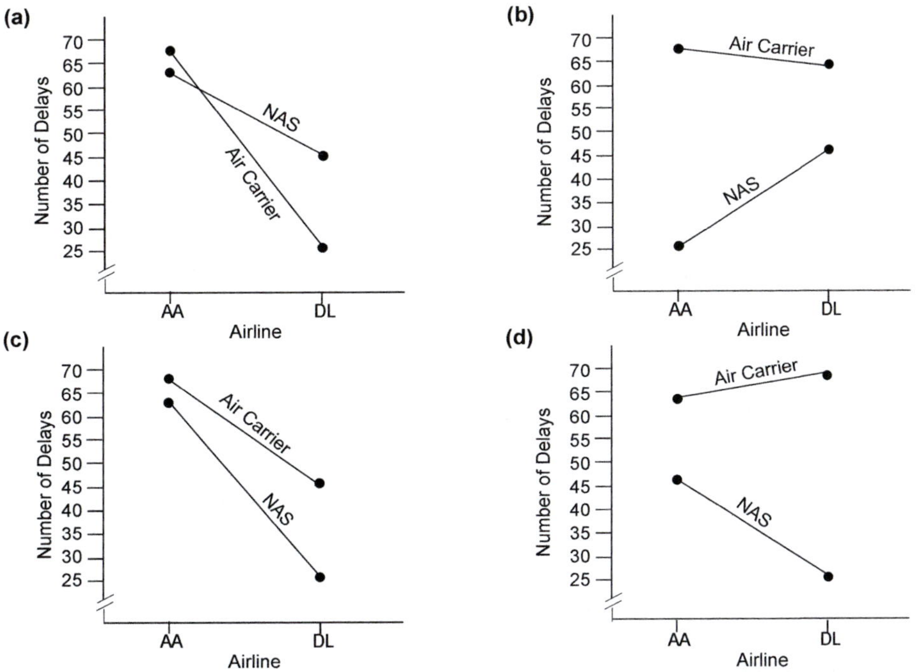

FIGURE 9.11 Interaction plots for Review Exercises Part A-6.

Use the following research description to answer Items 7–10.

Manalo and Wolfe (2000) conducted an ANCOVA to compare scores on the written essay component of the Test of English as a Foreign Language (TOEFL) between two groups of examinees who self-selected their composition medium: those who prepared their essays using a word-processor vs. those who hand-wrote their essays. Following guidance from the literature, Manalo and Wolfe treated examinees' English language proficiency scores as a covariate. As part of their presentation, Manalo and Wolfe reported the following (p. 14):

(a) With respect to the slopes:"...there is a larger difference for examinees with low levels of English language proficiency and a smaller difference for examinees with higher levels of English language proficiency between computer-based and handwriting scores" (p. 8).

(b) F(Covariate) = 98381.36, $p < .0001$; F(Main Model) = 98381.36, $p < .0001$; and F(Interaction) = 1486.61, $p < .0001$ (p. 13).

(c) Handwriting: $M_{\text{Unadjusted}} = 4.0$, $M_{\text{Adjusted}} = 4.18$;
Word Processor: $M_{\text{Unadjusted}} = 4.1$, $M_{\text{Adjusted}} = 3.89$.

7. Which of the following statements is accurate?
   a. The interaction plot most likely is disordinal.
   b. The interaction plot most likely is ordinal.
   c. The interaction plot most likely consists of parallel lines.
   d. There is not enough information given in the research description to answer this question.
8. The results suggest
   a. a valid ANCOVA model.
   b. a misspecified ANCOVA model.
   c. the increment in explained Y variance made by the group membership variable is significant.
   d. the inclusion of the covariate is unnecessary.
9. Controlling for the effect of examinees' English language proficiency
   a. had no effect on either group.
   b. favored the word-processor group.
   c. favored the handwriting group.
   d. confirmed that English language proficiency is not an appropriate covariate.
10. Manalo and Wolfe (2000, p. 8) also reported that N = 153,531 and that "the proportion of variance explained by (the) interaction is nearly zero." When this additional information is combined with what is reported in the given research description above, which course of action would be appropriate?
    a. Ignore the interaction and focus on the ANCOVA model because the significant interaction is most likely an artifact of the large sample size.
    b. Ignore the interaction and focus on the ANCOVA model but recommend for future research that subsequent or replication studies should examine the interaction.

Continued on next page ...

c. Analyze both the ANCOVA *and* interaction models and report the results from both analyses.
d. Conduct an ATI because even though the interaction is "nearly zero," the homogeneity of regression assumption was not met.

B. Applying Procedural and Conceptual Knowledge

Use the following research description and the corresponding data set to conduct the activities given in Parts B1–B7. See Section 9.5 for guidance.

An aviation researcher is asked by an airline to compare various treatments for treating depression among its pilot corps. The sample consists of pilots who volunteer to participate and who score above the 75th percentile on Goldberg and Williams' (1988) General Health Questionnaire (GHQ), which is administered as part of a pre-employment assessment by the airline. This 12-item instrument assesses a person's level of psychological distress (e.g., depression, anxiety, social dysfunction, and psychosomatic symptoms) and is measured on a 4-point Likert-type response scale where 0 = "Not at all," 1 = "No more than usual," 2 = "Rather more than usual," and 3 = "Much more than usual." Thus, scores could range from 0 to 36, with higher scores reflecting higher levels of psychological distress. The participating pilots self-select to receive one of three treatments: Group 1 receives counseling (cognitive therapy), Group 2 receives the drug Imipramine, and Group 3 is a control group, which neither participates in any professional counseling nor receives any medication for depression. These groups are dummy coded with Control as the reference group. After a 6-week study period, the pilots are administered the Patient Health Questionnaire (PHQ-9, Kroenke et al., 2001), which consists of nine items measured on a 4-point Likert-type response scale where 0 = "Not at all," 1 = "Several days," 2 = "More than half the days," and 3 = "Nearly every day." Thus, scores could range from 0 to 27, with higher scores reflecting higher levels of depression. Although the GHQ and PHQ are different instruments, the respective items are similar and therefore the researcher is concerned that the GHQ could confound the results and decides to control for its effect by treating it as a covariate. The data set, which consists of hypothetical data, is given in the file "Ch_9 Review Exercises Data-Part B."

B1. Research Questions and Hypotheses. Write the research questions and corresponding hypotheses.

B2. Sample Size Planning. Conduct a power analysis to see if the given data set is sufficiently large for the overall R^2 and sR_i^2 values for a medium effect size.

B3. Preliminary Data Screening. Perform regression diagnostics on the data set and confirm that the final model is compliant with the underlying regression assumptions, including homogeneity of regression. Assume that the reliability of the GHQ instrument is acceptable.

B4. Data Analysis. Based on the results of preliminary data screening, determine if the ANCOVA model is valid or misspecified, and then conduct either an ANCOVA or interaction analysis, whichever is appropriate. If ANCOVA is appropriate, then include a table similar to Table 9.7. If an interaction analysis is appropriate, then provide summary tables similar to Tables 9.11 and 9.12, derive the equations of the regression lines, and provide the corresponding interaction plots.

Continued on next page ...

B5. Interpretation of Results. Interpret the results relative to the research questions and corresponding null hypotheses.

B6. Post Hoc Considerations. Determine and comment on the effect sizes and power of the analysis.

B7. Commentary. Provide any additional commentary that is meaningful with respect to the results.

References

Adamson, M., Samarina, V., Xiangyan, X., Huynh, V., Kennedy, Q., Weiner, M., . . . Taylor, J. (2010). The impact of brain size on pilot performance varies with aviation training and years of education. *Journal of the International Neuropsychological Society, 16*(3), 412–423. https://doi.org/10.1017/S1355617710000111

Chen, G., Gully, S. M., & Eden, D. (2001). Validation of a new general self-efficacy scale. *Organizational Research Methods, 4*(1), 62–83. https://doi.org/10.1177/109442810141004

Cohen, J., Cohen, C., West, S. G., & Aiken., L. S. (2003). *Applied multiple regression/correlation analysis for the behavioral sciences* (3rd ed.). Lawrence Erlbaum Associates.

Goldberg, D., & Williams, P. (1988). *A user's guide to the general health questionnaire*. Windsor, Berkshire: NFER-Nelson.

Hofer, F., Hardmeier, D., & Schwaninger, A. (2006, November 27–December 1). *Increasing airport security using the X-Ray ORT as effective pre-employment assessment tool* [Symposium]. The 4th International Aviation Security Technology Symposium, Washington, DC, United States. https://doi.org/10.13140/RG.2.1.3658.6487

Hunter, D. R. (2006). *Risk perception among general aviation pilots*. International Journal of Aviation Psychology, 16(2), 135–144.

Jaccard, J., & Turrisi, R. (2003). *Interaction effects in multiple regression* (2nd ed.). Sage.

Johnson, P. O., & Fay, L. C. (1950). The Johnson-Neyman technique: Its theory and application. *Psychometrika, 15*, 349–367. https://doi.org/10.1007/BF02288864

Johnson, P. O., & Neyman, J. (1936). Tests of certain linear hypotheses and their applications to some educational problems. *Statistical Research Memoirs, 1*, 57–93.

Manalo, J. R., & Wolfe, E. W. (2000, April 24–28). *A comparison of word-processed and handwritten essays written for the test of English as a foreign language* [Conference session]. Annual Meeting of the American Educational Research Association, New Orleans, LA, United States. https://files.eric.ed.gov/fulltext/ED443845.pdf

Kroenke, K., Spitzer, T. L., & Williams, J. B. W. (2001). The PHQ-9. *Journal of General Internal Medicine,, 16*, 606–613. https://doi.org/10.1046/j.1525-1497.2001.016009606.x (see also https://pmc.ncbi.nlm.nih.gov/articles/PMC1495268/)

Potthoff, R. F. (1964). On the Johnson-Neyman technique and some extensions thereof. *Psychometrika, 29*, 241–256.

Soper, D. (2025a). *F-value calculator for hierarchical multiple regression*. https://www.danielsoper.com/statcalc/calculator.aspx?id=14

Soper, D. (2025b). *Critical F-value calculator*. https://www.danielsoper.com/statcalc/calculator.aspx?id=4

Soper, D. (2025c). *p-value calculator for an F-test*. https://www.danielsoper.com/statcalc/calculator.aspx?id=7

Soper, D. (2025d). *A-priori sample size calculator for multiple regression*. https://www.danielsoper.com/statcalc/calculator.aspx?id=1

Soper, D. (2025e). *A-priori sample size calculator for hierarchical multiple regression*. https://www.danielsoper.com/statcalc/calculator.aspx?id=16

Soper, D. (2025f). *Post-hoc statistical power calculator for multiple regression*. https://www.danielsoper.com/statcalc/calculator.aspx?id=9

Soper, D. (2025g). *Post-hoc statistical power calculator for hierarchical multiple regression*. https://www.danielsoper.com/statcalc/calculator.aspx?id=17

Yerkes, R. M., & Dodson, J. D. (1908). The relation of strength of stimulus to rapidity of habit-formation. *Punishment: Issues and Experiments*, 27–41.

Chapter 10
Other Regression Models: Mediation and Binary Logistic Regression

Student Learning Outcomes

After studying this chapter, you will be able to do the following:

1. Distinguish between full and partial mediation.
2. Conduct and interpret the results of an MRC analysis involving an IV, a mediator variable, and a DV.
3. Given a dichotomous DV and IV, prepare a corresponding contingency table and determine the probability, odds, and odds ratio for a specific event.
4. Interpret the partial regression coefficients of a logistic regression equation from both logit and odds ratios perspectives.
5. Conduct a binary logistic regression analysis for a research study that involves a dichotomous DV and multiple IVs by engaging in the activities presented in Section 10.4.

10.1 Chapter Overview

This chapter introduces two additional statistical strategies that can be performed via a regression analysis: *mediation* and *logistic regression*. With respect to the former, we introduce the concept of mediation and distinguish between *full* (or complete) *mediation* and *partial mediation*. We then present an example of a full mediation analysis followed by an example of a partial mediation analysis. We conclude this section with some additional commentary related to mediated models. We next introduce the concept of *binary logistic regression* and explain how the *logit* relates to predicting probabilities, odds, and odds ratios for comparing two groups or sets of conditions. We also comment on the statistical aspects of logistic regression and the interpretation of the logit through transformation of the regression coefficient. We conclude the chapter with a guided example of a research study that involves binary logistic regression.

10.2 Mediation

The Concept of Mediation

In the most general sense, the concept of *mediation* from a societal perspective involves an independent third party, called a *mediator*, who comes between two opposing factions with the intent of negotiating an acceptable resolution to whatever issues are at the heart of the factions' disagreement. When applied to statistics, mediation also involves an independent third party that comes between two factions. This third party consists of a variable, *M*, called a mediating variable, and the

DOI: 10.4324/9781003506935-14

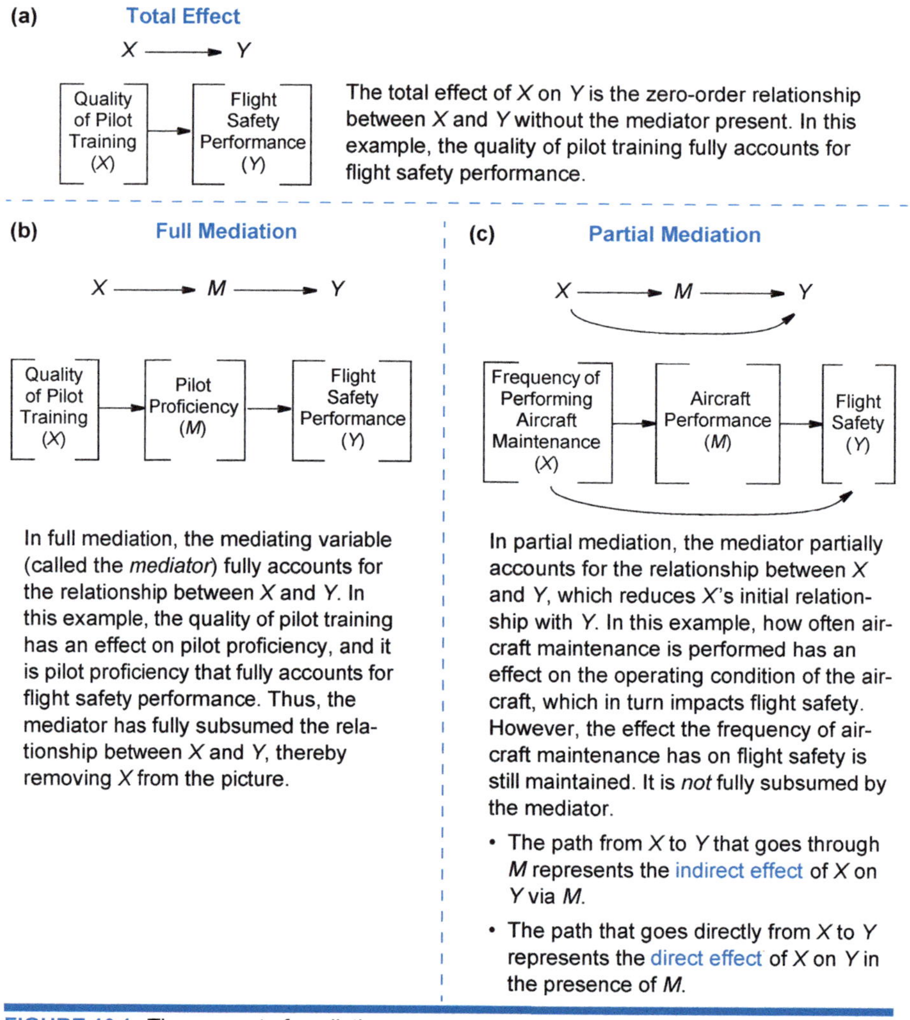

FIGURE 10.1 The concept of mediation.

two factions are an independent variable, *X*, and a dependent variable, *Y*. Although a mediating variable does not try to negotiate an amicable settlement between *X* and *Y*, it can nevertheless affect the relationship between *X* and *Y*. More specifically, it is possible for *M* to mediate the relationship between *X* and *Y* such that it (a) *completely accounts* for *X*'s relationship with *Y*, called full or complete mediation, or (b) *partially accounts* for *X*'s relationship with *Y*, called partial mediation. With respect to the former, *M* usurps *X*'s relationship with *Y* effectively rendering the relationship to zero, whereas in the latter, *X*'s relationship with *Y* is maintained, but it is diminished because of *M*'s presence.

The concepts of full and partial mediation are illustrated in Figure 10.1 using a concrete examle where we first consider the initial relationship between *X* = Quality of Pilot Training and *Y* = Flight Safety Performance. It is reasonable to conclude that *X* has a positive relationship with *Y*: As the quality of pilot training increases, flight safety performance also increases. This initial relationship is diagrammed in Figure 10.1(a) and represents the zero-order Pearson *r*. This relationship also can be represented by the corresponding zero-order coefficient of determination, r^2, the zero-order unstandardized regression coefficient, B_{XY}, and the zero-order standardized regression coefficient, β_{XY}. Regardless of its representation, the relationship reflects the total effect *X* has on *Y*.

We now introduce a second IV, namely, *M* = Pilot Proficiency. It is plausible that as the quality of pilot training increases (*X*), so too will pilot proficiency (*M*), which in turn leads to improved flight safety performance (*Y*). This is illustrated in Figure 10.1(b). As a result, the introduction of *M* changes the relationship between *X* and *Y*. More specifically, *M* takes over or fully assumes *X*'s relationship with *Y*. In other words, the initial relationship *X* has with *Y* is now completely assumed by *M*. When viewed from this perspective, *M* is a mediating variable, and the example illustrates the concept of full mediation: *M* completely eliminates *X*'s relationship with *Y* and now fully accounts for *X*'s initial relationship with *Y*. With respect to our example, the quality of pilot training (*X*) accounts for the level of pilot proficiency (*M*), which in turn accounts for a pilot's flight safety performance (*Y*). Thus, the mediator has now fully removed the influence the quality of pilot training has on flight safety performance.

Let's augment this example by considering the situation where *X* = Aircraft Maintenance Frequency, *Y* = Flight Safety, and *M* = Aircraft Performance. It can be posited that routine maintenance (*X*) has a direct effect on flight safety (*Y*). It also is reasonable to conjecture that routine maintenance (*X*) keeps aircraft operating in prime condition and thereby improving aircraft performance (*M*), which also can impact flight safety (*Y*). Unlike our previous example, we now have a situation where both *X and M* have a relationship with *Y*. This situation is depicted in Figure 10.1(c) and illustrates the concept of partial mediation. In partial mediation, the mediating variable accounts for *some* of the relationship between *X* and *Y*, but not all, and the presence of the mediating variable diminishes *X*'s initial relationship with *Y*, but it does not completely eliminate it. Because *M* does not completely account for the full relationship between *X* and *Y*, the *direct path* from *X* to *Y* represents *X*'s direct effect on *Y* in the presence of *M*, and the path from *X* to *Y* that goes *through M* is the indirect effect of *X* on *Y* via *M*. When considered from the given practical perspective, note that how often aircraft maintenance is performed (*X*) can impact both the operating condition of the aircraft (*M*) as well as flight safety (*Y*), but the operating condition of the aircraft also can impact flight safety. So, we now have a situation where *X*'s relationship with *Y* is *not* subsumed by *M*, but instead is maintained in the presence of *M*, which leads to *X* having both a direct and indirect (through *M*) relationship with *Y*.

A Numerical Example of Mediation

According to Baron and Kenny (1986), to determine if the relationship between *X* and *Y* is mediated by *M*, four conditions must be satisfied.

1. The first step is to confirm that the zero-order relationship between the targeted IV and the DV is significant. If not, then there is no relationship to mediate.
2. The second step is to confirm that the zero-order relationship between the targeted IV and the mediator is significant.

3. The third step is to confirm that the mediator has a significant effect on the DV when controlling for *X*.
4. The last step is to examine the effect the mediator has on the relationship between *X* and *Y*. If this relationship is *not* significant, then this implies that *X*'s initial relationship with *Y* has been completely subsumed by *M* and therefore there is full mediation. If, however, this relationship is significant, then this implies that *X* maintains its relationship with *Y* in the presence of *M* (albeit smaller) and therefore there is partial mediation.

We now apply this procedure to a concrete example. Consider the data set given in the file, "Ch_10 Mediation Example Data," which consists of hypothetical data relative to Tokarski (2021) who examined the effect of several factors on maintenance errors made by aviation maintenance technicians (AMTs) at a regional maintenance, repair, and overhaul (MRO) facility. A description of the variables for our example follows:

- *Y* = Maintenance Errors are the number of errors AMTs recall making on the job during a 1-year period, including self-detected errors as well as those reported by their supervisors.
- X_1 = Safety concerns are AMTs' perceptions of the extent to which they believe their organization has a strong concern for safety-related issues. This is measured using a 10-item instrument scored on a traditional 5-point Likert response scale. Thus, scores could range from 10 to 50, with higher scores reflecting higher perceptions (more positive) of safety concerns.
- X_2 = Training reflects the extent to which AMTs feel they were adequately trained for their job and if their organization provides opportunities for on-the-job training. This is measured using a 5-item instrument scored on a traditional 5-point Likert response scale. Thus, scores could range from 5 to 25, with higher scores reflecting higher perceptions (more positive) of training.
- *M* = Stress is the amount of on-the-job stress AMTs perceive to experience when performing their job. This is measured using a 9-item instrument scored on a traditional 5-point Likert response scale. Thus, scores could range from 9 to 45, with higher scores reflecting higher perceived levels of on-the-job stress.

Using the given data set, we present two examples of mediation—full and partial—and in both cases, the mediating variable is *M* = Stress. The reader is informed that we will use unstandardized regression coefficients (B_i values) to determine the significance of the targeted relationships.

An Example of Full Mediation. As noted earlier, full mediation occurs when an IV's relationship with the DV is fully accounted for by the mediating variable, as diagrammed in Figure 10.1(b). An example of full mediation is with respect to X_1's effect on *Y*. A summary of the results of this mediation analysis is provided in Table 10.1, and a step-by-step description follows. (See also Figure 10.2a.)

Step 1: Examine the zero-order relationship between X_1 and Y. As reported in Table 10.1, when *Y* is regressed on X_1, $B_{Y1} = -0.27$, $t(88) = -2.26$, $p = .0261$. Thus, the initial zero-order relationship between X_1 and *Y* is significant, and the *total effect* is equal to –0.27.

TABLE 10.1 *Summary of Results of a Mediation Analysis That Demonstrates Full Mediation*

Path	*B*	*SE*	*t*	*p*	95% CI
Zero-Order Relationships					
• Path *c*: $X_1 \rightarrow Y$ (Total effect) [a]	–0.27	0.12	–2.26	.0261	[–0.52, –0.03]
• Path *a*: $X_1 \rightarrow M$	–0.28	0.10	–2.78	.0067	[–0.49, 0.08]
• $M \rightarrow Y$	0.47	0.11	4.09	< .0001	[0.24, 0.70]
Full Model: *Y* Regressed on X_1 and *M*: $\hat{Y} = -0.15X_1 + 0.43M + 27.29$					
• Path *b*: $M \rightarrow Y$ [b]	0.43	0.12	3.57	.0006	[0.19, 0.66]
• Path *c′*: $X_1 \rightarrow Y$ (Direct effect) [c]	–0.15	0.12	–1.30	.1985	[–0.39, 0.08]
• Indirect effect [d]	–0.12	0.05	–2.22	.0260	[–0.29, –0.03]

Note. $N = 90$. The data set is given in the file "Ch_10 Mediation Example Data," where X_1 = Safety Concern, X_2 = Training, M = Stress, and Y = Maintenance Errors. See Figure 10.2(a). [a] Total effect = *Y* regressed on X_1 in the absence of the mediator, *M*. [b] This is $B_{Y(M \cdot 1)}$. [c] This is $B_{Y(1 \cdot M)}$. [d] $B_{\text{Indirect}} = B_{\text{Total}} - B_{\text{Direct}}$; SE_{Indirect} = Bootstrap *SE* (1000 cases); 95% $\text{CI}_{\text{Indirect}}$ = Bootstrap CI (1000 cases).

Step 2: Examine the zero-order relationship between X_1 and M. As reported in Table 10.1, when *M* is regressed on X_1, $B_{M1} = -0.28$, $t(88) = -2.78$, $p = .0067$. Thus, the zero-order relationship between X_1 and *M* is significant.

Step 3: Examine the effect M has on Y when controlling for X_1. When *Y* is regressed on X_1 and *M* simultaneously, the regression equation is $\hat{Y} = -0.15X_1 + 0.43M + 27.29$. Thus, the effect *M* has on *Y* in the presence of X_1 is 0.43. More specifically, as reported in Table 10.1's Full Model, $B_{Y(M \cdot 1)} = 0.43$, $t(87) = 3.57$, $p = .0006$. Therefore, the effect *M* has on *Y* when controlling for X_1 is significant, and this effect is equal to 0.43.

Step 4: Examine the effect M has on the X_1–Y Relationship. The effect the mediator has on the X_1–Y relationship is represented by the partial regression coefficient for X_1 from the full model's regression equation, $\hat{Y} = -0.15X_1 + 0.43M + 27.29$, which is –0.15. More specifically, as reported in Table 10.1's Full Model, $B_{Y(1 \cdot M)} = -0.15$, $t(87) = -1.30$, $p = .1985$. As a result, the effect X_1 has on *Y* in the presence of *M* is *not* significant, which indicates full mediation. Also note that this regression coefficient is "smaller" than the zero-order coefficient reported in Step 1.

Interpreting these results in the context of the given research setting, participants' perceptions of their organization's concern for safety-related issues initially has a significant negative effect on maintenance errors when examined in the absence of *M*. More specifically: As safety concern scores increase (i.e., become more positive), maintenance errors decrease ($B_1 = -0.27$). When stress enters the analysis, though, the effect of participants' safety concerns on maintenance errors becomes contextually "smaller," $B_{Y(1 \cdot M)} = -0.15$, which is not significant. Thus, stress fully mediates the effect of safety concerns on maintenance errors: As stress increases, maintenance errors also increase ($B_M = 0.42$).

The reader will note that the zero-order regression coefficient reported in Step 1 is $B_{Y1} = -0.27$, and the corresponding partial regression coefficient reported in Step 4 is $B_{Y(1 \cdot M)} = -0.15$. When considered with respect to the number line, the latter actually represents an *increase* in numerical value and not a reduction. However, when examined in the context of the given research setting, it actually is a *decrease*. More specifically, the zero-order B value indicates that as participants' safety concern scores increase (i.e., become more positive), maintenance errors decrease by 0.27 units. So, for every 10-point increase in safety concern scores, there are, on average, about 3 fewer reported maintenance errors. In the presence of stress, though, for every 10-point increase in safety concern scores, there are, on average, 1.5 fewer reported maintenance errors, which is a 50% reduction. This is why we placed quotes around the word "smaller" in Step 4's analysis. One plausible explanation for this reduction in maintenance errors in the presence of stress could be related to the theoretical model of Yerkes-Dodson (1908; see Figure 9.6 in Chapter 9), which posits that a certain amount of stress is needed for optimal performance.

An Example of Partial Mediation. As illustrated in Figure 10.1(c), X's relationship with Y in partial mediation takes two separate paths. In the first path, X leads to Y *indirectly* by going through M, and in the second path, X leads to Y *directly* by circumventing M. A numerical example of partial mediation based on the given data set is with respect to X_2's effect on Y. A summary of the results of this mediation analysis is provided in Table 10.2, and a step-by-step description follows.

Step 1: Examine the zero-order relationship between X_2 and Y. As reported in Table 10.2, when Y is regressed on X_2, $B_{Y2} = -0.73$, $t(88) = -3.54$, $p = .0006$. Thus, the initial effect X_2 has on Y is significant, and the *total effect* is equal to –0.73.

Step 2: Examine the zero-order relationship between X_2 and M. As reported in Table 10.2, when M is regressed on X_2, $B_{M2} = -0.53$, $t(88) = -2.96$, $p = .0040$. Thus, the zero-order relationship between X_2 and M is significant.

Step 3: Examine the effect M has on Y when controlling for X_2. When Y is regressed on X_2 and M simultaneously, the regression equation is $\hat{Y} = -0.53X_2 + 0.38M + 35.14$. Thus, the effect M has on Y in the presence of X_2 is 0.38. More specifically, as reported in Table 10.2's Full Model, $B_{Y(M \cdot 2)} = 0.38$, $t(87) = 3.25$, $p = .0016$. Therefore, the effect M has on Y when controlling for X_2 is significant, and this effect is equal to 0.38.

TABLE 10.2 *Summary of Results of a Mediation Analysis That Demonstrates Partial Mediation*

Path	*B*	*SE*	*t*	*p*	95% CI
Zero-Order Relationships					
• Path *c*: $X_2 \rightarrow Y$ (Total effect) [a]	–0.73	0.20	–3.54	.0006	[–1.14, –0.32]
• Path *a*: $X_2 \rightarrow M$	–0.53	0.18	–2.96	.0040	[–0.88, –0.17]
• $M \rightarrow Y$	0.47	0.11	4.09	< .0001	[0.24, 0.70]
Full Model: *Y* Regressed on X_2 and *M*: $\hat{Y} = -0.53X_2 + 0.38M + 35.14$					
• Path *b*: $M \rightarrow Y$ [b]	0.38	0.12	3.25	.0016	[0.15, 0.61]
• Path *c′*: $X_2 \rightarrow Y$ (Direct effect) [c]	–0.53	0.20	–2.58	.0115	[–0.93, 0.12]
• Indirect effect [d]	–0.20	0.09	–2.22	.0270	[–0.38, –0.02]

Note. $N = 90$. The data set is given in the file "Ch_10 Mediation Example Data," where X_1 = Safety Concern, X_2 = Training, M = Stress, and Y = Maintenance Errors. See Figure 10.2(b).
[a] Total effect = Y regressed on X_2 in the absence of the mediator, M. [b] This is $B_{Y(M \cdot 2)}$. [c] This is $B_{Y(2 \cdot M)}$. [d] $B_{\text{Indirect}} = B_{\text{Total}} - B_{\text{Direct}}$; SE_{Indirect} = Bootstrap SE (1000 cases); 95% $\text{CI}_{\text{Indirect}}$ = Bootstrap CI (1000 cases).

Step 4: Examine the effect M has on the X_2–Y Relationship. The effect the mediator has on the X_2–Y relationship is represented by the partial regression coefficient for X_2 from the full model's regression equation, $\hat{Y} = -0.53X_2 + 0.38M + 35.14$, which is –0.53. More specifically, as reported in Table 10.2's Full Model, $B_{Y(2 \bullet M)} = -0.53$, $t(87) = -2.58$, $p = .0115$. As a result, the effect X_2 has on Y in the presence of M is significant, and when coupled with the result from Step 3, this indicates partial mediation. Note further that this regression coefficient is contextually "smaller" than the zero-order coefficient reported in Step 1 (see the previous discussion about what we mean by "smaller"). This finding also indicates that the *direct effect* X_2 has on Y is –0.53, and the *indirect effect* is –0.2, which is the difference between the total and direct effects.

Concluding Comments

Following are a few closing comments relative to conducting a mediation analysis via MRC.

1. For our two examples, we summarized the results of the respective mediation analyses in Tables 10.1 and 10.2. Many researchers, however, report their findings via path diagrams, which populate the arrows of a mediation diagram with the numerical values that represent the relationship between variables. As noted earlier, these relationships may be reported as correlation coefficients, coefficients of determination, and unstandardized or standardized regression coefficients. Figure 10.2 contains the respective path diagrams for our two examples with the relationships being represented by unstandardized regression coefficients. The paths are denoted as follows: c = the total effect, which is the zero-order relationship between the IV and DV; a = the zero-order relationship between the IV and M; b = the relationship between M and the DV when controlling for the IV; and c' = the direct effect, which is the relationship between the IV and DV in the presence of M.
2. According to Baron and Kenny (1986), the concept of mediation has an inherent temporal precedence such that it is plausible and meaningful for X to first cause M, and then for M to cause Y. In other words, the notion that X precedes M in time, and that M in turn precedes Y in time, must have a theoretical or logical basis. This suggests that an *experimental research design* is appropriate for positing a mediation model or hypothesis because such designs permit manipulation and control of the variables. Therefore, the reader is advised that for *observational studies*, which lack manipulation and control, any claims of full or partial mediation that emerge from a mediation analysis are to be considered tentative or hypothetical at best.
3. Given a significant mediation model, the strength of the mediated relationship between the IV and DV is equal to the product of the a and b paths, where the respective correlation coefficients are reported.
4. The examples we provided reflect simple mediation, which involves three variables: an IV, DV, and M. There are several other forms of mediation. For example, in *sequential mediation*, several mediators are linked together and flow serially (i.e., in the same direction). For instance, in addition to stress, we could have included fatigue, anxiety, and depression as additional mediating variables

(a) Final Path Diagram for Full Mediation Example (see Table 10.1), Where X_1 = Safety Concern, M = Stress, and Y = Maintenance Errors

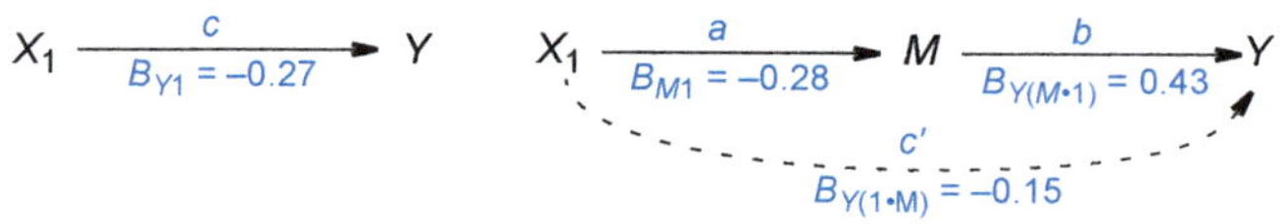

Path *c* = the *total effect* X_1 has on Y without the mediator, and it is significant, $t(88) = -2.26$, $p = .0261$.

Path *a* = zero-order effect between X_1 and M, and it is significant, $t(88) = -2.78$, $p = .0067$.

Path *b* = the effect M has on Y in the presence of X_1, and it is significant, $t(87) = 3.57$, $p = .0006$.

Path *c′* = the *direct effect* X_1 has on Y in the presence of M, and it is *not* significant, $t(87) = -1.30$, $p = .1985$. (We show it for completeness but use a dotted line to indicate it is not significant.)

(b) Final Path Diagram for Partial Mediation Example (see Table 10.2), Where X_2 = Training, M = Stress, and Y = Maintenance Errors

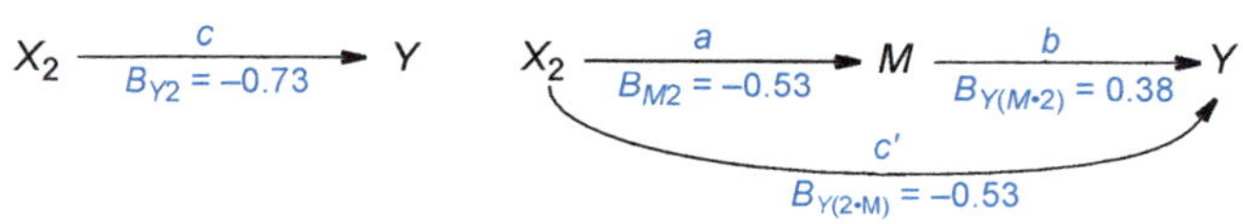

Path *c* = the *total effect* X_2 has on Y without the mediator, and it is significant, $t(88) = -3.54$, $p = .0006$.

Path *a* = zero-order effect between X_2 and M, and it is significant, $t(88) = -2.96$, $p = .0040$.

Path *b* = the effect M, has on Y in the presence of X_2, and it is significant, $t(87) = 3.25$, $p = .0016$.

Path *c′* = the *direct effect* X_2 has on Y in the presence of M, and it is significant, $t(87) = -2.58$, $p = .0115$.

FIGURE 10.2 Path diagrams for the two mediation examples.

in our two examples. Doing so would then lead to the following hypothetical diagram where M_i represents these additional mediators:

$$X \rightarrow M_1 \rightarrow M_2 \rightarrow M_3 \rightarrow M_4 \rightarrow Y$$

It also is possible to have *parallel mediation* where the mediators are "stacked" above each other and the path between X and Y goes through each mediator separately. This is illustrated in Figure 10.3. These and other forms of mediation are best examined using a statistical strategy called *structural equation modeling*, or *SEM*.

5. There are several online calculators that can be used to determine if a mediation model is significant. Two of the most frequently used tools are Soper (2025), which provides the "Sobel Test Calculator for the Significance of Mediation," and Preacher and Leonardelli (2025), which includes the Sobel test as well as the Aroian test and Goodman test. Readers are encouraged to consult these online calculators.

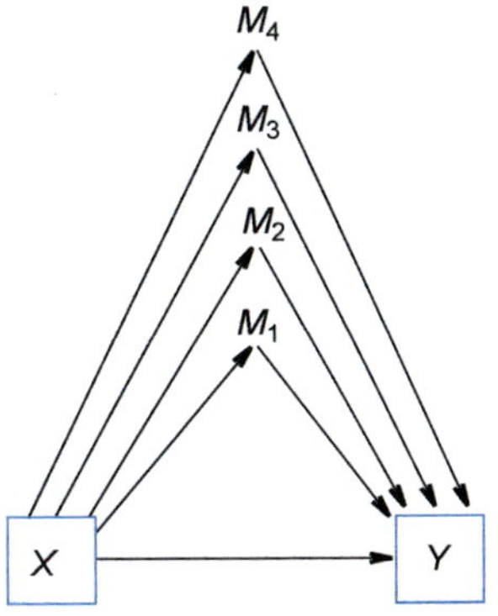

FIGURE 10.3 Example of parallel mediation.

10.3 Binary Logistic Regression

We now turn our attention to another special application of MRC, *logistic regression*, which is a statistical strategy for analyzing data when the dependent variable is nominal (categorical). Although logistic regression fundamentally is an extension of MRC, it differs from MRC in that it is not based on ordinary least squares (OLS) regression but on *maximum likelihood estimation* (MLE). In MRC the strength of the prediction equation is assessed by the multiple R, and the effect of the IVs on the DV is assessed by the regression coefficients. In logistic regression, we assess the regression coefficients and provide statistics related to *probability*, *odds*, and *odds ratios*, which represent the effect the IVs have on the group membership variable represented by the DV. In this section, we introduce the concept of logistic regression—more specifically, *binary logistic regression* where the DV is dichotomous—and then build an understanding of a logistic regression analysis by presenting the concepts of probability, odds, and odds ratios. We follow this discussion with a guided example of a binary logistic regression analysis for a research study.

The Concept of Logistic Regression

There are many instances of research studies where the dependent variable is not continuous but instead represents group membership. This is especially relevant in the medical field. For example, consider a study that examines the outcome of cancer patients relative to their marital status. Here, the research question might be, "What is the effect of marital status on cancer patients' survival?" Note that in this example the DV is a dichotomous variable (patient lives vs. dies), and the IV is also a dichotomous variable (married vs. not married). As another example, medical researchers might be interested in factors that distinguish between whether a man develops or does not develop prostate cancer. Once again, the DV is dichotomous, but the IV could be continuous or nominal and might include a man's age, diet, exercise routine, physical environment, or geographical location.

There are also numerous circumstances in aviation research where the dependent variable is not continuous. For example, consider a pilot who opts to abort a takeoff when presented with a hot engine warning light vs. another pilot who opts not to abort a takeoff when presented with the same situation. When examined from a research perspective, a corresponding research question might be, "What factors distinguish between pilots who opt to abort takeoff vs. pilots who opt not to abort takeoff when presented with a hot engine warning light?" Notice that in the given research setting, the group membership variable is now associated with the DV, not the IV, and the DV is binary or represents a dichotomy.

From a research design perspective, the given examples represent an *ex post facto study* (see Appendix A), and as noted in Appendix A, ex post facto studies are considered *retrospective* because the effects on the DV have already occurred. In an ex post facto study, the focus is on searching for presumed factors of a preexisting effect. More specifically, the researcher's objective is to identify possible factors of group membership by examining the potential effect various IVs have on the group membership of the DV. From a generic perspective, the primary research question is, "What are the possible factors that led to group membership?" When applied to our aviation example, pilots belong to one of two pre-existing groups: those who chose to abort takeoff and those who chose not to abort takeoff, and the focus of the study is to find factors that distinguish between the two groups of pilots. Some potential factors might be pilot experience (e.g., number of flight hours and years as PIC), physiological (e.g., biological sex at birth and age), psychological (e.g. self-efficacy and hazardous attitudes), or physical (e.g., remaining runway length available and atmospheric conditions). This is illustrated in Figure 10.4.

As another example, consider an airport where the visibility is obscured by radiation fog from time-to-time delaying departure operations in the morning. In this case, the research goal might be to predict the occurrence or non-occurrence of radiation fog at the airport. For this situation, note that the focus is on physical conditions and the corresponding RQ would be, "What factors predict the daily occurrence or non-occurrence of radiation fog at an airport?" Thus, daily weather reports at an airport might include one of two conditions in the morning: "fog" vs. "no fog," which is a binary DV. We would then search for possible factors that led to these conditions.

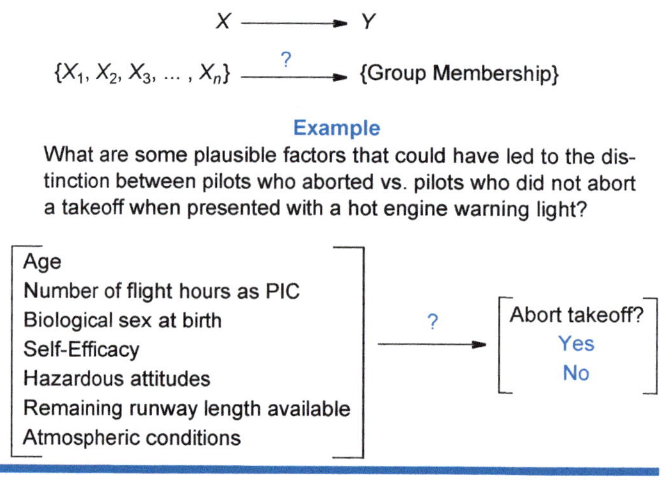

FIGURE 10.4 Illustration of an ex post facto study where group membership is on the DV.

Examples might include meteorological and geographical factors such as the wind speed at specific altitudes, dew point temperatures, and the airport's proximity to a cold body of water.

To analyze data from an ex post facto study where group membership is on the DV, we use a statistical strategy called logistic regression. When the DV consists of exactly two groups such as those we presented above, then this statistical strategy is referred to as binary logistic regression, which is the topic of discussion for the current section (10.3) and the next section (10.4). Thus, our presentation is restricted to those cases that involve a dichotomous DV. Other examples of aviation-based research studies that involve a dichotomous DV include (a) identifying factors that determine if an airline enters or does not enter a new market, (b) identifying factors that distinguish between commercial airline passengers who trust vs. do not trust aircraft automation, and (c) exploring factors that determine if GA pilots continue or do not continue with an inadvertent VFR-to-IMC flight. Note once again from these examples that the grouping is on the DV, the DV implies a preexisting condition based on group membership, and we are searching for possible factors that could have led to the group's effect.

Developing An Understanding of Logistic Regression

To understand the statistical strategy of logistic regression, we first present an example of a study that represents a chi-square test for independence, which is also called two-way chi-square because it involves two categorical variables. The reader might recall from a previous statistics course that a chi-square test for independence tests the null hypothesis that the proportion of frequencies in the categories that represent the levels of one variable are the same for those at all levels of the other variable. For our example, we will consider a hypothetical study involving aviation maintenance technicians (AMTs) at a local aircraft MRO where the objective is to examine the relationship between AMTs' marital status (Y) and their work-related stress level (X).

For this study, the DV reflects the dichotomy, Y = Married vs. Not Married, and the IV is X = AMTs' perceptions of the amount of stress they feel while working at the MRO, where perceptions are scored on an 8-item traditional 5-point Likert response scale. Thus, scores could range from 8 to 40 with higher scores reflecting higher perceived levels of stress. To perform a chi-square analysis, though, we must first express X as a categorical variable. To do this, we change the data type for X from continuous to nominal in our statistics program, and then create two groups: Low Stress and High Stress. Because the mean stress level for the continuous IV is $M_X = 26.4$, we define Low Stress as scores ≤ 26 and High Stress as scores > 26. A copy of the data set is provided in the file, "Ch_10 Numerical Example Data with 1 IV," and a copy of the corresponding contingency table is given in Table 10.3. The results of this chi-square analysis are $\chi^2(1) = 4.525$, $p = .0334$. Note that the adjusted Pearson residuals confirm that statistical significance is with respect to Married with High Stress vs. Married with Low Stress.

Focusing on Probability. The results presented in Table 10.3 can be used to compare the probability of AMTs being married or not married relative to their stress levels. Some readers might recall from a previous course that for simple probability, the probability event A will occur is represented by the following ratio:

TABLE 10.3 *Contingency Table for Stress Level and Martial Status of AMTs*

	Y = Marital Status		
X = Stress Dichotomy [a]	Married (Y = 1)	Not Married (Y = 0)	Total
High Stress (X = 1)			
Count (N)	38	6	44
Percent (%) [b]	86.4	13.6	
Residuals [c]	2.127	–2.127	
Low Stress (X = 0)			
Count (N)	31	15	46
Percent (%) [b]	67.4	32.6	
Residuals [c]	–2.127	2.127	
Total	69	21	90

Note. $N = 90$. $\chi^2(1) = 4.525$, $p = .0334$ (two-tailed test).
[a] High stress represents scores greater than 26 on the questionnaire, whereas low stress represents scores of 26 or less. [b] Percentages are based on the total counts per group with respect to rows (e.g., 38/44 = 86.4%). [c] Residuals are adjusted Pearson residuals, which follow a normal distribution. Thus, cells with absolute values greater than the critical value of 1.96 are significant at $\alpha = .05$ (two-sided test).

$$P(A) = \frac{\text{The number of successful outcomes } (S)}{\text{The total number of possible outcomes } (T)} = \frac{\text{Success } (S)}{\text{Total } (T)} = \frac{S}{T}$$

For example, note from Table 10.3 that of the 44 participants with high stress levels, 38 are married. Thus, the proportion of AMTs with a high stress level who are married is 38/44, or equivalently, the *probability* is 38/44 = .864, which means there is an 86.4% chance of randomly selecting from this subgroup a high-stressed AMT who is married. Similarly, if the focus in on AMTs with low stress levels, then the corresponding proportion is 31/46, the probability is .674, and there is a 67.4% chance of randomly selecting a low-level stressed AMT who is married from this subgroup. Given these results, we can infer that AMTs with high stress levels have a higher probability of being married, or equivalently, a lower probability of *not* being married. Note also from Table 10.3 that this result is statistically significant, $\chi^2(1) = 4.525$, $p = .0334$, which means that the two variables, level of stress and marital status, are not independent of each other.

From Probability to Odds. Let's now change the focus of this discussion from probability to *odds*. Most readers have an intuitive understanding of the concept of odds. For example, most readers know that if the odds on a bet are 3 to 1 and we bet a dollar and win, then we will get $3. When applied to the current example, the focus is on predicting the odds of an AMT being married so that we can generate probabilities and provide a comparison of the effects of stress levels on group membership.

Some readers might recall that odds can be presented in two ways: the odds in favor of an event occurring and the odds against an event occurring. Furthermore, odds are defined as the ratio between probabilities. With respect to odds in favor, the ratio is the probability of an event occurring, $p(A)$, vs. the probability of the event not occurring, $1 - p(A)$, and the odds against is the reciprocal of odds in favor.

$$\text{Odds in favor of } A = \frac{\text{Probability that } A \text{ will occur}}{\text{Probability that } A \text{ will not occur}} = \frac{p(A)}{1 - p(A)}$$

$$\text{Odds against } A = \frac{\text{Probability that } A \text{ will not occur}}{\text{Probability that } A \text{ will occur}} = \frac{1 - p(A)}{p(A)}$$

Let's apply this concept to our current example with respect to determining the odds that an AMT is married or not married. As reported in Table 10.3, of the $N = 90$ AMTs, 69 are married and 21 are not married. Therefore:

- $P(\text{Married}) = \frac{69}{90} = .767$
- $P(\text{Not Married}) = \frac{21}{90} = .233$
- $\text{Odds (Married)} = \frac{.767}{.233} = 3.29 \text{ to } 1$
- $\text{Odds (Not Married)} = \frac{.233}{.767} = 0.304 \text{ to } 1$

As a result, the odds of being married are greater than the odds of not being married. This means that if we were to randomly select an AMT from the given sample, there is a greater likelihood this person is married than not married. In general:

- When the odds in favor of an event occurring are greater than 1, then the event is *more likely* to occur than not occur. Thus, the higher the odds, the more likely the event will occur.
- When the odds in favor of an event occurring are less than 1, then the event is *less likely* to occur.
- If the odds are 1:1, then this is known as even odds and the likelihood that an event will occur is equal to the likelihood that the event will not occur.

When calculating odds by dividing the numerator by the denominator, it is understood that the odds are being expressed with respect to 1 and therefore this part of the expression is generally not included. So, for our example, we would express the odds of being married as 3.29 and the odds of not being married as 0.304.

In addition to converting probability to odds, we also can go in the opposite direction and convert odds to probability using the following formula:

$$P(A) = \frac{\text{Odds of } (A)}{[1 + \text{Odds of } (A)]}$$

For example, given that the odds of an AMT being married are 3.29, the corresponding probability is

$$P(\text{Married}) = \frac{\text{Odds of (Married)}}{[1 + \text{Odds of (Married)}]} = \frac{.329}{.429} = .767$$

The Concept of Odds Ratios. Another aspect of odds is the concept of odds ratios (*OR*). Just as odds represent a ratio between probabilities, we also can examine the *ratio of the odds*. Odds ratios are often more useful in research than probabilities and odds because they enable us to compare the occurrence of an event given two sets of conditions. For example, if a study finds that 3 of every 4 pilots with low self-efficacy abort a takeoff when presented with a hot engine warning light, then the odds a pilot with low self-efficacy aborts a takeoff are 3:1. If the same study finds that 2 of every 3 pilots with high self-efficacy abort a takeoff when presented with a hot engine warning light, then the odds a pilot with high self-efficacy aborts a takeoff are 2:1. As a result, the corresponding *OR*, which consists of the odds in favor of the first group (low self-efficacy pilots) vs. the odds in favor of the second group (high self-efficacy pilots) is

$$3{:}1/2{:}1 = \frac{\left(\frac{3}{1}\right)}{\left(\frac{2}{1}\right)} = \frac{3}{1} \times \frac{1}{2} = \frac{3}{2} = 3{:}2$$

This implies that pilots with low self-efficacy are 1.5 times more likely to abort a takeoff than pilots with high self-efficacy when presented with a hot engine warning light. Applying this concept to our current example, let's compare the status of being married with respect to AMTs' perceived stress levels at work. Note the following from Table 10.3:

- 38 of 44 AMTs with *high stress* levels are married and therefore

$$\text{Odds(Married and High Stress)} = \frac{38}{6} = 6.33$$

- 31 of 46 AMTs with *low stress* levels are married and therefore

$$\text{Odds(Married and Low Stress)} = \frac{31}{15} = 2.07$$

- So, the corresponding odds ratio with respect to stress and being married is

$$\text{Odds Ratio(Stress)} = \frac{6.33}{2.07} = 3.06$$

This last result indicates that AMTs with high work-related stress levels are 3 times more likely to be married than AMTs with low work-related stress levels. This result confirms our previous inference that AMTs with high work-related stress levels have a higher probability of being married. It also provides a comparison between the two sets of conditions of high versus low stress. The ability to compare the odds for and odds against an event occurring between groups is a powerful research tool.

Note that we also could report the reciprocal of this result, which is 0.33. In doing so, the interpretation would then be that AMTs with low work-related stress are one-third as likely to be married than AMTs with high work-related stress. Although it does not matter which form of the *OR* to report because the nature of the relationship does not change, readers generally appreciate when authors report *OR* values that are greater than 1 as opposed to less than 1 because the interpretation is more easily understood.

The Logit Function and the Logistic Regression Equation. With the concepts of probability, odds, and odds ratios in our background, we now introduce the *logit function* and corresponding *logistic regression equation*. The reader is informed that our presentation here is purposely abridged and designed to simply provide a working understanding of the concepts presented. For a more detailed or exhaustive exposition, the reader is directed to Cohen et al. (2003, Chapter 13) or Warner (2008, Chapter 21).

Because logistic regression involves a categorical dependent variable, the interpretation of the results of a logistic regression analysis involves predicted probabilities or odds, including *OR*. This poses an issue, though, because, without going into any detail, odds are not compliant with the concept of linearity. However, this issue can be resolved by taking the natural logarithm (ln) of odds, which is the *log*istic probability un*it*, or logit function:

$$\text{logit} = \ln\left[\frac{p_i}{1-p_i}\right]$$

By taking the natural logarithm of odds in favor, the resulting logit function better represents the characteristics of a nominal dependent variable. Furthermore, because scores on the logit function are compliant with a linear model, the logistic regression equation may be represented as a facsimile of the general linear regression equation:

$$\text{logit} = \ln\left[\frac{p_i}{1-p_i}\right] = B_1X_1 + B_2X_2 + B_3X_3 + \ldots + B_iX_i + B_0$$

As a result, the partial regression coefficients (B_i) and regression constant (B_0) of this equation may be interpreted in a similar manner as in linear regression. More specifically, each B_i reflects the facsimile of a "slope" and is interpreted as the amount of change in the logit for each unit change in the corresponding predictor, X_i. For example, when we conduct a logistic regression analysis for our earlier example involving AMTs' marital status (*Y*) and work-related stress levels (*X* where *X* is a dichotomized continuous IV; see the data set given in the file, "Ch_10 Numerical Example Data with 1 IV"), the regression equation is

$$\text{logit(Married)} = 1.12X_1 + 0.73$$

Here, $B_1 = 1.12$ is interpreted as "for every 1-unit change in X_1, there is a 1.12-unit change in the logit. Because X_1 = Low Stress (coded 0) vs. High Stress (coded 1), the 1-unit change in X_1 represents a change from either Low Stress to High Stress or from High Stress to Low Stress. Similarly, when $X_1 = 0$ (Low Stress), the logit is equal to $B_0 = 0.73$.

Notice how unwieldy or awkward this interpretation is of the partial regression coefficient. A more practical interpretation would be to equate the unit change in B_i to a unit change in *odds*. To do this, we can re-express the natural logarithm of odds to odds via the natural exponential function, *e*. For example, some readers might recall from a previous mathematics course that

$$\text{Log}_{\text{Base}}\ \text{Power} = \text{Exponent is equivalent to } \text{Base}^{\text{Exponent}} = \text{Power}$$

Therefore, if we know that ln (odds) = *B*, where the base of the natural logarithm is the natural exponential function, *e*, then we can re-express this result as e^B = odds. When applied to logistic regression, e^B represents odds ratio, and the corresponding regression coefficient is interpreted from the perspective of a 1-unit change in X_i that corresponds to a change in the odds ratio. Thus, for the current example, if $B_1 = 1.12$, then $e^{1.12} = 3.06$. Because High Stress is coded 1, AMTs with high work-related stress levels are 3 times more likely to be married than AMTs with low work-related stress levels. Note that this is exactly what we derived earlier.

A few comments:

1. When applied to a *specific IV*, the odds ratio is equal to e^{B_i}, where B_i represents the corresponding partial regression coefficient for the predictor, X_i.
2. When applied to a *specific case* from the data set, the odds ratio is equal to

$$e^{(B_1X_1 + B_2X_2 + B_3X_3 + \ldots + B_iX_i + B_0)}$$

3. To determine the percentage of change in the odds ratio that equates to a 1-unit change in a targeted IV, subtract 1 from e^{B_i} and then multiply the result by 100. So, for the current example where $e^{1.12} = 3.06$, the percentage of change from High Stress to Low Stress with respect to being married or not is

$$(3.06 - 1) \times (100) = 206$$

Therefore, the odds of being married increase by 206% for AMTs with high work-related stress vs. AMTs with low work-related stress.
4. Most statistics programs report odds ratios as part of their output for a logistic regression analysis.

As another example, consider the data set given in the file, "Ch_10 Numerical Example with 3 IVs," which relates to a previous example that compared pilots who abort vs. pilots who do not abort takeoff when presented with a hot engine warning light (see Figure 10.4). For this example, the predictors are the first three IVs listed in Figure 5.4 and are defined as follows:

- X_1 = Age is a dichotomous variable where "Older" represents pilots 50 years old or older (coded 1) and "Younger" represents pilots younger than 50 years old (coded 0). Thus, the comparison for X_1 is Older vs. Younger aged pilots.
- X_2 = Number of Flight Hours as PIC is a dichotomous variable where "High" represents pilots with more than 6,070 hours (coded 1) and "Low" represents pilots with at most 6,070 hours (coded 0). Thus, the comparison for X_2 is High vs. Low flight hours as PIC.
- X_3 = Biological Sex at Birth, where Female is coded 1 and Male is coded 0. Thus, the comparison for X_3 is Female vs. Male pilots.

For the dichotomous dependent variable, Y = "Yes" represents pilots who aborted takeoff (coded 1) and Y = "No" represents pilots who did not abort takeoff (coded 0). Thus, the comparison for the DV is Yes vs. No with respect to aborting takeoff. A summary of the results from conducting a logistic regression analysis for the given data set is provided in Table 10.4, and a brief review of the results follows.

A Note of Caution: For readers who prepare separate contingency tables for the IVs similar to Table 10.3 and then manually calculate the odds ratios for each IV, the respective calculated odds ratios will *not* be the same as the results from the logistic regression analysis. This is because the analysis involves regressing Y on the three IVs simultaneously, which means that the respective odds ratios are in the presence of the other IVs. However, the manually calculated odds ratio will be the same if separate logistic regression analyses are conducted for each IV.

1. The linear form of the logistic regression equation is

$$\text{logit(Abort Takeoff = Yes)} = 0.89X_1 - 0.26X_2 + 2.03X_3 - 2.18$$

TABLE 10.4 *Summary of Logistic Regression Analysis Results for Chapter 10's Numerical Example with 3 IVs*

Model Summary

Model	Log-Likelihood	*df*	Chi-Square	*p*
Difference	8.83	3	17.66	.0005
Full	39.53			
Reduced	48.36			

Note. Pseudo R^2 = Difference/Reduced = 8.83/48.36 = .1826.

Coefficients (Parameter Estimates)

						95% CI [b]	
Term [a]	*B*	*SE*	Chi-Square	*df*	*p*	*LL*	*UL*
Constant	−2.18	0.81	7.16	1	.0075	−3.99	−0.73
X_1 = Age	0.89	0.58	2.33	1	.1272	−0.25	2.06
X_2 = Flight Hours as PIC	−0.26	0.57	0.22	1	.6395	−1.37	0.86
X_3 = Biological Sex at Birth	2.03	0.70	8.40	1	.0038	0.76	3.59

Note. $N = 71$.
[a] Age is dichotomous: Younger is < 50 years old and Older is ≥ 50 years old. Flight Hours is dichotomous: Low is ≤ 6,070 hours and High is > 6,070 hours. [b] CIs are likelihood-based.

Odds Ratios

			95% CI	
Term	Odds Ratio	Reciprocal	*LL*	*UL*
X_1 = Age	2.43	0.41	0.78	7.84
X_2 = Flight Hours as PIC	0.77	1.30	0.25	2.37
X_3 = Biological Sex at Birth	7.61	0.13	2.15	36.25

As noted earlier, each term is a partial regression coefficient and is interpreted in a similar manner as in linear regression:

- $B_1 = 0.89$. Holding all other IVs in the model constant, the logit increases on average by 0.89 units for "Older" pilots compared to "Younger" pilots. This is not significant, $\chi^2(1) = 2.33$, $p = .1272$.
- $B_2 = -0.26$. Holding all other IVs in the model constant, the logit decreases on average by 0.26 units for pilots with "High" flight hours as PIC compared to pilots with "Low" flight hours as PIC. This is not significant, $\chi^2(1) = 0.22$, $p = .6395$.
- $B_3 = 2.03$. Holding all other IVs in the model constant, the logit increases on average by 2.03 units for Female pilots compared to Male pilots. This is significant, $\chi^2(1) = 8.40$, $p = .0038$.
- $B_0 = -2.19$ is the predicted logit when all $X_i = 0$. For the current example, this applies to Younger (X_1) Male pilots (X_3) with Low flight hours as PIC (X_2).

2. As reported in the Odds Ratios subtable of Table 10.4, the predicted odds ratios for each IV are as follows:
   - X_1. The odds ratio is $e^{0.89} = 2.43$. Holding all other IVs in the model constant, Older pilots are approximately 2.5 times more likely than Younger pilots to abort takeoff when a hot engine warning light is presented.

- X_2. The odds ratio is $e^{-0.26} = 0.77$. Holding all other IVs in the model constant, the odds of aborting takeoff are approximately three-fourths as high for pilots with High flight hours as PIC than it is for pilots with Low flight hours as PIC. Because $OR < 1$, it is more intuitive to interpret the reciprocal, which is $OR = 1.30$: Holding all other IVs in the model constant, pilots with Low flight hours as PIC are 1.3 times more likely than pilots with High flight hours as PIC to abort takeoff when a hot engine warning light is presented.
- X_3. The odds ratio is $e^{2.03} = 7.61$. Holding all other IVs in the model constant, Female pilots are approximately 7.6 times more likely to abort takeoff when a hot engine warning light is presented than Male pilots.

3. As noted earlier, in addition to determining the odds ratios for the predictors, we also can determine the odds ratio for a particular participant. For example, consider Case 4 from the given data set. Note that this participant is an Older Female with Low flight time as PIC. Therefore, $X_1 = 1$, $X_2 = 0$, and $X_3 = 1$, and the odds ratio is

$$\begin{aligned} & e^{(B_1X_1 + B_2X_2 + B_3X_3 + B_0)} \\ &= e^{[(0.89)(1) + (-0.26)(0) + (2.03)(1) - 2.18]} \\ &= e^{[(0.89) + (0) + (2.03) - 2.18]} \\ &= e^{0.74} \\ &= 2.09 \end{aligned}$$

Therefore, this particular participant is approximately 2 times more likely to abort takeoff than not abort takeoff when presented with a hot engine warning light.

10.4 Binary Logistic Regression in Research: A Guided Example

We now present a guided example of how to conduct a logistic regression analysis for a research study that involves a dichotomous DV. This Guided Example is based on a hypothetical study, motivated by Jones-McKinney (2015), and involves two groups of pilots: Pilots who were involved in a hazardous event during the past 5 years (the "Yes" group, coded 1), and Pilots who were not involved in a hazardous event during the past 5 years (the "No" group, coded 0). (*Note*: We leave it to the reader to operationally define "hazardous event.") The predictor variables are:

- X_1 = Age is a continuous variable measured in years.
- X_2 = Biological Sex is a categorical variable coded Female = 1 and Male = 0.
- X_3 = Level of Education is a categorical variable that represents pilots' highest level of education with respect to four categories: high school diploma, 2-year college, 4-year college, and graduate degree (master's or doctoral).
- X_4 = Total Flight Hours is a continuous variable that represents pilots' total flight time.
- X_5 = Perceived Airport Experience reflects pilots' perceptions of their experience with various airfield layouts, navigational and lighting aids, and signage. This is measured using an 8-item instrument scored on a traditional 5-point Likert response scale. Thus, scores could range from 8 to 40, with higher scores reflecting greater perceptions of airport experience.

The structure of our presentation consists of seven parts: Research Questions and Statistical Hypotheses, Sample Size Planning, Preparing the IVs for Analysis, Preliminary Data Screening, Data Analysis, Interpretation of Results, and Post Hoc Considerations. The corresponding data set is provided in the file, "Ch_10 Guided Example Data."

Research Question and Statistical Hypotheses

Research Question. To what extent do X_1 = Age, X_2 = Biological Sex, X_3 = Level of Education, X_4 = Total Flight Hours, and X_5 = Perceived Airport Experience distinguish between pilots who were involved vs. pilots who were not involved in a hazardous event during the past 5 years?

Statistical Hypotheses. We express the corresponding null and alternative hypotheses in terms of odds ratios via the partial regression coefficient. Instead of presenting separate sets of null and alternative hypotheses for each IV, we instead present a single set with subscript *i* to reflect each IV. Note that if $B_i = 0$, then $e^0 = 1$, which means "equal odds."

- H_0: $B_i = 0$. None of the IVs has a significant predictive effect in distinguishing between the pilot groups. Thus, pilots have an equal chance of belonging to the "Yes" or "No" group with respect to each predictor.
- H_1: $B_i \neq 0$. At least one IV has a significant predictive effect in distinguishing between the pilot groups. Thus, the odds of pilots belonging to the Yes group will change significantly in the presence of at least one IV.

Sample Size Planning

The literature is replete with various sample size guidelines for logistic regression. For example, according to Bujang et al. (2018), approaches to determining minimum sample sizes vary considerably and include: (a) a "flat" minimum of $N = 500$; (b) $N = xk$, where x is a multiple of 10 between 10 and 50 and k = the number of IVs; and (c) $N = 100 + xk$. Based on these respective guidelines, minimum sample sizes for the given Guided Example with five IVs are (a) $N = 500$, (b) N ranging between 50 and 250, and (c) N ranging between 150 and 350. The given data set has N = 109 cases, which is consistent with approach (b).

Preparing the IVs for Analysis

One of the assumptions of logistic regression is that a linear relationship exists between the IVs and the logit of the DV. This assumption of linearity of the logit is violated when interaction terms between a continuous IV and its natural logarithm are statistically significant within the regression model (Tabachnick & Fidell, 2013). One approach to ensuring the linearity of the logit is to transform all IVs into dichotomies because, mathematically, the linearity of the logit cannot be violated by binary data because ln(0) is undefined and ln(1) = 0. Using dichotomous independent variables in logistic regression also aids in the interpretation of the model's results. For example, MacKinnon and Dwyer (1993) argue that because the regression coefficients in a logistic regression analysis may be measured on different scales, the interpretation of any mediation effects can be problematic if the IVs are not dichotomous. Thus, partitioning the IVs into dichotomies also provides a format suitable for measuring mediation effects.

TABLE 10.5 *Assignment of Variables into Dichotomies*

Variables	Assigned Dichotomy [a]	Basis for Dichotomy
Independent Variables		
X_1 = Age	45 or older vs. Younger than 45	Midrange = 45.5
X_2 = Biological Sex	Female vs. Male	Self-evident
X_3 = Level of Education	Graduate degree vs. Less than graduate degree	Contingency analysis (see Table 10.6)
X_4 = Total Flight Hours	1500 or more vs. Fewer than 1500	FAA level for ATPL
X_5 = Airport Experience	Scores of 32 or less vs. Scores greater than 32	Mean (M = 32.3)
Dependent Variable		
Y = Hazardous Event	Yes vs. No	Group membership

Note. The splits for the dichotomies were tested and validated through the development of contingency tables.
[a] The assigned dichotomies are relative to the group coded 1 vs. the group coded 0.

For the data set given in the file, "Ch_10 Guided Example Data," we provide both raw and corresponding dichotomous data for each IV. We also provide in Table 10.5 the basis we used for creating the respective dichotomies. When creating dichotomies for a continuous IV, a common approach is to base the dichotomy on a measure of central tendency such as the mean or median of the distribution. However, it is also appropriate to consider the context of the targeted variable. For example, the dichotomy we created for X_4 = Total Flight Hours was based on the FAA's requirement for holding an Air Transport Pilot's License (ATPL) because it is an important consideration in the study.

When creating dichotomies for IVs, it also is important to confirm there is at least one case per group. This can be accomplished via a contingency table for each IV such as Table 10.6, which is an example of a contingency table using Y = Hazardous Events and X_4 = Level of Education based on the dichotomy Graduate Degree versus Less than Graduate Degree. Although not presented, we also conduct a contingency analysis of the other IVs to confirm that each cell has at least one frequency and that there is a relatively balanced number of counts in each group ("Yes" vs. "No") for each dichotomy. As reported in Table 10.6, there are 21 "No" cases and 14 "Yes" cases for Y, and 25 "No" cases and 46 "Yes" cases for pilots with a graduate degree and less than a graduate degree, respectively. Warner (2008) recommends a minimum five counts per cell and the choice of using the mean, median, or midrange as shown in Table 10.5 was predicated on these objectives.

Preliminary Data Screening

Preliminary data screening for logistic regression is similar to that of MRC and includes confirming that the data set is compliant with the assumptions of logistic regression and performing diagnostics on the data set to ensure that the model is appropriate.

Assumptions. The assumptions for logistic regression are less restrictive than those for MRC and are fairly straightforward: (a) a dichotomous DV; (b) independent scores on the DV from an assessment that served as the basis for group membership, for example, administering a self-efficacy instrument and then using the mean

TABLE 10.6 *Contingency Table for Graduate versus Less than Graduate Degree and Occurrence of Hazardous Events*

Y = Hazardous Event	No	Yes	Total
Graduate	22	15	37
Less than a Graduate	25	47	72
Total	47	62	109

Note. $N = 109$. $\chi^2(1) = 5.80$, $p = .015$. The split for the targeted dichotomy (graduate vs. less than a graduate degree) is valid because there is at least one occurrence and more than five for each cell in the table.

score to partition the group into low vs. high self-efficacy; (c) a correctly specified model, which presumes that the targeted predictors are grounded in theory, the literature, preliminary studies, or from personal experience/expertise; and (d) mutually exclusive categories for the DV. For the Guided Example, these assumptions are met as described below.

- The DV consists of two groups: pilots who have vs. pilots who have not been involved in a hazardous event during the past 5 years.
- Group membership is not the result of participants' scores on some assessment but instead they are based on participants' personal background with respect to whether or not they were involved in a hazardous event. Therefore, this assumption is not applicable.
- The targeted predictors are based on a past study (Jones-McKinney, 2015).
- The two categories of participants are presumed to be mutually exclusive based on pilots' self-reported involvement in a hazardous event, as defined by the researcher. We say "presumed" because we have no control over the veracity of participants' response to the question, "Were you involved in a hazardous event over the past 5 years?" Pilots who reported being involved in at least one hazardous event are assigned to the "Yes" group, and pilots who reported being involved in zero hazardous events are assigned to the "No" group.

Diagnostics. Two key diagnostics associated with logistic regression are cell sizes and outliers.

- Similar to chi-square, logistic regression requires a minimum cell size of 5. For the given data set, when we examine the respective contingency tables for each IV (see Table 10.6 as an example), the frequency of each cell is at least 5. Therefore, no further action is needed.
- An outlier analysis should be performed for all continuous IVs used in logistic regression to ensure there are no extreme cases included in the model (Warner, 2008). Although the given data set for the current Guided Example contains three continuous IVs, X_1 = Age, X_4 = Total Flight Hours, and X_5 = Perceived Airport Experience, the reader will recall that we transformed these IVs into dichotomous categories and therefore we do not check for outliers as we would for quantitative IVs. Instead, the focus is on ensuring that the size of the respective categories for the IVs are balanced as presented below.

- X_1 = Age: Under 45 (n = 53); 45 or older (n = 56).
- X_2 = Biological Sex: Female (n = 15); Male (n = 94).
- X_3 = Level of Education: Less than Graduate degree (n = 72); Graduate degree (n = 37).
- X_4 = Flight Hours: Fewer than 1500 (n = 48); 1500 or more (n = 61).
- X_5 = Perceived Airport Experience Scores: 32 or less (n = 54); greater than 32 (n = 55).

Of the five IVs, there is an imbalance for X_2. This imbalance, however, is consistent with the current state of affairs of ATPs within the aviation industry, and therefore we will not make any attempts to rectify this imbalance.

Data Analysis

For logistic regression, data analysis involves conducting two separate analyses. The first is for the null model, which is obtained by regressing the dichotomous DV on *none* of the IVs. The second is for the full model, which is obtained by regressing the dichotomous DV on the five dichotomous IVs simultaneously. A summary of the results for each model is provided in Table 10.7.

Interpretation of Results

Assessing the Overall Model. Prior to interpreting the overall logistic regression model, it is imperative to confirm that the model is correctly specified. This means that the model should only include relevant IVs. The log-likelihood (LL) function, which is comparable to the sum of the squared residuals in multiple regression, can be used to assess the overall goodness of fit of the model from a post-hoc perspective (Warner, 2008). To do this, we compare the respective chi-square values for the full and null models. To be judged statistically significant, the chi-square statistic for the full model should be sufficiently larger than that of the null model.

Null Model. As reported in Table 10.7, the log-likelihood of the null model is 74.52. Furthermore, the corresponding regression equation is logit(Abort = Yes) = 0.277, $\chi^2(0) = 2.05$, $p = .1521$. This model represents the absence of any information provided by the IVs. More specifically, the odds of a pilot belonging to the Yes group are $e^{B_0} = e^{0.277} = 1.319$, which equates to a probability of $\frac{1.319}{2.319} = .5688$. Thus, 56.9% of pilots in the given sample data were involved in at least one hazardous event. This can be confirmed from the data set: Of the N = 109 cases, 62 are in the Yes group, and $\frac{62}{109} = .5688$.

Full Model. As reported in the Model Summary subtable of Table 10.7, the log-likelihood of the Full model is 67.23, the log-likelihood of the Reduced model, which corresponds to the Null model, is 74.52, and the difference between the models is 7.29, which is significant, $\chi^2(5) = 14.58$, $p = .0123$. Thus, the full model provides a significant increase in the chi-square statistic over the null model. Because the full model, which includes all the IVs, produces a significantly better fit than the null model, we may deduce that the full model is correctly specified.

Interpreting the Predictors. As reported in the Coefficients subtable of Table 10.7, the logistic regression equation for the full model is logit(Abort = Yes) = $-0.98X_1 - 0.45X_2 - 0.77X_3 + 0.98X_4 - 0.08X_5 + 0.63$, and of the five IVs, two are significant:

TABLE 10.7 *Summary of Results for the Logistic Regression Analysis for Chapter 10's Guided Example*

NULL MODEL

Model Summary

Model	Log-Likelihood	*df*	Chi-Square	*p*
Difference	0.00	0	0	—
Full	74.52			
Reduced	74.52			

Note. N = 109.

Coefficients (Parameter Estimates)

Term	*B*	*SE*	χ^2	*df*	*p*	95% CI
Constant	0.277	0.19	2.05	0	.1521	[–0.10, 0.66]

Note. N = 109.

FULL MODEL

Model Summary

Model	Log-Likelihood	*df*	Chi-Square	*p*
Difference	7.29	5	14.58	.0123
Full	67.23			
Reduced	74.52			

Note. N = 109. Pseudo R^2 = Difference/Reduced = 7.29/74.52 = .0978.

Coefficients (Parameter Estimates)

Term [a]	*B*	*SE*	χ^2	*df*	*p*	95% CI
Constant	0.63	0.53	1.43	1	.2311	[–0.39, 1.69]
X_1 = Age	–0.98	0.49	4.01	1	.0453	[–1.97, –0.04]
X_2 = Biological Sex at Birth	–0.45	0.63	0.51	1	.4767	[–1.70, 0.79]
X_3 = Level of Education	–0.77	0.49	2.52	1	.1124	[–1.74, 0.18]
X_4 = Total Flight Hours	0.98	0.50	3.85	1	.0496	[0.02, 2.00]
X_5 = Airport Experience	–0.08	0.48	0.03	1	.8599	[–1.03, 0.86]

Note. N = 109.
[a] All IVs represent dichotomies and reflect the group coded 1 vs. the group coded 0. X_1 represents 45 years old or older vs. Younger than 45 years old; X_2 represents Female vs. Male; X_3 represents Graduate degree vs. Less than graduate degree; X_4 represents ≥ 1500 hours vs. < 1500 hours; X_5 represents Scores ≤ 32 vs. Scores > 32.

X_1 = Age and X_4 = Total Flight Hours. A brief description of the respective regression coefficients follows beginning with the two significant ones. (*Note*: Recall that group comparisons are made from the perspective of the group coded 1 to the group coded 0, which is the reference group.)

B_1 = –0.98. Holding all other IVs in the model constant, the logit decreases on average by 0.98 units for pilots 45 years old or older vs. pilots younger than 45 years old. This is significant, $\chi^2(1) = 4.01$, $p = .0453$. Furthermore, $e^{-0.98} = 0.3753$, which is the corresponding odds ratio. Focusing on the reciprocal, *OR* = 2.66, which means that pilots younger than 45 years old are approximately 2.66 times more likely than their older counterparts to be involved in a hazardous event. Therefore, we reject the corresponding null hypothesis because X_1 = Age yields an *OR* that differs significantly from 1.

B_4 = 0.98. Holding all other IVs in the model constant, the logit increases on average by 0.98 units for pilots who have 1500 or more total flight hours vs. pilots who have fewer than 1500 total flight hours. This is significant, $\chi^2(1) = 3.85$, p = .0496. Furthermore, $e^{0.98} = 2.66$, which is the corresponding odds ratio. Thus, more experienced pilots (i.e., those with at least 1500 hours total flight time) are approximately 2.66 times more likely than their less experienced counterparts to be involved in a hazardous event. Therefore, we reject the corresponding null hypothesis because X_4 = Total Flight Hours yields an *OR* that differs significantly from 1.

All Remaining Predictors. As reported in Table 10.8, which contains a summary of the odds ratios for the full model: (a) male pilots are approximately 1.5 times more likely to be involved in a hazardous event than female pilots, (b) pilots with less than a graduate degree are approximately 2 times more likely to be involved in a hazardous event than pilots with a graduate degree, and (c) the odds of being involved in a hazardous event are approximately the same for pilots with perceived airport experience scores of 32 or less vs. those with scores greater than 32. None of these regression coefficients are significant, though, and therefore we fail to reject the respective null hypotheses for the corresponding IVs. As a side note, X_1 = Age and X_3 = Level of Education, respectively, are significant in the absence of any other variable. However, when both IVs are in the model together but in the absence of the other IVs, neither is significant. This could imply that the two IVs might be redundant or there is a possible mediating effect. Further investigation of these variables is left to the reader.

Post Hoc Considerations

We now examine various aspects of the findings of the logistic regression analysis. This includes discussing and interpreting the *pseudo* R^2 and the 95% CIs,.

Pseudo R^2. As an analog to R^2 in multiple linear regression, Cohen et al. (2003, p. 502) recommend reporting the pseudo R^2, denoted R_L^2 , for logistic regression:

TABLE 10.8 *Summary of Odds Ratios for the IVs in the Full Model for Chapter 10's Guided Example*

Term	Odds Ratio	95% CI	*p*
X_1 = Age			
45 or older vs. Younger than 45	0.3766	[0.14, 0.96]	.0453
Reciprocal	2.66	[1.04, 7.14]	
X_2 = Biological Sex at Birth			
Female vs. Male	0.6408	[0.19, 2.19]	.4767
Reciprocal	1.56	[0.46, 5.26]	
X_3 = Level of Education			
Graduate degree vs. Less than graduate	0.4621	[0.18, 1.19]	.1124
Reciprocal	2.16	[0.84, 5.56]	
X_4 = Total Flight Hours			
1500 or more vs. Fewer than 1500	2.6672	[1.00, 7.37]	.0496
Reciprocal	0.37	[0.14, 1.00]	
X_5 = Airport Experience			
Scores of 32 or less vs. Scores greater than 32	0.9189	[0.36, 2.37]	.8599
Reciprocal	1.09	[0.42, 2.78]	

Note. N = 109. Tests and CIs for odds ratios are likelihood ratio based.

$$R_L^2 = \frac{\text{Log-Likelihood for Difference between Full and Reduced Models}}{\text{Log-Likelihood for the Reduced Model}},$$

where the Full Model includes all the IVs and the Reduced Model is the Null Model, which contains none of the IVs. As reported in the Model Summary subtable of the Full Model in Table 10.7, the Difference between the Full and Reduced models is 7.29, and the Null model is 74.52. Thus, R_L^2 = .0978. The reader is informed that the interpretation of R_L^2 is not a proportion of variance accounted for by the IVs as in multiple linear regression. Instead, what it represents is essentially the gain in prediction obtained from adding variables to a baseline model. Therefore, the full model provides a significant predictive gain of approximately 9.8% above the null model for our guided example.

95% CIs. As with other MRC applications, it is important to examine the confidence intervals of the *OR*s for each IV in the full model. Table 10.8 contains the 95% CIs for the *OR*s for the Guided Example. The interpretations, which follow those of CIs presented in previous chapters except for the different context, are left for the reader.

Chapter 10 Summary

10.2 Mediation

- Simple mediation occurs when a third variable (*M*) affects the relationship between *X* and *Y*. When *M* fully accounts for *X*'s relationship with *Y* it is referred to as full mediation and when *M* does not fully account for the relationship between *X* and *Y* it is referred to as partial mediation.
- Four conditions must be satisfied in mediation: (a) there is a significant zero-order relationship between *X* and *Y*, (b) there is a significant zero-order relationship between *X* and *M*, (c) *M* has a significant effect on *Y* when controlling for *X*, and (d) *M* has an effect on the relationship between *X* and *Y*. If this relationship is not significant, which implies that *X*'s initial relationship with *Y* has been completely assumed by *M*, then it is full mediation. If, however, this relationship is significant, which implies that *X* maintains its relationship with *Y* in the presence of *M* (albeit smaller), then it is partial mediation.
- Path diagrams are used to represent the relationship between variables. These paths are denoted as follows: c = the total effect, which is the zero-order relationship between *X* and *Y*; a = the zero-order relationship between *X* and *M*; b = the relationship between *M* and *Y* when controlling for *X*; and c' = the direct effect, which is the relationship between *X* and *Y* in the presence of *M*.
- There are other forms of mediation including sequential mediation where mediators are linked together and flow serially, as well as parallel mediation where mediators are stacked above each other and the path between *X* and *Y* goes through each mediator separately.

10.3 Binary Logistic Regression

- Logistic regression is a statistical strategy used to analyze data that consists of a categorical dependent variable. If the DV is dichotomous, then the analysis is called binary logistic regression. The research design that requires a logistic regression strategy is ex post facto.
- Binary logistic regression is predicated on four main assumptions: (a) a dichotomous DV; (b) independent scores on the DV; (c) a correctly specified model; and (d) mutually exclusive categories on the DV, meaning each case or person belongs to one group or another but not both.
- A logistic regression analysis is grounded in probability, odds, and odds ratios. Simple probability is the ratio of the number of successful outcomes to total number of possible outcomes. Odds in favor of an event occurring are defined as the ratio between the probability an event will occur vs. the probability the event will not occur; and odds against is the reciprocal of odds in favor. Odds ratios are defined as the ratio of odds in favor for one group vs. the odds in favor of a second group.
- Logit is an abbreviation for logistic probability unit, and the logit function is defined as the natural logarithm of the odds in favor. The logit function enables the logistic regression equation to be of the same form of a linear regression equation and the corresponding partial regression coefficients to be interpreted in a similar manner as linear regression.
- The odds ratios for a specific IV from a logistic regression equation may be determined by treating the partial regression coefficients (B_i) as the exponent to the natural exponential function; that is, e^{B_i}. These odds ratios also may be acquired from most statistics programs.
- It is often advantageous to dichotomize continuous IVs prior to conducting a logistic regression analysis.
- A logistic regression analysis compares the results of a full model, which contains all the IVs, to the results of a null model, which contains no IVs. If the difference is significant, then the full model provides a significant increase in the chi-square statistic over the null model and is therefore considered correctly specified.
- A pseudo R^2, denoted R_L^2, may be calculated from a logistic regression analysis as evidence of model fit. R_L^2, though, does *not* represent a proportion of variance accounted for by IVs but instead reflects the gain in prediction obtained from adding variables to a baseline model.

Key Terms and Concepts

Binary logistic regression
Chi-square test for independence
Complete mediation
Contingency table
Direct effect
e^{B_i}
Even odds
Mediating variable
Natural exponential function (e)
Natural logarithm (ln)
Odds in favor
Odds against
Odds ratio (OR)
Partial mediation

Full mediation
Indirect effect
Logistic regression
Logistic regression equation
Logit function
Log-likelihood (LL) function
Path diagrams
Probability (simple)
Pseudo R^2 (R_L^2)
Simple mediation
Total effect
Two-way chi-square

Review Exercises

A. Checking Conceptual Knowledge

In 1–10, choose the best answer among the choices provided.

1. In a study examining the impact of pilot training on flight safety outcomes, researchers investigate whether stress levels mediate the relationship between training and safety performance. Which of the following statements best describes the role of mediation in this context?
   a. Mediation occurs when pilot training directly affects safety performance without any influence from stress levels.
   b. Mediation indicates that stress levels are a moderator that strengthens the relationship between training and safety performance.
   c. Mediation implies that pilot training influences stress levels, which in turn affects safety performance.
   d. Mediation shows that flight safety outcomes are solely determined by pilot training, independent of stress levels.
2. In a research study assessing the effects of flight simulator training on pilots' decision-making skills, researchers found that the relationship between training and improved decision-making was entirely explained by changes in pilots' confidence levels. This scenario exemplifies which of the following concepts?
   a. Full mediation, where the effect of training on decision-making is completely mediated by confidence levels.
   b. Direct effect, where flight simulator training impacts decision-making skills without mediation.
   c. Moderation, where confidence levels alter the strength of the relationship between training and decision-making skills.
   d. Partial mediation, where both training and other factors contribute to decision-making skills.
3. The term "logit" is commonly used in logistic regression models. What does the logit function represent?
   a. The likelihood of an event not occurring.
   b. The standard deviation of event occurrences.
   c. The probability of an event occurring.
   d. The logarithm of the odds of an event occurring.
4. If the probability of an event occurring such as an aircraft's flaps system malfunction during a flight is .2, what are the odds in favor of that event occurring?

   a. 1:5 b. 1:4 c. 4 :1 d. 5:1

5. In a study exploring the effects of cockpit resource management (CRM) training on flight crew performance, researchers discovered that CRM training improves team communication and, in turn, enhances performance. However, they also found that CRM training has a direct positive effect on performance, independent of communication. This scenario illustrates which of the following concepts?
   a. Full mediation, where communication entirely accounts for the relationship between CRM training and performance.
   b. No mediation, indicating that CRM training has no effect on performance.
   c. Direct effect, where CRM training influences performance without any mediation.
   d. Partial mediation, where communication partially explains the relationship between CRM training and performance.
6. In an aviation safety study examining the relationship between pilot training (X), safety performance (M), and accident rates (Y), what would be considered the direct effect of pilot training on accident rates?
   a. The effect of pilot training on safety performance.
   b. The effect of safety performance on accident rates.
   c. The effect of pilot training on accident rates, controlling for safety performance.
   d. The indirect effect of safety performance on accident rates due to pilot training.
7. In an aviation safety study examining the relationship between pilot training (X), safety performance (M), and accident rates (Y), what would be considered the indirect effect of pilot training on accident rates?
   a. The direct relationship between pilot training and accident rates.
   b. The effect of safety performance on accident rates after pilot training.
   c. The effect of pilot training on accident rates, controlling for safety performance.
   d. The effect of pilot training on accident rates through safety performance as a mediator.
8. If the odds of an aircraft experiencing an engine failure are 1:10, and the odds of an engine failure on a different aircraft type are 1:20, what is the odds ratio comparing these two aircraft types?
   a. 2:1 b. 1:5 c. 10:1 d. 5:1
9. Which of the following is the key difference between logistic regression and linear regression when modeling the likelihood of an aviation event?
   a. Logistic regression predicts continuous outcomes, whereas linear regression predicts binary outcomes.
   b. Linear regression models the odds of an event, whereas logistic regression models the probability of an event.
   c. Logistic regression models the probability of an event, whereas linear regression models continuous outcomes.
   d. Logistic regression can only handle categorical data, whereas linear regression handles only numerical data.

10. Which of the following is an assumption made in logistic regression when analyzing data?
    a. The relationship between the dependent and independent variables is linear.
    b. The dependent variable is continuous and normally distributed.
    c. The residuals follow a normal distribution.
    d. The dependent variable is binary, and the observations are independent of each other.

B. Applying Procedural and Conceptual Knowledge

Use the following research description and the corresponding data set to conduct the activities given in Parts B1–B7. See Section 10.4 for guidance.

Consider a hypothetical study designed to will help inform a targeted airline's marketing program by determining what factors distinguish between passengers who are part of the airlines' loyalty program and those who are not. Data are obtained from a random sample of N = 268 airline passengers. The sample is restricted to adult travelers 18 years or older, including leisure, business, and student travelers, who flew domestically and/or internationally via the airline and not its regional subsidiary. Researchers conducted a survey of these participants with respect to the following variables:

- Y = Loyalty program status. Participants reported if they did or did not belong to the airline's loyalty program so they can accumulate miles/points that could be redeemed for subsequent air travel.
- X_1 = Age. Participants were asked to self-report their age in years.
- X_2 = Airline amenities. Participants were asked to self-report if they took advantage of any airline amenities such as food or beverages offered on board an aircraft, in-flight entertainment, and seat upgrades.
- X_3 = Passenger satisfaction. Participants were asked to complete a 5-item airline satisfaction questionnaire scored on a traditional 5-point Likert response scale. Thus, scores could range from 5 to 25, with higher scores indicating greater satisfaction with the airline.

The file "Ch_10 Review Exercises Data-Part B" contains a set of hypothetical data relative to this study.

B1. Research Question and Statistical Hypotheses. Write the research question and corresponding statistical hypotheses that pertain to the study.

B2. Sample Size Planning. Conduct an a priori power analysis to determine the minimum sample size needed for the logistic regression. Compare this result to the size of the given data set and explain what impact the size of the given sample will have on the results relative to the minimum size needed.

B3. Prepare the IVs for Analysis. Dichotomize the IVs and prepare a table similar to Table 10.5 that presents the basis for the dichotomies.

B4. Conduct Preliminary Data Screening. Confirm that the given data set is compliant with the assumptions of logistic regression and perform appropriate data diagnostics.

Continued on next page ...

B5. Data Analysis. Conduct a hypothesis test of the odds ratios for the given IVs and summarize the results in tables similar to Tables 10.4, 10.7, and 10.8.

B6. Interpretation of Results. Interpret the results of the study relative to the research question and corresponding null hypothesis stated in Part B1.

B7. Post Hoc Considerations. Calculate and comment on the pseudo R^2 and the 95% CIs. Present plausible explanations for the results.

References

Baron, R. M., & Kenny, D. A. (1986). The moderator-mediator variable distinction in social psychology research: Conceptual, strategic, and statistical considerations. *Journal of Personality and Social Psychology, 51*(6), 1173–1182. https://doi.org/10.1037/0022-3514.51.6.1173

Bujang, M. A., Sa'at, N., Sidik, T. M. I. T. A. B., & Joo, L. C. (2018). Sample size guidelines for logistic regression from observational studies with large population: Emphasis on the accuracy between statistics and parameters based on real life clinical data. *The Malaysian Journal of Medical Science*, *25*(4), 122–130. https://pmc.ncbi.nlm.nih.gov/articles/PMC6422534/

Cohen, J., Cohen, C., West, S. G., & Aiken., L. S. (2003). *Applied multiple regression/correlation analysis for the behavioral sciences* (3rd ed.). Lawrence Erlbaum Associates.

Jones-McKinney, U. O. (2015). *Factors related to hazardous events: The precursors of runway incursions classified as pilot deviations.* [Doctoral dissertation, Florida Institute of Technology].

MacKinnon, D. P., & Dwyer, J. H. (1993). Estimating mediated effects in prevention studies. *Evaluation Review*, *17*(2), 144–158. https://doi.org/10.1177/0193841X9301700202

Preacher, K. J., & Leonardelli, G. J. (2025). *Calculation for the Sobel test: An interactive calculation tool for mediation tests.* https://quantpsy.org/sobel/sobel.htm

Soper, D. (2025). *Sobel test calculator for the significance of mediation.* https://www.danielsoper.com/statcalc/calculator.aspx?id=31

Tabachnick, B. G., & Fidel, L. S. (2013). *Using multivariate statistics* (6th ed.). Pearson Education Inc. Upper Saddle River, NJ.

Tokarski, R. V. (2021). *Examining the safety climate of a general aviation maintenance, repair, and overhaul (MRO) organization: A replication study and application of Bandura's reciprocal causation model.* [Doctoral dissertation, Florida Institute of Technology]. Scholarship Repository. https://repository.fit.edu/etd/41/

Warner, R. M. (2008). *Applied statistics: From bivariate through multivariate techniques.* Sage Publications Inc.

Yerkes, R. M., & Dodson, J. D. (1908). The relation of strength of stimulus to rapidity of habit-formation. *Punishment: Issues and Experiments*, 27–41.

Appendixes

Appendix A
Key Concepts in Research and Statistics

A.1 Appendix Overview

This appendix contains a review of some of the basic aspects of research and statistics. Most readers will recognize this material from previous courses they completed in research design and statistics. Other readers, however, might require some additional review. Regardless of the reader's academic background, it is important to recognize that the material presented here is strictly a review of the salient topics and concepts of research and statistics and is not to be considered exhaustive. The first part of this appendix focuses on *research fundamentals*. Although a previous course in research methods is not needed to understand the material in this textbook, some of the fundamental concepts of research, including *research questions* and *operational definitions*, *research designs*, and *measurement scales* are highlighted as part of the discussion. As a result, the research fundamentals section contains a review of: (a) the *research process*, which includes writing research questions and operational definitions; (b) a review of the different *quantitative research methodologies*; and (c) the *concept of measurement*, including the different measurement scales used in research. The second part of this appendix focuses on *statistics basics* and includes a review of: (a) how to organize and summarize quantitative data using distributions and histograms; (b) descriptive statistics, including measures of central tendency, variability, and position; and (c) inferential statistics, including the concepts of normal, *t*, and *F* distributions, parameter estimation, and hypothesis testing.

A.2 Research Fundamentals

The General Nature of Research

The purpose of research is to gain a better understanding of our world by focusing on key topics or issues. For example, in the world of aviation, we might be interested in understanding (a) factors that are related to runway incursions, (b) the extent to which pilot fatigue impacts flight safety, (c) how air traffic control could use artificial intelligence to optimize route management, (d) the impact noise pollution near airports has on academic achievement or home prices, and (e) the effectiveness of SMS in preventing aircraft incidents. Regardless of the objective, all research studies begin with a research question (RQ) and operational definitions, which specify how key terms and variables are defined and measured. As an example, consider Nakushian's (2020) research proposal:

Research Question. What is the difference in the assessment of susceptibility to cybersecurity threats between GA pilots and ATPs? (p. 5)

Operational Definitions. *GA pilots* are pilots involved in any civilian aircraft operation that is *not* operated under Part 121 of the Federal Aviation Regulations (Title 41 CFR) and are restricted to those who fly single-engine and multi-engine land airplanes. *Airline transport pilots* (ATPs) are defined as pilots operating under Part 121 of the Federal Aviation Regulations. A *cybersecurity threat* is defined as any hacking, virus, or malware that obtains unauthorized access to information used by pilots in flight such as ransomware, which is software that locks up a computer system until a ransom is paid. *Assessment of susceptibility* is measured by a researcher-modified version of Hadlington's (2018) 25-item Attitudes Toward Cyber-security in Business (ATC-IB) instrument, which is measured on a 4-point Likert-type scale where 1 = Strongly Disagree and 4 = Strongly Agree; higher scores reflect a higher awareness of cybersecurity threats. (p. 2)

After the framework of a study is established, the researcher begins data collection. This involves identifying the *target population*, selecting a *representative sample* from the population, and *collecting data* from the sample. The researcher then *analyzes the data*, *interprets the results* of the analysis in the context of the RQ, and presents information about the likelihood the results from the sample can be generalized to the population. This process is summarized in Figure A.1.

Types of Quantitative Research Methodologies

As illustrated in Figure A.2, quantitative research methodologies fall into two primary categories: experimental and observational. The primary difference between the two is grounded in the concept of *manipulation and control*. In experimental research, the researcher manipulates variables and observes within a controlled environment the effect this manipulation has on an outcome variable (the DV). In observational research, the researcher simply observes the effect of a variable on a DV without any manipulation or control. A brief description of the various types of experimental and observational studies given in Figure A.2 follows.

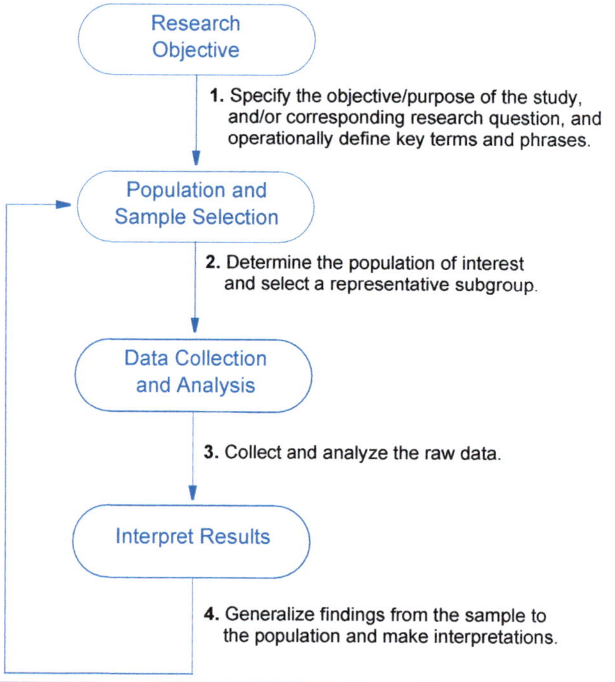

FIGURE A.1 The research process. (*Source*: Gallo et al., 2023, p. 5. Reprinted with permission.)

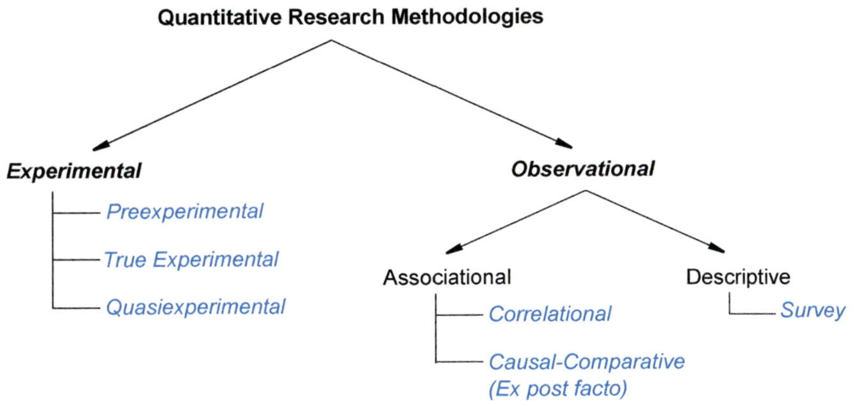

FIGURE A.2 Types of quantitative research methodologies. (Adapted from Gallo et al., 2023, p. 13.)

Experimental Studies. Experimental research is an *intervention* methodology used to demonstrate a *cause-and-effect* relationship. As noted above, the distinguishing characteristic of experimental research is the concept of manipulation and control whereby the researcher deliberately and systematically manipulates the treatment condition, which is the IV, and then controls for the effect of all other factors that might influence the DV. Experimental studies are generally classified as *preexperimental*, *true experimental*, and *quasiexperimental*.

Preexperimental. A preexperimental study is so named because it lacks sufficient control on potentially *confounding variables*, which could provide an alternative explanation for the results. For example, consider a study that consists of two groups of airline pilots. Group A (Treatment) receives flight simulator training focused specifically on dealing with emergency procedures, and Group B (Control) does not receive this training but has received previous routine emergency procedure training. Both groups are then assessed on the time it takes to correctly respond to a simulated emergency. Note that this study lacks *random assignment* to groups and therefore has a *group equivalency* problem. Because the only group comparison occurs post treatment, there might be preexisting differences among the participants with respect to relevant extraneous factors such as their age, years of experience, and type of emergency situations they have encountered previously, which could provide alternative explanations for the results. Because of this lack of control, preexperimental research is considered a *weak experimental design*.

True Experimental. A true experimental study is what most readers think of when they hear the term "research." Although there are several variations, true experimental research involves two groups—one treatment and one control—where participants are randomly assigned to one of the groups. With random assignment, participants have an equal and independent chance of being assigned to either the treatment or control group, which addresses the group equivalency issue of preexperimental studies. When random assignment is used to form groups, the groups are considered to be *statistically equivalent* because any differences that might exist between them would be due to chance. As a result, true experimental studies are often referred to as *randomized designs*. Once the groups are created, one group is then administered the experimental treatment and the other group is administered a "pseudo treatment," or *placebo*, which is an imitation treatment. At the end of the study period, both groups are then assessed to see if the experimental treatment was more effective. Although randomization makes it unnecessary to confirm group equivalency a priori, most randomized designs involve administering some sort of pre-assessment prior to administering treatment to confirm that the groups are indeed "equivalent" on key variables that could impact the dependent measure.

Quasiexperimental. A quasiexperimental study is similar to a true experimental study except instead of randomly assigning participants to groups, *intact groups* are used for treatment and control conditions. For example, let's suppose we want to determine if there is any difference in flight students' achievement in an aviation meteorology course between two different instructional approaches: traditional (large-group lecture) vs. simulation. Both instructional approaches are taught by the same instructor, use the same instructional material, and consist of 150 minutes of instruction per week for 10 weeks. The difference is that students in the simulation group are presented various weather situations via a simulator to augment their lecture—for example, this might occur during the last 15 or 20 minutes of class—whereas the traditional group does not receive this augmentation. Thus, the traditional approach is the control group and the simulation approach is the treatment group. In true experimental research, flight students are randomly assigned to one of the two groups. However, in most educational settings, assigning students to different groups is not feasible because students self-select the classes they take and the times they are offered. For the current example, let's assume one section of the course is offered on MW from 10 A.M. to 11:15 A.M. and is designated the treatment group, and another section is offered on TR from 3 P.M. to 4:15 P.M. and is designated the control group. Thus, students within these groups will respectively receive the treatment condition and the control condition. Because using intact groups lacks random assignment, there is a group equivalency problem. Therefore, unlike true experimental research where it is not necessary to confirm group equivalency, quasiexperimental research must establish group equivalency via pre-assessments prior to beginning the study.

Observational Studies. As noted earlier, observational research involves observing the effect a variable has on a DV without any manipulation or control. Three such types of studies are *correlational*, *causal-comparative*, and *survey*.

Correlational. A correlational study is used when the objective is to examine the relationship between two (or more) variables consisting of the same group of participants. For example, let's say a researcher is interested in determining if there is a relationship between X = the number of training days flight students need to become ready to take the private pilot check ride and Y = the total number of flight hours students complete to become "check-ride ready." To conduct this study, the researcher acquires a sample of flight students whose CFIs deemed they were check-ride ready and records the data for the two variables. The researcher then conducts a correlation analysis (see Chapter 1) to assesses the strength and direction of the relationship. Note there is no manipulation of variables or control of extraneous variables that could provide an alternative explanation for the results and therefore the researcher cannot claim a cause-and-effect relationship.

Causal-Comparative. A causal-comparative study is similar to experimental research: Both involve a group membership variable consisting of two (or more) groups. The difference, though, is with causal-comparative research a researcher has no direct control over the independent variable because either (a) its effect on the dependent variable has already occurred or (b) because it is impossible or unethical to manipulate the independent variable. Given one or both of these two situations, the only way a researcher can observe the effects of an IV is by comparing groups that were exposed to different conditions in the past. As a result, causal-comparative research is often referred to as ex post facto ("after the fact") studies or retrospective studies.

As an example, consider a study designed to assess the effect of "time of flight" on flight performance among Part 121 pilots, where time of flight consists of two groups: night flight pilots who contractually fly during nighttime hours vs. day flight pilots who contractually fly during daylight hours. We define flight performance as any operational error that is documented during flights, including miscommunications between pilots and ATC, navigation errors, and pilot deviation runway incursions. Note that in this example, because pilots are contractually bound, they cannot be randomly assigned to groups, which rules out a true experimental study. Further note there is no treatment or post-assessment being administered, which rules out a quasiexperimental study. So, what we have is a study involving two groups that we have no control in forming where the effect of group membership has already occurred. One way to conduct this study is to (a) form the groups by consulting pilots' personnel records to determine who are under contract for nighttime vs. daytime flights and (b) acquire pilots' flight performance data from the Flight Data Analysis program recorded during their flights. We would then conduct a *t* test of independent samples to determine if there is a significant difference in performance between the groups. So, although a causal-comparative study might appear similar to an experimental study, it lacks manipulation and control, and the effect of the IV on the DV has already occurred.

Survey. Survey research is a *descriptive methodology* because it involves describing various characteristics of people such as their age, education background, marital status, race/ethnicity, gender identification, and annual earned income. Surveys are also used to assess peoples' opinions and attitudes about some issue. For example, an airport director might conduct survey research to (a) acquire passenger demographic information to help inform decisions regarding what airport services to provide, (b) assess passengers' satisfaction with airport services, or (c) assess passengers' experiences with TSA security procedures. Readers are cautioned not to confuse "survey research," which is a research methodology vs. "survey," which is a data collection instrument. To mitigate this potential confusion, we refer to the latter as a *questionnaire*.

Measurement and Measurement Scales

The concept of *measurement* involves assigning a numerical value to a specific attribute of an object or event (Glass & Stanley, 1970, p. 7). As an example, consider the measurement of a Boeing 747's wingspan. The object of the measurement is the wingspan, the specific attribute is the wingspan's length measured in feet or meters, and the numerical value assigned is 224 ft. 5 in. or 68.4 m.

In general, there are five key elements to any measurement:

- The properties being measured, which in research are the variables of interest.
- The units of measurement.
- The instrument being used for the measurement.
- The measurement scale.
- The accuracy and precision of the measurement.

For an example grounded in research, consider Nakushian's (2020) proposed study, which we presented earlier:

- The property being measured is participants' assessment of susceptibility to cybersecurity threats.
- The unit of interest is aggregate scores, where higher scores reflect a higher awareness of cybersecurity threats and hence less susceptibility.
- The measuring instrument is the researcher-modified version of Hadlington's (2018) ATC-IB.
- The measuring instrument is scored on a 4-point Likert-type response scale where responses can range from 1 = Strongly Disagree to 4 = Strongly Agree.
- The accuracy and precision of the measurement is assessed by the instrument's *validity*, which is determined via face, content-, or criterion-related validity, and *reliability*, which often is acquired via Cronbach's alpha or McDonald's omega.

When collecting data for a research study, the variable of interest most likely will consist of either numerical values or alphabetical characters. For example, the variable "flight hours" consists of numerical values, but the variable "FAA ratings" consist of alphabetical characters. In addition to a variable's *data type*, research variables are also measured using specific scales of measurement. These measurement scales, which initially were presented by Stevens (1951), are *nominal*, *ordinal*, *interval*, and *ratio* scales. A brief summary of each measurement scale follows.

Nominal Scales. Variables measured on a nominal scale consist of either numerical values or alphabetical characters. Researchers use nominal scales to collect data on variables that represent specific attributes of participants. These data are then organized into categories that are reflective of the variable and are often called categorical data. For example, a variable measured on a nominal scale is *marital status*, and the different categories of this variable might include "single–never married," "married," "divorced," and "widowed." Thus, the attribute is "marital status," and participants respond by selecting the category that represents their current marital status. Other examples of variables measured on a nominal scale (and sample categories) include: (a) biological sex assigned at birth (female or male), (b) employment status (full-time, part-time, unemployed), and (c) highest level of education (high school only, some college, 2-year college degree, 4-year college degree, post-graduate degree). Notice that for each attribute, the respective categories are *mutually exclusive*, which means that participants' responses indicate they either identify or do not identify with a specific category of the given attribute. For example, participants who respond to employment status as "full-time" are declaring that they "belong" to the full-time employment category and "do not belong" to the other employment categories. This infers that variables measured on a nominal scale have the property of *distinctness*, which means that they differ in kind only.

When analyzing data collected from nominal scale variables, the categories are usually assigned numerical labels. For example, with biological sex, we might assign 1 to "female" and 0 to "male." For variables with more than two categories, some online survey sites assign positive integers to the categories. As an example, consider the three categories of employment status cited above. Some sites represent these categories numerically and in alphabetical order as 1 = full-time, 2 = part-time, and 3 = unemployed. It is important to recognize that these numerical values are simply labels for the categories and have no quantitative sense, which means they cannot be arithmetically manipulated. For example, we know that 1 + 2 = 3, but when applied to the current context, full-time employment (1) plus part-time employment (2) does not equal unemployment (3). These labels also should not be interpreted as one category having more (or less) of the attribute being measured than another category. For example, part-time (2) does not reflect a higher level of "employment status" than full-time (1). As noted in the foregoing paragraph, these numerical values simply reflect distinct outcomes.

Ordinal Scale. Variables measured on an ordinal scale may consist of either numerical values or alphabetical characters. Researchers use ordinal scales to collect data on variables that represent rankings of specific attributes. Examples include anything that lends itself to a ranked order such as airline rankings based on customer satisfaction surveys or annual gross revenue.

As another example, in his study on aviation professionalism, Alhallaf (2016, p. 200) presented participants with the sentence, *I believe professionalism is based on or related to ...* followed by a list of 10 items given below.

... being compliant with procedures and policies.
... being ethical.
... being competent.
... being qualified and reliable.
... demonstrating excellence via a person's behavior, personal appearance, and quality of work.
... the number of certificates/licenses obtained.
... the number of ratings obtained.
... the number of years of experience.
... the level of formal education.
... earning professional certificates from professional aviation organizations.

Alhallaf asked participants to rank each item from 1 to 10, where 1 = the phrase that most closely reflects their perception of professionalism, and 10 = the phrase that least reflects their perception of professionalism. In this example, the variable is "perceptions of professionalism," the categories are the 10 phrases that complete the given sentence, and the outcomes are ranked relative to how much of the attribute they possess. For example, if the phrase "being competent" is ranked 1 and the phrase "being ethical" is ranked 2, then "being competent" has more of the attribute associated with aviation professionalism than "being ethical." Lastly, as is the case with nominal scale variables, the numbers participants assign to the categories simply indicate a category's ranked order and therefore have no quantitative sense, which means they cannot be arithmetically manipulated. As a result, the data from variables that are measured on an ordinal scale are treated as categorical data.

Interval Scale. Variables measured on an interval scale consist solely of numerical values and are considered continuous variables. Researchers use interval scales to collect data on variables that reflect *equal differences,* or *intervals,* between successive categories of the attribute being measured. A commonly used example of a variable measured on an interval scale is temperature, which is measured in degrees (F or C). Note that a temperature scale consists of equal intervals between each numerical value. For example, the difference between temperatures of 80° F vs. 70° F is the same as the difference between temperatures of 27° F vs. 37° F. Each comparison represents a difference of 10° F. As another example, consider classroom test scores that vary between 1 and 100. Similar to the temperature example, the difference between scores of 80 and 90 is the same as the difference between scores of 27 and 37. Notice in both examples, though, that a measurement of 90 vs. a measurement of 45 does not mean that (a) 90° F is twice as warm as 45° F or (b) a person who scores 90 on the test has twice as much knowledge of what was being tested than the person who scores 45. This is because interval scales do not have a well-defined zero point that represents the absence of the attribute being measured. With respect to our two examples, recognize that a temperature of 0° F does not mean the absence of temperature, and an algebra test score of 0 does not mean the absence algebra knowledge. In addition to equal differences, interval scales also have properties of distinctness and order.

One of the most common interval scales used in social science research, which includes aviation, is the traditional 5-point Likert response scale, with categories 1 = Strongly Disagree, 2 = Disagree, 3 = Neutral, 4 = Agree, and 5 = Strongly Agree. Some in the research community argue that this scale has the same properties as an ordinal scale because it reflects a ranking (e.g., "5" has a higher ranking than "3"). Others, however, argue that the intervals between categories are "theoretically" equal, which implies, for example, that a score of Agree (4) is three units more than a score of Strongly Disagree (1). Still others, like to classify a traditional Likert scale or Likert-type scale as *quasi-interval*, which is somewhere between ordinal and interval. Regardless of the perspective, for our purposes, it is accepted practice to treat data acquired from a Likert-response scale as interval data and the corresponding variables are considered continuous.

Ratio Scale. Variables measured on a ratio scale consist solely of numerical values and are considered continuous variables. Researchers use ratio scales to collect data on variables that have a "true" or well-defined zero as well as equal intervals between units. For example, the variable "flight hours" is measured on a ratio scale, as is the variable "number of years as Pilot-in-Command." Focusing on flight hours, note that when comparing a flight student with 100 flight hours vs. a flight student with 200 flight hours, the second student has 100 *more hours* as well as *twice as many hours* as the first student. Similarly, a pilot with 10 years as PIC vs. one with 5 years as PIC has 5 more years as well as twice as many years as PIC than the other pilot. Also notice that 0 flight hours and 0 years as PIC are meaningful and reflect the absence of the respective attribute being measured. Examples of other variables measured on a ratio scale include age, distance, height, income, and weight. Variables measured on a ratio scale have the highest level of measurement and include a true zero point as well as properties of equal differences, distinctness, and order.

A.3 Statistics Basics

This section contains a brief review of the fundamental concepts of statistics, including key concepts associated with distributions, descriptive statistics, and inferential statistics. It is assumed that the reader has completed at least one first-level statistics course, and therefore the material presented here is designed to serve as a refresher and is not to be considered exhaustive.

Organizing and Summarizing Research Data: Frequency Distributions and Histograms

Research data initially are collected in *raw form*, which makes it difficult to understand or interpret the data. For example, the raw data table in Figure A.3 reflects the amount of on-the-job stress aviation maintenance technicians (AMTs) at a local MRO perceive to experience when performing their job. Higher scores represent higher perceived levels of on-the-job stress. Note that when the data are presented in raw form as shown in the table, they are not easy to interpret. However, although the details provided by the raw data are lost, when organized in a grouped frequency distribution table, which summarizes the data, and histogram, which graphically illustrate the shape of the distribution, the data are easier to describe and interpret. Therefore, an early post-data collection activity involves organizing the data in a meaningful way and two commonly used methods are summary tables and graphs.

Raw Data Table of AMTs' Perceived Stress Levels

Case #	Stress Level	Case #	Stress Level	Case #	Stress Level	Case #	Stress Level	Case #	Stress Level	Case #	Stress Level
1	30	16	25	31	33	46	23	61	26	76	37
2	36	17	22	32	27	47	28	62	20	77	30
3	33	18	29	33	24	48	23	63	31	78	33
4	34	19	28	34	45	49	34	64	21	79	33
5	25	20	30	35	14	50	23	65	26	80	20
6	23	21	25	36	25	51	20	66	27	81	28
7	32	22	28	37	21	52	28	67	25	82	24
8	11	23	33	38	29	53	18	68	37	83	21
9	34	24	36	39	27	54	17	69	19	84	14
10	21	25	24	40	26	55	25	70	40	85	24
11	12	26	24	41	33	56	27	71	29	86	20
12	26	27	29	42	21	57	40	72	32	87	22
13	23	28	20	43	25	58	15	73	20	88	22
14	25	29	21	44	27	59	34	74	31	89	9
15	30	30	26	45	24	60	26	75	33	90	32

Grouped Frequency Distribution Table

Stress Levels	Frequency
8–15	6
16–23	23
24–31	40
32–39	18
40–47	3

Histogram

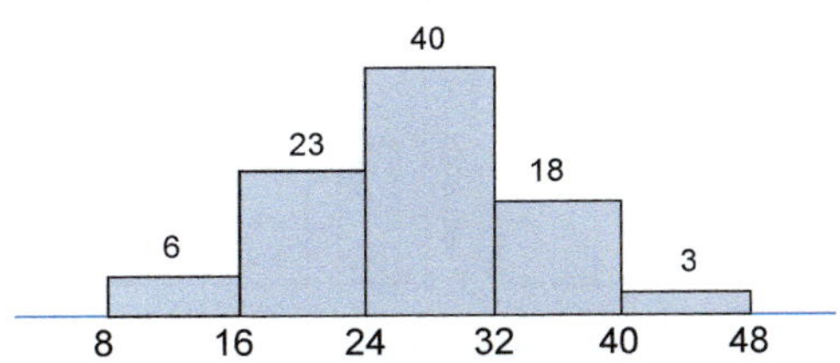

FIGURE A.3 Illustration of how data can be organized as a frequency distribution and histogram.

Note from Figure A.3 how the histogram and grouped frequency distribution table are aligned to each other. For example: (a) The numerical labels of the histogram (8, 16, 24, 32, 40, and 48) correspond to the *lower limits* of the *class intervals* given in the grouped frequency distribution table (e.g., 8–15, 16–23, 24–31, etc.), and (b) because the difference between any two consecutive lower limits is 8 (e.g., 16 – 8 = 8, 24 – 16 = 8, 32 – 24 = 8, etc.), the *upper limits* of the class intervals respectively are 15, 23, 31, 39, and 47 (e.g., 8–15, 16–23, 24–31, etc.). (*Note*: We do not show the class interval 48–55 because there are no data within this interval.) This also means that when interpreting the histogram, working from left-to-right, the first bar represents the number of AMTs whose stress level scores are from 8 to *less than* 16, the second bar represents the number of AMTs whose stress level scores are from 16 to *less than* 24, and so forth. Three additional points:

- When examining distributions, most statistics programs provide a histogram but not the corresponding grouped frequency distribution table. However, this table can be created manually by understanding the relationship between the two forms as noted above.
- Many statistics programs provide an option for users to change histogram's *bin width*. This width, which is also called the *class width*, is the difference between any two consecutive lower limits of the histogram. For our example, we used a bin width of 8.
- In addition to describing a distribution using a grouped frequency distribution table or histogram, distributions are often described by their *shape*. For example, the distribution given in Figure A.3 is *bell-shaped*. As shown in Figure A.4, although there are several different shapes of distributions, they all may be categorized as being either *symmetrical* and *skewed*.

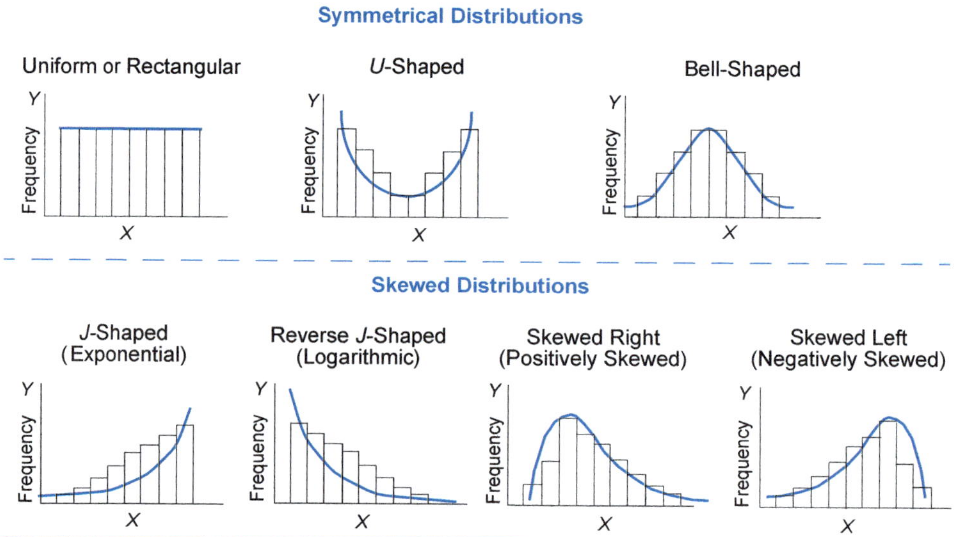

FIGURE A.4 Common shapes of distributions partitioned into two main categories: symmetrical and skewed. (*Source*: Gallo et al., 2023, p. 33. Reprinted with permission.)

Descriptive Statistics

Organizing data into grouped frequency distribution tables, presenting them graphically via histograms, and describing the shape of the resulting distribution are different ways of summarizing data and are part of a statistical strategy called descriptive statistics. Descriptive statistics consist of a set of procedures that *describe* data. In addition to organizing data via tables and graphs, these procedures also include calculating a single metric that describes specific characteristics of data. Collectively referred to as summary statistics, these include *measures of central tendency*, *measures of variability*, and *measures of position*. Brief summaries of each follow.

Measures of Central Tendency. A measure of central tendency describes the center of a distribution, and the three most common measures are mode, median, and mean.

Mode. The mode of a distribution is the score or category that occurs most frequently. Its advantages are (a) it is easy to determine and interpret, (b) it is not impacted by outliers, and (c) it can be used for qualitative data. Its major disadvantage is that it has limited mathematical properties and therefore is not used in advanced statistical analyses involving quantitative data. A distribution with two modes is called bimodal, and a distribution with more than two modes is called multimodal.

Median. The median of a distribution is the score that divides a ranked distribution in half so that 50% of the scores are above the median and 50% are below the median. In addition to being a measure of central tendency, the median is also a *measure of position* and is equivalent to the 50th percentile. The median's primary advantages are that it truly reflects the middle of a distribution, and it is not impacted by outliers. It also is the preferred measure of central tendency vs. the mean for skewed distributions. Its major disadvantage is the same as the mode, namely, it has limited mathematical properties and therefore is not used in advanced statistical analyses involving quantitative data. We denote the median of a sample *Mdn*.

Mean. The mean of a distribution is the (unweighted) arithmetic average, which is calculated by dividing the sum of the scores by the total number of scores, *N*. The mean is the most stable measure of central tendency because it uses all the scores of a distribution. This feature, though, is also a disadvantage because it makes the mean sensitive to outliers, which "pull" it in the direction of the outliers. As a result, the mean is not representative of skewed distributions. Nevertheless, the mean is the most important measure of central tendency and serves as the foundation of all statistical analyses. With respect to inferential statistics, the mean of a sample, denoted *M*, is distinguished from the mean of the population, which is represented by the Greek letter *mu*, μ (pronounced "mew").

In a symmetrical distribution, the mode, median, and mean are all approximately equal to each other. In a skewed distribution, if the mean is less than the median, then the distribution is skewed left, but if the mean is greater than the median, then the distribution is skewed right.

Measures of Variability. A measure of variability, which is also referred to as a measure of dispersion, refers to how spread out the scores in a distribution are about the mean. If the scores are widely spread out then the variation is large, whereas if the scores are clustered together, then the variation is small. The most commonly reported measures of variability in descriptive statistics are the *range*, *interquartile range*, *five-number summary*, and *standard deviation*.

Range. The range of a distribution is the difference between the highest and lowest observations ($H - L$). Although the summary statistics reported by statistics programs report the range as this numerical difference, it is more informative to readers to report the high and low scores. For example, if the scores of a distribution range from a low of 10 to a high of 90, instead of reporting the range as 80, it is more informative to report it as "10 to 90." The major advantages of the range are that it is intuitive and easy to determine and interpret. Major disadvantages are it is based on two scores and does not involve every observation, and therefore does not always provide an accurate description of a distribution's variation.

Interquartile Range. The interquartile range (*IQR*) of a distribution represents the middle 50% of a distribution and is often called "the middle fifty." More specifically, the *IQR* is the distance between the first quartile (Q_1) and the third quartile (Q_3), which are measures of position. The *IQR* is an improvement over the range because it is not influenced by outliers and therefore is more stable than the range. For example, given the distribution {2, 3, 4, 6, 7, 8, 10, 40}, the range is 40 – 2 = 38 but the *IQR* is 6.25. The reason for this disparity is that 40 is an outlier. Because there are several ways to determine quartiles, the *IQR* reported by different statistics programs will not always be the same.

Five-Number Summary. Unlike the range and *IQR*, which provide a single metric that represents the variability of a distribution, the five-number summary describes a distribution's variability by providing five key statistics that help readers get a sense of the degree of spread there is among the scores. This summary consists of the L, Q_1, *Mdn*, Q_3, and H. A five-number summary is often represented graphically via *boxplots*.

Standard Deviation. The standard deviation of a distribution is a measure of the average distance the scores of a distribution are from the mean. Unlike the other measures of variability, the calculation of standard deviation involves every observation, which makes it a stable measure. The smaller the standard deviation, the less spread there is in the data. Standard deviation is derived from another measure of variability called *variance*, which is not meaningful as a descriptive statistic, but has vast importance in inferential statistics. With respect to inferential statistics, the standard deviation of a sample, denoted *SD*, is distinguished from the standard deviation of the population, which is represented by the Greek letter *sigma* σ.

Measures of Position. A measure of position represents a specific point that is used to describe where a score is located within a distribution relative to all the other scores in the distribution. Commonly reported measures of position include *percentiles*, *quartiles*, and *standard scores*.

Percentiles. A percentile partitions a distribution of scores into 100 equal parts. As a measure of position, the *P*th percentile represents the point in a distribution where *P* percent of the scores lie below it. For example, if a distribution consists of 200 scores and 80 of these scores are less than 45, then the percentile is 80/200 = 0.40 = the 40th percentile, denoted P_{40}, which means that 40% of the scores in the distribution are located below the point 45 (and 60% of the scores lie above 45). Thus, P_{40} = 45 is the measure of position that separates the scores in the distribution into two parts: scores that lie below 45 and scores that lie above 45. Note that the score that represents a measure of position (in this case 45,) is not necessarily an actual score in the distribution. It is simply a specific location within the distribution. Further note that based on the definition of a percentile, it is meaningless to report P_{100} because this would mean that 100% of the scores lie below a certain point. Therefore, the highest percentile possible is the 99th percentile, P_{99}.

Quartiles. Percentiles are commonly partitioned into quartiles, where (a) $Q_1 = P_{25}$ is the 25th percentile and represents the position where 25% of the scores in a distribution lie below it (and 75% lie above it); (b) $Q_2 = P_{50}$ is the 50th percentile and represents the position where 50% of the scores in a distribution lie below it and 50% lie above it (this is also the median); and (c) $Q_3 = P_{75}$ is the 75th percentile and represents the position where 75% of the scores in a distribution lie below it (and 25% lie above it).

Standard Scores. Because each distribution has its own mean and standard deviation, it is not possible to make direct comparisons between scores of different distributions. To reconcile this problem, we *standardize* different distributions by converting raw scores to standard scores, and the most widely used standard scores are *z* scores. A *z* score allows us to describe the location of a raw score by indicating how far it deviates from the distribution's mean in standard deviation units. The raw scores of a distribution from a sample can be transformed to *z* scores via the formula

$$z = \frac{x - \mu}{\sigma}$$

where x is the raw score, μ is the mean of the distribution, and σ is the standard deviation of the distribution. When a raw score is transformed into a *z* score, the location of the raw score is now relative to the distribution's mean and measured in standard deviation units. When all the scores of a distribution are transformed into *z* scores, the transformed distribution is called a *z* distribution with mean 0 and standard deviation 1.

Other Measures of Variability: Skewness and Kurtosis. Two other descriptive statistics that are commonly reported as part of summary statistics are *skewness* and *kurtosis*. Skewness refers to the extent to which a distribution is asymmetrical. If the skewness statistic of a distribution is zero, then the distribution is symmetrical; if the skewness statistic is negative, then the distribution is skewed left; and if the skewness statistic is positive, then it is skewed right. Kurtosis refers to the extent to which a distribution "peaks" or "flattens out." The kurtosis statistic for a *normal distribution* is 3; the kurtosis statistic for "peaked" distributions is greater than 3 and becomes greater in magnitude with higher peaks; and the kurtosis statistic for "flat" distributions is greater than 0 but less than 3.

Inferential Statistics

As previously illustrated in Figure A.1, the research process involves selecting a representative sample from the population, collecting data from the sample, analyzing and interpreting the results of data analysis, and then generalizing the results from the sample to the population. This latter part of the process is the heart of inferential statistics, which consists of a set of statistical procedures that help us make informed decisions or conclusions about a population based on sample data.

Because these inferences about a population are being made from a sample, they are not absolute and therefore we cannot be certain that they are accurate. As a result, inferential statistics is often considered "the science of making reasonable decisions based on limited information" (Gallo et al., 2023, p. 77). Two approaches to statistical inference are *parameter estimation* and *hypothesis testing*, and underlying these inferential statistical procedures are various theoretical distributions, including the *normal distribution*, the *t distribution*, and the *F distribution*. (*Note*: We do not include the *chi-square distribution* because it is not a topic of discussion in this book.)

Theoretical Distributions. As noted above, the theoretical distributions we present are the normal, *t*, and *F* distributions. Although we do not discuss the normal distribution in this book, we present a brief summary here because both the *t* and *F* distributions build on it. Our presentations are brief because we assume the reader has a sufficient background of these distributions from a previous statistics course.

The Normal Distribution. The normal distribution is based on a mathematical equation derived by DeMoivre (see Glass & Stanley, 1970, pp. 95–97). As it turned out, this distribution also is representative of many different variables in the natural and social sciences. For example, height, weight, blood pressure, test scores, IQ, and income all approximate a normal distribution. Augmenting this empirical revelation is the *Central Limit Theorem*, which was proven mathematically: If we randomly select all possible samples of the same size from the same population, then the distribution of the sample means from these samples will approximate a normal distribution even if the population distribution is not normal.

The graph of DeMoivre's equation is a bell-shaped, symmetric curve called a normal curve, where its mean, μ, is the mean of a large frequency distribution, and its standard deviation, σ, is the standard deviation of the distribution. Although there are different normal curves for different values of μ and σ, all will maintain the same shape. When a normal distribution is *standardized* using *z* scores, the mean of the distribution is $\mu = 0$, the standard deviation is $\sigma = 1$, and the curve of this *standard normal distribution* is called the unit normal curve or the standard normal curve. Furthermore, as illustrated in Figure A.5, the area under the curve is 1, with approximately 99.7% of the area located within 3 standard deviations of the mean.

Also observe from Figure A.5 that each interval represents a proportion of the total area. For example, the interval between 0 and 1 contains approximately 34% of the total area. When applied to a frequency distribution of scores, we can use the unit normal curve to determine the proportion of observations that fall within a particular interval. For example, if we have a sample distribution of $N = 200$ scores with $M = 20$ and $SD = 5$, then approximately $0.68 \times 200 = 136$ of the 200 scores are located within one standard deviation of the mean. This equates to the interval between $20 - 5 = 15$, which is 1 *SD* below *M*, and $20 + 5 = 25$, which is 1 *SD* above *M*. Furthermore, because we can represent proportions as probabilities, this result also can be expressed with respect to a probability. For example, the probability of randomly selecting a score from the given distribution that is within one standard deviation of the mean is .68. In other words, there is a 68% chance of selecting such a score. The unit normal table, or more simply, the *z* table, is used to determine the area

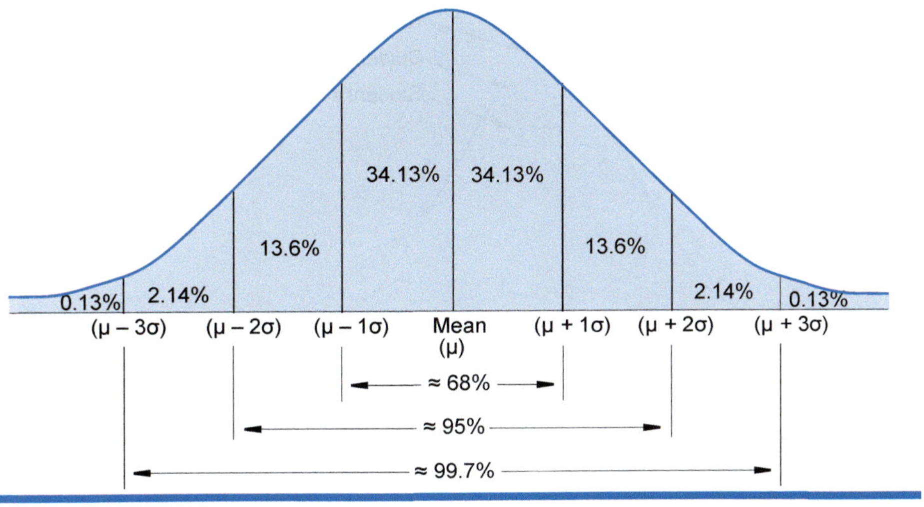

FIGURE A.5 The unit normal curve. (*Source*: Gallo et al., 2023, p. 59. Reprinted with permission.)

under the curve for any interval of the unit normal curve. We do not include a *z* table in this book, but one can be found in most first-level statistics books.

The t Distribution. The *t* distribution is derived from the normal distribution and is used when the population standard deviation, σ, is unknown. This is evident in the corresponding *t* statistic, which is derived from the *z* formula and is based on a sampling distribution of sample means. Thus, the numerator of the *z* statistic is now the difference between the mean of the sample means (*M*) and the population mean (μ), and the denominator is the standard error of the mean ($SE = \sigma_M = \frac{\sigma}{\sqrt{n}}$. However, because a *t* distribution is used when σ is unknown, we use *SD*:

$$z = \frac{M - \mu}{\frac{\sigma}{\sqrt{n}}} \rightarrow t = \frac{M - \mu}{\frac{SD}{\sqrt{n}}}$$

There is no single *t* distribution but instead there is a family of *t* distributions based on sample size and degrees of freedom (*df*). As a result, the shape of the *t* distribution and area under the curve are indexed to *df*. This is illustrated in both Figure A.6(a) as well as in the *t* table given in Appendix B, Table B.1. Note from Figure A.6(a) that the shape of the *t* distribution is similar to the normal distribution and approximates a normal distribution for larger sample sizes. In fact, from a theoretical perspective, if *df* is infinite, then the corresponding *t* distribution is equivalent to the normal distribution. We assume that the reader is familiar with the concept of degrees of freedom as well as how to use a *t* table and therefore do not provide any further discussion about the *t* distribution. Furthermore, examples of using the *t* distribution are provided in the chapters of this book when discussing confidence intervals and hypothesis testing.

The F Distribution. Similar to the *t* distribution, the shape of the *F* distribution is based on degrees of freedom, but unlike *t*, the *F* ratio has a pair of *df*:

$$F = \frac{\text{Variance between groups}}{\text{Variance within groups}} = \frac{SS_{\text{Between}}}{df_{\text{Between}}} \times \frac{df_{\text{Within}}}{SS_{\text{Within}}}$$

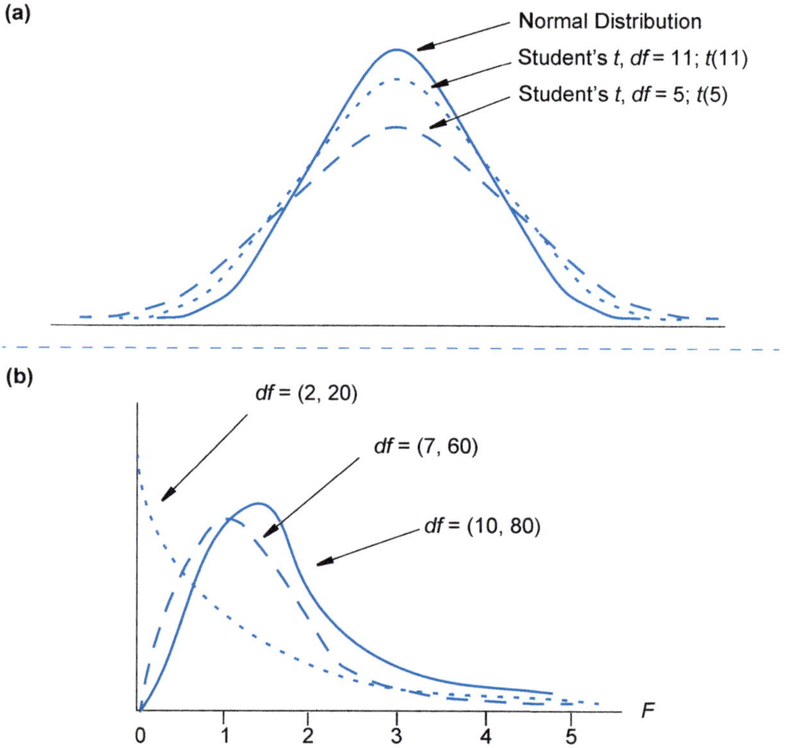

FIGURE A.6 (a) Graphs of various t distributions and (b) graphs of various F distributions. (*Source*: Gallo et al., 2023, p. 116 and p. 224. Reprinted with permission.)

As a result, there are different F distributions for each pair of *df* associated with the respective sample variances. Examples are illustrated in Figure A.6(b). Notice that the shape of the F distribution is positively skewed. However, as the pairs of *df* become larger, the F distribution approximates a normal distribution. To visualize this, the reader is encouraged to consult StatDistributions.com (n.d.) and enter $df = (60, 60)$. Some readers also might recall that $F = t^2$, or alternatively, $t = \sqrt{F}$. In general, a t distribution with $df = (n - 2)$ is equal to the square root of the F distribution with $df = (1, n - 2)$. An F table is provided in Appendix B, Tables B.4a and B.4b.

Parameter Estimation. A direct approach to statistical inference is parameter estimation, which uses sample statistics to approximate the corresponding population parameter. Two examples are *point estimates* and *confidence intervals* (*CIs*). An example of a point estimate is using a sample mean as a direct estimate of the population mean. So, if $M = 25$, then we would state that $\mu = 25$. Although a point estimate might approximate the true population parameter, its accuracy is suspect because it does not provide any information about how "close" the estimate is to the population parameter.

An improvement to a point estimate is an *interval estimate*, which consists of a range of probable values where we believe the parameter is located. Better yet, is to associate a probability factor to the interval estimate that provides the degree of confidence that the population parameter is indeed within the given interval. This is the concept of a confidence interval. Thus, a confidence interval (CI) represents a range of probable values, paired with a specific degree of certainty, that is assigned to a population parameter. In general, a CI states, "With *P*% certainty, the population parameter falls within the interval bounded by *a* and *b*." For example, in estimating the mean of a population, if the lower bound of the interval is $a = 16$, and the upper bound of the interval is $b = 21$, and the degree of certainty is 95%, then we would report that with 95% certainty, the true mean of the population lies within the interval [16, 21].

The accuracy of a CI, which is formally referred to as accuracy in parameter estimation (AIPE), is based on the width of the interval, called precision. The smaller the width of a CI, the more precise it is and the higher is its AIPE. The width of an interval is the difference between the lower and upper limits ($b - a$), and the calculations of these limits are based on the point estimate and standard error associated with the sample statistic. Underlying these calculations is sample size: The larger the sample, the more precise the interval and the higher is the AIPE. For additional information about CIs, including how they are determined and interpreted, the reader is directed to the chapters in this textbook.

Hypothesis Testing. An indirect approach to statistical inference is hypothesis testing. It is considered an indirect approach because a researcher must first state two hypotheses about a population parameter. The *null hypothesis* (H_0) maintains the status quo: "Whatever effect or relationship observed in the sample occurred by chance and therefore unless there is compelling evidence to the contrary, there is no change, no treatment effect, no difference, no relationship; nada." The *alternative hypothesis* (H_1), on the other hand, is a prediction of the researcher's expectation, which generally is grounded in theory, the literature, or preliminary studies. Once the hypotheses are established, a sample is selected from the population, and data are collected from the sample and analyzed. Based on the results of data analysis, the researcher either rejects or fails to reject H_0. This decision is then generalized to the population.

Although the decision resulting from a hypothesis test is grounded in sound statistics, there is always a chance that it could be wrong. In all total, there are four possible outcomes related to the decision made about the null hypothesis:

- If the decision, based on sample data, is "reject" H_0, and if the effect observed in the sample truly exists in the population, then the decision is correct. This is called power.
- If the decision, based on sample data, is "reject" H_0, but the effect observed in the sample does *not* truly exist in the population, then the decision is incorrect. This is referred to as a Type I error, which is also known as a *false positive*.
- If the decision, based on sample data, is "fail to reject" H_0, but the effect observed in the sample truly exists in the population, then the decision is incorrect. This is referred to as a Type II error, which is also known as a *false negative*.
- If the decision, based on sample data, is "fail to reject" H_0, and the effect observed in the sample truly does *not* exist in the population, then the decision is correct, but it also is considered a waste of time and resources.

For additional information about hypothesis testing, the reader is directed to the chapters in this textbook.

References

Alhallaf, H. F. A. (2016). *In pursuit of aviation professionalism: Examining the relationship between factors affecting the aviation profession and the concept of aviation professionalism.* (Publication No. 10055349) [Doctoral dissertation, Florida Institute of Technology]. ProQuest Dissertations and Theses Global. https://www.proquest.com/openview/4986608f6380eadea370781248a92f1f/1?pq-origsite=gscholar&cbl=18750

Gallo, M. A., Wheeler, B. E., & Silver, I. M. (2023). *Fundamentals of statistics for aviation research.* Routledge.

Glass, G. V., & Stanley, J. C. (1970). *Statistical methods in education and psychology.* Prentice-Hall.

Hadlington, L. (2018). Employees attitude towards cyber security and risky online behaviours: An empirical assessment in the United Kingdom. *International Journal of Cyber Criminology*, *12*(1), 269–281. https://doi.org/10.5281/zenodo.1467909

Nakushian, A. (2020). *Airline pilots vs. general aviation pilots assessment of cybersecurity threats* [Unpublished research proposal]. College of Aeronautics, Florida Institute of Technology.

StatDistributions.com (n.d.). *F-distribution.* https://www.statdistributions.com/f/

Stevens, S. S. (Ed.). (1951). *Handbook of experimental psychology.* Wiley. https://psycnet.apa.org/record/1951-07758-000

Appendix B
Statistics Tables

TABLE B.1 *Critical Values of Student's t Distribution*

			One-Tailed (α)			
	.10	**.05**	**.025**	**.01**	**.005**	**.0005**
			Two-Tailed (α)			
df	**.20**	**.10**	**.05**	**.02**	**.01**	**.001**
1	3.078	6.314	12.71	31.82	63.66	636.62
2	1.886	2.920	4.303	6.965	9.925	31.599
3	1.638	2.353	3.182	4.541	5.841	12.924
4	1.533	2.132	2.776	3.747	4.604	8.610
5	1.476	2.015	2.571	3.365	4.032	6.869
6	1.440	1.943	2.447	3.143	3.707	5.959
7	1.415	1.895	2.365	2.998	3.499	5.408
8	1.397	1.860	2.306	2.896	3.355	5.041
9	1.383	1.833	2.262	2.821	3.250	4.781
10	1.372	1.812	2.228	2.764	3.169	4.587
11	1.363	1.796	2.201	2.718	3.106	4.437
12	1.356	1.782	2.179	2.681	3.055	4.318
13	1.350	1.771	2.160	2.650	3.012	4.221
14	1.345	1.761	2.145	2.624	2.977	4.140
15	1.341	1.753	2.131	2.602	2.947	4.073
16	1.337	1.746	2.120	2.583	2.921	4.015
17	1.333	1.740	2.110	2.567	2.898	3.965
18	1.330	1.734	2.101	2.552	2.878	3.922
19	1.328	1.729	2.093	2.539	2.861	3.883
20	1.325	1.725	2.086	2.528	2.845	3.850
21	1.323	1.721	2.080	2.518	2.831	3.819
22	1.321	1.717	2.074	2.508	2.819	3.792
23	1.319	1.714	2.069	2.500	2.807	3.768
24	1.318	1.711	2.064	2.492	2.797	3.745
25	1.316	1.708	2.060	2.485	2.787	3.725
26	1.315	1.706	2.056	2.479	2.779	3.707
27	1.314	1.703	2.052	2.473	2.771	3.690
28	1.313	1.701	2.048	2.467	2.763	3.674
29	1.311	1.699	2.045	2.462	2.756	3.659
30	1.310	1.697	2.042	2.457	2.750	3.646
32	1.309	1.694	2.037	2.449	2.738	3.622
34	1.307	1.691	2.032	2.441	2.728	3.601
36	1.306	1.688	2.028	2.434	2.719	3.582
38	1.304	1.686	2.024	2.429	2.712	3.566
40	1.303	1.684	2.021	2.423	2.704	3.551
60	1.296	1.671	2.000	2.390	2.660	3.460
80	1.292	1.664	1.990	2.374	2.639	3.416
100	1.290	1.660	1.984	2.364	2.626	3.390
120	1.289	1.658	1.980	2.358	2.617	3.373
200	1.286	1.653	1.972	2.345	2.601	3.340
250	1.285	1.651	1.969	2.341	2.596	3.330
1000	1.282	1.646	1.962	2.330	2.581	3.300
∞	1.282	1.645	1.960	2.326	2.576	3.291

Note. For unlisted *df* values, consult the online Student's *t* calculator at https://good-calculators.com/student-t-value-calculator/.

TABLE B.2 *Critical Values for Pearson's r*

	α (One-Tailed)			α (Two-Tailed)		
df = (*N* – 2)	.10	.05	.01	.10	.05	.01
1	.951	.988	.9999	.988	.997	.9999
2	.800	.900	.980	.900	.950	.990
3	.687	.805	.934	.805	.878	.959
4	.608	.729	.882	.729	.811	.917
5	.551	.669	.833	.669	.755	.875
6	.507	.621	.789	.621	.707	.834
7	.472	.582	.750	.582	.666	.798
8	.443	.549	.715	.549	.632	.765
9	.419	.521	.685	.521	.602	.735
10	.398	.497	.658	.497	.576	.708
11	.380	.476	.634	.476	.553	.684
12	.365	.457	.612	.457	.532	.661
13	.351	.441	.592	.441	.514	.641
14	.338	.426	.574	.426	.497	.623
15	.327	.412	.558	.412	.482	.606
16	.317	.400	.542	.400	.468	.590
17	.308	.389	.529	.389	.456	.575
18	.299	.378	.515	.378	.444	.561
19	.291	.369	.503	.369	.433	.549
20	.284	.360	.492	.360	.423	.537
21	.277	.352	.482	.352	.413	.526
22	.271	.344	.472	.344	.404	.515
23	.265	.337	.462	.337	.396	.505
24	.260	.330	.453	.330	.388	.496
25	.255	.323	.445	.323	.381	.487
26	.250	.317	.437	.317	.374	.479
27	.245	.311	.430	.311	.367	.471
28	.241	.306	.423	.306	.361	.463
29	.237	.301	.416	.301	.355	.456
30	.234	.296	.409	.296	.349	.449
31	.229	.291	.403	.291	.344	.442
32	.225	.287	.397	.287	.339	.436
33	.222	.283	.392	.283	.334	.430
34	.219	.279	.386	.279	.329	.424
35	.216	.275	.381	.275	.325	.418
36	.213	.271	.376	.271	.320	.413
37	.210	.267	.371	.267	.316	.408
38	.207	.264	.367	.264	.312	.402
39	.204	.261	.362	.261	.308	.398
40	.202	.257	.358	.257	.304	.393
45	.190	.243	.338	.243	.288	.338
50	.181	.231	.322	.231	.273	.354
55	.172	.220	.307	.220	.261	.339
60	.165	.211	.295	.211	.250	.325
70	.153	.195	.274	.195	.232	.302
80	.143	.183	.257	.183	.217	.283
90	.135	.173	.242	.173	.205	.267
100	.128	.164	.230	.164	.195	.254

Note. For unlisted *df* values, $r_{Critical} = \frac{a}{\sqrt{b+n-2}}$, where $a = t_{Critical}$ (see Table B.1) and $b = t^2_{Critical}$. A complete table of *df* values between 1 and 100 is available online at http://commres.net/wiki/_media/correlationtable.pdf.

TABLE B.3 *Power Table for Pearson's r*

	r								
	.10	**.20**	**.30**	**.40**	**.50**	**.60**	**.70**	**.80**	**.90**
Power	***One-Tailed Test* (α = .05)**								
.25	96	26	13	8	6	5	4	4	3
.50	271	68	31	18	12	8	6	5	4
.60	360	90	40	23	14	10	7	6	4
.70	470	117	51	29	18	12	9	6	5
.75	537	133	59	32	20	14	19	7	5
.80	616	153	67	37	23	15	11	8	5
.85	716	177	77	42	26	17	12	8	6
.90	853	211	92	50	31	20	14	10	6
.95	1077	266	115	63	38	25	17	11	8
.99	1569	386	167	91	55	36	24	16	10
Power	***Two-Tailed Test* (α = .05)**								
.25	167	42	20	12	8	6	5	4	3
.50	385	96	42	24	15	10	7	6	4
.60	490	122	53	29	18	12	9	6	5
.70	616	153	67	37	23	15	10	7	5
.75	692	172	75	41	25	17	11	8	6
.80	783	194	85	46	28	18	12	9	6
.85	895	221	97	52	32	21	14	10	6
.90	1047	259	113	62	37	24	16	11	7
.95	1294	319	139	75	46	30	19	13	8
.99	1828	450	195	105	64	40	27	18	11

TABLE B.4a *Critical F Values (α = .05)*

df	*df* Num.																
Den.	**1**	**2**	**3**	**4**	**5**	**6**	**7**	**8**	**9**	**10**	**15**	**20**	**30**	**40**	**50**	**150**	**9999**
1 [a]	161.4	199.5	215.7	224.6	230.2	234.0	236.8	238.9	240.5	241.9	246.0	248.0	250.1	251.1	251.8	253.5	254.3
2	18.51	19.00	19.16	19.25	19.30	19.33	19.35	19.37	19.38	19.40	19.43	19.45	19.46	19.47	19.48	19.49	19.50
3	10.13	9.55	9.28	9.12	9.01	8.94	8.89	8.85	9.81	8.79	8.70	8.66	8.62	8.59	8.58	8.54	8.53
4	7.71	6.94	6.59	6.39	6.26	6.16	6.09	6.04	6.00	5.96	5.86	5.80	5.75	5.72	5.70	5.65	5.63
5	6.61	5.79	5.41	5.19	5.05	4.95	4.88	4.82	4.77	4.74	4.62	4.56	4.50	4.46	4.44	4.39	4.36
6	5.99	5.14	4.76	4.53	4.39	4.28	4.21	4.15	4.10	4.06	3.94	3.87	3.81	3.77	3.75	3.70	3.67
7	5.59	4.74	4.35	4.12	3.97	3.87	3.79	3.73	3.68	3.64	3.51	3.44	3.38	3.34	3.32	3.26	3.23
8	5.32	4.46	4.07	3.84	3.69	3.58	3.50	3.44	3.39	3.35	3.22	3.15	3.08	3.04	3.02	2.96	2.93
9	5.12	4.26	3.86	3.63	3.48	3.37	3.29	3.23	3.18	3.14	3.01	2.94	2.86	2.83	2.80	2.74	2.71
10	4.96	4.10	3.71	3.48	3.33	3.22	3.14	3.07	3.02	2.98	2.85	2.77	2.70	2.66	2.64	2.57	2.54
11	4.84	3.98	3.59	3.36	3.20	3.09	3.01	2.95	2.90	2.85	2.72	2.65	2.57	2.53	2.51	2.44	2.40
12	4.75	3.89	3.49	3.26	3.11	3.00	2.91	2.85	2.80	2.75	2.62	2.54	2.47	2.43	2.40	2.33	2.30
13	4.67	3.81	3.41	3.18	3.03	2.92	2.83	2.77	2.71	2.67	2.53	2.46	2.38	2.34	2.32	2.24	2.21
14	4.60	3.74	3.34	3.11	2.96	2.85	2.76	2.70	2.65	2.60	2.46	2.39	2.31	2.27	2.24	2.17	2.13
15	4.54	3.68	3.29	3.06	2.90	2.79	2.71	2.64	2.59	2.54	2.40	2.33	2.25	2.20	2.18	2.10	2.07
20	4.35	3.49	3.10	2.87	2.71	2.60	2.51	2.45	2.39	2.35	2.20	2.12	2.04	1.99	1.96	1.88	1.84
30	4.17	3.32	2.92	2.69	2.53	2.42	2.33	2.27	2.21	2.16	2.01	1.93	1.84	1.79	1.76	1.67	1.62
40	4.08	3.23	2.84	2.61	2.45	2.34	2.25	2.18	2.12	2.08	1.92	1.84	1.74	1.69	1.66	1.56	1.51
50	4.03	3.18	2.79	2.56	2.40	2.29	2.20	2.13	2.07	2.03	1.87	1.78	1.69	1.63	1.60	1.50	1.44
150	3.90	3.05	2.66	2.43	2.27	2.16	2.07	2.00	1.94	1.89	1.73	1.64	1.54	1.58	1.44	1.31	1.22
9999	3.84	3.00	2.60	2.37	2.21	2.10	2.01	1.94	1.88	1.83	1.67	1.57	1.46	1.39	1.35	1.20	1.03

Note. For critical values not listed, consult Soper's critical *F* calculator at https://www.danielsoper.com/statcalc/calculator.aspx?id=4, which was used to create this table.

[a] *F* values in this row are rounded to one decimal place.

TABLE B.4b *Critical F Values (α = .01)*

df	*df* Num.																
Den.	**1**	**2**	**3**	**4**	**5**	**6**	**7**	**8**	**9**	**10**	**15**	**20**	**30**	**40**	**50**	**150**	**9999**
1 [a]	4052	5000	5403	5625	5764	5860	5928	5981	6023	6056	6157	6208	6261	6287	6303	6345	6366
2	98.50	99.00	99.17	99.25	99.30	99.33	99.36	99.37	99.39	99.40	99.41	99.45	99.47	99.47	99.45	99.49	99.50
3	34.12	30.82	29.46	28.71	28.24	27.91	27.67	27.49	27.35	27.23	26.87	26.69	26.50	26.41	26.35	26.20	26.13
4	21.20	18.00	16.69	15.98	15.52	15.21	14.98	14.80	14.66	14.55	14.20	14.02	13.84	13.75	13.69	13.54	13.46
5	16.26	13.27	12.06	11.39	10.97	10.67	10.46	10.29	10.16	10.05	9.72	9.55	9.38	9.29	9.24	9.09	9.02
6	13.75	10.92	9.78	9.15	8.75	8.47	8.26	8.10	7.98	7.87	7.56	7.40	7.23	7.14	7.09	6.95	6.88
7	12.25	9.55	8.45	7.85	7.46	7.19	6.99	6.84	6.72	6.62	6.31	6.16	5.99	5.91	5.86	5.72	5.65
8	11.26	8.65	7.59	7.01	6.63	6.37	6.18	6.03	5.91	5.81	5.52	5.36	5.20	5.12	5.06	4.93	4.86
9	10.56	8.02	6.99	6.42	6.06	5.80	5.61	5.47	5.35	5.26	4.96.	4.81	4.65	4.57	4.52	4.38	4.31
10	10.04	7.56	6.55	5.99	5.64	5.39	5.20	5.06	4.94	4.85	4.56	4.41	4.25	4.17	4.12	3.98	3.91
11	9.65	7.21	6.22	5.67	5.32	5.07	4.89	4.74	4.63	4.54	4.25	4.10	3.94	3.86	3.81	3.67	3.60
12	9.33	6.93	5.95	5.41	5.06	4.82	4.64	4.50	4.39	4.30	4.01	3.86	3.70	3.62	3.57	3.43	3.36
13	9.07	6.70	5.74	5.21	4.86	4.62	4.44	4.30	4.19	4.10	3.82	3.66	3.51	3.43	3.38	3.24	3.17
14	8.86	6.51	5.56	5.04	4.69	4.46	4.28	4.14	4.03	3.94	3.66	3.51	3.35	3.27	3.22	3.08	3.00
15	8.68	6.36	5.42	4.89	4.56	4.32	4.14	4.00	3.89	3.80	3.52:	3.37	3.21	3.13	3.08	2.94	2.87
20	8.10	5.85	4.94	4.43	4.10	3.87	3.70	3.56	3.46	3.37	3.09	2.94	2.78	2.69	2.64	2.50	2.42
30	7.56	5.39	4.51	4.02	3.70	3.47	3.30	3.17	3.07	2.98	2.70	2.55	2.39	2.30	2.24	2.09	2.01
40	7.31	5.18	4.31	3.83	3.51	3.29	3.12	2.99	2.89	2.80	2.52	2.37	2.20	2.11	2.06	1.90	1.80
50	7.17	5.05	4.20	3.72	3.41	3.19	3.02	2.90	2.78	2.70	2.42	2.26	2.10	2.00	1.95	1.78	1.68
150	6.81	4.75	3.91	3.45	3.14	2.92	2.76	2.63	2.53	2.44	2.16	2.00	1.83	1.73	1.66	1.46	1.33
9999	6.63	4.61	3.78	3.32	3.02	2.80	2.64	2.51	2.41	2.32	2.04	1.88	1.70	1.59	1.52	1.25	1.00

Note. For critical values not listed, consult Soper's critical *F* calculator at https://www.danielsoper.com/statcalc/calculator.aspx?id=4, which was used to create this table.
[a] *F* values in this row are rounded to the nearest whole number.

Appendix C
Answers to Part A Review Exercises

	Chapter									
Item	**1**	**2**	**3**	**4**	**5**	**6**	**7**	**8**	**9**	**10**
1.	a	b	c	c	d	d	c	b	c	c
2.	c	d	d	b	b	c	a	c	b	a
3.	c	b	c	a	a	d	b	c	d	d
4.	b	b	a	d	c	c	c	d	c	b
5.	a	c	c	c	c	c	b	c	b	d
6.	c	d	b	a	c	a	a	b	a	c
7.	b	a	c	b	b	d	d	d	b	d
8.	b	b	a	c	c	a	d	d	b	a
9.	b	c	b	d	c	b	b	a	c	c
10.	d	d	c	b	d	a	a	b	d	d

Index

Note: **Bold** page numbers refer to tables; *italic* page numbers refer to figures, and page numbers followed by "n" denote endnotes.

S

T